Dr. Sanjay Kumar

Universal Gravitation

by

Dr. Sanjay Kumar
M.Tech, PhD

Managing Director: Quanta Classes Lucknow, Mo. 9453763058
Ex. Sr. Faculty of Physics: Imagine Point Kanpur, Jain Classes Jhansi, Bansal Classes MP, TATA Aarambh Engineering and Medical Simplified Lucknow.

K 423 A Sector K Ashiyana Colony Lucknow

January 4, 2023

SK Education
Quanta Classes: K 423 A Sector K Ashiyana Colony Lucknow (UP)
JEE (Main & Advanced) / SAT / NEET / Foundation
Mo. +919453763058

Email: spphysicsworld@gmail.com
Concepts and Problems in Physics
Title: Universal Gravitation

First edition 2023
ISBN: 9798372239081

To JEE (MAIN & ADVANCED), NEET & SAT aspirants
with the hope that this work will stimulate
an interest in physics and provide an acceptable guide to its understanding.

Indian astronomer-mathematician Bhaskaracharya (c. 1114–1185), expanded on Aryabhata's heliocentric model in his astronomical treatise "Siddhanta Shiromani", where he mentioned the law of gravity, discovered that the planets don't orbit the Sun at a uniform velocity, and accurately calculated many astronomical constants based on this model, such as the solar and lunar eclipses, and the velocities and instantaneous motions of the planets.

Preface

This physics book is the product of more than eighteen years of teaching and innovation experience in physics to engineering (JEE main and Advanced) and medical (NEET and AIIMS) aspirants. Our main goals in writing this book are-

- to present the basic concepts and principles of physics that students need to know for JEE (main and advanced), NEET and other related competitive exams.
- to provide a balance of quantitative reasoning and conceptual understanding, with special attention to concepts that have been causing difficulties to student in understanding the concepts.
- to develop students problem-solving skills and confidence in a systematic manner.
- to motivate students by integrating real-world examples that build upon their everyday experiences.

What's New?
Lots! Much is new and unseen before. Here are the big five:

1. Every concept is given in student friendly language with various category based solved problems. The solution is provided with problem solving approach and discussion.

2. Checkpoint questions have been added to applicable sections of the text to allow students to pause and test their understanding of the concept explored within the current section. The answer keys and solutions are given at the end of the book, in "answer keys and solutions", so that students can confirm their knowledge without jumping too quickly to provided answer.

3. Special attention is given to all tricky topics (like - applications of Newton's law of gravitation, calculation of gravitational field from gravitational potential and vice versa, gravitational field and potential due to a continuous mass distribution, variation in g in a rotatory frame, Gauss's law of gravitation, motion of a satellite, circular motion of a multistar system, energy consideration and applications of Kepler's laws etc.) so that students can easily solve them with fun.

4. At the end of the theory part, there are miscellaneous solved examples which involve the application of multiple concepts of the chapter.

5. To test the understanding level of students, multiple choice questions, conceptual questions, practice problems with previous years JEE Main and Advanced problems are provided at the end of the whole discussion. Number of dots indicates level of problem difficulty. Straightforward problems (basic level) are indicated by single dot ($\bullet$), intermediate problems (JEE mains and NEET level) are indicated by double dots ($\bullet\bullet$), whereas challenging problems (advanced level) are indicated by thee dots ($\bullet\bullet\bullet$). Answer keys with hints and solutions are provided at the end of the book.

We have kept these goals in mind while developing the main themes of our physics book.

Dr. Sanjay Pandey

Online Physics Classes
by
Dr. Sanjay Kumar

JEE (Main & Advanced) / SAT / NEET / International Physics Olympiad / Foundation (IX - XII)
Quanta Classes: K 423 A Sector K Ashiyana Colony Kanpur Road Lucknow
Mo. +919453763058
Email: spphysicsworld@gmail.com

Contents

GRAVITATION

So far, in previous chapters, we have discussed various forces: pushes and pulls, elastic forces, friction, and other forces that act when one body is in contact with another. In this chapter we study the properties of one particularly important non-contact force, gravitation, which is one of the fundamental and (we believe) universal forces of nature. The law that describes the gravitational force between any two bodies was discovered by Newton in 1665, and it has had spectacular success in accounting for the gravitational forces exerted on objects on Earth as well as for the motions of the planets in the solar system. A modern theory of gravitation, "Einstein's general theory of relativity", is necessary to account for effects in strong gravitational fields which is beyond the scope of this book.

Please note that, many of the basic concepts of dynamics find application here. In particular, we shall use Newton's force laws, dynamics of circular motion, potential energy, and conservation of energy and angular momentum. So, the knowledge of these topics is prerequisite.

1.1 Gravitation From the Ancients to Kepler

1.1.1 Heliocentric (Sun-centred) Model of Hinduism

The study of the structure of the universe is called cosmology. The term gravity was first used in "Rig veda" which was probably written between 15th-12th century BC. Rig veda is the earliest of the four vedas and one of the most important texts of Hindu religion.

According to Rig veda "the gravitational effect of solar system makes the earth stable"

Several Vedic Sanskrit texts written in Hinduism, always supported heliocentric (Sun-centred) scheme, in which the Earth (along with the other planets) moves about the Sun, which is at the centre of the solar system. Yajnavalkya (c. 9th-8th century BC) recognized that the Earth is spherical and believed that the Sun was at the common centre of the spheres as described in the vedas at the time. In his astronomical text Shatapatha Brahmana (87.3.10) he states that "The sun attracts these bodies-the earth, and the other planets towards it's centre and these bodies revolve in their orbits around the Sun". He recognized that the Sun was much larger than the Earth, which would have influenced this early heliocentric concept.

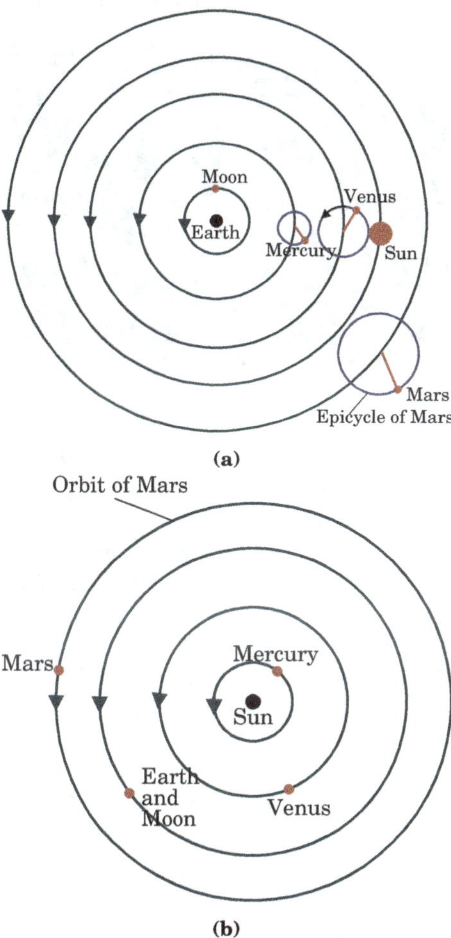

Figure 1.1: (*a*) The Ptolemaic view of the solar system. The Earth is at the centre and the Sun and planets move around it. The planets move in small circles (epicycles), whose centers travel along large circles. (*b*) The Copernican view of the solar system. The Sun is at the centre, and the planets move around it.

He also accurately measured the relative distances of the Sun and the Moon from the Earth. He also described a solar calendar in the Shatapatha Branmana.

The Vedic Sanskrit text Aitareya Brahmana (2.7) (c. 9th-8th century BC) also states that, Sun never sets nor rises that's right. When people think the sun is setting it is not so; they are mistaken. This indicates that the Sun is stationary (hence the Earth is moving around it), which is elaborated later in vishnu Purana (2.8) (which was probably written between 400

BC- 900 CE), which states that "the Sun is stationed for all the time at the same place".

The Indian astronomer-mathematician Aryabhata (476-550 AD), in his text Aryabhatiya, propounded a heliocentric model in which the Earth was taken to be spinning on its axis and the periods of the planets were given with respect to a stationary Sun. He was also the first to discover that the planets follow an elliptical orbit around the Sun, and thus propounded an eccentric elliptical model of the planets, on which he accurately calculated many astronomical constants, such as the times of the solar and lunar eclipses, and the instantaneous motion of the Moon (expressed as a differential equation).

Bhaskaracharya (1114-1185) expanded on Aryabhata's helio-

Figure 1.2: Indian astronomer-mathematician Aryabhata (476-550 AD)

centric model in his astronomical treatise "Siddhanta Shiromani", where he mentioned the law of gravity, discovered that the planets don't orbit the Sun at a uniform velocity, and accurately calculated many astronomical constants based on this model, such as the solar and lunar eclipses, and the velocities and instantaneous motions of the planets. Arabic translations of Aryabhata's Aryabhatiya were available from the 8th century, while Latin translations were available from the 13th century, before Copernicus had written "De revolutionibus orbium coelestium", so it's quite likely that Aryabhata' s work had an influence on Copernicus' ideas.

Some scholars from Islamic and and Christian word rejected the heliocentric model proposed by Indian mathematicians and astronomers. They developed a new geocentric (Earth-centred) cosmological model, in which, as the name implies, the Earth remains stationary at the centre of the universe while the moon, the sun, the planets, and the stars were points of light turning about the earth on large "celestial spheres". This viewpoint was further expanded by the second-century Egyptian astronomer Ptolemy (the 'P' is silent). He developed an elaborate mathematical model of the solar system that quite accurately predicted the complex planetary motions.

The Earth seems to us to be a substantial body. Even today, in navigational astronomy we use a geocentric reference frame, and in ordinary conversation we use terms such as sunrise," which implies such a frame.

Because simple circular orbits cannot account for the complicated motions of the planets, Ptolemy had to use the concept of epicycles, in which a planet moves around a circle whose centre moves around another circle centred on the Earth (1.1a). He also had to resort to several other geometrical arrangements, each of which preserved the supposed sanctity of the

circle as a central feature of planetary motions.

We now know that it is not a circle that is fundamental but an ellipse, with the Sun at one focus, as we shall discuss.

In the 16th century Nicolaus Copernicus (1473-1543) proposed a heliocentric (Sun-centred) scheme, in which the Earth (along with the other planets) moves about the Sun (see Fig.1.1b). Even though the Copernican scheme seems much simpler than that of Ptolemy, it was not immediately accepted by Christian and Islamic world. Copernicus still believed in the sanctity of circles, and his use of epicycles and other arrangements (which are not shown in Fig. 1.1b) was about as great as that of Ptolemy. However, by putting the Sun at the centre of things, Copernicus proposed the correct reference frame, from which our modern view of the solar system could develop.

To resolve the conflict between the Copernican and Ptolemaic schemes, more accurate observational data were needed. Such data were compiled by great astronomer Tycho Brahe (1546-1601). For 30 years, from 1570 to 1600, Tycho compiled the most accurate astronomical observations the world had known. The invention of the telescope was still to come, but Tycho developed ingenious mechanical sighting devices that allowed him to determine the positions of stars and planets in the sky with unprecedented accuracy.

Tycho had a young mathematical assistant named Johannes Kepler. Kepler had become one of the first outspoken defenders of Copernicus, and his goal was to find evidence for circular planetary orbits in Tycho's records. To appreciate the difficulty of this task, keep in mind that Kepler was working before the development of graphs or of calculus—and certainly before calculators! His mathematical tools were algebra, geometry, and trigonometry, and he was faced with thousands upon thousands of individual observations of planetary positions measured as angles above the horizon.

Many years of work led Kepler to discover that the orbits are

Figure 1.3: Johannes Kepler (1571-1630)

not circles, as Copernicus claimed, but ellipses. Furthermore, the speed of a planet is not constant but varies as it moves around the ellipse (as proposed by Bhaskaracharya).

Kepler's laws, as we call them today, state that-

1. Planets move in elliptical orbits, with the sun at one focus of the ellipse.

2. A line drawn between the sun and a planet sweeps out equal areas during equal intervals of time.

3. The square of a planet's orbital period is proportional to the cube of the semi-major axis length.

   Later, in section 1.20, we derive Kepler's second and third law mathematically.

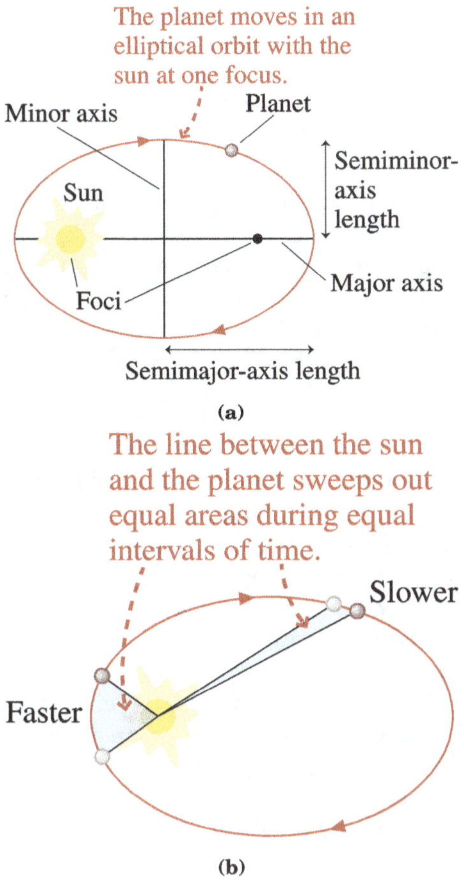

Figure 1.4: The elliptical orbit of a planet about the sun.

Fig.1.4a shows that an ellipse has two foci (plural of focus), and the sun occupies one of these. The long axis of the ellipse is the major axis, and half the length of this axis is called the semimajor-axis length. As the planet moves, a line drawn from the sun to the planet "sweeps out" an area. Fig. 1.4b shows two such areas. Kepler's discovery that the areas are equal for equal Δt implies that the planet moves faster when near the sun, slower when farther away.

All the planets except Mercury have elliptical orbits that are only very slightly distorted circles. As Fig.1.5 shows, a circle is an ellipse in which the two foci move to the center, effectively making one focus, and the semimajor-axis length becomes the radius. Because the mathematics of ellipses is difficult, this chapter will focus on circular orbits.

Kepler discovered the third law by carefully examining, over a period of many years, countless combinations of planetary data.

Kepler published the first two of his laws in 1609, the same year in which Galileo first turned a telescope to the heavens. Through his telescope Galileo could see moons orbiting Jupiter, just as Copernicus had suggested the planets orbit the sun. He could *see* that Venus has phases, like the moon, which implied its orbital motion about the sun. By the time of Galileo's death in 1642, the Copernican revolution was complete.

> Because Galileo supported the idea of a heliocentric solar system, the Roman Catholic Church placed him under house arrest.The conflict between the Roman Catholic Church and Galileo was complicated. His house arrest was likely influenced by his personality and by global factors such as the Protestant Reformation. In the 1990s, Pope John Paul II expressed regret for the church's response to Galileo and acquitted him.

1.2 Newton's Law of Universal Gravitation

There is a popular story about Newton that he got the idea of gravity after an apple fell on his head. Newton himself said that the "notion of gravitation" came to him as he "sat in a contemplative mood" and "was occasioned by the fall of an apple." It occurred to him that, perhaps, the apple was attracted to the centre of the earth but was prevented from getting there by the earth's surface. And if the apple was so attracted, why not the moon?

In the case of the apple the motion is linear as it accelerates

Figure 1.6: Sir Issac Newton (1642-1726)

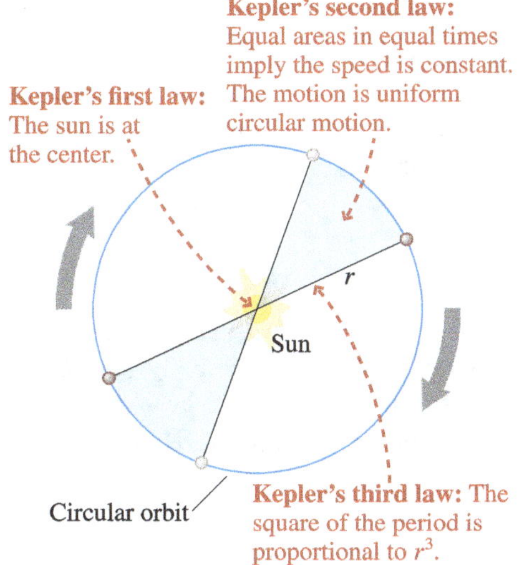

Figure 1.5: A circular orbit is a special case of an elliptical orbit.

downward toward the centre of the Earth. In the case of the Moon the motion is circular with constant speed. As we know that, an object in uniform circular motion accelerates toward the center of the circle. It follows, therefore, that the Moon also accelerates toward the center of the Earth. In fact, the force responsible for the Moon's centripetal acceleration is the Earth's gravitational attraction, the same force responsible for the fall of the apple.

So, gravitation is a universal force between all objects in the universe.

Newton recognized the importance of Kepler's laws. Galileo's observations supported the heliocentric model, and Kepler's laws fit observations of planetary orbits. Newton was thus able to focus on finding a general theory to explain all three laws. One of his remarkable insights was his hypothesis that *terrestrial* and *celestial* physical laws are the same. Starting from this idea, Newton theorized that gravity is not limited to objects near the Earth but instead applies to all objects in the Universe. Today, this theory is known as *Newton's law of universal gravitation*.

Newton used Kepler's second and third laws to derive two properties of gravity. From Kepler's second law, he concluded that the force of gravity must be directed straight toward the Sun, i.e., it must be central.

Using Kepler's third law, he mathematically reasoned that the magnitude of the gravitational force F_g felt by any planet, is inversely proportional to the square of distance r between the Sun and the planet.

According to Newton's law of universal gravitation-

every particle with non-zero mass in the Universe exerts a gravitational force on every other particle with non-zero mass in the Universe. The gravitational force is always attractive. For example, the Sun pulls the Earth toward the centre of Sun, and the Earth pulls the Sun toward the centre of Earth.

The gravitational force is always attractive in nature. The magnitude of gravitational force (F_g) acting between the particles of masses m_1 and m_2 separated by distance r [Fig.1.7] is -

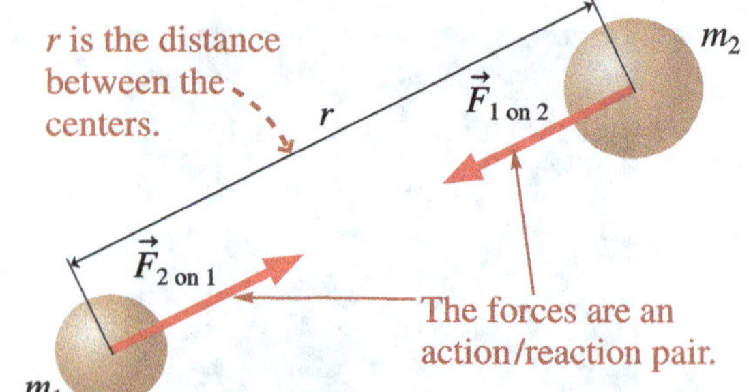

Figure 1.7: The gravitational forces on masses m_1 and m_2.

1. directly proportional to the product of their masses, i.e.,
$$F_g \propto m_1 m_2$$

2. inversely proportional to the square of the distance between the particles, i.e.,
$$F_g \propto \frac{1}{r^2}$$

On combining these two statements, we get-
$$F_g \propto \frac{m_1 m_2}{r^2}$$
or
$$\boxed{F_g = G \frac{m_1 m_2}{r^2}} \tag{1.1}$$

The proportionality constant G is called the gravitational constant and has a value of $G = 6.674 \times 10^{-11}$ Nkg^{-2}m^2.

Here, it is important to note that the gravitational force does not depend upon the medium in which objects are placed.

Dimensional formula of G:
$$[G] = \left[\frac{Fr^2}{m_1 m_2}\right] = \frac{[MLT^{-2}][L^2]}{[M^2]} = [M^{-1}L^3T^{-2}]$$

☞ Newton's law of gravity [Eq.1.1] applies to point objects. For calculating the force of gravity for an object of finite size we divide the finite object into a collection of small mass elements, then use superposition and the methods of calculus to determine the net gravitational force.

Having figured out these properties of gravity, Newton's next step was an important test of the validity of his thinking. He used his laws of motion and these two properties of gravity that he had inferred from Kepler's second and third laws to derive Kepler's first law. His success in deriving Kepler's first law showed that all three of Kepler's planetary laws stemmed from one physical theory: The Sun exerts a gravitational force on all the planets.

Fig.1.8 is a graph of the gravitational force as a function of the distance between the two masses. As you can see, an inverse-square force decreases rapidly.

Strictly speaking, Equation 1.1 is valid only for particles.

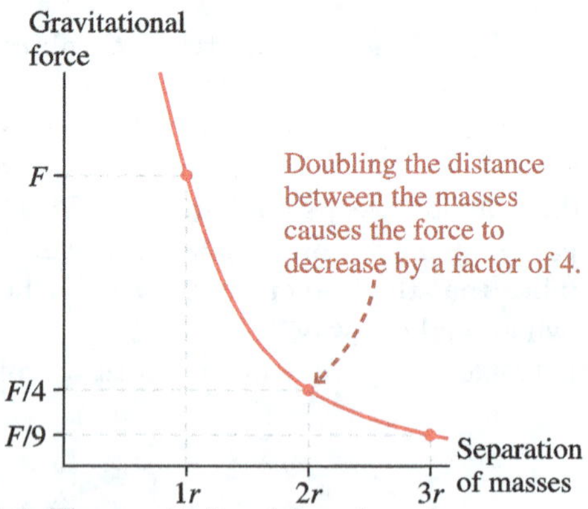

Figure 1.8: The gravitational force is an inverse-square force.

However, Newton was able to show that this equation also applies to spherical objects, such as planets, if r is the distance between their centres.

1.2.0.1 Vector Form of Law of Universal Gravitation

Suppose, position vectors of masses m_1 and m_2 with respect to origin O are $\vec{r}_1$ and $\vec{r}_2$ respectively [see Fig.1.9]. If the magnitude of gravitational force between m_1 and m_2 is $F_g \left(= G\frac{m_1 m_2}{r^2}\right)$, then gravitational force on mass m_1 in vector form, can be written as-
$$\vec{F}_{12} = F_g \, \hat{r}_{12}$$
Here, $\hat{r}_{12}$ is the unit vector from the centre of mass m_2 to m_1. Since, $\hat{r}_{12} = \frac{\vec{r}_2 - \vec{r}_1}{|\vec{r}_2 - \vec{r}_1|}$, therefore, above equation gives-

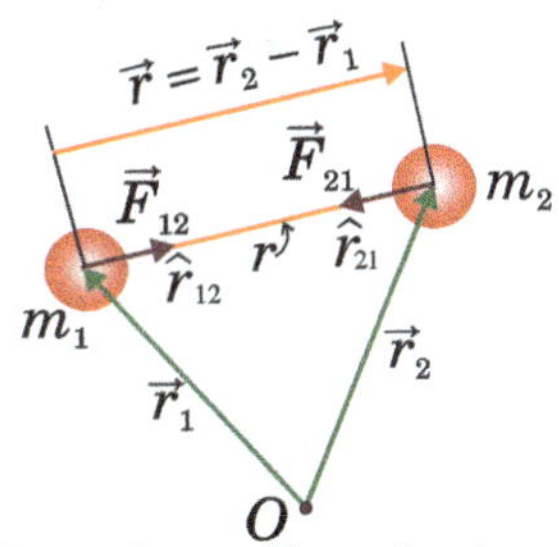

Figure 1.9: Vector form of law of universal gravitation

$$\vec{F}_{12} = F_g\,\hat{r}_{12} = F_g\frac{\vec{r}_2 - \vec{r}_1}{|\vec{r}_2 - \vec{r}_1|} = G\frac{m_1 m_2}{|\vec{r}_2 - \vec{r}_1|^2}\frac{\vec{r}_2 - \vec{r}_1}{|\vec{r}_2 - \vec{r}_1|}$$
$$= G\frac{m_1 m_2}{|\vec{r}_2 - \vec{r}_1|^3}(\vec{r}_2 - \vec{r}_1)$$

Similarly, the gravitational force on mass m_2 due to mass m_1, is

$$\vec{F}_{21} = G\frac{m_1 m_2}{|\vec{r}_1 - \vec{r}_2|^3}(\vec{r}_1 - \vec{r}_2) = -G\frac{m_1 m_2}{|\vec{r}_2 - \vec{r}_1|^3}(\vec{r}_2 - \vec{r}_1) = -\vec{F}_{21}$$

i.e., in vector form $\vec{F}_{12} = -\vec{F}_{12}$.

Thus, the force exerted by object 1 on object 2 is equal and opposite to the force exerted by object 2 on object 1, i.e., gravitational forces always form action and reaction pair.

☞ Gravitational forces always act along the line joining the two particles, and they form an action–reaction pair.

1.2.1 Gravitation and Spherically Symmetric Bodies

We have stated the law of gravitation in terms of the interaction between two particles. It turns out that the gravitational interaction of any two bodies having spherically symmetric mass distributions (such as solid spheres or spherical shells) is the same as if we concentrated all the mass of each at its centre. For the spherical bodies shown in Figure 1.10a, we have

$$F_g = \frac{m_1 m_2}{r^2}$$

Now, if we model the earth as a spherically symmetric body with mass M and radius R, the force it exerts on a particle or a spherically symmetric body with mass m, at a distance r between centres, is

$$\boxed{F_g = \frac{GMm}{r^2}} \tag{1.2}$$

provided that the body lies outside the earth. A force of the same magnitude is also exerted on the earth by the body.
If the body is placed on the surface of earth, then $r = R$, so Eq.1.2, becomes

$$\boxed{F_g = \frac{GMm}{R^2}} \tag{1.3}$$

At points inside the earth the situation is different. If we could drill a hole to the centre of the earth and measure the gravitational force on a body at various depths, we would find that toward the centre of the earth the force decreases, rather than increasing as $1/r^2$. As the body enters the interior of the earth (or other spherical body), some of the earth's mass is on the

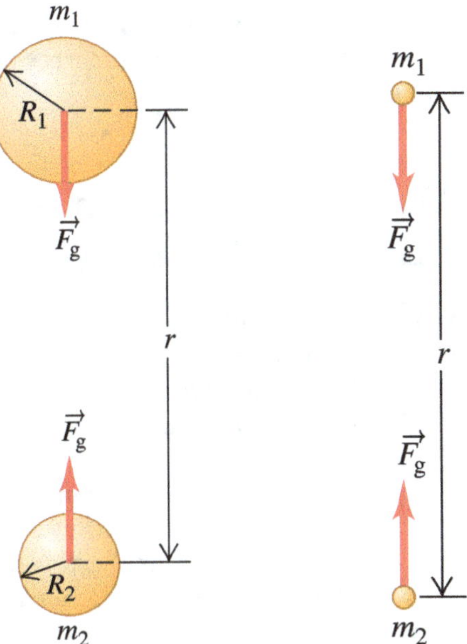

(a) The gravitational force between two spherically symmetric masses m_1 and m_2 ...

(b) ... is the same as if we concentrated all the mass of each sphere at the sphere's center.

Figure 1.10

side of the body opposite from the centre and pulls in the opposite direction. Exactly at the centre, the earth's gravitational force on the body is zero.

Spherically symmetric bodies are an important case because moons, planets and stars all tend to be spherical. Since all particles in a body gravitationally attract each other, the particles tend to move to minimise the distance between them. As a result, the body naturally tends to assume a spherical shape, just as a lump of clay forms into a sphere if you squeeze it with equal forces on all sides.

Now, consider a uniform sphere of radius R and mass M, as shown in Figure 1.11. A point object of mass m is brought near the sphere, though still outside it at a distance r from its center. The object experiences a relatively strong attraction from mass near the point A, and a weaker attraction from mass near point B. In both cases the force is along the line connecting the mass m and the center of the sphere—that is, along the x axis. In addition, masses at the points C and D exert a net force that is also along the x axis. Thus, the symmetry of the sphere guarantees that the net force it exerts on m is directed toward the sphere's center.

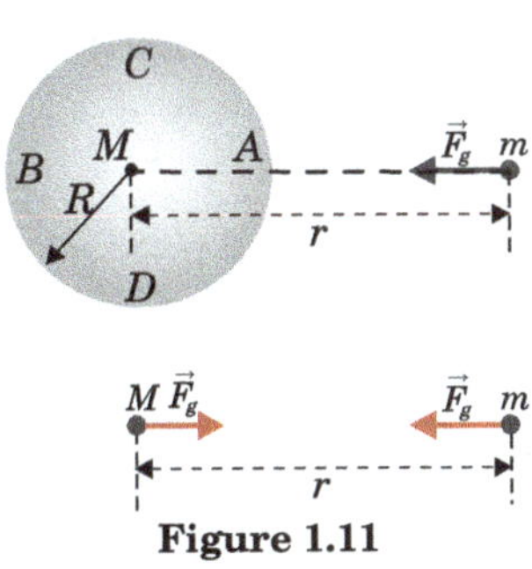

Figure 1.11

☞ **Conclusion:** Newton's law of universal gravitation [Eq.1.1] can be directly applied to point masses and spherical masses only.

1.3 The Cavendish Experiment (Determining the Value of G)

Eq.1.2,has a proportionality constant "G". Newton did not know the value of G. He could only say that the gravitational force is proportional to the product $m_1 m_2$ and inversely proportional to r^2, but he had no means of knowing the value of the proportionality constant.

Determining G requires a direct measurement of the gravitational force between two known masses at a known separation. Since, the gravitational force between ordinary-size objects, is very weak so this work is not so easy. Yet the English scientist Henry Cavendish came up with an ingenious way of doing so with a device called a *torsion balance*. Two fairly small masses m, typically about 10 gram, are placed on the ends of a lightweight rod. The rod is hung from a thin fiber, as shown in Fig.1.12a, and allowed to reach equilibrium. If the rod is then rotated slightly and released, a restoring force will return it to equilibrium. This is analogous to displacing a spring from equilibrium, and in fact the restoring force and the angle of displacement obey Hooke's law: $F_{\text{restore}} = \kappa \Delta\theta$. The "torsion constant" κ can be determined by knowing the period of oscillations[1]. Once κ is known, a force that twists the rod slightly away from equilibrium can be measured by the product $\kappa \Delta\theta$. It is possible to measure very small angular deflections, so this device can be used to determine very small forces.

Two larger masses M (typically lead spheres with $M \approx 10$ kg) are then brought close to the torsion balance, as shown in Fig 1.12b. The gravitational attraction that they exert on the smaller hanging masses causes a very small but measurable twisting of the balance, enough to measure F_g. Because m, M, and the distance between their centers, r, are all known, Cavendish was able to determine G from

$$G = \frac{F_g r^2}{Mm}$$

Cavendish found $G = 6.754 \times 10^{-11} N.m^2/kg^2$, in good agreement with the currently accepted value of $G = 6.67 \times 10^{-11} N.m^2/kg^2$

To see why Cavendish is said to have weighed the Earth, recall that the force of gravity on the surface of the Earth, mg, can be written as follows:

$$mg = \frac{GMm}{R^2}$$

Cancelling m and solving for M yields

Mass of earth $M = \dfrac{gR^2}{G}$

The value of g at the earth's surface is known with great accuracy from kinematics experiments. The earth's radius R is

---

[1] Period of oscillation of two ball system about an axis, is given by- $T = 2\pi\sqrt{I/\kappa}$. Here, I is the moment of inertia of the two ball system about the axis of rotation and κ is the torsion constant of

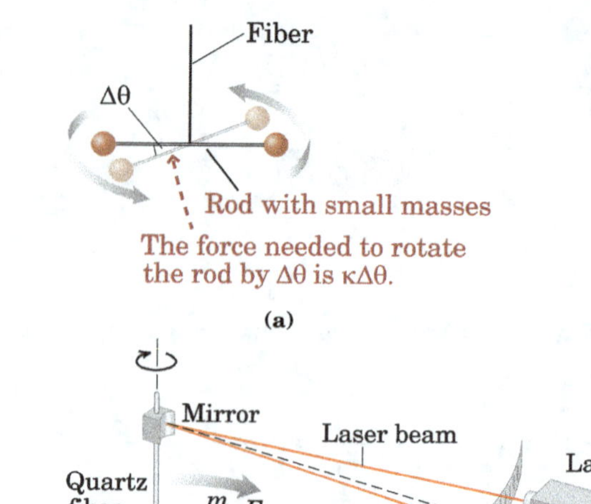

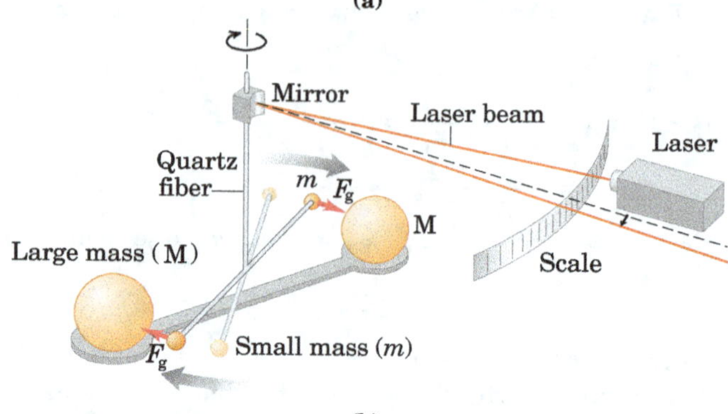

Figure 1.12: The principle of the Cavendish balance, used for determining the value of G. The angle of deflection has been exaggerated here for clarity.

determined by surveying techniques. Combining our knowledge from these very different measurements has given us a way to determine the mass of the earth (M).

When Cavendish measured G, he didn't actually "weigh" the Earth, of course. Instead, he calculated its mass, M.

1.3.0.1 Calculation of mass of Earth

Mass of earth,

$$M = \frac{gR^2}{G} = \frac{\left(9.81\frac{m}{s^2}\right)\left(6.37 \times 10^6 m\right)^2}{6.67 \times 10^{11} \frac{N.m^2}{kg^2}} = 5.97 \times 10^{24} kg$$

EXAMPLE 1. Two solid sphere of same size of a metal are placed in contact by touching each other [Fig.1.13]. Prove that the gravitational force acting between them is directly proportional to the fourth power of their radius.

APPROACH Since spherical geometries are symmetrical

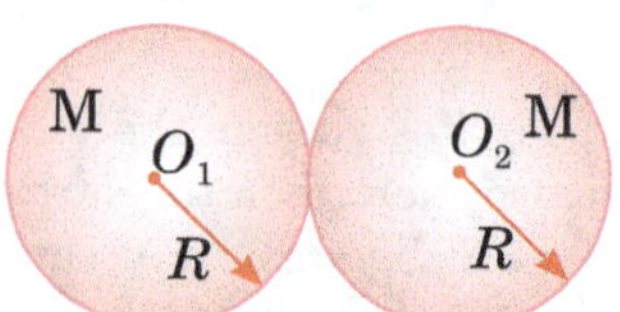

Figure 1.13

about their center, therefore their mass can be considered at their center. Furthermore, Newton's formula (Eq.1.1) also applicable for spherical bodies. So, apply Eq.1.1 and calculate gravitational force F acting between the spherical bodies.

SOLUTION The masses of the spheres may be assumed to be concentrated at their centres. So,

$$F = \frac{G\left[\frac{4}{3}\pi R^3 \rho\right] \times \left[\frac{4}{3}\pi R^3 \rho\right]}{(2R)^2} = \frac{4}{9}\left(G\pi^2\rho^2\right)R^4$$

Clearly, $F \propto R^4$

EXAMPLE 2. Three particles, each of mass m, are situated at the vertices of an equilateral triangle of side 'a'. The only forces acting on the particles are their mutual gravitational forces. It is desired that each particle moves in a circle while maintaining their original separation 'a'. Determine the initial velocity that should be given to each particle and time period of circular motion.

APPROACH To move all three particles on a circular path, the net force on each particle must be directed towards the center of the circle and it's magnitude must also provide the required centripetal force for circular motion. So, to find the speed of each particle, we first find the net gravitational force on any one particle due to other two and then equate it to the required centripetal force for that particle.

SOLUTION The resultant force on particle at A due to other two particles is [Fig.1.14]

$$F_A = \sqrt{F_{AB}^2 + F_{AC}^2 + 2F_{AB}F_{AC}\cos 60^\circ} = \sqrt{3}\frac{Gm^2}{a^2} \qquad \ldots (1)$$

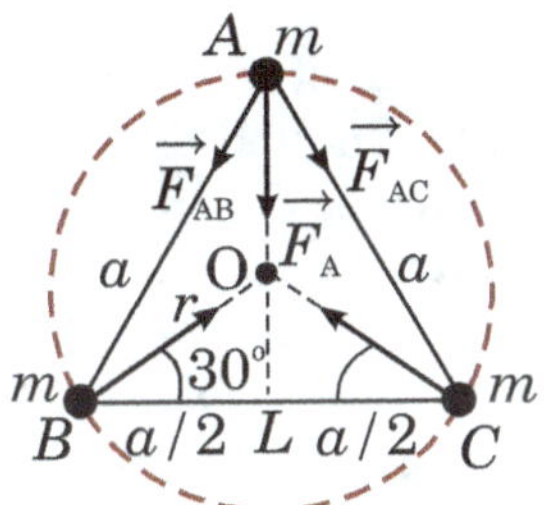

Figure 1.14

$$\left[\because F_{AB} = F_{AC} = \frac{Gm^2}{a^2} \right]$$

In ΔOLB, we have

$$\cos 30^\circ = \frac{a/2}{r}$$

$$\Rightarrow \quad r = \frac{a}{2\cos 30^\circ} = \frac{a}{\sqrt{3}}$$

Therefore, radius of the circle $r = \frac{a}{\sqrt{3}}$

If each particle is given a tangential velocity v, so that F acts as the centripetal force, then

$$\frac{mv^2}{r} = \sqrt{3}\frac{mv^2}{a} \qquad \ldots (2)$$

From Eq.(1) and (2) $\quad \sqrt{3}\frac{mv^2}{a} = \frac{Gm^2\sqrt{3}}{a^2} \Rightarrow v = \sqrt{\frac{Gm}{a}}$

Time period,

$$T = \frac{2\pi r}{v} = \frac{2\pi a}{\sqrt{3}}\sqrt{\frac{a}{Gm}} = 2\pi\sqrt{\frac{a^3}{3Gm}}$$

EXAMPLE 3. Four identical particles are arranged in a square, as shown in Figure 1.15. If the mass of each particle is M and the edge length of the square is a, then find net force on any one of the particle.

APPROACH Let us calculate the net gravitational force on the particle placed at point A, due to all other particles [Fig.1.16]. The particles at points B and D are equidistant from A, so gravitational forces applied by each of them on

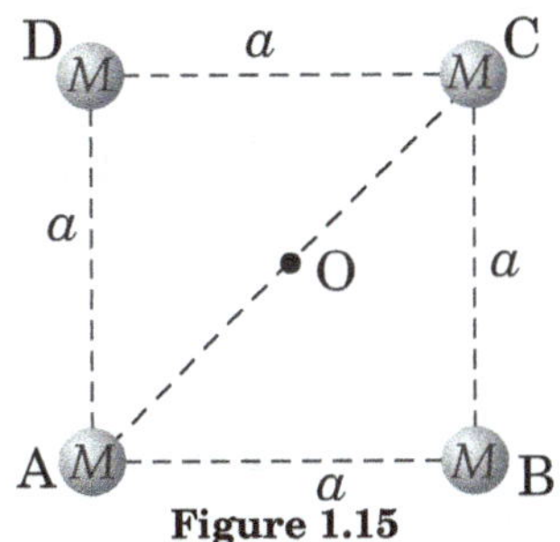

Figure 1.15

A, will have equal magnitude. Let us denote it by F_1. The gravitational force applied by particle C on the particle A, is F_2 and it is directed from A to C i.e., it is towards center 'O' of the square. These values can be obtained by applying Eq.1.1:

$$F_g = G\frac{m_1 m_2}{r^2} \qquad \ldots (1)$$

Due to symmetry, the resultant of forces of B and D on A, will also be towards the centre 'O' of the square.
Now, use vector algebra to get resultant gravitational force on the particle at A.

SOLUTION From Fig.1.16, the resultant of forces applied

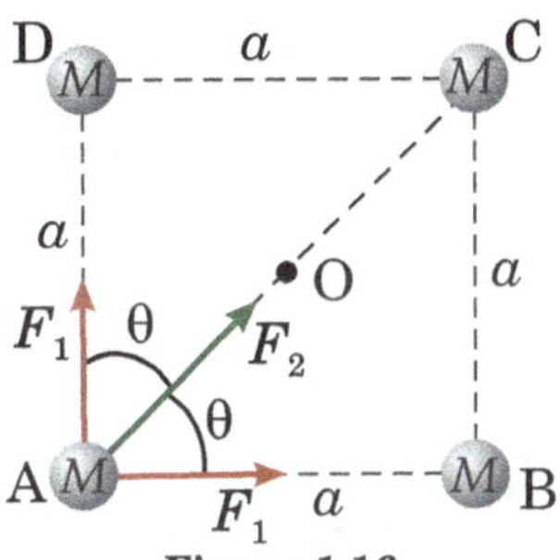

Figure 1.16

by particles B and D, is

$$= \sqrt{F_1^2 + F_1^2 + 2F_1 F_1 \cos 90^\circ} = F_1\sqrt{2}$$

It is directed towards the center of square, 'O'.
Since, both $F_1\sqrt{2}$ and F_2, are directed towards the centre 'O', therefore, net force on the particle at 'A', is given by

$$F_{net} = F_1\sqrt{2} + F_2 \qquad \ldots (2)$$

From Eq.(1), $F_1 = GM^2/a^2$ and $F_2 = GM^2/(a\sqrt{2})^2$
On substituting these values in Eq.(2), we get

$$F_{net} = \frac{GM^2}{a^2}\sqrt{2} + \frac{GM^2}{(a\sqrt{2})^2}$$
$$= \frac{GM^2}{a^2}\left(\sqrt{2} + \frac{1}{2}\right)$$

Aliter
From Fig.1.16, the magnitude of net gravitational force on the particle of mass M, at A, can also be given by-

$$F_{net} = 2F_1\cos\theta + F_2 \qquad \ldots (3)$$

Here, $\theta = 45^\circ$ $F_1 = GM^2/a^2$ and $F_2 = GM^2/(a\sqrt{2})^2$
On substituting these values in Eq.(3), we get

$$F_{net} = \frac{2GM^2}{a^2}\cos 45^\circ + \frac{GM^2}{(a\sqrt{2})^2}$$
$$= \frac{2GM^2}{a^2}\frac{1}{\sqrt{2}} + \frac{GM^2}{2a^2}$$
$$= \frac{GM^2}{a^2}\left(\sqrt{2} + \frac{1}{2}\right)$$

EXAMPLE 4. Five particles each of mass 'm' are kept at five vertices of a regular pentagon. A sixth particle of mass 'M' is kept at centre of the pentagon 'O'. Distance between 'M'

and '*m*' is '*a*'. Find:

(a) net force on '*M*'.

(b) magnitude of net force on '*M*', if any one particle is removed from one of the vertices.

APPROACH To get gravitational force applied by each particle on M, apply Eq.1.1. Net force on the particle at centre O, will be equal to the vector sum of all the forces on M.

SOLUTION (a) Let A, B, C, D, and E are the vertices of a regular pentagon, the angle between any two successive edges of the pentagon, is $(360/5)° = 72°$. Gravitational forces applied by particles at A, B, C, D and E on the particle at O are respectively- F_A, F_B, F_C, F_D, and F_E. Since, in a regular pentagon, all vertices are equidistant from O, therefore

$$F_A = F_B = F_C = F_D = F_E = \frac{GMm}{a^2} = F \text{ (say)}$$

Angle between two successive force vectors is

$$\theta = \frac{360°}{5} = 72°.$$

When these five force vectors are added as per polygon law of vector addition we get another closed regular polygon as shown in Fig.1.17b.

Therefore, net resultant force on '*M*' is zero.

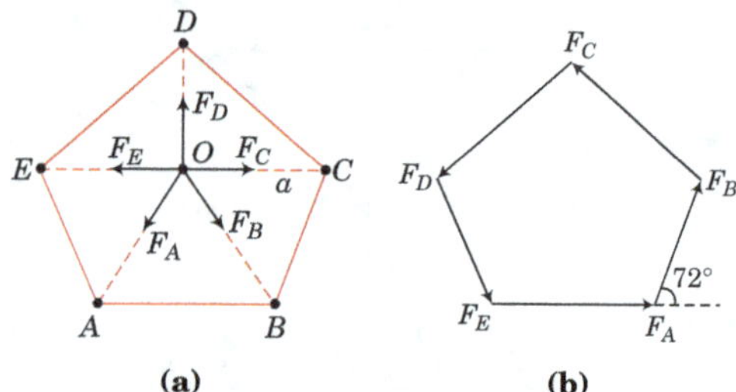

Figure 1.17

Note that, above result ($F_{net} = 0$) holds for any number of identical particles placed at vertices of a regular polygon.

(b) Removal of particle at A means removal of F_A. Now, you think like that- if there was force F_A, then the resultant of all the forces was zero. It means, F_A is equal and opposite to resultant of remaining forces. So, the magnitude of net force in absence of F_A will be equal to F_A, i.e.,

$$F_{net} = F_A = \frac{GMm}{a^2}$$

EXAMPLE 5. Six particles each of mass '*m*' are placed at six vertices A, B, C, D, E and F of a regular hexagon of side '*a*'. A seventh particle of mass '*M*' is kept at center '*O*' of the hexagon.

(a) Find net force on '*M*'.

(b) Find net force on '*M*' if particle at A is removed.

(c) Find net force on '*M*' if particles at A and C are removed.

APPROACH To get gravitational force applied by each particle on M, apply Eq.1.1. Now, the net force can be obtained by applying vector algebra as used in previous examples.

SOLUTION (a) Since, A, B, C, D, E and F are the vertices of a regular hexagon, therefore the angle between any two successive forces is 60°. Since, in a regular hexagon, all

vertices are equidistant from O, therefore

$$F_A = F_B = F_C = F_D = F_E = F_F = \frac{GmM}{a^2} = F \text{ (say)}$$

Here, corresponding to each above force there is an equal and

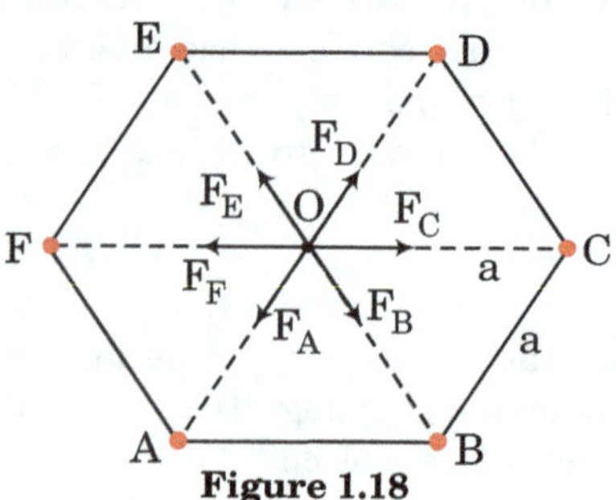

Figure 1.18

opposite force. So, net force on the seventh particle, placed at O, will be zero.

(b) Removal of the particle at A, means removal of the force F_A. In this case, there is no force to cancel F_D and hence this is the only force in excess on the particle at O.

$$\therefore \qquad F_{net} = F_D = \frac{GMm}{a^2} \text{ (towards } D \text{)}$$

(c) Removal of particles at A and C means removal of F_A and F_C. F_B is still get cancelled by F_E. Now, the remaining un-cancelled forces are F_D and F_F. So, net force is the resultant of these two forces F_D and F_F. The magnitude of these forces are equal and the angle between them is 120°. Also, the resultant of these forces, lies between them and divides the angle between these forces in two equal parts, i.e., it would be along $\overrightarrow{OE}$. Magnitude of this resultant is

$$F_{net} = \sqrt{F^2 + F^2 + 2(F)(F)\cos 120°}$$

$$= F = \frac{GMm}{a^2}(\text{ towards } E)$$

EXAMPLE 6. Two particles of masses 1 kg and 2 kg are placed at a separation of 50 cm. Assuming that the only forces acting on the particles are their mutual gravitation, find the initial acceleration of heavier particle.

APPROACH It is a problem based on Newton's second law of motion where gravitational force is the only force acting between the particles. Let us denote $m_1 = 1$ kg, $m_2 = 2$ kg and distance between the particles $r = 50$ cm $= 0.50$ m

Now, according to the problem, acceleration of mass m_2 is required.

Newton's second law of motion for particle of mass m_2 can be written as-

$$\Sigma F = m_2 a \qquad \qquad ... (1)$$

Here, ΣF is the net force acting on the particle of mass m_2 and a is the acceleration produced due to this force.

Since, there is only gravitational force between the particles, so from Eq.1.1, we have

$$\Sigma F = F_g = G\frac{m_1 m_2}{r^2} \qquad \qquad ... (2)$$

Now, calculate the value of net force on particle m_2 by using Eq.(2) and then use it in Eq.(1) and simplify for acceleration a

SOLUTION From above Ee.(2), the net force acting on heavier particle, of mass m_2, is

$$\Sigma F = \frac{Gm_1 m_2}{r^2} = \frac{6.67 \times 10^{-11} \times 1 \times 2}{(0.5)^2} = 5.34 \times 10^{-10} \text{N}$$

Therefore, from Eq. (1), the acceleration of heavier particle,
$$a = \frac{\Sigma F}{m_2} = \frac{5.3 \times 10^{-10}}{2} = 2.67 \times 10^{-10} ms^{-2}$$
This example shows that although gravitational force is very weak, it binds our solar system and also this universe of all galaxies and other interstellar system.

EXAMPLE 7. Two stationary particles of masses M_1 and M_2 are at a distance 'd' apart[Fig.1.19a]. A third particle lying on the line joining the particles, experiences no resultant gravitational force. What is the distance of this particle from M_1?

APPROACH If at distance r from M_1, the gravitational

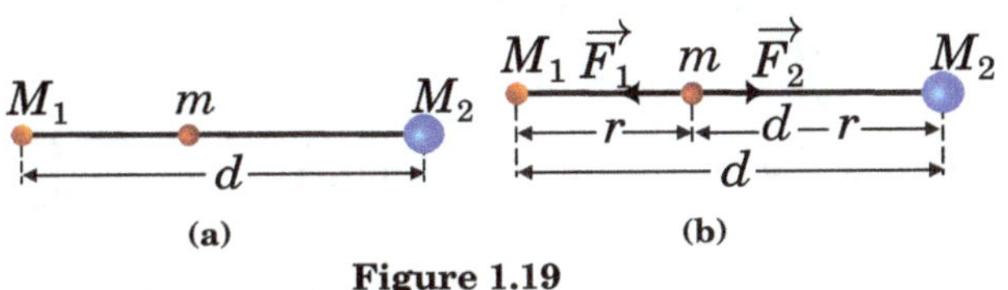

(a)   (b)

Figure 1.19

forces acting on m due to masses M_1 and M_2 are $\vec{F}_1$ and $\vec{F}_2$ respectively, then net force on m [Fig.1.19b]:
$$\Sigma \vec{F} = \vec{F}_1 + \vec{F}_2$$
Force $\vec{F}_1$ is directed towards M_1 and $\vec{F}_2$ towards M_2
Therefore, net gravitational force on mass m due to M_1 and M_2, in scalar form is
$$\Sigma F = F_2 - F_1$$
Since, mass m experiences no net force i.e., it is in translational equilibrium, therefore,
$$\Sigma F = 0 \quad \Rightarrow \quad F_2 - F_1 = 0$$
or $\qquad F_2 = F_1$ ... (1)
Now, to get the value of r, calculate the values of F_1, F_2 by using Eq.1.1 and substitute it in Eq. (1).
SOLUTION The system is shown in Fig.1.19a. The force on m towards M_1 is, $F_1 = \dfrac{GM_1 m}{r^2}$

The force on m towards M_2 is, $F_2 = \dfrac{GM_2 m}{(d-r)^2}$
According to question net force on m is zero, i.e.,
$$F_1 = F_2$$
$$\Rightarrow \quad \frac{GM_1 m}{r^2} = \frac{GM_2 m}{(d-r)^2} \quad \Rightarrow \left(\frac{d-r}{r}\right)^2 = \frac{M_2}{M_1}$$
$$\Rightarrow \quad \frac{d}{r} - 1 = \frac{\sqrt{M_2}}{\sqrt{M_1}} \quad \Rightarrow \quad r = \left[\frac{\sqrt{M_1}}{\sqrt{M_1} + \sqrt{M_2}}\right]$$
EXAMPLE 8. Two particles of equal mass (m) each move in a circle of radius (r) under the action of their mutual gravitational attraction [Fig.1.20]. Find the speed of each particle.

APPROACH The gravitational force acting between any two

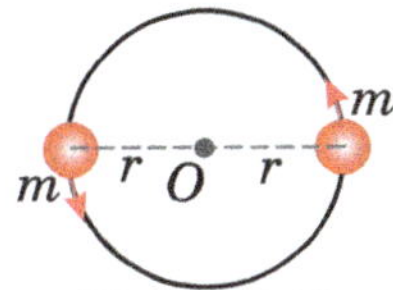

Figure 1.20

particles is always attractive, so, the direction of this force on

each particle is toward the other particle.

Now, for the motion of a particle in a circular path, the force must directed towards the centre of circle and it must always provide the required centripetal force. So, for motion of two particles in a circular path, due to their mutual gravitational attraction only, both particles must always remain diametrically opposite to each other, because this is the only position of the particles in which the gravitational force acting between the particles passes through the centre of the circle. To remain diametrically opposite to each other, both particles must also have equal speeds.

In such a case, the gravitational force acting on each particle, will be towards the centre of the circle which provides the required centripetal force for circular motion of particles.

To calculate the speed of each particle, find the gravitational force between the particles and then put it equal to required centripetal force for circular motion.

SOLUTION Given mass of each particle is m and radius of circular path is r. So, for particles at diametrically opposite to each other, the separation between them will be $2r$.

Now, the gravitational force acting between the particles is given by
$$F_g = \frac{Gmm}{(2r)^2} \qquad\qquad ... (1)$$
Required centripetal force, for each particle, is given by
$$F_C = \frac{mv^2}{r} \qquad\qquad ... (2)$$
Since, required centripental force is provided by gravitational foce, therefore from Eq. (1) and (2), we have
$$F_C = F_g$$
or $\quad \dfrac{mv^2}{r} = \dfrac{Gmm}{(2r)^2} \quad \Rightarrow \quad v^2 = \dfrac{Gm}{4r}$
$$\Rightarrow \qquad v = \frac{1}{2}\sqrt{\frac{Gm}{r}}$$

EXAMPLE 9. A large spherical mass M is fixed at one position and two identical point masses m are kept on a line passing through the center of M (see Fig.1.21). The point masses are connected by a rigid massless rod of length l and this assembly is free to move along the line connecting them. All three masses interact only through their mutual gravitational interaction. If the point mass nearer to M is at a distance $r = 3l$ from M, then find the value of m corresponding to which the tension in the rod becomes zero.

APPROACH Here, the important point is that, the two

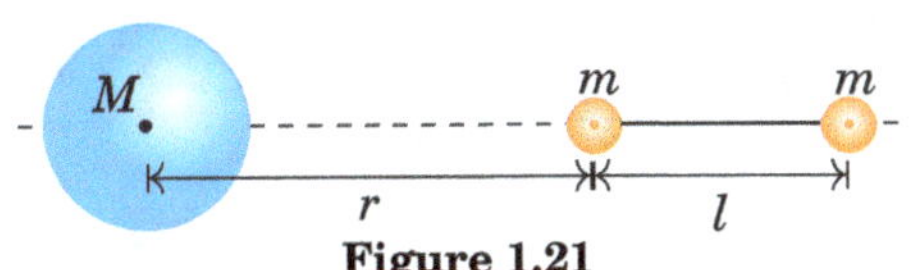

Figure 1.21

point masses are connected with a massless rod. So, the acceleration is same for both point masses. Also note that the tension in the light rigid rod is zero, therefore, it won't apply any force on point masses.

Draw free body diagrams (FBDs) of both point masses attached with light rod [Fig.1.22] and then, apply Newton's second law of motion on each particle. Here, you will get two equations. Now, set tension force equal to zero (given) for both the particles, and then simplify for mass of each particle (it is equal for both the particles)

SOLUTION In Fig.1.22, the distances of point masses at A and B from the centre of larger mass M, are r and $r + l$ respectively.

Free body diagrams (FBDs) of both the particles are shown in single diagram 1.22. A and B are the positions of the particles. The acceleration $\vec{a}$ of both point masses are equal because they are connected by a massless rigid rod of length l.

Forces acting on each particle, are shown in Fig.1.22.

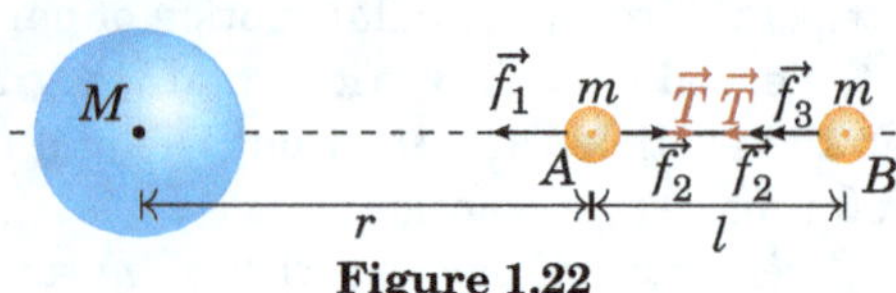

Figure 1.22

Gravitational forces acting on left and right particles due to large mass M are f_1 and f_3 respectively.

From Eq.1.1, we have $f_1 = GMm/r^2$ and $f_3 = GMm/(r+l)^2$. Gravitational force, due to mutual attraction between the two point masses, is $f_2 = Gmm/l^2$.

Suppose, the magnitude of tension in the rod is T (Here we have taken T as a general case, later we will set, $T = 0$.)

Applying Newton's second law on the point mass at A, gives

$$f_1 - f_2 - T = ma \qquad \ldots (1)$$

Similarly, applying Newton's second law on the point mass at B, we get

$$f_3 + f_2 + T = ma \qquad \ldots (2)$$

Now, corresponding to zero tension, i.e., $T = 0$, both equations (1) and (2) change to following Eq. (1a) and (2a) respectively.

$$f_1 - f_2 = ma \qquad \ldots (1a)$$
$$f_3 + f_2 = ma \qquad \ldots (2a)$$

On substituting the values of f_1, f_2 and f_3 in equations (1a) and (2a), we get-

$$\frac{GMm}{r^2} - \frac{Gmm}{l^2} = ma \qquad \ldots (1b)$$
$$\frac{GMm}{(r+l)^2} + \frac{Gmm}{l^2} = ma \qquad \ldots (2b)$$

Now, we eliminate a by subtracting Eq. (1b) in (2b):

$$\frac{GMm}{(r+l)^2} + \frac{Gmm}{l^2} - \frac{GMm}{r^2} + \frac{Gmm}{l^2} = 0 \qquad \ldots (3)$$

Given that: $r = 3l$, therefore, Eq.(3) gives:

$$\frac{GMm}{16l^2} + \frac{Gmm}{l^2} - \frac{GMm}{9l^2} + \frac{Gmm}{l^2} = 0$$

or $\quad \dfrac{M}{16} + 2m - \dfrac{M}{9} = 0$

or $\quad m = \dfrac{7M}{288}$

EXAMPLE 10. A mass M is splitted into two parts m and $(M - m)$, which are then separated by a certain distance. What ratio (m/M) maximises the gravitational force between the parts.

APPROACH Here, we have to use the concept of maxima and

minima.[2] Since, total mass M and the distance between the splitted parts, are constants, therefore the gravitational force between the splitted parts depends on m only.

So, for force F to be maximum: $\dfrac{dF}{dm} = 0$ and $\dfrac{d^2F}{dm^2} < 0$

SOLUTION If r is the distance between m and $(M - m)$, then from Eq.1.1, the gravitational force acting between the particles

$$F = G\frac{m(M-m)}{r^2} = \frac{G}{r^2}(mM - m^2) \qquad (1)$$

Here, r and M are constants. So, gravitational force F depends on m only.

On differentiating Eq. (1) with respect to m, we get

$$\frac{dF}{dm} = \frac{d}{dm}\left(\frac{G}{r^2}\left(mM - m^2\right)\right)$$

or $\qquad \dfrac{dF}{dm} = \dfrac{G}{r^2}(M - 2m) \qquad \ldots (2)$

Corresponding to maximum/minimum force, $\qquad \dfrac{dF}{dm} = 0$

$\Rightarrow \qquad M - 2m = 0 \qquad\qquad \left[\because \dfrac{G}{r^2} \neq 0\right]$

or $\qquad m = M/2$

Again differentiating Eq.(2), with respect to m, we get

$$\frac{d^2F}{dm^2} = -\frac{2G}{r^2}$$

which is negative and independent on m. So, it will also be negative for $m = M/2$. Therefore, force will be maximum for $m = M/2$

Hence, corresponding to maximum force between the two parts, $m/M = 1/2$.

EXAMPLE 11. (a) Write an expression for the force exerted by the Moon, mass M, on a particle of water, mass m, on the Earth at A, directly under the Moon, as shown in Fig. 1.23. The radius of the Earth is R, and the centre-to-centre Earth-Moon distance is r. (b) Suppose that the particle of water was at the centre of the Earth. What force would the Moon exert on it there? (c) Show that the difference in these forces is given by $\qquad F_T = \dfrac{2GMmR}{r^3}$

and represents the tidal force, the force on water relative to the Earth. What is the direction of the tidal force? (d) Repeat for a particle of water at B, on the far side of the Earth from the Moon. What is the direction of this tidal force? (e) Explain why there are two tidal bulges in the oceans (and solid Earth), one pointing toward the Moon and the other away from it.

SOLUTION (a) The magnitude of the gravitational force from the Moon on a particle at A is

$$F_A = \frac{GMm}{(r-R)^2},$$

where the denominator is the distance from the centre of the moon to point A.

(b) At the centre of the Earth the gravitational force of the moon on a particle of mass m is

---

[2] If $f(x)$ be any given function of x, then corresponding to maximum or minimum value of $f(x)$, the first order derivative of $f(x)$ is zero, i.e.,

$$\frac{df(x)}{dx} = 0 \qquad \ldots (1)$$

If for any value of x obtained from Eq. (1), the value of $\frac{d^2f(x)}{dx^2} > 0$, then $f(x)$ will be minimum and if $\frac{d^2f(x)}{dx^2} < 0$, then $f(x)$ will be maximum for that value of x.

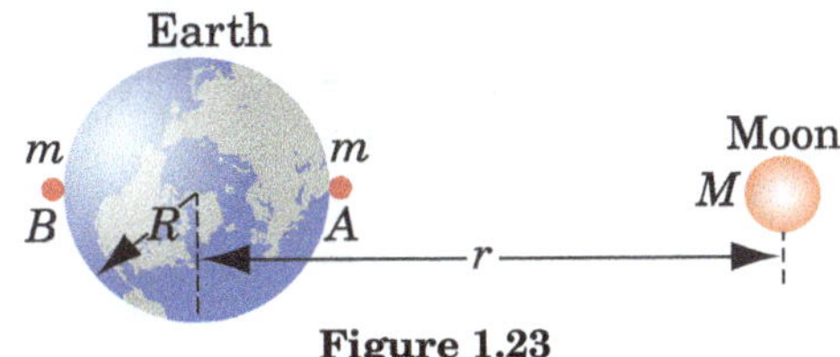

Figure 1.23

$$F_C = GMm/r^2$$

(c) Now we want to know the difference between these two expressions:

$$F_A - F_C = \frac{GMm}{(r-R)^2} - \frac{GMm}{r^2}$$

$$= GMm\left(\frac{r^2}{r^2(r-R)^2} - \frac{(r-R)^2}{r^2(r-R)^2}\right)$$

$$= GMm\left(\frac{r^2 - (r-R)^2}{r^2(r-R)^2}\right)$$

$$= GMm\left(\frac{R(2r-R)}{r^2(r-R)^2}\right)$$

To simplify assume $R \ll r$ and then substitute $(r-R) \approx r$, $2r-R \approx 2r$. The force difference simplifies to

$$F_T = GMm\frac{R(2r)}{r^2(r)^2} = \frac{2GMmR}{r^3}$$

(d) Repeat part (c) except we want $r+R$ instead of $r-R$. Then

$$F_A - F_C = \frac{GMm}{(r+R)^2} - \frac{GMm}{r^2}$$

$$= GMm\left(\frac{r^2}{r^2(r+R)^2} - \frac{(r+R)^2}{r^2(r+R)^2}\right)$$

$$= GMm\left(\frac{r^2 - (r+R)^2}{r^2(r+R)^2}\right)$$

$$= GMm\left(\frac{-R(2r+R)}{r^2(r+R)^2}\right)$$

To simplify assume $R \ll r$ and then substitute $(r+R) \approx r$, $2r + R \approx 2r$. The force difference simplifies to

$$F_T = GMm\frac{-R(2r)}{r^2(r)^2} = -\frac{2GMmR}{r^3}$$

The negative sign indicates that this "apparent" force points away from the moon, not toward it.

(e) Consider the directions: the water is effectively attracted to the moon when closer, but repelled when farther.

1.3.0.2 Calculation of Gravitational Force on a Body of a Continuous Mass Distribution

So for, we have dealt with only massive particles and spherical objects because Newton's law of gravitation applies only to massive particles and spherical objects. however, most objects - like a rod, circular ring , disk etc., are not particles. Instead, they are extended objects. Although every macroscopic object consists of very large number of massive particles but it is not feasible to calculate the gravitational force on each particle and then add them vectorially. Instead, we shall treat any macroscopic object as having a continuous mass distribution and divide it in to infinitesimally small elements. The mass of each such element is "dm" and the force on each such element

is assumed to be $d\vec{F}$. The resultant force on the complete body is then found from the superposition principle by integrating (i.e., adding) the force contributions due to all mass elements, i.e.,

$$\boxed{\vec{F} = \int d\vec{F}} \tag{1.4}$$

1.3.1 Gravitational force due to a rod

Let us consider a uniform rod of mass M and length l. We have to calculate the gravitational force on this rod due to a point particle of mass m placed at a distance r from one of its ends as shown in Fig.1.24a

To find the required gravitational force on the rod, choose a

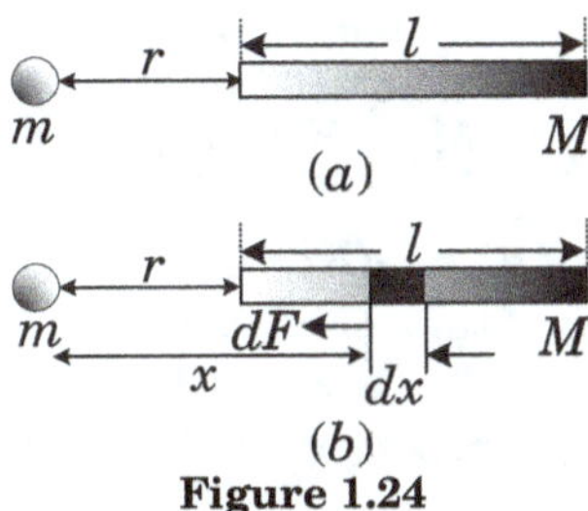

Figure 1.24

mass element dm of the rod of thickness dx at a distance x from the mass m as shown in Fig.1.24b. All such elements of the rod experience a gravitational force dF due to presence of the point mass m. We can find the total gravitational force of attraction experienced by the rod by integrating dF from $x = r$ to $x = r + l$.

Calculation of force on the rod: If λ be the linear mass density of the rod,i.e., $\lambda = M/l$, then mass of the element of width dx on the rod, (Fig.1.24b) at a distance x from the point mass m, is given by

$$dm = \lambda dx = \frac{M}{l}dx$$

The force on mass dm due to m is,

$$dF = \frac{Gm\,dm}{x^2}$$

$$dF = \frac{GmM}{lx^2}dx$$

To find the net force on rod we integrate the above expression for the whole rod, i.e., from a distance $x = r$ to $x = r + l$, so net force on the rod:

$$F = \int dF = \int_r^{r+l} \frac{GMm}{l}\frac{1}{x^2}dx$$

or $\qquad F = \frac{GMm}{l}\left[\frac{1}{r} - \frac{1}{r+l}\right] = \frac{GMm}{r(r+l)}$

EXAMPLE 12.   A thin rod of mass M and length L is bent in a semicircle (a) What is it's gravitational force (both in magnitude and direction) on a particle with mass m at O, the centre of curvature? (b) What would be the force on m if the rod is, in the form of a complete circle?(c) If mass of the rod is 20 kg, length is 5.0 m and the mass of a particle located at the centre of curvature is, 0.10 kg. Find the gravitational force exerted by the rod on it.

(a) APPROACH The semi-circular rod is shown in the Fig.1.25. We'll use an element of length $dl = Rd\theta$ whose mass

dM is $\lambda dl = (M/L)(Rd\theta)$. We'll first find dF_x and dF_y and then integrate over θ from 0 to π [By symmetry, we can write $F_x = 0$]. Now, the net force on a particle of mass m, placed at origin O, will be given by: $F = \sqrt{F_x^2 + F_y^2}$

SOLUTION Considering an element of rod of length dl as

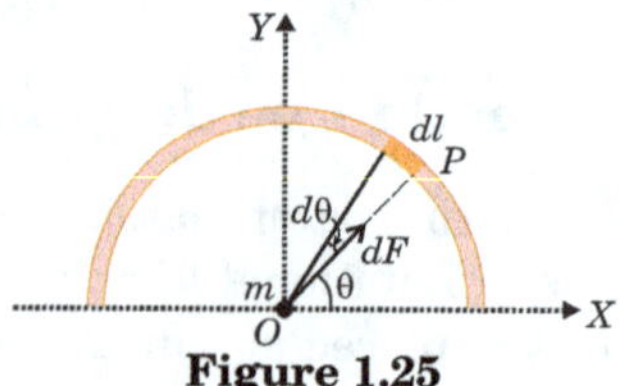

Figure 1.25

shown in figure and treating it as a point of mass $(M/L)dl$ situated at a distance R from P, the gravitational force due to this element on the particle will be

$$dF = \frac{GmdM}{R^2} = \frac{Gm(M/L)(Rd\theta)}{R^2}$$

along OP [as $dM = \lambda dl = \frac{M}{L}(Rd\theta)$]
So, the components of this force along x and y-axis will be
$$dF_x = dF\cos\theta = \frac{GmM\cos\theta d\theta}{LR}; dF_y = dF\sin\theta = \frac{GmM\sin\theta d\theta}{LR}$$
So that, $F_x = \frac{GmM}{LR}\int_0^\pi \cos\theta d\theta = \frac{GmM}{LR}[\sin\theta]_0^\pi = 0$

and $F_y = \frac{GmM}{LR}\int_0^\pi \sin\theta d\theta = \frac{GmM}{LR}[-\cos\theta]_0^\pi$
$= \frac{2\pi GmM}{L^2}$ [as $R = \frac{L}{\pi}$]
So, $F = \sqrt{F_x^2 + F_y^2} = F_y = \frac{2\pi GmM}{L^2}$ [as $L = \pi R$, i.e., F_x is zero]
i.e., the resultant force is along the y-axis and has magnitude $(2\pi GmM/L^2)$
(b) If the rod was bent into a complete circle, then
$F_x = \frac{GmM}{LR}\int_0^{2\pi} \cos\theta d\theta = 0$ and also $F_y = \frac{GmM}{LR}\int_0^{2\pi} \sin\theta d\theta = 0$
i.e, the resultant force on m at O due to the ring is zero.
(c) From part (a), the force on the particle is
$$F = \frac{2\pi GmM}{L^2}$$
On substituting the given values, we get
$$F = \frac{2(3.14)(6.67 \times 10^{-11})(0.10 \text{ kg})(20 \text{ kg})}{(5 \text{ m})^2}$$
$$= 3.351008 \times 10^{-11}\text{N} \approx 33.5 \text{ pN}$$

☞ This result is reasonable. As corresponding to each element there is an opposite element which applies equal and opposite force on 'm'.

EXAMPLE 13. A uniform ring of mass m and radius a is lying at a distance $\sqrt{3}a$ from the centre of a sphere having same radius but mass M. Find the magnitude of gravitational force between them.

APPROACH Consider two opposite very small segments each of mass dm at A and B positions of the ring (Fig.1.27). The whole ring can be considered as the sum of such symmetric segments. From Fig.1.27,
$$\tan\angle O'OB = \frac{a}{\sqrt{3}a} = \frac{1}{\sqrt{3}}$$
or $\angle O'OB = 30°$
Distance $AO = BO = \sqrt{a^2 + (\sqrt{3}a)^2} = 2a$
Since, each segment of the ring is at equal distance from the center of the sphere, therefore, the magnitude of force applied by sphere on each segment of the ring will be same.

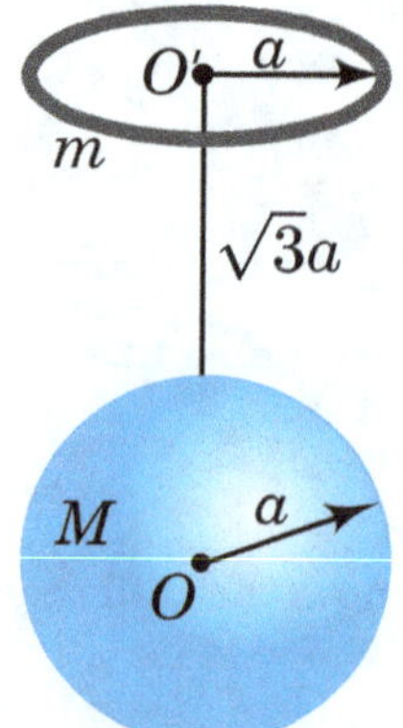

Figure 1.26

From Fig.1.27, the gravitational force applied by sphere on each segment at A and B is
$$dF = G\frac{M\,dm}{(2a)^2} = G\frac{M\,dm}{4a^2}$$
At each points A and B, the gravitational forces can be resolved into two components $dF\cos 30°$ and $dF\sin 30°$ Components $dF\sin 30°$ at A and B are toward the center of the ring i.e., opposite to each other and hence get cancelled. Similarly corresponding to each segment of the ring these components get cancelled. Now, components $dF\cos 30°$ at each point A and B is downward and so get added. This is true for all other segments of the ring. Net gravitational force on the ring will be the integration of $dF\cos 30°$ for complete ring.

SOLUTION Net force on ring $= \oint dF\cos 30°$

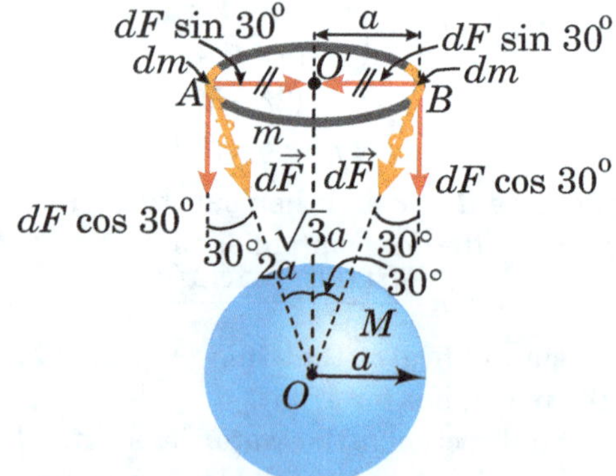

Figure 1.27

Here, the symbol '$\oint$', represents the integration for whole ring.

$$= \oint \frac{GM(dm)}{(2a)^2}\frac{\sqrt{3}}{2}$$
$$= \frac{\sqrt{3}GMm}{8a^2} \qquad \left[\because \oint dm = m\right]$$

1.3.2 Anti or Negative Gravitational Mass

In Newtons law of gravitation, there is always a gravitational attraction between two massive particles(Eq.1.1). These masses are generally said to be positive. Now, if we consider a particle having a hypothetical mass such that it is pushed away by a gravitational force of a positive mass, then this mass is said to be a negative mass.

For a negative-mass particle the force F and the acceleration a in $F = ma$ point in opposite directions, i.e., when you push it, it doesn't accelerate in the direction it was pushed. It accelerates backwards.

When a negative gravitational mass is placed in an external gravitational field, unlike positive gravitational mass, it experiences a gravitational force opposite to the external gravitational field.

If a negative gravitational mass $-m$ is placed in external gravitational field g, the gravitational force on $-m$, will be

$$\vec{F}_g = -m\vec{g}$$

In this chapter, we use the concept of negative mass and negative density (negative density = negative mass/volume), when we calculate the gravitational field of sphere or disc having hole inside it. **Note:** Recently, a fluid with negative mass have been created by Washington State University physicists. It can be used to explore some of the more challenging concepts of the cosmos.

EXAMPLE 14. Consider a 'mass dipole' consisting of two particles having opposite masses $m(>0)$ and $-m$. Describe its motion when the dipole is initially at rest in empty inertial space.

APPROACH As given above, opposite-mass particles repel each other with force: $F = \frac{Gm_1m_2}{r^2}$

SOLUTION (a) In empty inertial space the only force acting on neutral particles at a distance d apart is the gravitational force $F = Gm^2/d^2$. Since, we have considered opposite mass pairs, therefore, this force is repulsive. In response, each particle accelerates at the same rate $a = Gm/d^2$ but in the opposite direction.

Remarks

Despite its strange dynamical properties, a mass dipole would not violate any of the laws of physics. For example, despite the acceleration in empty space, energy is conserved because the total kinetic energy $\frac{1}{2}mv^2 + \frac{1}{2}(-m)v^2$ is always zero.

1.3.3 Properties of Gravitational Forces

Gravitational forces show following properties-

- Gravitational force between two masses of same nature (i.e., either both are positive or both are negative), is always attractive.
- Gravitational force between to masses of opposite nature is repulsive.
- Gravitational forces are developed in the form of action and reaction pair. Hence, they obey Newton's third law of motion.
- Unlike electrostatic force, it is independent of nature of medium in between two masses and presence or absence of other bodies.
- Gravitational forces are central forces as they act along the line joining the centres of two bodies.
- The gravitational forces are conservative forces so work done by gravitational force, on any object, does not depends upon path followed by the object.
- If any particle moves along a closed path under the action of gravitational force then the work done by this force is always zero.
- Gravitational force is weakest force of nature.
- Force developed between any two masses is called gravitational force and force between Earth and any body is called force of gravity.
- The net gravitational force acting on any particle due to number of particles is the vector sum of gravitational forces of attraction exerted on the given particle due to each particles i.e.,

$$\vec{F} = \vec{F}_1 + \vec{F}_2 + \vec{F}_3 + \dots$$

  It means the principle of superposition is valid.
- Gravitational force holds good over a wide range of distances. Infact, it is found true from interplanetary distances to interatomic distances.
- It is a two body interaction i.e. gravitational force between two particles is independent of the presence or absence of other particles.

1.3.4 Check Point 1

1. ••Two particles of masses m_1 and m_2, initially at rest at infinite distance from each other, move under the action of mutual gravitational pull. Without using the principle of conservation of mechanical energy, show that at any instant their relative velocity of approach is $\sqrt{2G(m_1+m_2)/R}$, where R is their separation at that instant.

2. ••Two lead balls of mass m and $2m$ are placed at a separation d. A third ball of mass m is placed at an unknown location on the line joining the first two balls such that the net gravitational force experienced by the first ball is $6Gm^2/d^2$. What is the location of the third ball?

3. Three point masses 'm' each are placed at the three vertices of an equilateral triangle of side 'a'. Find net gravitational force on any one point mass.

4. Three identical point masses m each are kept at the vertices A, B, and C of a cube having side length 'a' (fig.1.28). Another identical mass is placed at the center point D of the cube. (a) Where will you place a fifth identical mass so that the net gravitational force acting on mass at D becomes zero?

   (b) Calculate the net gravitational force acting on the mass at D.

5. •• Two point masses m and M are held at rest at a large distance from each other. When released, they begin moving under their mutual gravitational pull. (a) Find their relative acceleration (a) when separation between them becomes x.

   (b) Calculate the relative velocity of the two masses when their separation is x.

   (c) Write the velocity of centre of mass of the system when separation between them is x.

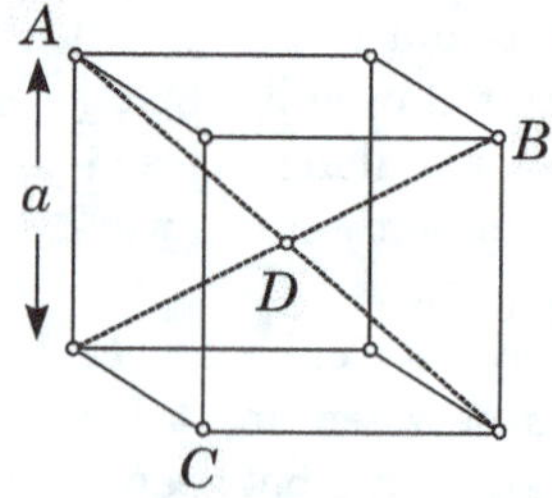

Figure 1.28

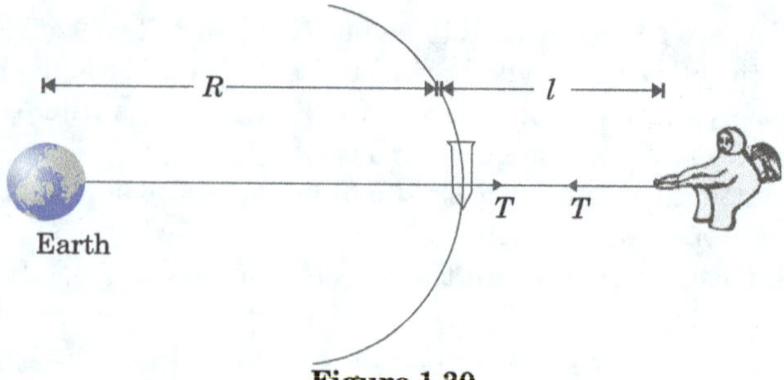

Figure 1.30

6. •• A small asteroid is at a large distance from a planet and its velocity makes an angle $\phi(\neq 0)$ with line joining the asteroid to the centre of the planet [Fig.1.29]. Prove that such an asteroid can never fall normally on the surface of the planet.

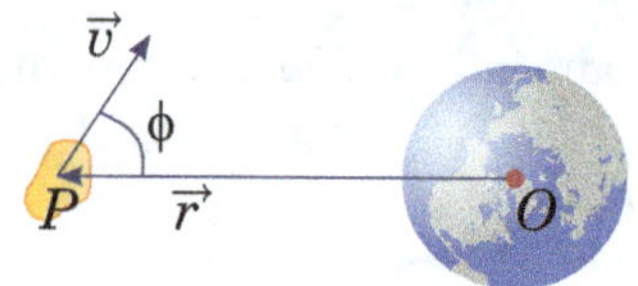

Figure 1.29

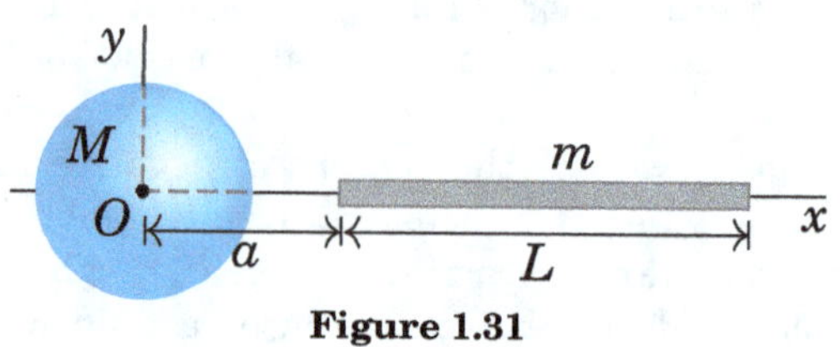

Figure 1.31

7. ••• Three identical particles, each of mass m, are located in space at the vertices of an equilateral triangle of side length a. They are revolving in a circular orbit under mutual gravitational attraction.
(a) Find the speed of each particle.
(b) Find the acceleration of the centre of mass of a system comprising of any two particles.
(c) Assume that one of the particles suddenly loses its ability to exert gravitational force. Find the velocity of the centre of mass of the system of other two particles after this.

8. •• A spaceship is orbiting the earth in a circular orbit at a height equal to radius of the earth ($R_e = 6400$ km) from the surface of the earth. An astronaut is on a space walk outside the spaceship. He is at a distance of $l = 200$ m from the ship and is connected to it with a simple cable which can sustain a maximum tension of 10 N . Assume that the centre of the earth, the spaceship and the astronaut are in a line. Mass of astronaut along with all his accessories is 100 kg.
(a) Do you think that a weak cable that can only take a load of 10 N, can prevent him from drifting in space? Make a guess.
(b) Estimate the tension in the cable.
(Acceleration due to gravity on the surface of Earth $g = 9.8$ m/s^2)

9. •• A uniform sphere of mass M is located near a thin, uniform rod of mass m and length L, as in Figure 1.31. Find the gravitational force of attraction exerted by the sphere on the rod.

Multiple Choice Questions

10. •• Two astronauts are floating in gravitational free space after having lost contact with their spaceship. The two will
(A) move towards each other
(B) move away from each other
(C) will become stationary
(D) keep floating at the same distance between them.

11. •• Two spheres of masses m and M are situated in air and the gravitational force between them is F. The space around the masses is now filled with a liquid of specific gravity 3 . The gravitational force will now be
(A) $3F$ (B) F (C) $F/3$ (D) $F/9$

12. •• Gravitational force is required for
(A) stirring of liquid   (B) convection
(C) conduction   (D) radiation

13. •• A body of weight 72 N moves from the surface of earth at a height half of the radius of earth, then gravitational force exerted on it will be
(A) $36N$ (B) $32N$ (C) $144N$ (D) $50N$

14. •• Two particles of equal mass m go around a circle of radius R under the action of their mutual gravitational attraction. The speed v of each particle is
(A) $\frac{1}{2}\sqrt{\frac{Gm}{R}}$ (B) $\sqrt{\frac{4Gm}{R}}$
(C) $\frac{1}{2R}\sqrt{\frac{1}{Gm}}$ (D) $\sqrt{\frac{Gm}{R}}$

15. •• The earth (mass $= 6\times 10^{24}$kg) revolves around the sun with an angular velocity of 2×10^{-7}rad/s in a circular orbit of radius 1.5×10^8km. The force exerted by the sun on the earth, in newton, is
(A) 36×10^{21} (B) 27×10^{39}
(C) zero   (D) 18×10^{25}

16. •• If the gravitational force between two objects were proportional to $1/R$ (and not as $1/R^2$), where R is the distance between them, then a particle in a circular path (under such a force) would have its orbital speed v, proportional to
(A) R (B) $1/R^2$
(C) R^0 (independent of R)   (D) $1/R$

1.4 Celestial terminology

Some important terms used in astrophysical situations, are defined here.

Star: A star is a celestial body that is massive enough to be held together by its own gravity and to be luminous (usually through nuclear reactions in the core). Stars range in mass from less than 10% of the mass of the Sun to about 150 times the mass of the Sun.

Planet: A planet is a significantly smaller object in orbit about a star, but one that is large enough to be approximately spherical and to have gravitationally "cleared" the space around it from similarly sized objects. By current convention the term planet is reserved for objects in orbit about our Sun. A planet in orbit about a star other than our Sun, is called an *exoplanet*. By the definition of planet, there are eight planets in orbit about our Sun. A *dwarf planet* is an object that would be considered a planet, but is not large enough to gravitationally clear the area of other similar sized objects. Pluto was designated a dwarf planet in 2006, along with similar sized objects in its region, as was Ceres, which orbits in the region between Mars and Jupiter.

Satellite: Celestial objects called satellites are in gravitational orbits not directly about the Sun but rather around a planet (and indirectly move about the Sun through the planet's orbit). You may think of satellite as a term for artificial objects in orbit about the Earth, such as weather and communications satellites. Satellite is a general term for objects in orbit about planets, and while including these artificial objects, it also includes our Moon and the many similar objects in orbit around other planets. For example, the planet Jupiter has at least 67 satellites, 4 of which were large enough to have been discovered by Galileo.

Asteroid: There is a much larger number of smaller solid objects in orbit, like planets, directly around the Sun. But these objects are not large enough to have achieved an approximately spherical shape and are called asteroids. It is estimated that there are about 2 million asteroids larger than 1 km in diameter, and almost 10 times that number larger than 100 m in diameter. Asteroids are generally of rocky composition. Many asteroids reside in the main asteroid belt, located between Mars and Jupiter.

Comet: A comet is of a size comparable to many asteroids, but the composition is a mix of ices and rocky materials. Comets were formed farther out in the early Solar System than asteroids. As comets approach the Sun, the ices vaporize and dust and gas are released.

Meteoroids: Smaller solid objects in orbit about the Sun are called meteoroids. There is not yet formal agreement on the boundary between small asteroids and large meteoroids, but usually, objects larger than 10m diameter, are called asteroids and those smaller than 1 m meteoroids, although most meteoroids are much smaller than that, extending down to dust-sized objects. The vast majority of meteoroids have origins in either comets or asteroids, although a few come from the surfaces of other planets or satellites.

Meteor: The term meteor refers to the light and ionization phenomena when a meteoroid ablates in a planetary atmosphere.

Meteorite: A meteorite refers to those meteoroids that reach the surface of Earth or another planet or satellite. Most meteors vaporize entirely and never become meteorites.

Solar System and Galaxy: We refer to the combination of the Sun and the objects directly, or indirectly, in orbit about it, as the Solar System. Stars, along with the planets, clouds of gas and dust, and smaller bodies in orbit around them, are part of much larger structures called galaxies. A typical galaxy contains millions to hundreds of billions of stars and is held together with the attractive gravitational force.

tidal Forces: The term tidal force refers to the difference in gravitational forces across any body. For example, if a spacecraft were near an extremely dense star, the difference between the gravitational forces on the side facing the star and the side facing away would produce a tidal force strong enough to tear the spacecraft apart.

This difference in the gravitational force on water on the different sides of Earth causes the tides. However, tides are by no means restricted to Earth. On Io, a natural satellite of Jupiter (i.e., the moon of Jupiter), the tidal forces actually change the solid surface of the satellite.

1.5 Latitude and Longitude

Any point on Earth can be defined by the intersection of its lines of latitude and longitude.

Latitude is measured as the angle from the equator, to the Earth's centre, to your position on the Earth's surface (Figure 1.32). It is expressed as degrees north or south of the equator (0°), with the poles at a latitude of 90°. Thus the poles are referred to as high latitude, while the equatorial region is considered low latitude. Lines of equal latitude are always the same distance apart, and so they are called parallels of latitude; they never converge. However, the circles created by the parallels of latitude do get smaller as they approach the poles.

Longitude measures the distance east or west of an imag-

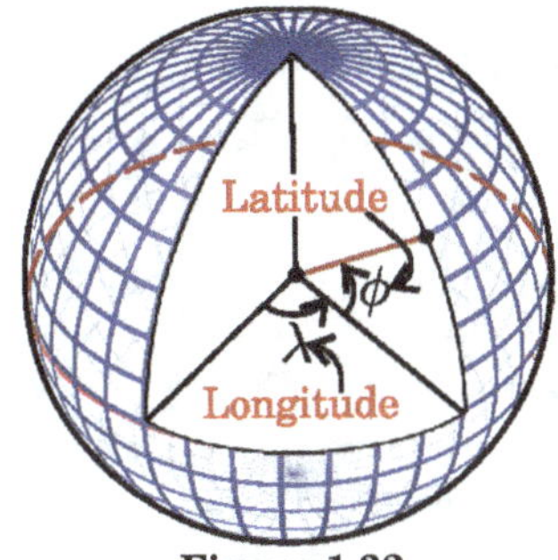

Figure 1.32

inary reference point, the prime meridian (0°), which is now defined as the line passing through Greenwich, England (although throughout history the prime meridian has also been located in Rome, Copenhagen, Paris, Philadelphia, the Canary Islands, and Jerusalem; unlike the equator, the prime meridian's location is fairly arbitrary). Your longitude represents the angle east or west between your location, the center of the Earth, and the prime meridian (Fig. 1.32).

As you move east and west from the prime meridian, eventually you reach 180° E and W on the opposite side of the globe from Greenwich. This point is the International Date Line. Lines of longitude are called meridians of longitude, or great circles. All circles of longitude are the same length, and are not parallel like lines of latitude; they converge as they near the poles.

1.6 The Force of Gravity on the Surface of the Earth (Weight) and Gravitational Acceleration

The weight of a body is the total gravitational force exerted on the body by all other bodies in the universe.

When the body is near the surface of the earth, we can neglect all other gravitational forces and consider the weight as just the earth's gravitational attraction. At the surface of the moon we consider a body's weight to be the gravitational attraction of the moon, and so on.

If we again model the earth as a spherically symmetric body with radius R and mass M, the weight w of a small body of mass m at the earth's surface (a distance R from its centre) is

$$w = F_g = \frac{GMm}{R^2} \qquad (1.5)$$

(weight of a body of mass m at the earth's surface)

But we also know that the weight w of a body is the force that causes the acceleration g of free fall, so by Newton's second law $w = mg$. Equating this with Eq.1.5, gives

$$mg = \frac{GMm}{R^2}$$

or $\qquad \boxed{g = \frac{GM}{R^2}} \qquad (1.6)$

(acceleration due to gravity at the earth's surface)

If the mass density of Earth is uniform throughout it's volume and suppose it is ρ, then

$$M = \frac{4}{3}\pi R^3 \rho$$

Substituting this value of M, in Eq.1.6, we get

$$\boxed{g = \frac{4}{3}\pi GR\rho \quad (g \text{ in terms of density of Earth})} \qquad (1.7)$$

So, for uniform mass density of Earth, $g \propto R$

The acceleration due to gravity g is independent of the mass m of the body because m doesn't appear in this equation.

Numerical value of g on the surface of earth:

$$g = \frac{GM}{R^2} = \frac{(6.67 \times 10^{-11} \frac{N.m^2}{kg^2})(5.97 \times 10^{24} kg)}{(6.37 \times 10^6 m)^2}$$

$$= 9.81 \ m/s^2$$

The apparent weight of a body on earth differs slightly from the earth's gravitational force because the earth rotates and is therefore not precisely an inertial frame of reference.

If we assume a spherical earth, the volume is

$$V = \frac{4}{3}\pi R^3 = \frac{4}{3}\pi \ (6.38 \times 10^6 m)^3 = 1.09 \times 10^{21} m^3$$

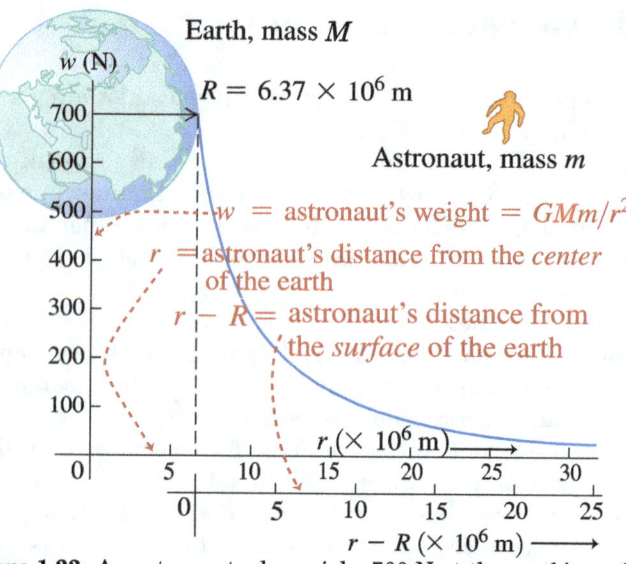

Figure 1.33: An astronaut who weighs 700 N at the earth's surface experiences less gravitational attraction when above the surface. The relevant distance r is from the astronaut to the centre of the earth (not from the astronaut to the earth's surface).

Therefore, the average density ρ (the Greek letter rho) of the earth is

$$\rho = \frac{\text{Total mass of earth (M)}}{\text{Total volume of earth (V)}}$$

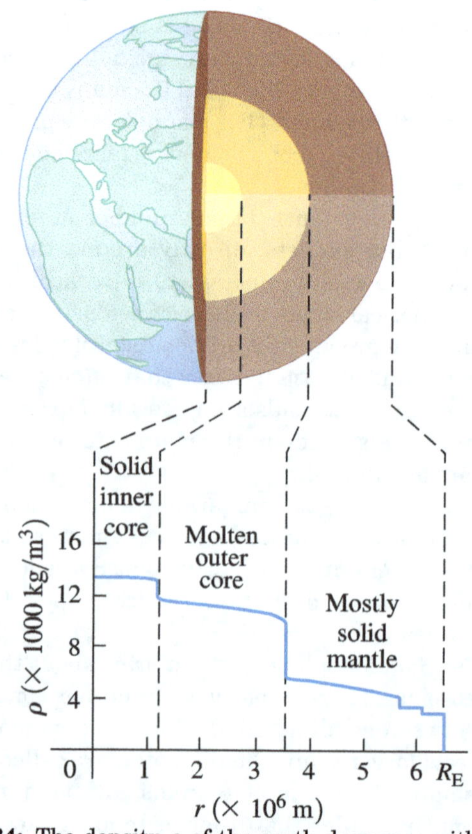

Figure 1.34: The density ρ of the earth decreases with increasing distance r from its center.

$$= \frac{M}{V} = \frac{5.97 \times 10^{24} \text{kg}}{1.09 \times 10^{21} \text{m}^3} = 5500 \text{ kg/m}^3 = 5.5 \text{ g/cm}^3$$

If the earth were uniform, we would expect the density of individual rocks near the earth's surface to have this same value. In fact, the density of surface rocks is substantially lower, ranging from about 2000 kg/m³ = 2 g/cm³ for sedimentary rocks to about 33000 kg/m³ = 3.3 g/cm³ for volcanic rock. So, the earth cannot be uniform, and the interior of the earth must be much more dense than the surface in order that the *average* density be 55000 kg/m³ = 5.5 g/cm³.

1.6.0.1 Relative Density (RD) or Specific Gravity (SG) of Earth

The relative density (RD) of any substance is defined as:

$$\text{RD} = \frac{\text{Density of substance}}{\text{Density of water at } 4°C}$$

Relative density is also called specific gravity (SG). It is a dimensionless quantity.

Since, the density of water at 4°C is 1000 kg/m³), therefore,

$$\text{RD}_{\text{earth}} = \frac{\text{Density of earth}}{\text{Density of water at } 4°C} = \frac{5500\text{kg/m}^3}{1000\text{kg/m}^3} = 5.5$$

1.6.0.2 Calculation of Error in Measurement of g

In form of density, ρ, the gravitational acceleration can be written as

$$g = \frac{GM}{R^2} = \frac{G}{R^2} \times \frac{4}{3}\pi R^3 \times \rho \quad \left[\because \quad M = \frac{4}{3}\pi R^3 \rho\right]$$

$$\Rightarrow \qquad g = \frac{4}{3}\pi GR\rho \qquad\qquad (1.8)$$

If the density of earth (ρ) is considered uniform throughout the volume of Earth, then $g \propto R$

Since, total mass of earth (M) is constant, therefore, $g \propto \dfrac{1}{R^2}$

In this case, % variation in 'g' (upto 5%), can be calculated as follows:

$$g \propto \frac{1}{R^2}$$

or $\qquad g = k\dfrac{1}{R^2}, \qquad k$ is a proportionality constant.

Taking natural log of both sides, we get

$$\ln g = \ln k + \ln \frac{1}{R^2} = \ln k + \ln 1 - 2\ln R$$

Differentiating both sides, yields-

$$\frac{dg}{g} = -2\frac{dR}{R}$$

Therefore, relative or fractional error in measurement of g:

$$\frac{\Delta g}{g} = -2\left(\frac{\Delta R}{R}\right)$$

So, the maximum possible fractional error in g, is given by

$$\left|\frac{\Delta g}{g}\right|_{\text{max}} = 2\left(\frac{\Delta R}{R}\right) \qquad\qquad \ldots(A)$$

If small change occurs in (M) and (R) then by

$$g = \frac{GM}{R^2}$$

we have: $\ln g = \ln G + \ln M - 2\ln R$

Differentiating both sides, yields-

$$\frac{dg}{g} = \frac{dM}{M} - 2\frac{dR}{R}$$

So, relative or fractional error in measurement of g:

$$\frac{\Delta g}{g} = \frac{\Delta M}{M} - 2\frac{\Delta R}{R} \qquad\qquad \ldots(B)$$

Therefore, the maximum possible fractional error in g, is given by

$$\left|\frac{\Delta g}{g}\right|_{\text{max}} = \frac{\Delta M}{M} + 2\left(\frac{\Delta R}{R}\right) \qquad\qquad \ldots(C)$$

Now, if R is constant, then

$$\left|\frac{\Delta g}{g}\right|_{\text{max}} = \frac{\Delta M}{M} \qquad\qquad \ldots(D)$$

EXAMPLE 15. If by keeping the mass same, the radius of the earth were shrink by one percent, then find the change in acceleration due to gravity.

APPROACH The acceleration due to gravity on the earth's surface is given by $\qquad g = GM/R^2 \qquad \ldots(1)$

Thus, for constant M, the value of g increases when radius R is decreased.

On taking natural logarithm of both sides of Eq.(1), we get

$$\ln g = \ln GM - 2\ln R \qquad\qquad \ldots(2)$$

Now, differentiating Eq.(2), to get the percentage in g.

SOLUTION Differentiate equation (2) and simplify to get percentage increase in the acceleration due to gravity,

$$\Delta g/g = -2(\Delta R/R) = -2(-1) = 2\%.$$

EXAMPLE 16. Two point-like objects, each with mass m, are connected by a massless rope of length l. The objects are suspended vertically near the surface of Earth, so that one object is hanging below the other. Then the objects are released. Show that the tension in the rope is

$$T = \frac{GMml}{R^3}$$

where M is the mass of the Earth and R is its radius.

APPROACH Let us denote the lower point mass by A and upper by B. Suppose, these are at distances r_1 and r_2 from the centre of earth. When the system is released, each point mass will have same acceleration because they are connected with a massless rope of length l.

Forces acting on point mass m at A are

1. Downward gravitational force $\vec{F}_1 = \frac{GMm}{r_1^2}$,

2. Upward tension force $\vec{T}$.

Forces acting on point mass m at B are

1. Downward gravitational force $\vec{F}_2 = \frac{GMm}{r_2^2}$, with $r_2 = r_1 + l$

2. Downward tension force $\vec{T}$.

Now apply Newton's second law of motion on both the masses and simplify for T

SOLUTION Equation of motion for point mass at A:

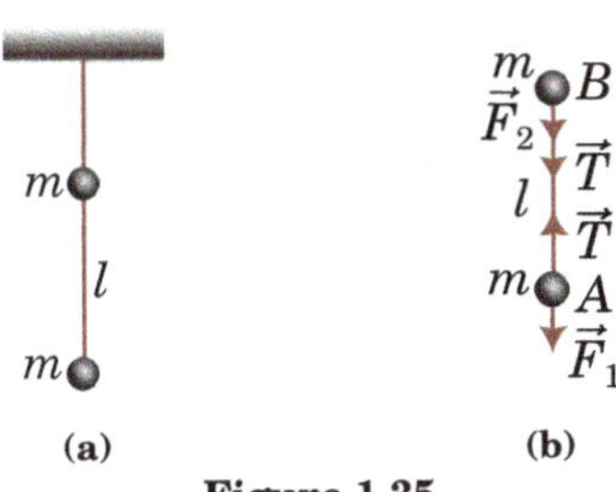

Figure 1.35

$$\frac{GMm}{r_1^2} - T = ma \qquad \qquad \dots (1)$$

Equation of motion for point mass at B:

$$\frac{GMm}{r_2^2} + T = ma \qquad \qquad \dots (2)$$

On eliminating ma from above equations, we get

$$\frac{GMm}{r_1^2} - T = \frac{GMm}{r_2^2} + T$$

or
$$2T = \frac{GMm}{r_1^2} - \frac{GMm}{r_2^2}$$

$$= \frac{GMm}{2}\left(\frac{1}{r_1^2} - \frac{1}{r_2^2}\right)$$

$$= \frac{GMm}{2}\frac{r_2^2 - r_1^2}{r_1^2 r_2^2}$$

Now $r_1 \approx r_2 \approx R$ in the denominator, but $r_2 = r_1 + l$, so $r_2^2 - r_1^2 \approx 2Rl$ in the numerator. So,

$$T \approx \frac{GMml}{R^3}$$

EXAMPLE 17. A plumb bob (a mass m hanging on a string) is deflected from the vertical by an angle θ due to a massive mountain nearby (Fig. 1.36). (a) Find an approximate formula for θ in terms of the mass of the mountain, m_M, the distance to its centre, D_M, and the radius and mass of the Earth. (b) Make a rough estimate of the mass of Mt. Everest, assuming it has the shape of a cone 4000 m high and base of diameter 4000 m. Assume its mass per unit volume is 3000 kg per m³. (c) Estimate the angle θ of the plumb bob if it is 5 km from the centre of Mt. Everest.

APPROACH Since, the plumb bob is in equilibrium, so draw

Figure 1.36

free body diagram (FBD) and and apply Newton's second law of motion for translational equilibrium.

SOLUTION (a) Free-body diagram (FBD) for the plumb bob, is shown in Fig.1.37. The attractive gravitational force on the plumb bob is $F_M = G\frac{mm_M}{D_M^2}$. Since the bob is not accelerating, the net force in all direction will be zero. Write the net force for both vertical and horizontal directions. Use $g = G\frac{M_{\text{Earth}}}{R_{\text{Earth}}^2}$.

$$\sum F_{\text{vertical}} = F_T \cos\theta - mg = 0$$

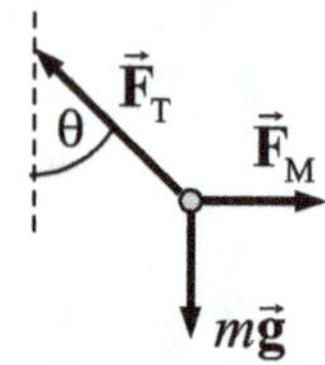

Figure 1.37

$$\Rightarrow \qquad F_T = \frac{mg}{\cos\theta}$$

$$\sum F_{\text{horizontal}} = F_M - F_T \sin\theta = 0$$
$$\Rightarrow \qquad F_M = F_T \sin\theta = mg\tan\theta$$
$$G\frac{mm_M}{D_M^2} = mg\tan\theta$$

$$\Rightarrow \qquad \theta = \tan^{-1}G\frac{m_M}{gD_M^2} = \tan^{-1}\frac{m_M R_{\text{Earth}}^2}{M_{\text{Earth}} D_M^2}$$

(b) We estimate the mass of Mt. Everest by taking its volume times its mass density. If we approximate Mt. Everest as a cone with the same size diameter as height, then its volume is $V = \frac{1}{3}\pi r^2 h = \frac{1}{3}\pi (2000m)^2 (4000m) = 1.7 \times 10^{10}m^3$. The density is $\rho = 3 \times 10^3 kg/m^3$.

Find the mass by multiplying the volume times the density.

$$M = \rho V = \left(3 \times 10^3 kg/m^3\right)\left(1.7 \times 10^{10} m^3\right) = 5 \times 10^{13} kg$$

(c) With $D = 5000m$, use the relationship derived in part (a).

$$\theta = \tan^{-1}\frac{M_M R_{\text{Earth}}^2}{M_{\text{Earth}} D_M^2} = \tan^{-1}\frac{\left(5 \times 10^{13} kg\right)\left(6.38 \times 10^6 m\right)^2}{\left(5.97 \times 10^{24} kg\right)(5000m)^2}$$

$$= 8 \times 10^{-4}\ \text{degrees}$$

1.6.1 Variation in g with Height Above the Surface of Earth

Let us consider a mass m at point P at a distance r form the centre of earth [see Fig.1.38]. The gravitational force between earth and mass m is

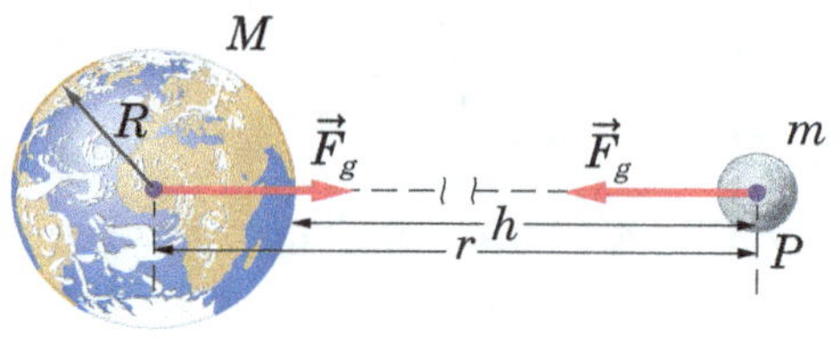

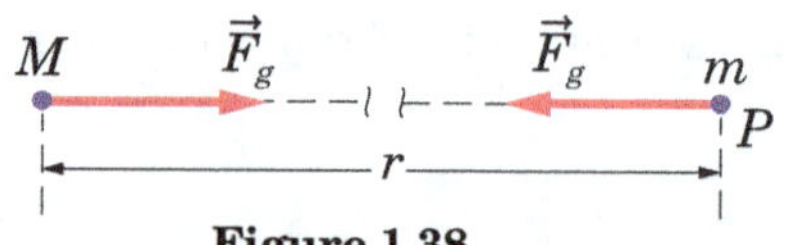

Figure 1.38

$$F_g = \frac{GMm}{r^2}$$

If g' is the effective gravitational acceleration of m at P, then

$$F_g = mg' = \frac{GMm}{r^2}$$

$$\Rightarrow \qquad \boxed{g' = \frac{GM}{r^2}} \qquad \qquad (1.9)$$

Since, $g = \dfrac{GM}{R^2}$, therefore, on substituting the value of GM from this to above equation gives-

$$\boxed{g' = \frac{gR^2}{r^2}} \qquad \qquad (1.10)$$

Since, $r > R$, therefore from Eq.1.10, $g' < g$

Eq.1.10 shows that the gravitational field outside a solid sphere is the same as if all the mass of the sphere were concentrated at it's centre. (This statement is correct only if the mass density of the sphere is uniform or if it varies only with distance from the center of the sphere.)

If h is the height of P from the surface of earth, then $r = R + h$. On substituting this value of r in Eq.1.10, we get

$$g' = \left(\frac{R}{R+h}\right)^2 g$$

$$\boxed{g' = \frac{g}{\left(1 + \frac{h}{R}\right)^2}} \qquad (1.11)$$

or $\qquad \boxed{g' = g\left(1 + \frac{h}{R}\right)^{-2}} \qquad (1.12)$

If $h \ll R$, then by binomial theorem, we can write

$$\boxed{g' = g\left(1 - \frac{2h}{R}\right)} \qquad (1.13)$$

From either Eq.1.9, 1.11 or 1.12 it is clear that-

(i) as we go above the surface of the earth, the value of 'g' decreases.

$$g' \propto \frac{1}{r^2} \quad \text{for} \quad r > R$$

(ii) if $r \to \infty$, then, $g' \to 0$. So, at infinite distance from the centre of earth, the value of 'g' becomes zero.

(iii) At the surface of earth, $r = R$, therefore from Eq.1.10,

$$g' = g = \frac{GM}{R^2}$$

Note: From Eq. 1.13, $\dfrac{g - g'}{g} = \dfrac{2h}{R}$

Therefore, percentage decrease in g on going at height h above the surface of earth is,

$$\boxed{\frac{\Delta g}{g} \times 100 = \frac{g - g'}{g} \times 100 = \frac{2h}{R} \times 100\%} \qquad (1.14)$$

So, at a height of $h = 100\ km = 10^5\ m$ above the surface of earth, the acceleration due to gravity decreases by a fraction of

$$\frac{2h}{R} = \frac{2 \times 100}{6400} \times 100 \cong 3\%$$

EXAMPLE 18. Two identical objects each of mass 'm' are hung from a balance whose scale pans differ in vertical height by 'h'[Fig.1.39]. If 'h' is sufficient to change the value of gravity, then determine the error in weighing in terms of average density of the Earth ρ.

APPROACH First of all write the expressions of gravitational accelerations at heights h_1 and h_2 and then find the weights of particles at these heights by applying relation $W = mg$. Now calculate the difference in their weights and substitute the mass of earth in terms of it's average density ρ.

SOLUTION From Eq. 1.13, the value of gravity at height h is given by,

$$g_h = g\left[1 - \frac{2h}{R}\right]$$

If g_1 and g_2 are the accelerations due to gravity at heights h_1 and h_2 respectively, then

weight at height h_1, $W_1 = mg_1 = mg\left[1 - \dfrac{2h_1}{R}\right]$

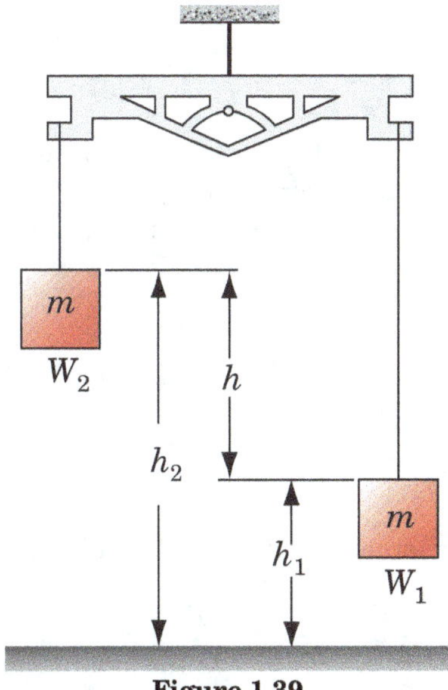

Figure 1.39

weight at height h_2, $W_2 = mg_2 = mg\left[1 - \dfrac{2h_2}{R}\right]$

Therefore, difference in weight

$$W_2 - W_1 = mg_2 - mg_1 = -2mg\left[\frac{h_2}{R} - \frac{h_1}{R}\right]$$

$$= -2\,m\frac{GM}{R^2} \times \frac{h}{R}$$

here, $h = h_2 - h_1$ [Fig.1.39]

$-$ve sign shows that weight decreases with height. Now, we consider just magnitude of this weight difference.

Therefore, error in weighing $\Delta W = |W_2 - W_1| = 2m\dfrac{GM}{R^2} \times \dfrac{h}{R}$

If ρ is the density of earth, then it's mass $M = \dfrac{4}{3}\pi R^3 \rho$, therefore,

$$g = \frac{GM}{R^2} = \frac{G\left(4\pi R^3 \rho/3\right)}{R^2} = \frac{4}{3}\pi R \rho G$$

So, error in weighing $\Delta W = \dfrac{8}{3}\pi \rho G m h$

EXAMPLE 19. A simple pendulum has a time period T_1 when on the earth's surface and T_2 when taken to a height R above the earth's surface, where R is the radius of the earth. What is the value of T_2/T_1?

APPROACH Time period of a simple pendulum at any position depends on the gravity at that position. If l is the length of string of simple pendulum and g is the gravity at the given position, then time period of simple pendulum at that point, can be obtained by relation

$$T = 2\pi\sqrt{\frac{l}{g}}$$

So, if g' is gravitational accelerations at height h above earth's surface, then time periods of simple pendulum at height h, will be

$$T' = 2\pi\sqrt{\frac{l}{g'}}$$

Therefore, $\dfrac{T'}{T} = \sqrt{\dfrac{g}{g'}}$ $\qquad\qquad$ ... (1)

Here, it is given that, the time period of simple pendulum at earth's surface is T_1 and at height R, it is T_2, therefore Eq.(1), changes to

$$\frac{T_2}{T_1} = \sqrt{\frac{g}{g'}} \qquad \ldots (2)$$

SOLUTION Now, from Eq.1.10, find the value of g/g' and substitute it in Eq.(2) and then simplify for T_2/T_1.

SOLUTION From Eq.1.10, the value of gravity at distance r from the center of earth is,

$$g' = \frac{gR^2}{r^2}$$

For height R, we have $r = R + R = 2R$, therefore

$$g' = \frac{gR^2}{(2R)^2}$$

So, $\qquad \dfrac{g}{g'} = 4 \qquad \ldots (3)$

SOLUTION On substituting the value of g'/g, from Eq.(3) in Eq.(2), we get

$$\frac{T_2}{T_1} = \sqrt{4} \quad \Rightarrow \quad \frac{T_2}{T_1} = 2$$

1.6.2   Variation in g With Depth Below the Surface of Earth

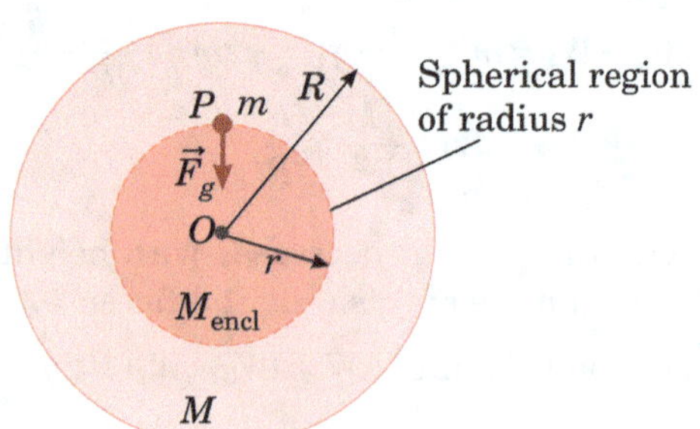

Figure 1.40

Let us consider any point P inside the earth at distance $r(<R)$ from the center of earth. The gravitational force of earth on a particle of mass m at P

$$F_g = G\frac{M_{encl}\, m}{r^2},$$

Here, $M_{encl} \to$ the portion of Earth's mass that is enclosed in a sphere of radius r.

Now, volume mass density of earth, $\rho = \dfrac{M}{\frac{4}{3}\pi R^3}$,

Volume of enclosed earth,

$$V = \frac{4}{3}\pi r^3$$

Therefore, $M_{encl} = \rho V = \rho\left(\dfrac{4}{3}\pi r^3\right) = \dfrac{M}{\frac{4}{3}\pi R^3} \times \left(\dfrac{4}{3}\pi r^3\right) = \dfrac{M}{R^3} r^3$

$$\therefore \quad F_g = G\frac{M_{encl}\, m}{r^2} = G\frac{\left(\frac{M}{R^3}r^3\right)m}{r^2} = G\frac{Mm}{R^3}r$$

If g' is the acceleration due to gravity at P, then $F_g = mg'$

$$\therefore \qquad mg' = G\frac{Mm}{R^3}r$$

or $\qquad g' = \dfrac{GM}{R^3}r$

Since, $g = \dfrac{GM}{R^2}$, therefore

$$g' = \frac{GM}{R^2}\frac{r}{R} = g\frac{r}{R}, \text{ i.e.,}$$

$$\boxed{g' = \frac{GM}{R^3}r = g\frac{r}{R}} \qquad (1.15)$$

$\because \quad r < R, \qquad\qquad \therefore \quad g' < g$

☞ At the center of earth, $r = 0$, therefore, from Eq.1.15, $g' = 0$

If h is the depth of point P from the surface of earth, then, $r = R - h$, therefore, from Eq.1.15, we can write-

$$\boxed{g' = \frac{g}{R}(R-h) = g\left(1 - \frac{h}{R}\right)} \qquad (1.16)$$

If we substitute the value of g in terms of density of Earth from Eq.1.7 in Eq.1.16, we get

$$\boxed{g' = \frac{4}{3}\pi G\rho(R-h)} \qquad (1.17)$$

So, at depth h, the % decrease in acceleration due to gravity, is

$$\boxed{\frac{g-g'}{g} \times 100 = \frac{h}{R} \times 100\%} \qquad (1.18)$$

Therefore, at a depth of 10 km below the earth's surface, the % decrease in the acceleration due to gravity is approximately, $\frac{10}{6400} \times 100 \approx 0.16$ **EXAMPLE 20.** At which depth from Earth

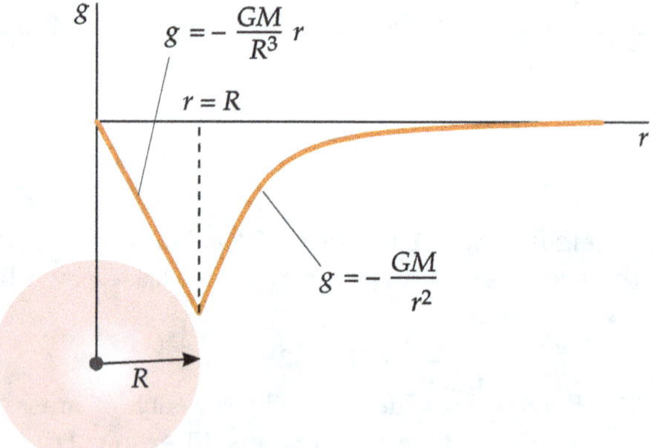

Figure 1.41

surface, acceleration due to gravity is decreased by 1%

APPROACH From Eq.1.18, the percentage change in gravity with depth h, is given by

$$\frac{g-g'}{g} \times 100 = \frac{h}{R} \times 100\% \qquad \ldots(1)$$

Now, substitute the given values and simplify for h.

SOLUTION Here, it is given that $\dfrac{g-g'}{g} \times 100 = 1$, therefore Eq.(1) gives

$$1 = \frac{h}{6400} \times 100 \quad \Rightarrow \quad h = 64\text{ km}$$

EXAMPLE 21. Imagine a tunnel dug along a diameter of the Earth [Fig.1.42]. Show that a particle (mass $= m$) dropped from one end of the tunnel executes simple harmonic motion. What is the time period of this motion? Assume the Earth to be a sphere of uniform mass density (equal to its known average density $= 5520$ kg/m^3) and $G = 6.67 \times 10^{-11}$N.m^2/kg^2. Neglect all damping forces.

APPROACH First of all, find the the acceleration of the par-

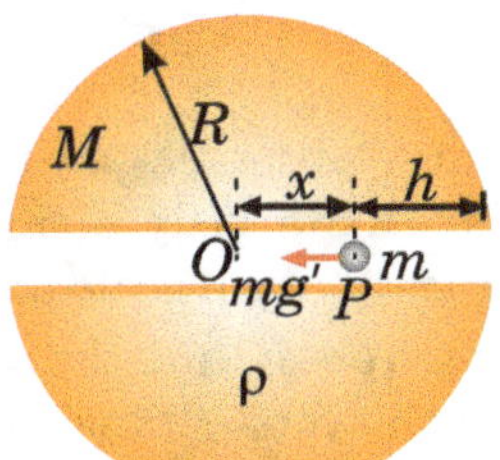

Figure 1.42

ticle at any general point P inside the earth. If it is proportional to $-ve$ of displacement from the centre of the Earth, then particle will execute a Simple Harmonic Motion. Now, transform the acceleration in the form of
$$\frac{d^2x}{dt^2} = -\omega^2 x$$

It is the form of SHM with time period, $T = 2\pi\sqrt{\dfrac{l}{g}}$

SOLUTION Let M be the mass of the Earth, R its radius and G, the universal gravitational constant.

If at any instant t, the particle is at point P at depth h from the surface of Earth, then from Eq.1.17, acceleration due to gravity is given by
$$g' = \frac{4\pi}{3}(R-h)\rho G \qquad \ldots (1)$$

$\therefore$ Gravitational force on the particle at P
$$F = mg' = m\frac{4\pi}{3}(R-h)\rho G \qquad \ldots (2)$$
Putting $(R-h) = x$, distance of P from the centre of the Earth, in Eq. (2), we get
$$F = -m\frac{4\pi}{3}x\rho G \qquad \ldots (3)$$
here, $-ve$ sign in RHS shows that the force is oppositely directed to position x from the center of Earth.

If d^2x/dt^2 is the acceleration of particle at P, then above equation can also be written as
$$m\frac{d^2x}{dt^2} = -m \times \frac{4\pi}{3}x\rho G$$
or $\qquad \dfrac{d^2x}{dt^2} = -\dfrac{4\pi G\rho}{3}x \qquad \ldots (4)$

Equation (4) is of the form of simple harmonic motion (SHM) given by equation:
$$\frac{d^2x}{dt^2} = -\omega^2 x \text{ (Differential form of SHM}^3) \qquad \ldots (5)$$
here, $\omega = \sqrt{\dfrac{4\pi G\rho}{3}}$, is the angular frequency of the SHM. Since, the acceleration of the particle at $P \propto (-x)$, i.e., $-ve$ of displacement from the centre of the Earth, the particle will execute a Simple Harmonic Motion.

The time period of the particle,
$$T = \frac{2\pi}{\omega} = 2\pi\sqrt{\frac{3}{4\pi\rho G}}$$
$$= \sqrt{\frac{3\pi}{\rho G}} = \sqrt{\frac{3 \times 3.1416}{5520 \times 6.67 \times 10^{-11}}}$$

---

^{3}It will be discussed in chapter "Simple Harmonic Motion"

$= 0.5059 \times 10^4$ seconds

$= 5059$ seconds $= 84.32$ minutes.

EXAMPLE 22.   To alleviate the traffic congestion between two cities A and B, engineers have proposed building a rail tunnel along a chord line connecting the cities (Fig. 1.43). A train, unpropelled by any engine and starting from rest, would fall through the first half of the tunnel and then move up the second half. Assuming Earth is a uniform sphere and ignoring air drag and friction, find the city-to-city travel time.

APPROACH First of all, find the the acceleration of the train

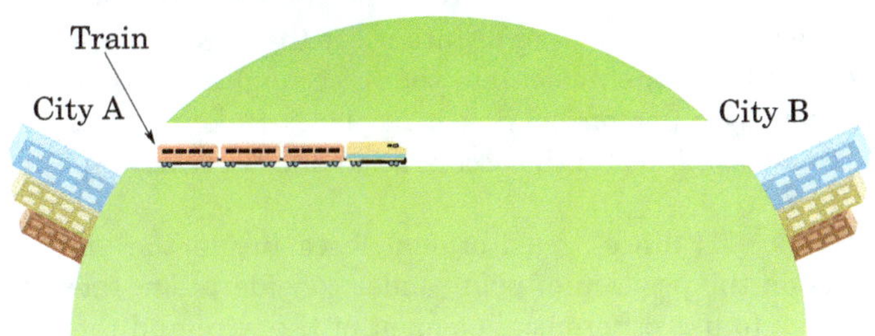

Figure 1.43

at any general point P inside the earth. If it is proportional to $-ve$ of displacement from the centre of the tunnel, then motion of the train would be simple harmonic. Now, transform the acceleration in the form of
$$\frac{d^2x}{dt^2} = -\omega^2 x$$

It is the form of SHM with time period, $T = 2\pi\sqrt{\dfrac{l}{g}}$

The travel time required from city A to city B is only half of it.

SOLUTION The gravitational force at a radial distance r inside Earth (e.g., point P in the Fig.1.44) is
$$F_g = -\frac{GMm}{R^3}r$$

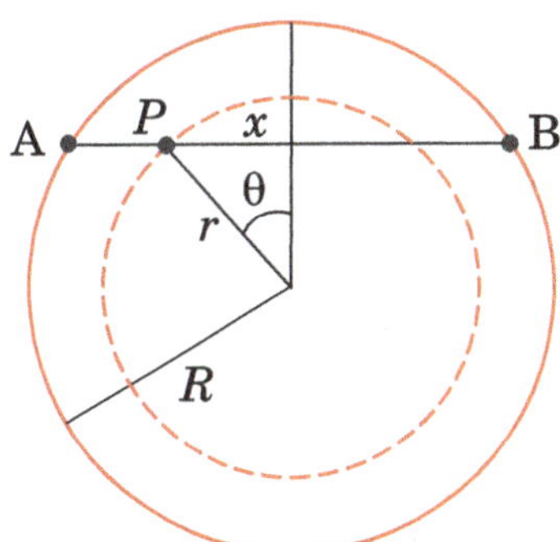

Figure 1.44

The component of the force along the tunnel is
$$F_x = F_g\sin\theta = \left(-\frac{GMm}{R^3}r\right)\frac{x}{r} = -\frac{GMm}{R^3}x$$
which can be rewritten as
$$a_x = \frac{d^2x}{dt^2} - \frac{GM}{R^3}x = -\omega^2 x \qquad \text{(Equation of SHM)}$$
where $\omega^2 = GM/R^3$. The equation is similar to Hooke's law, in that the force on the train is proportional to the displacement of the train but oppositely directed. Without exiting the tunnel, the motion of the train would be simple harmonic with time period given by $T = 2\pi/\omega$. The travel time required from city A to city B is only half of that (one-way):

$$\Delta t = \frac{T}{2} = \frac{\pi}{\omega} = \pi \sqrt{\frac{R^3}{GM}}$$

$$= \pi \sqrt{\frac{\left(6.37 \times 10^6 \, m\right)^3}{\left(6.67 \times 10^{-11} m^3/kg \cdot s^2\right)\left(5.98 \times 10^{24} kg\right)}}$$

$$= 2529 \, s = 42.1 \, min$$

Note that the result is independent of the distance between the two cities.

EXAMPLE 23. A planet of radius $R = \frac{1}{10} \times$ (radius of Earth) has the same mass density as Earth. Scientists dig well of depth $\frac{R}{5}$ on it and lower a wire of the same length and of linear mass density 10^{-3} kg/m into it. If the wire is not touching anywhere, then calculate the force applied at the top of the wire by a person holding it in place. [take the radius of Earth $= 6 \times 10^6$ m and the acceleration due to gravity on Earth $= 10$ m/s^2.]

APPROACH Since, gravitational force inside the well depends on the position of point under consideration, therefore we have to use differential element of the wire and then calculate the gravitational force on it. Now, integration of it from $r = \frac{4}{5}R$ to $r = R$ will give us the net gravitational force applied by earth. This force will be equal to the force applied by the person to hold it in place.

SOLUTION Given $R_e = 10R = 6 \times 10^6$ m and $g_e = 10$ m/s^2. Consider a wire element of length dr placed at a distance r from the centre O [Fig.1.45].

Let ρ be the common mass density of the planet and the earth

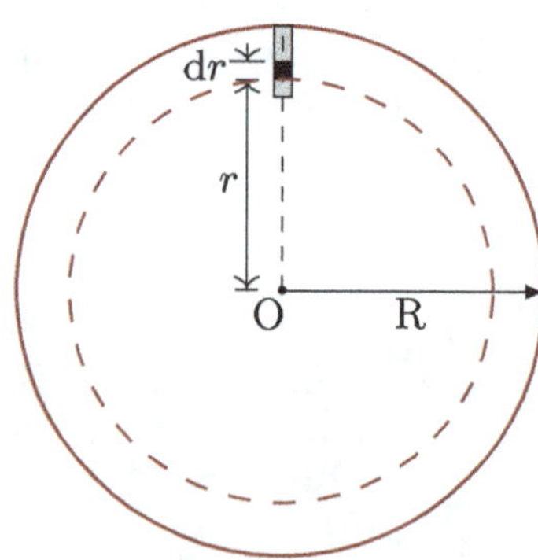

Figure 1.45

and m be the mass of the planet inside the sphere of radius r i.e., $m = \frac{4}{3}\pi r^3 \rho$. The gravitational force on the wire element by the planet is equal to the force by a point mass of magnitude m placed at the centre O. Thus, force on the wire element is

$$dF = \frac{Gm\lambda \, dr}{r^2} = \frac{4}{3}G\pi\rho\lambda r \, dr$$

Integrate dF from $r = \frac{4}{5}R$ to $r = R$ to get

$$F = \frac{4}{3}G\pi\rho\lambda \int_{\frac{4}{5}R}^{R} r \, dr = \frac{4}{3}G\pi\rho\lambda \left(\frac{9R^2}{50}\right)$$

$$= \frac{\frac{4}{3}\pi\rho R_e^3 G}{R_e^2}\left(\frac{9\lambda R_e}{5000}\right)$$

$$= g_e \frac{9\lambda R_e}{5000} = \frac{(10)(9)\left(10^{-3}\right)\left(6 \times 10^6\right)}{5000} = 108 \, N.$$

1.6.2.1 Variation with Latitude

Because the earth rotates on its axis, so it is a non-inertial frame of reference and we have to consider the pseudo force on all objects in this frame of reference, due to angular velocity (ω, say) of earth and hence of objects stationary with respect to earth.

Figure 1.46 shows three observers on the earth. Each one holds a spring scale with a body of mass m (true weight $\vec{w}_0 = m\vec{g}$) hanging from it. Each scale applies a tension force $\vec{F}$ to the body hanging from it, and the reading on each scale is the magnitude F of this force. If the observers are unaware of the earth's rotation, each one *thinks* that the scale reading equals the weight of the body because he thinks the body on his spring scale is in equilibrium. So, each observer thinks that the tension $\vec{F}$ must be opposed by an equal and opposite force $\vec{w}$, which we call the **apparent weight**. From Fig.1.46,

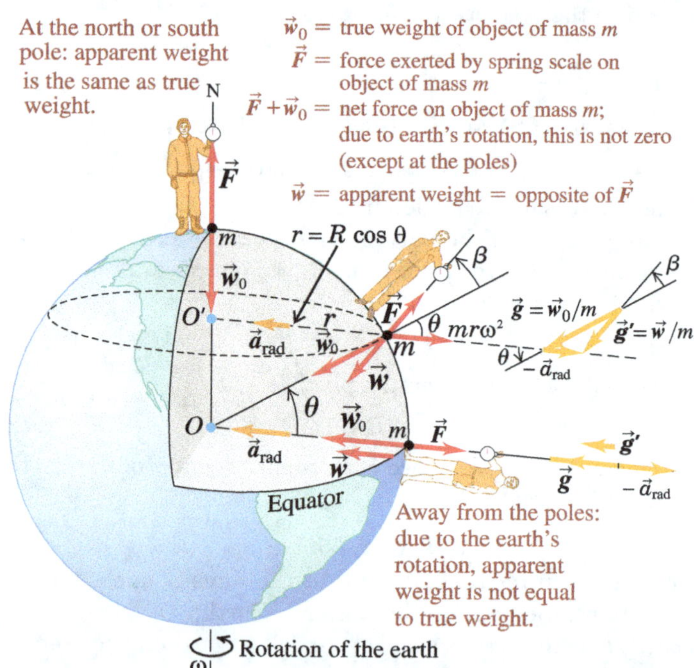

Figure 1.46: Except at the poles, the reading for an object being weighed on a scale (the *apparent weight*) is less than the gravitational force of attraction on the object (the *true weight*). The reason is that a net force is needed to provide a centripetal acceleration as the object rotates with the earth. For clarity, the illustration greatly exaggerates the angle β between the true and apparent weight vectors.

the pseudo force acting on the body $= mr\omega^2$

True weight $= w_0$

The angle between these two forces $= \pi - \theta$

Therefore, the resultant force, i.e., apparent weight of the body

$$w = \sqrt{w_0^2 + \left(mr\omega^2\right)^2 + 2w_0\left(mr\omega^2\right)\cos(\pi - \theta)}$$

[see Figure 1.46]

or $mg' = \sqrt{m^2g^2 + \left(mR\cos\theta\omega^2\right)^2 - 2mg\left(mR\cos\theta\omega^2\right)\cos\theta}$

$[\because \quad r = R\cos\theta, \, \cos(\pi - \theta) = -\cos\theta]$

or $\quad g' = \sqrt{g^2 + (R\cos\theta\omega^2)^2 - 2g(R\cos^2\theta)\omega^2}$

or $\quad g' = \sqrt{g^2 + (R^2\cos^2\theta\omega^4) - 2g(R\cos^2\theta)\omega^2}$

The value of ω is 7.2921159×10^{-5} rad/s. Therefore, the term containing ω^4 is much smaller than g^2 so, can be neglected. Therefore, above equation can be written as-

$$g' = [g^2 - 2g(R\cos^2\theta)\omega^2]^{1/2}$$

or $\quad g' = g\left[1 - \frac{2(R\cos^2\theta)\omega^2}{g}\right]^{1/2}$

On using binomial theorem and neglecting higher order terms, we get

or $\quad g' = g\left[1 - \frac{2\left(R\cos^2\theta\right)\omega^2}{2g}\right]$

$$\left[\because (1+x)^n \approx 1+nx \text{ for } x << 1\right] \text{ or}$$

$$g' = g\left[1 - \frac{(R\cos^2\theta)\omega^2}{g}\right] \quad g' = g - \omega^2 R\cos^2\theta$$

or $\qquad \boxed{g' = g - \omega^2 R\cos^2\theta} \qquad (1.19)$

or $\qquad \boxed{g' = g\left(1 - \frac{\omega^2 R}{g}\cos^2\theta\right)} \qquad (1.20)$

Note that, $\omega = 2\pi/86400$ s^{-1}, $R \cong 6400$ km, g = 9.8 m/s^2

So, $\omega^2 R/g \cong 0.34\%$, which is really a very small effect

At poles: $\theta = 90°$, therefore, from Eq.1.20, $g' = g$

$\Rightarrow$ the rotation of the earth has no effect on the gravity at poles.

At equator: $\theta = 0°$, therefore, Eq. 1.20, gives

$$\Rightarrow g' = g\left(1 - \frac{\omega^2 R}{g}\right)$$

Therefore, change in 'g', only due to rotation,

$$\Delta g_{\text{rot}} = g_{\text{pole}} - g_{\text{equator}}$$
$$= g - g' = \omega^2 R$$
$$= (2\pi/86400 \ s^{-1})^2 (6.4 \times 10^{-7} \text{ m}) \approx 0.03 \text{ m/s}^2$$

Hence, it is clear that, if rotation of Earth suddenly stops then acceleration due to gravity increases at all places on Earth except the poles.

Condition of weightlessness on the Surface of Earth

If apparent weight of body is zero then angular speed of Earth can be calculated as $mg' = mg - mR\omega^2\cos^2\theta$

$$0 = mg - mR\omega^2\cos^2\theta \Rightarrow \omega = \frac{1}{\cos\theta}\sqrt{\frac{g}{R}}$$

Conclusion

1. Due to rotational motion of earth, the effective gravity vector $\vec{g}$ does not pass through the centre of earth.
2. Rotational motion of earth reduces the effective value of gravity g.
3. At equator the effective value of gravity is minimum while at poles, it is maximum.

1.6.2.2 Weight of a Body moving on Earth's Surface from west to East at latitude θ

From Eq.1.20, the acceleration due to gravity at latitude θ, is

$$g' = g\left(1 - \frac{\omega^2 R}{g}\cos^2\theta\right)$$

If mass of the stationary body on the surface of earth, is m, then it's weight at latitude θ will be-

$$\boxed{W = mg - m\omega^2 R\cos^2\theta} \qquad (1.21)$$

Note that the Eq.1.21 holds for a body of mass m placed at rest on the surface of earth. In Eq.1.21, ω is the absolute angular velocity of the body placed on the earth surface about the axis of rotation of earth.

If $\vec{\omega}_e$ is the angular velocity of earth and $\vec{\omega}_b$ is the angular velocity of the body with respect to the earth, then absolute angular velocity of the body with respect to the axis of rotation of earth, is

$$\vec{\omega} = \vec{\omega}_{\text{earth}} + \vec{\omega}_{\text{body/earth}} = \vec{\omega}_e + \vec{\omega}_b$$

Note that earth rotates from west to east as seen by an observer over the northern hemisphere.

If the body also revolves in the same sense as earth, then

$$\omega = \omega_e + \omega_b$$

But, if the body revolves in opposite sense to the rotation of earth, then

$$\omega = \omega_e - \omega_b$$

Now, suppose, the body moves at speed v from west to east relative to earth, then absolute angular speed of the body with respect to axis of rotation of earth:

$$\omega_{\text{body}} = \omega_e + \omega_b, \text{ i.e.,-}$$
$$\omega = \omega_e + \frac{v}{R\cos\theta}$$

Using this value of ω in Eq.1.21, we get

$$W = mg - m\left[\omega_e + \frac{v}{R\cos\theta}\right]^2 R\cos^2\theta$$

or $\quad W = mg - m\left[\omega_e^2 + \frac{v^2}{R^2\cos^2\theta} + \frac{2m\omega_e v}{R\cos\theta}\right]R\cos^2\theta$

or $\quad W = mg - \left[m\omega_e^2 R\cos^2\theta + \frac{mv^2}{R} + 2\omega_e v\cos\theta\right]$

$$= mg\left[1 - \frac{\omega_e^2 R\cos^2\theta}{g} - \frac{2\omega_e v\cos\theta}{g}\right]$$

(Neglecting mv^2 as being very small)

1.6.2.3 Effect of the Shape of Earth

The equatorial radius is about 21 km longer than its polar radius. From Eq.1.6, we have,

$$g = \frac{GM}{R^2}$$

Hence, $g_{\text{pole}} > g_{\text{equator}}$.

So, the weight of the body increases as the body taken from the equator to the pole.

If R_p, R_e are respectively the polar, equatorial radii of earth and g_p, g_e are corresponding gravitational accelerations, then

$$g_p = \frac{GM}{R_p^2} \quad \text{and} \quad g_e = \frac{GM}{R_e^2}$$

with $R_e = R_p + 21$ km

here, $R_p = 6357$ km and $R_e = 6357 + 21 = 6378$ km

Therefore, difference in the values of gravity can be written as

$$g_p - g_e = \frac{GM}{R_p^2} - \frac{GM}{R_e^2}$$

By substituting the values of G, M and R_p, in above expression, we obtain:

$$\boxed{\Delta g_{\text{shape}} = g_p - g_e = 0.02 \text{ m/s}^2} \qquad (1.22)$$

So, due to shape of Earth, 'g_p' is greater than g_e by 0.02 m/s^2

Therefore, on considering shape and rotation both simultaneously, the net change in g:

$$[g_p - g_e]_{net} = (0.05 m/s^2) \longrightarrow < \begin{array}{ll} 0.03 m/s^2 & \text{(due to rotation)} \\ 0.02 m/s^2 & \text{(due to shape)} \end{array}$$

$$\text{(1.23)}$$

1.6.3 Condition of Not Flying Off the Object From the Surface of Earth

If we consider the rotation of earth, then the apparent weight of any object can be obtained by Eq.1.21

$$W = mg - m\omega^2 R\cos^2\theta \qquad \text{... (1)}$$

If object is at rest on the surface of earth, then ω will be equal to angular speed of earth.

From above equation, we can say that, if angular speed of earth keeps on increasing continuously, then apparent weight of the object keeps on decreasing. For a certain value of angular speed (ω), the apparent weight becomes zero and as soon as ω exceeds this value, the objects on its surface start flying off from it.

Let for $\omega = \omega_0$, the object feels the condition of weightlessness, i.e., $W = 0$, for $\omega = \omega_0$, then, Eq.(a) gives

$$0 = mg - m\omega_0^2 R\cos^2\theta$$

or $\qquad g = \omega_0^2 R\cos^2\theta$

or $\qquad \omega_0 = \dfrac{1}{\cos\theta}\sqrt{\dfrac{g}{R}}$

Since, ω_0 is inversely proportional to $\cos\theta$, therefore the needed value of ω_0 is minimum corresponding to maximum value of $\cos\theta$, i.e.,

$$\omega_{0,\,min} = \dfrac{1}{(\cos\theta)_{max}}\sqrt{\dfrac{g}{R}}$$

Since, $(\cos\theta)_{max} = 1$, which is corresponding to $\theta = 0$, i.e., a position of equator, therefore,

$$\omega_{0,\,min} = \sqrt{\dfrac{g}{R}}$$

For this value of angular speed, only the objects at equator feels weightlessness. Objects at other positions will still have some apparent weight.

If angular speed of earth exceeds this value of ω_0, then objects on the equator of earth's surface starts flying off.

So, objects will not flying off if angular velocity of earth is less than above value.

Thus, the condition for objects to not flying off the surface of earth is:

$$\omega \le \omega_0$$

If T is the time period of earth, then $\omega = 2\pi/T$, on substituting it in above equation, we get

$$\dfrac{2\pi}{T} \le \sqrt{\dfrac{g}{R}}$$

or $\qquad T \ge 2\pi\sqrt{\dfrac{R}{g}}$

So, the condition for rocks to not fly away from the equator of the earth becomes-

$$T \ge 2\pi\sqrt{\dfrac{R}{g}} = 2\pi\sqrt{\dfrac{R}{GM/R^2}}$$

$$= 2\pi\sqrt{\dfrac{R^3}{G \cdot \frac{4}{3}\pi R^3 \cdot \rho}} = \sqrt{\dfrac{3\pi}{G\rho}}$$

Key Points

1. Acceleration produced in a body due to the force of gravity is called the acceleration due to gravity.

2. The acceleration due to gravity is the rate of increase of velocity of a body freely falling towards the earth.

3. The acceleration due to gravity is equal to the force by which earth attracts a body of unit mass towards its centre.

4. Let 'm' be the mass of a body and 'F' be the force of attraction at a distance 'r' from the centre of earth, then acceleration due to gravity (g) at that place will be $g = \dfrac{F}{m} = \dfrac{GM}{r^2}$, where M = mass of earth.

5. The expression $g = GM/r^2$ is free from 'm' (mass of body). This means that the value of 'g' does not depend upon the shape, size and mass of the body. Hence if two bodies of different masses, shapes and sizes are allowed to fall freely, they will have the same acceleration. If they are allowed to fall from the same height, they will reach the earth simultaneously.

6. The acceleration of a body on the surface of the earth is $g = 9.81\ m/s^2$ or $981\ cm/s^2$.

7. Dimensional formula of g is $[M^0 L^1 T^{-2}]$.

8. The value of acceleration due to gravity depends on the following factors-
   (a) Height above the earth surface.
   (b) Depth below the earth surface.
   (c) Shape of the earth.
   (d) Axial rotation of the earth.

EXAMPLE 24. Determine the speed with which the earth has to rotate on its axis so that a person on the equator would weigh 20% as much as present. Take the equatorial radius as 6400 km.

APPROACH From Eq.1.21, the apparent weight of a person at latitude θ, due to rotation of the earth is given by

$$W' = W - m\omega^2 R\cos^2\theta$$

here, $W = mg$ is the actual wait at the pole of earth.

At equator $\theta = 0$

So, $\qquad W' = W - mR\omega^2 \qquad \text{... (1)}$

Now, substitute the given values in Eq.(1), and simplify for ω.

SOLUTION Substituting, $W = mg$, $W' = 20\%$ of mg, i.e., $W' = (20/100)mg = \frac{1}{5}mg$, in Eq.(1), we get

$$\tfrac{1}{5}mg = mg - mR\omega^2$$

$$\Rightarrow \qquad mR\omega^2 = \tfrac{4}{5}mg$$

$$\Rightarrow \qquad \omega = \sqrt{\left(\dfrac{4}{5}\dfrac{g}{R}\right)}, g = 9.8 m/s^2, R = 6400 \times 10^3\ m$$

$$\Rightarrow \qquad \omega = \sqrt{\left(\dfrac{4}{5} \times \dfrac{9.8}{6400 \times 10^3}\right)}\ \text{radian/sec.}$$

$$= 1.106 \times 10^3\ \text{rad/sec}$$

EXAMPLE 25. What should be the numerical value of the angular velocity of rotation of the earth and it's time period in order to make the effective acceleration due to gravity at the equator equal to zero?

APPROACH Use Eq.1.20 and simplify for ω.

SOLUTION From Eq.1.20, the effective acceleration at latitude θ is given by:

$$g' = g\left(1 - \frac{\omega^2 R}{g}\cos^2\theta\right) \qquad \ldots (1)$$

where $g = 9.8$ m/s^2 is a constant, ω is the angular speed of the earth and $R = 6400$ km $= 6.4 \times 10^6$ m is the radius of the earth. At equator, $\theta = 0$, therefore from above Eq. (1), we have-

$$g' = g - \omega^2 R,$$

when, $g' = 0$, then $g - \omega^2 R = 0$, so $\omega = \sqrt{g/R}$

Substitute the values to get ω:

$$\omega = \sqrt{g/R} = 1.24 \times 10^{-3}\,\text{rad/s}$$

Time period,

$$T = 2\pi/\omega = 2\pi/(1.24 \times 10^{-3}) = 5067.084 \text{ sec.}$$
$$= 84.45 \text{ min} = 1.4 \text{ hr}$$

Thus, to make effective acceleration equal to zero at equator, the new time period should be 1.4 h instead of 24h.

1.7 Gravitational Field Intensity (I_g or E_g or g)

The gravitational force exerted by a point particle of mass m_1 on a second point particle of mass m_2 a distance r_{12} away is given by

$$\vec{F}_{12} = -\frac{Gm_1 m_2}{r_{12}^2}\hat{r}_{12}$$

where $\hat{r}_{12} = \vec{r}_{12}/r_{12}$ is a unit vector directed away from particle 1 towards particle 2. The gravitational field at point P is determined by placing test mass[4] m_0 at P and calculating the gravitational force $\vec{F}_g$ on it due to all other particles. The gravitational force $\vec{F}_g$ divided by the mass m_0 is the gravitational field $\vec{g}$ at P

$$\boxed{\vec{g} = \frac{\vec{F}_g}{m_0}} \qquad (1.24)$$

The point P is called a field point.

For example, in Fig.1.47, we have a source point mass M at a point O and P is a point at distance r from O, where gravitational field due to this source point mass is required. To determine this field at P, we place a test point mass m at P and determine the gravitational force of M on m_0 at P. If F_g is the gravitational force of M on m_0, then from Eq.1.24, gravitational field at P will be given by

$$\vec{g} = \frac{\vec{F}_g}{m_0}$$

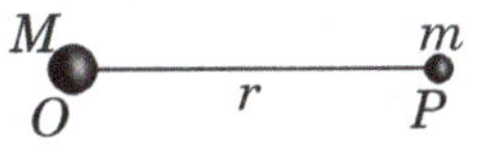

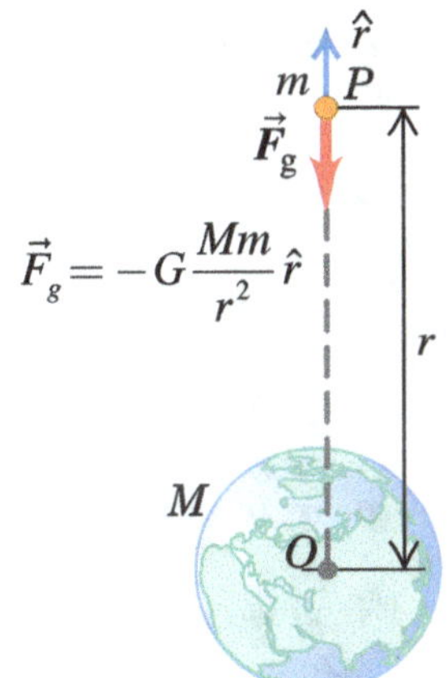

Figure 1.47

Since, $\vec{F}_g = -\dfrac{GMm_0}{r^2}\hat{r}$

Therefore, gravitational field $\vec{g} = \dfrac{\vec{F}_g}{m_0}$

---

[4]A very small mass, which cannot change the mass distribution of other nearby bodies, is called a test mass. For m_0 to be a test mass, $m_0 \to 0$

$$\text{or} \qquad \boxed{\vec{g} = -\frac{GM}{r^2}\hat{r}} \qquad (1.25)$$

The gravitational field at a field point due to the masses of a collection of point particles is the vector sum of the fields due to the individual masses:

$$\vec{g} = \sum_i \vec{g}_j \qquad (1.26)$$

The locations of these point particles are called source points. To find the gravitational field at a field point due to a continuous object, we find the field $d\vec{g}$ due to a small element of volume dV with mass dm and integrate over the entire mass distribution of the object (the entire set of source points).

$$\vec{g} = \int d\vec{g} \qquad (1.27)$$

The gravitational field of Earth (also called gravitational acceleration) at a point P at distance $r \geq R$ ($R =$ radius of Earth) from the center of Earth, points toward the center of Earth. It's magnitude $g(r)$ is given by Eq.1.9 [see Fig.1.48]

Figure 1.48

$$g(r) = \frac{F_g}{m} = \frac{GM}{r^2} \qquad (1.28)$$

In vector form, Eq.1.28, can be written as-

$$\vec{g} = -\frac{GM}{r^2}\hat{r} \qquad (1.29)$$

The intensity of gravitational field at a point is the net gravitational force exerted on a unit test mass kept at that point.

Dimensional formula of gravitational field intensity is-

$$[g] = \frac{[F]}{[m]} = \frac{[MLT^{-2}]}{[M]} = [M^0 LT^{-2}]$$

EXAMPLE 26. A 3.0-kg space probe experiences a gravitational force of 12 N$\hat{i}$ as it passes through point P. What is the gravitational field at point P ?

APPROACH The gravitational field at any point is defined by $\vec{g} = \vec{F}/m$.

SOLUTION Using its definition, express the gravitational field at a point in space:
$$\vec{g} = \frac{\vec{F}}{m} = \frac{(12N)\hat{i}}{30\,kg} = (4.0\ N/kg)\hat{i}$$

EXAMPLE 27.   The gravitational field at some point is given by $\vec{g} = 2.5 \times 10^{-6}$ N/kg$\hat{j}$. What is the gravitational force on a 0.0040 kg object located at that point?

APPROACH The gravitational force acting on an object of mass m where the gravitational field is $\vec{g}$ is given by $\vec{F} = m\vec{g}$.

SOLUTION The gravitational force acting on the object is the product of the mass of the object and the gravitational field:
$$\vec{F} = m\vec{g}$$
Substitute numerical values and evaluate $\vec{g}$:
$$\vec{F} = (0.0040 kg)\left(2.5 \times 10^{-6} N/kg\right)\hat{j}$$
$$= \left(1.0 \times 10^{-8} N\right)\hat{j}$$

EXAMPLE 28. A point particle of mass m is on the x axis at $x = L$ and an identical point particle is on the y axis at $y = L$. (a) What is the direction of the gravitational field at the origin? (b) What is the magnitude of this field?

APPROACH We can use the definition of the gravitational field due to a point mass to find the x and y components of the field at the origin and then add these components to find the resultant field.

APPROACH (a) The gravitational field at the origin is the sum of its x and y components:
$$\vec{g} = \vec{g}_x + \vec{g}_y \qquad \text{... (1)}$$
The gravitational field due to the point mass at $x = L$:
$$\vec{g}_x = \frac{Gm}{L^2}\hat{i}$$
The gravitational field due to the point mass at $y = L$:
$$\vec{g}_y = \frac{Gm}{L^2}\hat{j}$$
Substituting in equation (1), we get
$$\vec{g} = \vec{g}_x + \vec{g}_y = \frac{Gm}{L^2}\hat{i} + \frac{Gm}{L^2}\hat{j}$$
$\Rightarrow$ the direction of the gravitational field is along a line at $45°$ above the $+x$ axis.

(b) The magnitude of $\vec{g}$ is given by:
$$|\vec{g}| = \sqrt{g_x^2 + g_y^2}$$
Substituting the values of g_x and g_y, we get:
$$|\vec{g}| = \sqrt{\left(\frac{Gm}{L^2}\right)^2 + \left(\frac{Gm}{L^2}\right)^2} = \sqrt{2}\,\frac{Gm}{L^2}$$

EXAMPLE 29.   Five objects, each of mass M, are equally spaced on the arc of a semicircle of radius R, as in Figure 1.49. An object of mass m is located at the center of curvature of the arc. (a) If M is 3.0 kg, m is 2.0 kg, and R is 10 cm, what is the gravitational force on the particle of mass m due to the five objects? (b) If the object whose mass is m is removed, what is the gravitational field at the center of curvature of the arc?

APPROACH We can find the net force acting on m by super-position of the forces due to each of the objects arrayed on the circular arc. Once we have expressed the net force, we can find the gravitational field at the center of curvature from its defi-nition. Choose a coordinate system in which the $+x$ direction is to the right and the $+y$ direction is upward.

SOLUTION (a) Express the net force acting on the object whose mass is m :
$$\vec{F} = F_x\hat{i} + F_y\hat{j} \qquad \text{... (1)}$$
F_x is given by:
$$F_x = \frac{GMm}{R^2} - \frac{GMm}{R^2} + \frac{GMm}{R^2}\cos 45° - \frac{GMm}{R^2}\cos 45° = 0$$

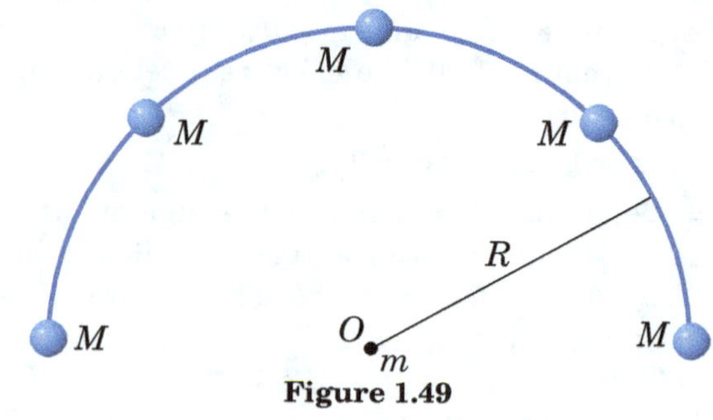

Figure 1.49

F_y is given by:
$$F_y = \frac{GMm}{R^2} + \frac{GMm}{R^2}\sin 45° + \frac{GMm}{R^2}\sin 45°$$
$$= \frac{GMm}{R^2}\left(2\sin 45° + 1\right)$$
Substitute numerical values and evaluate F_y :
$$F_y = \frac{(6.673 \times 10^{-11} N \cdot m^2/kg^2)(3.0 kg)}{(0.10 m)^2}(2.0 kg)(2\sin 45° + 1)$$
$$= 9.67 \times 10^{-8}\ \text{N}$$
Substitute in equation (1) to obtain:
$$\vec{F} = 0\hat{i} + (97 \times 10^8 N)\hat{j}$$

EXAMPLE 30.   A point particle of mass $m_1 = 2.0$ kg is at the origin and a second point particle of mass $m_2 = 4.0$ kg is on the x axis at $x = 6.0$ m. Find the gravitational field $\vec{g}$ at (a)$x = 2.0$ m, and (b) $x = 12$ m. (c) Find the point on the x axis for which $g = 0$.

APPROACH The configuration of point masses is shown in

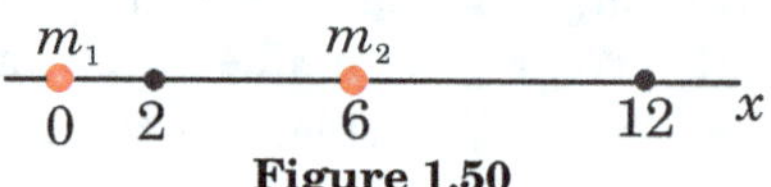

Figure 1.50

Fig.1.50. The gravitational field at any point can be found by superimposing the fields due to each of the point masses.

SOLUTION (a) The gravitational field at $x = 2.0\ m$ as the sum of the fields due to the point masses m_1 and m_2:
$$\vec{g} = \vec{g}_1 + \vec{g}_2 \qquad (1)$$
Here,
$$\vec{g}_1 = -\frac{Gm_1}{x_1^2}\hat{i} \quad \text{and} \quad \vec{g}_2 = \frac{Gm_2}{x_2^2}\hat{i}$$
On substituting in equation (1), we get
$$\vec{g} = -\frac{Gm_1}{x_1^2}\hat{i} + \frac{Gm_2}{x_2^2}\hat{i} = -\frac{Gm_1}{x_1^2}\hat{i} + \frac{Gm_2}{(2x_1)^2}\hat{i}$$
$$= -\frac{G}{x_1^2}\left(m_1 - \frac{1}{4}m_2\right)\hat{i}$$
Substituting numerical values, we get:
$$\vec{g} = -\frac{6.673 \times 10^{-11} N \cdot m^2/kg^2}{(2.0 m)^2}\left[2.0 kg - \frac{1}{4}(4.0 kg)\right]\hat{i}$$
$$= \left(-1.7 \times 10^{-11} N/kg\right)\hat{i}$$
(b) $\vec{g}_1 = -\frac{Gm_1}{x_1^2}\hat{i}$ and $\vec{g}_2 = -\frac{Gm_2}{x_2^2}\hat{i}$ Substituting in equation (1), we get
$$\vec{g} = -\frac{Gm_1}{x_1^2}\hat{i} - \frac{Gm_2}{x_2^2}\hat{i} = -\frac{Gm_1}{(2x_2)^2}\hat{i} - \frac{Gm_2}{x_2^2}\hat{i}$$
$$= -\frac{G}{x_2^2}\left(\frac{1}{4}m_1 + m_2\right)\hat{i}$$
On substituting numerical values, we get

$$\vec{g} = -\frac{6.673 \times 10^{-11} \text{N}\cdot\text{m}^2/\text{kg}^2}{(6.0m)^2} \left[\frac{1}{4}(2.0 \text{ kg}) + 4.0 \text{ kg}\right]\hat{i}$$

$$= \left(-8.3 \times 10^{-12} \text{N/kg}\right)\hat{i}$$

(c) For $\vec{g} = 0$, we have :

$$\frac{Gm_1}{x^2} - \frac{Gm_2}{(6.0-x)^2} = 0$$

or $\quad \dfrac{2.0}{x^2} - \dfrac{4.0}{(6.0-x)^2} = 0$

On simplifying, we get

$$x = 2.48 \text{ m and } x = -14.5 \text{ m}$$

From the diagram it is clear that the physically meaningful root is the positive one at $x = 2.5$ m

EXAMPLE 31. Two point particles, each of mass M, are fixed in position on the y axis at $y = +a$ and $y = -a$. (a) Find the gravitational field at all points on the x axis as a function of x. (b) Show that on the x axis, the maximum value of field g occurs at points $x = \pm a/\sqrt{2}$.

(a) APPROACH Make a sketch of the two particles and the

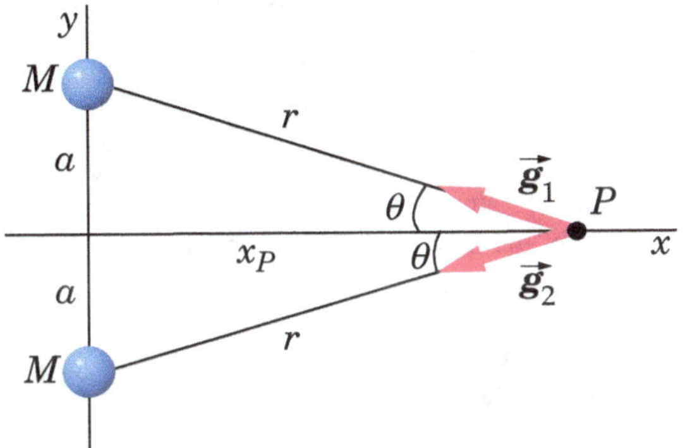

Figure 1.51

coordinate axis (Figure 1.51). Two particles of mass M each produce a gravitational field at point P located at $x = x_p$. The distance r between P and either particle is $\sqrt{x_P^2 + a^2}$. The resultant field $\vec{g}$ is the vector sum of the fields $\vec{g}_1$ and $\vec{g}_2$ due to each particle.

SOLUTION 1. Calculate the magnitude of $\vec{g}_1$ and $\vec{g}_2$:

$$g_1 = g_2 = \frac{GM}{r^2}$$

2. The y component of the resultant field, the sum of g_{1y} and $g_{2y'}$, is zero. The x component is the sum of g_{1x} and g_{2x} :

$$g_y = g_{1y} + g_{2y} = g_1 \sin\theta - g_2 \sin\theta = 0$$

$$g_x = g_{1x} + g_{2x} = g_1 \cos\theta + g_2 \cos\theta = 2g_1 \cos\theta$$

$$= \frac{2GM}{r^2}\cos\theta$$

3. Express $\cos\theta$ in terms of x_P and r from the Fig.1.51:

$$\cos\theta = \frac{x_P}{r}$$

4. Combining the last two results yields $\vec{g}$. To express $\vec{g}$ as a function of x_P, substitute $\left(x_P^2 + a^2\right)^{1/2}$ for r :

$$\vec{g} = g_x\hat{i} = -\frac{2GM}{r^2}\frac{x_P}{r}\hat{i} = -\frac{2GMx_P}{r^3}\hat{i}$$

$$= -\frac{2GMx_P}{\left(x_P^2 + a^2\right)^{3/2}}\hat{i}$$

5. x_p is the distance of an arbitrary point P on the x axis, from y axis. For simplicity, we replace it with x :

$$\vec{g} = -\frac{2GMx}{\left(x^2 + a^2\right)^{3/2}}\hat{i}$$

Comment: For $x < 0$, $\vec{g}$ is in the positive x direction and for $x > 0$, $\vec{g}$ is in the negative direction, as expected. If $x = 0$, we find that $\vec{g} = 0$; the fields $\vec{g}_1$ and $\vec{g}_2$ are equal and opposite at $x = 0$, and hence they cancel.

Note: For $x \gg a, \vec{g} \approx -\left(2GM/x^2\right)\hat{i}$. The field is the same as if a single particle of mass $2M$ were at the origin.

(b) APPROACH To show that, on the x axis, the maximum value of g occurs at the points $x = \pm a/\sqrt{2}$, we can differentiate g_x with respect to x and set the derivative equal to zero

. SOLUTION From part (a):

$$g_x = -\frac{2GMx}{\left(x^2 + a^2\right)^{3/2}}$$

Differentiate g_x with respect to x and set the derivative equal to zero to find extreme values:

$$\frac{dg_x}{dx} = -2GM\left[\left(x^2 + a^2\right)^{-3/2} - 3x^2\left(x^2 + a^2\right)^{-5/2}\right] = 0$$

$$\text{(for extrema)}$$

On simplifying for x, we get

$$x = \pm\frac{a}{\sqrt{2}}$$

Remarks: To establish that this value for x corresponds to a relative maximum, we need to either evaluate the second derivative of g_x at $x = \pm a/\sqrt{2}$ or examine the graph of $|g_x|$ at $x = \pm a/\sqrt{2}$ for concavity downward.

1.8   Gravitational Field of a Continuous Mass Distribution

So far we have dealt with only point mass masses or spherical bodies because Newton's law applies only to point masses and spherical bodies. However, most objects— like rod, plane sheet, solid cube etc. — are not point masses. Instead, they are extended bodies. Although every macroscopic object consists of very large numbers of tiny point masses (atoms)-it is not practical to calculate the individual field of each of these point masses and then add them vectorially. Instead, we shall treat any macroscopic object as having a continuous mass distribution(Fig.1.52). The gravitational field set up by a continuous mass distribution can be computed by dividing the distribution into infinitesimal elements dM. Each element of mass establishes a field $d\vec{g}$ at a given point P, and the resultant gravitational field at P is then found from the superposition principle by adding (that is, integrating) the field contributions due to all the mass elements, i.e.,

$$\vec{g} = \int d\vec{g} \tag{1.30}$$

In Fig.1.52, the gravitational field produced by a small mass element dM at point P is,

$$d\vec{g} = -G\frac{dM}{r^2}\hat{r} \tag{1.31}$$

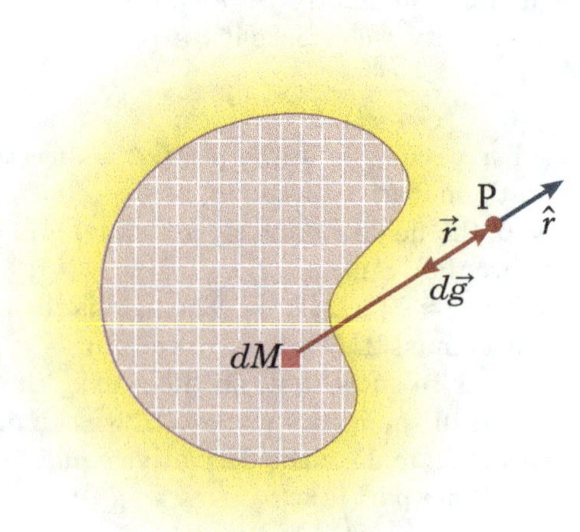

Figure 1.52

Therefore, net gravitational field at point P, will be given as-

$$\vec{g} = \int d\vec{g} = -G \int \frac{dM}{r^2}\hat{r} \qquad (1.32)$$

In order to evaluate this integral, we must express dM, $1/r^2$, and $\hat{r}$ in terms of the same coordinate(s). To do so, it is necessary to express the mass on the object in terms of a mass density. There are three types of mass densities:

1. Volume mass density ρ (lowercase Greek letter rho) is the amount of mass per unit volume (Fig. 1.53). For a uniform mass distribution,

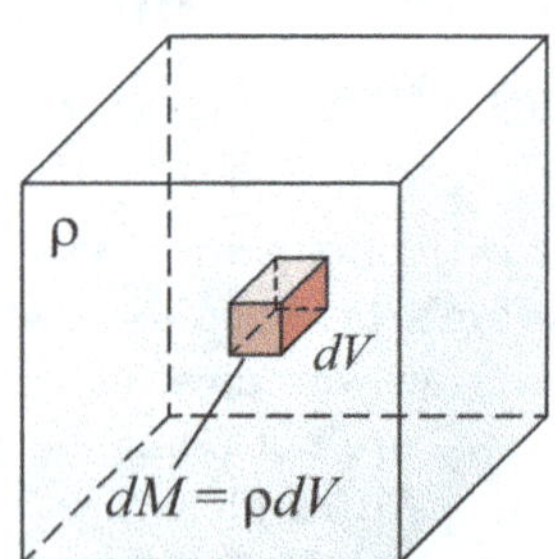

Figure 1.53

$$\rho \equiv \frac{M}{V} \qquad (1.33)$$

Volume mass density has the dimensions mass per volume, with SI units kg/m^3. The amount of mass dM in a small volume dV can be written in terms of the volume mass density:

$$\left(\begin{array}{c} \text{mass contained} \\ \text{in small volume} \end{array} \right) = \left(\begin{array}{c} \text{mass per} \\ \text{unit volume} \end{array} \right) \times (\,\text{volume}\,)$$

$$\boxed{dM = \rho \; dV} \qquad (1.34)$$

2. Surface mass density σ (lowercase Greek letter sigma) is the amount of mass per unit area (Fig. 1.54). For a uniform mass distribution,

$$\sigma \equiv \frac{M}{A}$$

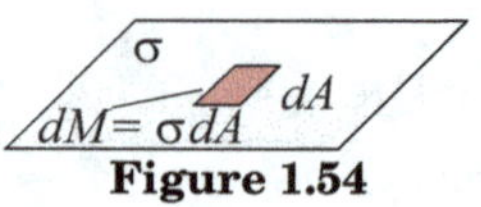

Figure 1.54

The dimensions of σ are mass per area, with SI units kg/m^2. The amount of mass dM in a small area dA can be written in terms of the surface mass density:

$$\left(\begin{array}{c} \text{mass contained} \\ \text{in small area} \end{array} \right) = \left(\begin{array}{c} \text{mass per} \\ \text{unit area} \end{array} \right) \times (\,\text{area}\,)$$

$$\boxed{dM = \sigma \; dA} \qquad (1.35)$$

3. Linear charge density λ (lowercase Greek letter lambda) is the amount of mass per unit length (Fig. 1.55). For a uniform mass distribution,

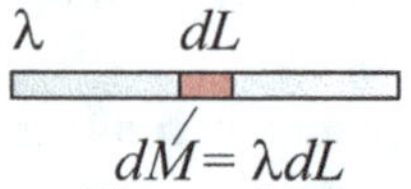

Figure 1.55

$$\lambda \equiv \frac{M}{L} \qquad (1.36)$$

The dimensions of λ are mass per length, with SI units Kg/m. The amount of mass dM in a small length element dL can be written in terms of the linear mass density:

$$\left(\begin{array}{c} \text{mass contained in} \\ \text{small length element} \end{array} \right) = \left(\begin{array}{c} \text{mass per} \\ \text{unit length} \end{array} \right) \times (\,\text{length}\,)$$

$$\boxed{dq = \lambda \; dL} \qquad (1.37)$$

Problem solving tactics for calculating the gravitational field from continuous charge distributions:

1. Identify the type of mass distribution and compute the mass density λ, σ or ρ.
2. Divide the mass distribution into infinitesimal masses dM, each of which will act as a tiny point mass.
3. The amount of mass dM, i.e., within a small element dL, dA or dV is
   $dM = \lambda dL$ (mass distributed along length)
   $dM = \sigma dA$ (mass distributed over a surface)
   $dM = \rho dV$ (mass distributed throughout a volume)
4. At point P, draw vector $d\vec{g}$ produced by the mass dm. The magnitude of $d\vec{g}$ is
$$dg = G\frac{dM}{r^2}$$
   Vector $d\vec{g}$ is opposite to radial line joining dM to P. $d\vec{g}$ is directed away for negative mass dM while directed towards dM for positive dM.
5. Resolve vector $d\vec{g}$ into its components. Identify any special symmetry features to show whether any component(s) of the field that are not get cancelled by other components.
6. Write the distance r and any trigonometric factors in terms of given coordinates and parameters.
7. The gravitational field is obtained by summing over all the infinitesimal contributions.

$$\vec{g} = \int d\vec{g} = -G \int \frac{dM}{r^2}\hat{r}$$

8. Perform the indicated integration over limit of integration that include all the source charges.

1.8.1 Gravitational Field Due to Finite Rod at Perpendicular Distance x from the Rod

Let us consider a rod of mass M and length L as shown in Fig.1.56. Suppose, the mass distribution of the rod is uniform throughout it's length. P is an arbitrarily positioned point, where the gravitational field $\vec{g}$ is to be calculated.

APPROACH Choose, the x axis so that the rod is on the x axis between points x_1 and x_2, and choose the y axis to be through the field point P. Let y be the perpendicular distance of P from the x axis. To calculate the gravitational field $\vec{g}$ at P, we separately calculate g_x and g_y. Using Equations (1.31) and (1.32), first find the gravitational field increment $d\vec{g}$ at P due to an arbitrary increment dM of the charge distribution. Then integrate each component of $d\vec{g}$ over the entire mass distribution. (Because M is distributed uniformly, the linear mass density λ equals M/L)

Stepwise Calculation

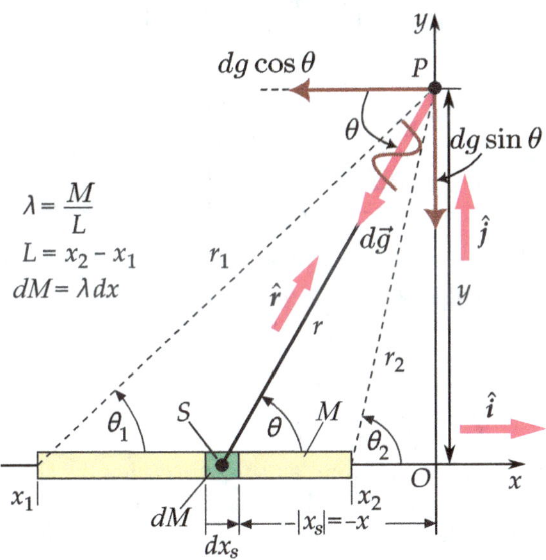

Figure 1.56: Geometry for the calculation of the gravitational field at field point P due to a uniform rod.

1. Sketch the mass configuration and the field point P. Include the x and y axes with the x axis lying along the length of rod and the y axis passing through P. In addition, sketch an arbitrary increment of the linear mass at point S (at $x = x_s$) that has a length dx_s and a mass dM, and the gravitational field at P due to dM. Sketch the gravitational field vector $d\vec{g}$ as if dM is positive (Figure 1.56)

2. $\vec{g} = -g_x\hat{i} - g_y\hat{j}$. Find expressions for dg_x and dg_y in terms of dE_r and θ, where dg_r, is the component of $d\vec{g}$ in the direction of $\overrightarrow{PS}$:
$$d\vec{g} = -dg_r\hat{r}$$
so $dg_x = dg_r\cos\theta$ and $dg_y = dg_r\sin\theta$

3. First we solve for g_x. Express dg_r using Equation (1.31),

where r is the distance from the source point S to the field point P. From Fig.1.56, $\cos\theta = -|x_s|/r = -x/r$. In addition, use $dM = \lambda dx$:
$$dg_r = \frac{GdM}{r^2} \quad \text{and} \quad \cos\theta = \frac{-x}{r}$$
so
$$dg_x = \frac{GdM}{r^2}\cos\theta = \frac{G\cos\theta\lambda dx}{r^2}$$

4. Integrate the step-3 result:
$$dg_x = \int_{x_1}^{x_2} \frac{G\cos\theta\lambda dx_s}{r^2} = G\lambda\int_{x_1}^{x_2} \frac{\cos\theta dx_s}{r^2}$$

5. Now, change the integration variable from x to θ. From Figure 1.56, find the relation between x and θ and between r and θ,
$$\tan\theta = \frac{y}{-|x_s|} = \frac{y}{-x}, \quad \text{so,} \quad x = -\frac{y}{\tan\theta} = -y\cot\theta$$
$$\sin\theta = \frac{y}{r}, \quad \text{so,} \quad r = \frac{y}{\sin\theta}$$

6. Differentiate the step 5 result to obtain an expression for dx (the field point P remains fixed, so y is constant):
$$dx = \frac{d}{d\theta}(-y\cot\theta)\,d\theta = y\operatorname{cosec}^2\theta d\theta$$

7. Substitute $y\operatorname{cosec}^2\theta d\theta$ for dx and $y/\sin\theta$ for r in the integral in step 4 and simplify:
$$\int_{x_1}^{x_2} \frac{\cos\theta\,dx}{r^2} = \int_{\theta_1}^{\theta_2} \frac{\cos\theta\,y\,\operatorname{cosec}^2\theta\,d\theta}{y^2}\sin^2\theta$$
$$= \frac{1}{y}\int_{\theta_1}^{\theta_2}\cos\theta\,d\theta \quad (y \neq 0)$$

8. Evaluate the integral and solve for g_x :
$$g_x = G\lambda\frac{1}{y}\int_{\theta_1}^{\theta_2}\cos\theta\,d\theta = \frac{G\lambda}{y}(\sin\theta_2 - \sin\theta_1)$$
$$= \frac{G\lambda}{y}\left(\frac{y}{r_2} - \frac{y}{r_1}\right)$$
$$= G\lambda\left(\frac{1}{r_2} - \frac{1}{r_1}\right) \quad (r_1 > 0 \text{ and } r_2 > 0)$$
Since, for $x > x_2$, we have, $r_1 > r_2$, therefore from above expression, it is clear that $g_x > 0$ at all points on the x axis in the region $x > x_2$.

9. Similar to g_x, use steps steps 3-7 to get g_y:
$$g_y = -\frac{G\lambda}{y}(\cos\theta_2 - \cos\theta_1)$$
$$= -G\lambda\left(\frac{\cot\theta_2}{r_2} - \frac{\cot\theta_1}{r_1}\right) \quad (y \neq 0)$$
If point P lies on the x-axis, then $y = 0$ and in this case, $\theta_1 = \theta_2 = 0$, therefore, $g_y = 0$

10. Combine steps 8 and 9 to obtain an expression for the gravitational field at P :
$$\vec{g} = -g_x\hat{i} - g_y\hat{j}$$

Making Sense of the Result: Consider the plane that is perpendicular to and bisecting the rod. At each point on this plane, symmetry dictates that $\vec{g}$ points towards the centre of the rod. That is, we expect that $g_x = 0$ throughout this plane. At all points on this plane $r_1 = r_2$. The step- 8 result gives $g_x = 0$ if $r_1 = r_2$, as expected.

Comment: The first expression for g_y in the step 9 result is valid everywhere in the xy plane but not on the x axis. The two cotangent functions in the expression for g_y are given by
$$\cot\theta_1 = \frac{-x_1}{y} \quad \text{and} \quad \cot\theta_2 = \frac{-x_2}{y}$$

and neither of these functions is defined on the x axis (where $y = 0$). The second expression for g_y in the step- 9 result is obtained using Equation (1.31). By recognizing that on the x axis $\hat{r} = \pm \hat{i}$, we can see that Equation (1.31) tells us that $d\vec{g} = \mp dg\hat{i}$, which implies $g_y = 0$.

Generalisation of Result: The gravitational field at point P due to a thin uniform rod (1.57) located on the z axis is given by $\vec{g} = -g_z\hat{k} - E_R\hat{R}$, where

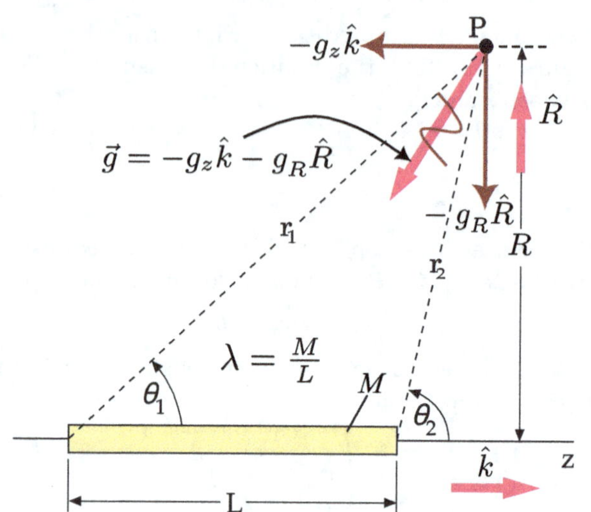

Figure 1.57: The gravitational field due to a uniformly charged thin rod.

$$\boxed{g_z = \frac{G\lambda}{R}(\sin\theta_2 - \sin\theta_1)} \qquad (1.38a)$$

$$\boxed{\text{or} \quad g_z = G\lambda\left(\frac{1}{r_2} - \frac{1}{r_1}\right) \quad (r_1 \neq 0) \text{ and } (r_2 \neq 0)} \qquad (1.38b)$$

$$\boxed{\text{and} \quad g_R = -\frac{k\lambda}{R}(\cos\theta_2 - \cos\theta_1)} \qquad (1.39a)$$

$$\boxed{\text{or} \quad g_R = -k\lambda\left(\frac{\cot\theta_2}{r_2} - \frac{\cot\theta_1}{r_1}\right) \quad (R \neq 0)} \qquad (1.39b)$$

The expressions for g_z (Equation (1.38)) are undefined at the end points of the thin uniform rod and the expressions for g_R (Equation (1.39) are undefined at all points on the z axis (where $R = 0$). However, $g_R = 0$ at all points where $R = 0$. Here, it is to be noted that θ_1 and θ_2 should to be used with proper sign.

1.8.1.1 Gravitational Field due to a Finite Rod For a Point on the Extension of it

EXAMPLE 32. A rod of mass M and length L is placed along z axis. It's centre is at $z = 0$ and mass distribution along it's length is uniform. Show that for large values of z the expression for the gravitational field of the rod on the z axis approaches the expression for the gravitational field of a point mass M at the origin.

APPROACH To find $\vec{g}$, at an axial point P, use Eq. (1.38).

Stepwise Calculation

1. The gravitational field on the z axis has only a z component, given by Equation (1.38):
$$g_z = G\lambda\left(\frac{1}{r_2} - \frac{1}{r_1}\right)$$

2. Sketch the rod. Include the z axis, the field point P, and r_1 and r_2 (Figure 1.58):

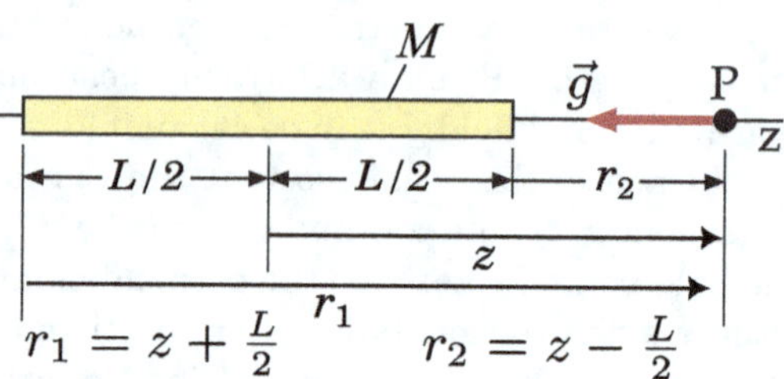

Figure 1.58: Geometry for the calculation of the gravitational field on the axis of a uniform rod.

3. Substitute with $r_1 = z + \frac{1}{2}L$ and $r_2 = z - \frac{1}{2}L$ into the step 1 result and simplify:

$$g_z = G\lambda\left(\frac{1}{z - \frac{1}{2}L} - \frac{1}{z + \frac{1}{2}L}\right)$$

$$= \frac{GM}{L}\left(\frac{L}{z^2 - \left(\frac{1}{2}L\right)^2}\right)$$

$$= \frac{GM}{z^2 - \left(\frac{1}{2}L\right)^2} \qquad \left(\text{here, } z > \frac{1}{2}L\right)$$

This is the required expression for gravitational field at an axial position of a uniform rod.

Note that the result is valid for the region $L/2 < z < \infty$ only. The result is not valid in the region $-L/2 < z < +L/2$?

For $z \gg L$, the above result gives-

$$g_z \approx \frac{GM}{z^2} \qquad (z \gg L)$$

This expression is the same as the expression for the gravitational field of a point mass M located at the origin.

EXAMPLE 33. What is the gravitational field at any point on the axis of a uniform rod length 'L' having linear mass density 'λ'? The point is separated from the nearer end by distance 'a' [Fig.1.59].

Note: Although this problem has been already discussed in

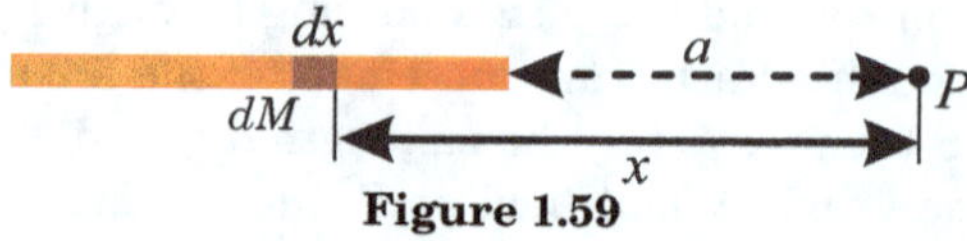

Figure 1.59

above article, here I am providing you an alternative method by using integration.

SOLUTION Consider an elementary length dx of the rod at distance x, from point P, where gravitational field g is to be determined.

The elemental mass, $dM = \lambda dx$

The small gravitational field at P due to this element,

$$dg = G\frac{\lambda\,dx}{x^2}$$

or $\quad g = G\lambda \int_a^{a+L}\frac{1}{x^2}\,dx = G\lambda\left[-\frac{1}{x}\right]_a^{a+L} = G\lambda\left[\frac{-1}{a+L}+\frac{1}{a}\right]$

Thus, $\qquad g = G\left[\frac{1}{a}-\frac{1}{L+a}\right]$

1.8.1.2 Gravitational Field Due to an Infinite Rod

In this section, we find the gravitational field due to a uniform rod extends to infinity in both directions and has linear mass density λ.

APPROACH If $L\to\infty$, then from the Figure 1.57, we see that $\theta_1\to 0$ and $\theta_2\to\pi$. Now, put these values in Equations (1.38) and (1.39) and solve for g_z and g_R. The resultant gravitational field at point P can be obtained by using the expression-

$$\vec{g} = -g_z\hat{k} - g_R\hat{R}$$

Stepwise Calculations

1. Choose the first expression for the gravitational field in each of Equations (1.38) and (1.39):

$$g_z = \frac{G\lambda}{R}(\sin\theta_2 - \sin\theta_1)$$
$$g_R = -\frac{G\lambda}{R}(\cos\theta_2 - \cos\theta_1)$$

2. Take the limit as both $\theta_1\to 0$ and as $\theta_2\to\pi$

$$g_z = \frac{G\lambda}{R}(\sin\pi - \sin 0) = \frac{G\lambda}{R}(0-0) = 0$$
$$g_R = -\frac{G\lambda}{R}(\cos\pi - \cos 0) = -\frac{G\lambda}{R}(-1-1) = 2\frac{G\lambda}{R}$$

3. Express the gravitational field in vector form:

$$\vec{g} = -g_z\hat{k} - g_R\hat{R} = -0\hat{k} - \frac{2G\lambda}{R}\hat{R} = -\frac{2G\lambda}{R}\hat{R}$$

$$\Rightarrow\qquad \boxed{\vec{g} = -\frac{2G\lambda}{R}\hat{R}}\qquad (1.40)$$

Thus, the magnitude of gravitational field decreases inversely with the radial distance from the rod.

Special Results: semi infinite rod

For a semi-infinite rod, as shown in Fig.1.60, we have -

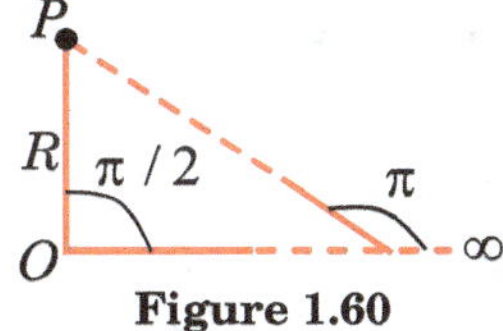

Figure 1.60

$$\theta_1 = \pi/2, \quad \theta_2 = \pi$$

Therefore,

$$g_z = \frac{G\lambda}{R}(\sin\theta_2 - \sin\theta_1)$$
$$= \frac{G\lambda}{R}\left(\sin\pi - \sin\frac{\pi}{2}\right)$$
$$= \frac{G\lambda}{R}(0-1) = -\frac{G\lambda}{R}$$

and $\qquad g_R = -\frac{G\lambda}{R}(\cos\theta_2 - \cos\theta_1)$

$$= -\frac{G\lambda}{R}\left(\cos\pi - \cos\frac{\pi}{2}\right)$$

$$= -\frac{G\lambda}{R}(-1-0) = \frac{G\lambda}{R}$$

In this case, net gravitational field,

$$g = \sqrt{g_z^2 + g_R^2}$$
$$= \sqrt{(-G\lambda/R)^2 + (G\lambda/R)^2}$$
$$= \frac{G\lambda}{R}\sqrt{2}$$

The angle between net gravitational field vector $\vec{g}$ and length of the rod, i.e., z-axis, is given by-

$$\theta = \tan^{-1}\frac{g_R}{g_z} = \tan^{-1}1 = \frac{\pi}{4}$$

1.8.1.3 Approximating Equations (1.38) and (1.39) on the Symmetry Plane

EXAMPLE 34. A mass M is uniformly distributed along the z axis, from $z = -\frac{1}{2}L$ to $z = +\frac{1}{2}L$. (a) Find an expression for the gravitational field on the $z = 0$ plane as a function of R, the radial distance of the field point from the z axis. (b) Show that for $R\gg L$, the expression found in Part (a) approaches that of a point charge at the origin of mass M. (c) Show that for $R\ll L$, the expression found in Part (a) approaches that of an infinitely long line mass on the z axis with a uniform linear mass density $\lambda = M/L$.

APPROACH The mass distribution is uniform and the linear mass density is $\lambda = M/L$. Sketch the line mass on the z axis and put the field point in the $z = 0$ plane. Then use Equations (1.38) and (1.39) to find the gravitational field expression for part (a). The gravitational field due to a point mass decreases inversely with the square of the distance from the mass. Examine the part (a) result to see how it approaches that of a point charge at the origin for $R\gg L$. The gravitational field due to a uniform line mass of infinite length decreases inversely with the radial distance from the line (Equation (1.40)). Examine the Part (a) result to see how it approaches the expression for the gravitational field of a line mass of infinite length for $R\ll L$.

SOLUTION (a) Stepwise solution is given below-

1. From Eq.(1.38) and (1.38), we have-

$$g_z = \frac{G\lambda}{R}(\sin\theta_2 - \sin\theta_1)$$
$$g_R = -\frac{G\lambda}{R}(\cos\theta_2 - \cos\theta_1)$$

2. Sketch the mass configuration with the line mass on the z axis from $z = -\frac{1}{2}L$ to $z = +\frac{1}{2}L$. Show the field point P in the $z = 0$ plane at a distance R from the origin (Figure 1.61):

3. From the Figure1.61, $r_1 = r_2$, therefore, the interior angle at the right end of the rod will also be θ_1. Therefore, $\theta_2 + \theta_1 = \pi$, so $\sin\theta_2 = \sin(\pi - \theta_1) = \sin\theta_1$ and $\cos\theta_2 = \cos(\pi - \theta_1) = -\cos\theta_1$. Substitution of these values in equations of step 1 gives:

$$g_z = \frac{G\lambda}{R}(\sin\theta_1 - \sin\theta_1) = 0$$

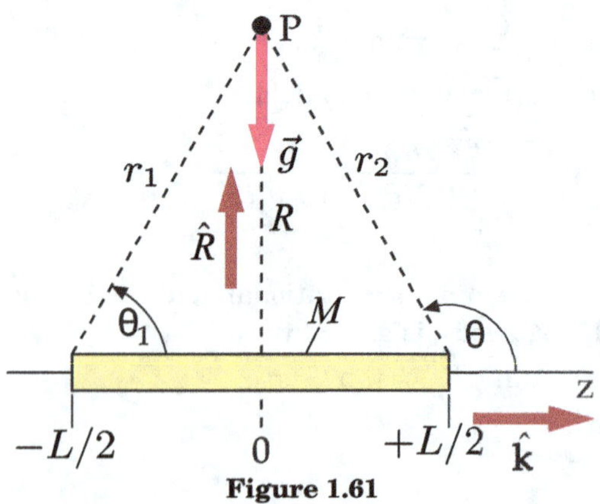

Figure 1.61

$$g_R = -\frac{G\lambda}{R}\left(-\cos\theta_1 - \cos\theta_1\right) = \frac{2G\lambda}{R}\cos\theta_1$$

4. Express $\cos\theta_1$ in terms of R and L and substitute into the step-3 result:

$$\cos\theta_1 = \frac{\frac{1}{2}L}{\sqrt{R^2 + \left(\frac{1}{2}L\right)^2}}$$

so, $$g_R = \frac{2G\lambda}{R}\frac{\frac{1}{2}L}{\sqrt{R^2 + \left(\frac{1}{2}L\right)^2}} = \frac{G\lambda L}{R\sqrt{R^2 + \left(\frac{1}{2}L\right)^2}}$$

5. Express the gravitational field in vector form, and substitute M for λL:

$$\vec{g} = -g_z\hat{k} - g_R\hat{R} = 0\hat{k} - g_R\hat{R}$$

so $$\vec{g} = -g_R\hat{R} = -\frac{GM}{R\sqrt{R^2 + \left(\frac{1}{2}L\right)^2}}\hat{R}$$

(b) The step wise solution is given below-

1. Examine the step-5 result of part (a). If $R \gg L$ then $R^2 + \left(\frac{1}{2}L\right)^2 \approx R^2$. Substitute R^2 for $R^2 + \left(\frac{1}{2}L\right)^2$:

$$\vec{g} \approx -\frac{GM}{R\sqrt{R^2}}\hat{R} = -\frac{GM}{R^2}\hat{R} \quad (R \gg L)$$

2. This (approximate) expression for the gravitational field decreases inversely with the square of the distance from the origin, just as it would for a point charge M at the origin.

$$\vec{g} - \approx \frac{GM}{R^2}\hat{R} \quad (R \gg L)$$

(c) Examine the Part (a), step - 5 result. If $R \ll L$ then $R^2 + \left(\frac{1}{2}L\right)^2 \approx \left(\frac{1}{2}L\right)^2$. Substitute $\left(\frac{1}{2}L\right)^2$ for $R^2 + \left(\frac{1}{2}L\right)^2$. This (approximate) expression for the gravitational field falls off inversely with the radial distance from the line mass, just as the exact expression for a rod of infinite length (1.40).

$$\vec{g} \approx -\frac{G\lambda L}{R\sqrt{\left(\frac{1}{2}L\right)^2}}\hat{R} = -\frac{2G\lambda}{R}\hat{R} \quad (R \ll L)$$

EXAMPLE 35. A thin uniform rod of mass M and length L is centred at the origin and lies along the x axis. Find the gravitational field due to the rod at all points on the x axis in the region $x > L/2$.

☞ This problem is similar to Example 32. Here, we will find a solution with the help of integration.

APPROACH Since, rod is an expanded mass distribution along the length, so we cannot directly apply Eq.1.25. To find

the net gravitational field intensity produced by complete rod, we have to divide it's complete length into elementary strips each of thickness dx and mass dm. If λ is the linear mass density of the rod, then $dm = \lambda\,dx$. Now, consider any one strip of the rod and find the intensity of gravitational field due to it at any general point P (say), on the axis of the rod. The direction of intensity of this gravitational field is towards the rod. Similarly, all elementary strips produce their gravitational field at P which will also be directed toward the rod. So, net intensity will be vector some of all these intensities, i.e., it can be obtained by simply integrating the intensity of one strip for complete length of the rod.

SOLUTION Intensity of gravitational field at P, due to an

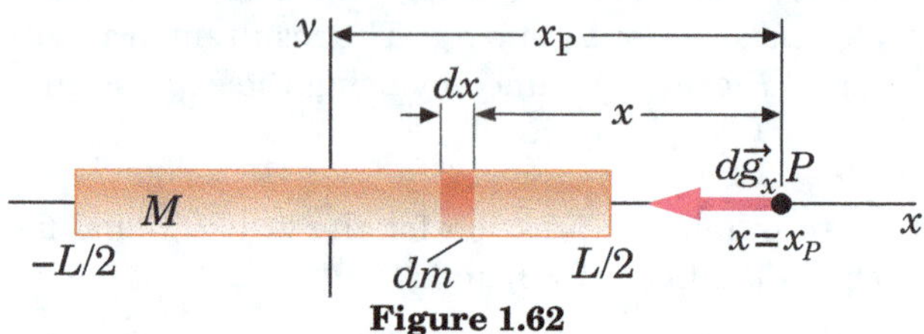

Figure 1.62

elementary strip of thickness dx and mass dm is.

$$dg_x = \frac{Gdm}{x^2} = \frac{G\,dx}{x^2}$$

$$g_x = \int dg_x = G\int_{x_P - L/2}^{x_P + L/2}\frac{dx}{x^2} = G\left[-\frac{1}{x}\right]_{x_P - L/2}^{x_P + L/2}$$

$$= -G\left[\frac{1}{x_P + L/2} - \frac{1}{x_P - L/2}\right]$$

$$= -G\frac{L}{x_P^2 - (L/2)^2} = -\frac{GM}{x_P^2 - (L/2)^2}$$

In vector form $\vec{g} = -\frac{GM}{x_P^2 - (L/2)^2}\hat{i}, \ x > L/2$.

Active and Passive Gravitational Masses

The active gravitational mass of a particle is an attribute that enables it to establish a gravitational field in space, whereas the passive gravitational mass is an attribute that enables the particle to respond to this field. The law of conservation of momentum states that active and passive gravitational mass should be identical. So, there is no need to distinguish between active and passive gravitational masses; it is sufficient to work with just gravitational masses.

1.9 Flux of a Gravitational Field and Gauss's Law for Gravity

1.9.1 Flux

There are many situations in physics where we deal with flows of matter or energy. For example, the flows of blood through a vein, water in a pipe or river, or air over an airplane's wing. It is measured in kg/s or m³/s. Each of these flows is a flux–a term that comes from the Latin *fluxus*, for "flow." In each case we can represent the flow by drawing lines that give the local direction of the flow.

We can also describe the flow by giving the flow rate per unit area, we call it, "the flux per unit area". For example, we consider a fluid passing through an area. In SI, the flux per unit area can be measured in kilograms per second per square meter i.e., "(kg/s·m²)". Now, if the flow is uniform across the cross-sectional area, then total flux can be obtained by multiplying the flux per unit area by the net cross-sectional area. For example, a river with cross-sectional area 25 m² and a flux per unit area of 100 kg/s·m² would carry a total flux of 2500 kg/s. If the flow per unit area weren't uniform, then we would have to consider small patches of area, calculate the flux over each, and sum the results.

1.9.1.1 Flux of Gravitational Field $\vec{g}$

In the case of gravitational fields, there's nothing "flowing." But direction of gravitational field lines gives the local direction of that field, and their closeness reflects the field strength. In the simplest case of a uniform gravitational field of magnitude g perpendicular to an area A, the flux of the gravitational field is $\Phi_g = gA$, in analogy with our finding the flux in a river by multiplying the flux per unit area by the river's cross-sectional area. Here we use Φ_g, the Greek phi with subscript g, as our symbol for gravitational flux.

Consider first the flux through a flat surface. Comparison

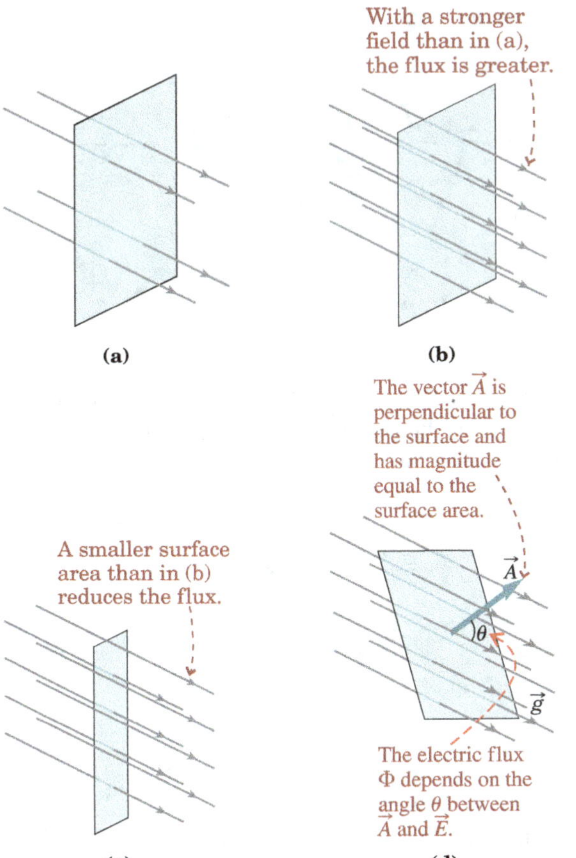

Figure 1.63: Gravitational flux through flat surfaces.

of Figs. 1.63a and 1.63b shows that the number of field lines crossing a given surface is larger when the field is stronger–a fact that's precisely captured in our formula $\Phi_g = gA$. Comparison of Figs. 1.63b and 1.63c shows further that, for a

given field strength, the number of field lines crossing a given surface is larger when the surface is larger. Again, $\Phi_g = gA$ captures this situation. Finally, Fig.1.63d shows that the number of field lines is reduced when the surface is tilted relative to the field. You can also see this effect in Fig. 1.64, which shows a water-flow analogy. Specifically, the flux is reduced by a factor $\cos\theta$, where θ is the angle between the vector field $\vec{g}$ and area vector $\vec{A}$ that's normal to the surface. So our flux expression generalizes to $\Phi_g = gA\cos\theta$. If we define the normal vector $\vec{A}$ as having magnitude equal to the surface area A, then we can write the gravitational flux through a flat surface in a uniform vector field using the dot product:

$$\boxed{\Phi_g = \vec{g}\cdot\vec{A}} \tag{1.41}$$

where the dot product, is the product of the two vector magnitudes with the cosine of the angle between them.

The surfaces in Fig. 1.63 are *open* surfaces, meaning it's pos-

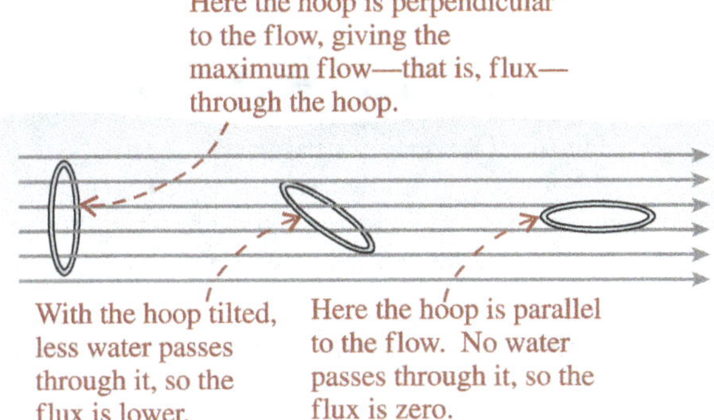

Figure 1.64: A water-flow analogy for gravitational flux, showing a circular hoop immersed in a flow of water.

sible to get from one side to the other without passing through the surface. For open surfaces there's an ambiguity in the sign of Φ, since we could have taken $\vec{A}$ in either of the two directions along the perpendicular to the surface. But for *closed* surfaces, we uniquely define the direction of $\vec{A}$ as the direction of the outward-pointing normal to the surface.

If a surface is curved and/or the field varies with position, then we divide the surface into patches, each small enough that it is essentially flat and that the field is essentially uniform over it (Fig.1.65). If a patch has area dA, then Equation 1.41 gives the flux through it: $d\Phi_g = \vec{g}\cdot d\vec{A}$, where the vector $d\vec{A}$ is normal to the patch. The total flux through the surface is then the sum over all the patches. If we make the patches arbitrarily small, that sum becomes an integral, and the flux is

$$\boxed{\Phi_g = \int_{\text{surface}} \vec{g}\cdot d\vec{A}} \tag{1.42}$$

The limits of the integral range over the entire surface, picking up contributions from all the patches $d\vec{A}$. The result is a surface integral–in general, the integral of a vector field over a surface, with the orientation of field and surface taken into account.

If the surface is closed, then net flux assosiated with the closed surface is written as-

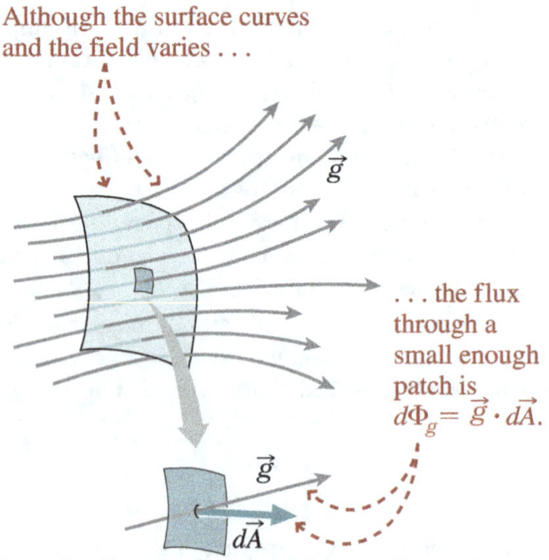

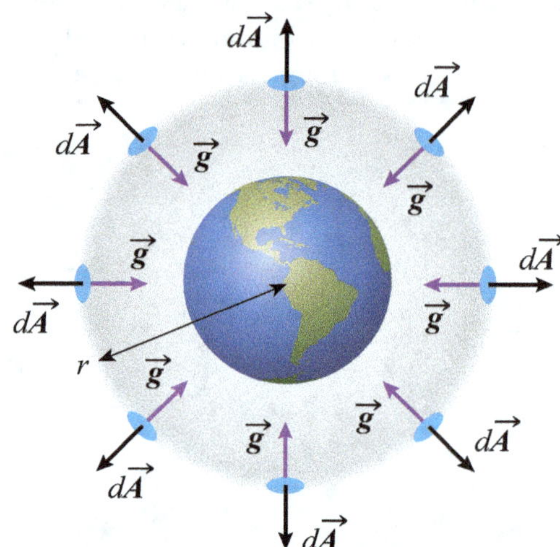

Figure 1.65: Finding the flux through a small area dA, so small it's essentially flat.

$$\boxed{\Phi_g = \oint_{\text{surface}} \vec{g} \cdot d\vec{A}} \qquad (1.43)$$

Here, the symbol $\oint_{\text{surface}}$ shows that the integral is over a closed surface, and it is a vector dot product between the gravitational field $\vec{g}$ and an outward pointing differential surface element $d\vec{A}$:

Note: If we replace gravitational field by electric field, then flux is called electric flux and if it is replaced by magnetic field, then the flux is called the magnetic flux.

1.10   Gauss's Law for Gravity

According to this theorem "the net gravitational flux through any closed hypothetical Gaussian surface is equal to $-4\pi G$ times the net enclosed mass M_{encl} within the hypothetical Gaussian surface."

$$\boxed{\Phi_g = \oint \vec{g} . d\vec{A} = -4\pi G M_{\text{encl}}} \qquad (1.44)$$

here $\vec{g}$ is the intensity of gravitational field. It can also be denoted by $\vec{E}$ or $\vec{I}$.

Proof: We can calculate the magnitude of $\vec{g}$ at any location due to a point mass M, by using Eq.(1.45):

$$\vec{g} = -\frac{GM}{r^2}\hat{r} \qquad (1.45)$$

where G is the universal gravitational constant, r is the radial distance of the location from the point mass, M.

This can be written in terms of the unit vector $\hat{r}$ that points radially outward from the centre of the mass, M and $\hat{r}$ is the unit radial vector pointing away from point mass M at the location of field point.

Above relation also holds for a large spherical mass.

In scalar form, above relation becomes

$$g = \frac{GM}{r^2} \qquad (1.46)$$

Now, from Eq.**??**, the gravitational flux, Φ_g, is given by -

$$\boxed{\Phi_g = \oint_{\text{surface}} \vec{g} \cdot d\vec{A}} \qquad (1.47)$$

The gravitational flux is a scalar quantity and it can be zero, positive or negative. Think $\vec{g}$ like a gravitational field here, rather than an acceleration, although they have the same numerical value. We call the magnitude of $\vec{g}$ the gravitational field strength, g.

We will now calculate the gravitational flux around Earth. For a radially symmetric planet, we have spherical symmetry, and our closed Gaussian surface is a sphere. In Fig.1.66 we illustrate Earth and a larger concentric Gaussian sphere. As long as the mass of Earth is distributed in a radially symmetric way, the gravitational field must have constant magnitude (but not direction) all around the closed Gaussian surface. Note that everywhere the gravitational field $\vec{g}$ and the differential surface vector $d\vec{A}$ are in exactly opposite directions.

The gravitational flux passing through the Gaussian surface

Figure 1.66: For the case of a Gaussian sphere around a radially symmetric Earth, the gravitational field $\vec{g}$ and the differential area vectors $d\vec{A}$ are directly opposite to each other.

is given by-

$$\Phi_g = \oint_{\text{sphere}} \vec{g} \cdot d\vec{A}$$

or

$$\Phi_g = \oint_{\text{sphere}} g \, dA \cos 180°$$

or

$$\Phi_g = -g \oint_{\text{sphere}} dA$$

or

$$\Phi_g = -g 4\pi r^2$$

here g is the scalar gravitational field strength at distance r from the centre of Earth.

On substituting the value of the gravitational field strength, g, at distance r, using Equation 1.46, we get:

$$\Phi_g = -\left(\frac{GM}{r^2}\right)4\pi r^2$$

$$\Rightarrow \quad \boxed{\Phi_g = -4\pi GM} \qquad (1.48)$$

Although, we have only developed this for the case of a spherically symmetric closed Gaussian surface, it holds for any shape of closed surface. We can generalize above result (Eq.1.45) to obtain Gauss's law for gravity: the gravitational flux through any closed surface is equal to $-4\pi G$ times of the mass enclosed (M_{encl}) by the closed surface, i.e.,

$$\Phi_g = \oint_{\text{surface}} \vec{g} \cdot d\vec{A} = -4\pi G M_{\text{encl}} \qquad (1.49)$$

For a spherical Gaussian surface of radius r, we have

$$\oint \vec{g}.d\vec{A} = \oint g\,dA\cos 180° = -g\oint dA = -g(4\pi r^2)$$

Therefore, Eq.1.49, becomes

$$g = \frac{GM_{\text{encl}}}{r^2} \qquad (1.50)$$

This is an important outcome of Gauss's law of gravitation and can also be directly applied without writing the complete integral form.

Gauss's law is only helpful in calculating the gravitational field strength when there is sufficient symmetry that we can choose a surface with constant field strength.

From Eq.1.49, it is clear that, to determine g at any point, we need only the mass enclosed (M_{encl}) within the hypothetical Gaussian surface passing through that point. We don't need the mass residing outside the Gaussian surface.

1.10.1 Applications of Gauss's law

Selecting a Gaussian Surface: While applying Gauss's law we are interested in evaluating the integral

$$\Phi_g = \oint \vec{g} \cdot \vec{dA} = \oint g\,dA\cos\theta$$

Now note that, this integration may be complicated if product of g and $\cos\theta$ is not constant. Therefore we always select a Gaussian surface in such a way that product of g and $\cos\theta$ remains constant. That is we have to select some symmetrical surfaces.

The types of symmetry are illustrated in Fig.1.67 and summarized in Table 1.1. If the object does not have any of these three types of symmetry, Gauss's law, even though still true, is unlikely to be helpful in calculating the gravitational field strength.

Table 1.1: Type of symmetry and the corresponding gaussian surface that should be used

Symmetry	Type of Gaussian Surface	Examples
spherical	sphere concentric with point	point mass, spherical mass, spherical shell
cylindrical	cylinder coaxial with line	linear mass, cylindrical mass, coaxial cylindrical shells
planar	cylinder or box perpendicular to plane	parallel plane(s), large flat object

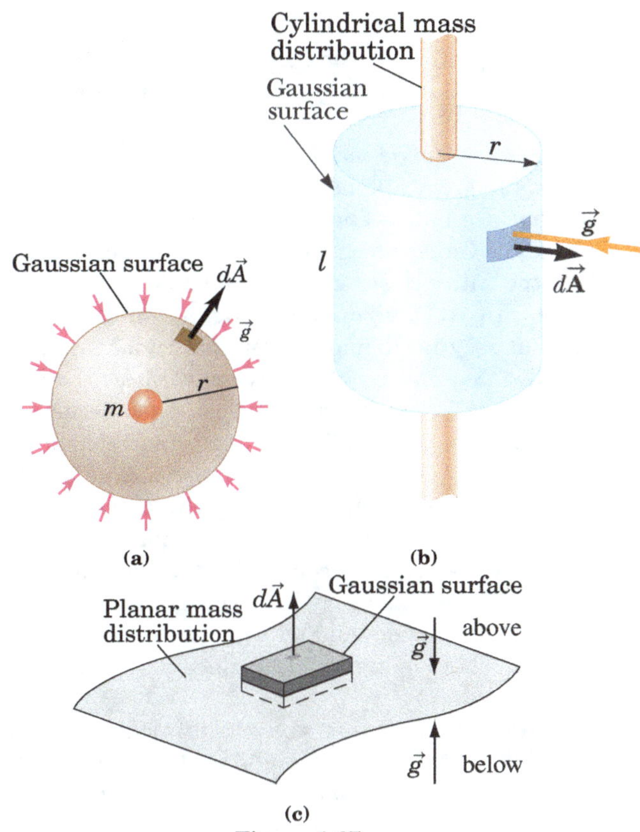

Figure 1.67

1.10.1.1 Gauss's Law: Problem Solving Approach

To apply Gauss's law, we simply follow the following steps-

1. Sketch the mass distribution, as well as the position(s) where the gravitational field is to be calculated.

2. Determine the symmetry of its gravitational field.

3. Draw the appropriate closed Gaussian surface (see Table 1.1) based on the symmetry, choosing the size of the surface according to where the gravitational field is to be determined. For example, in the case of a spherical mass, a Gaussian sphere is drawn concentric with the centre of the object and is of radius equal to the radial distance where the gravitational field is to be computed. Sketch the directions of $\vec{g}$ and $d\vec{A}$ on your diagram. You need not enclose all the masses within the Gaussian surface.

4. Be sure every part of the Gaussian surface satisfies one or more of the following conditions:

   (a) The value of the gravitational field can be argued by symmetry to be constant over the portion of the surface.

   (b) Gaussian surface is either tangent to or perpendicular to the gravitational field.

      i. If the gravitational field is everywhere tangent to a surface, the gravitational flux through the surface is $\Phi_g = \vec{g}.\vec{A} = 0$

      ii. If the gravitational field is everywhere perpendicular to a surface and has the same magnitude g at every point, the gravitational flux through the surface is $\Phi_g = \vec{g}.\vec{A} = gA$

   (c) The gravitational field is zero over the portion of the

surface.

Different portions of the Gaussian surface can satisfy different conditions as long as every portion satisfies at least one condition.

5. *If necessary, divide the closed surface into parts*. In some cases, the gravitational field is not constant over the entire surface but is known over each part. In this case, divide the Gaussian surface into different parts, ensuring that the entire closed surface is included.

6. For each part, *determine the orientation of $\vec{g}$ relative to $d\vec{A}$ and use that to find the dot product for that part*.

7. Finally *integrate over the Gaussian surface to compute the gravitational flux*. However, if you have a surface with uniform gravitational field strength (and constant direction relative to the surface normal) you can take g out of the integral and simply multiply the constant g by the surface area multiplied by the $\cos\theta$ value.

8. *Calculate the net mass enclosed within the Gaussian surface*. Remember that only the charge inside the surface needs to be considered, and when there are positive and negative charges inside the surface it is the net charge that is used.

9. *Use mathematical expression of Gauss's law* (with the expression for the gravitational flux (from step 6) and the enclosed mass *(step 7))* to solve for the gravitational field strength-

$$\Phi_g = \oint_{\text{surface}} \vec{g} \cdot d\vec{A} = -4\pi GM_{\text{encl}}$$

10. Finally, you can check that your answer makes sense with limiting cases(i.e., does your relationship make sense just outside a surface, or at infinite distance).

Following examples show some applications of Gauss's law.

EXAMPLE 36. Using Gauss's law, find the gravitational field at a distance r from a point charge M.

APPROACH By definition, a point is a sphere of negligible radius, i.e., a sphere of zero radius, therefore we can assume a point mass as spherical and draw a spherical Gaussian surface of radius r around it (Fig.1.68). From the symmetry of the Gaussian sphere, we know that at each point on it, the gravitational field $\vec{g}$ is perpendicular to the surface and directed towards the point mass i.e., towards the centre of the Gaussian sphere. Thus, $\vec{g}$, is directed opposite to $d\vec{A}$, therefore, $\vec{g}.d\vec{A} = -g\,dA$, with $M_{\text{encl}} = M$. Now, apply Gauss's law-

$$\Phi_g = \oint \vec{g} \cdot d\vec{A} = -4\pi GM_{\text{encl}} \qquad (1.51)$$

and solve for g

SOLUTION Substituting, $\vec{g}\cdot d\vec{A} = -g\,dA$ and $M_{\text{encl}} = M$ in Eq. (1.51), we get

$$-\oint g\,dA = -4\pi GM$$

$$\Rightarrow \qquad g\oint dA = 4\pi GM$$

$$\Rightarrow \qquad 4\pi r^2 g = 4\pi GM$$

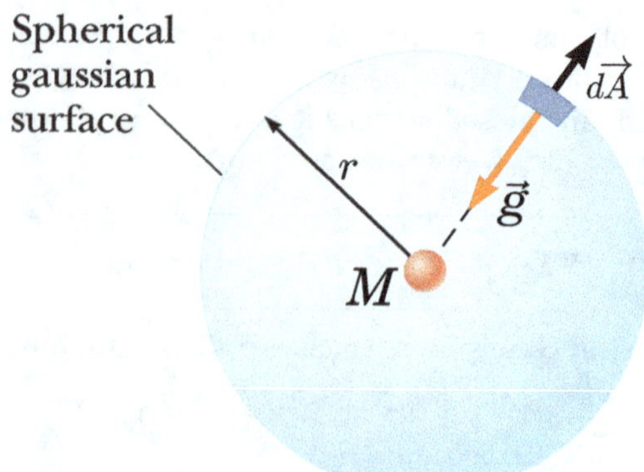

Figure 1.68

$$g = \frac{GM}{r^2} \qquad (1.52)$$

Eq.1.52 gives the gravitational field at a distance r from a point mass M.

If we consider the vector $\vec{r}$ away from point M, i.e. along the radius vector of the Gaussian surface, then Eq.1.52, in vector form, can be written as:

$$\vec{g} = -\frac{GM}{r^2}\hat{r} \qquad (1.53)$$

here $\hat{r}$ is the unit vector along $\vec{r}$.

$-$ ve sign shows that $\vec{g}$ is directed opposite to vector $\vec{r}$ at field point.

Eq.1.53 is same as Eq.1.25, which was obtained by using the basic definition of gravitational field.

EXAMPLE 37. Gravitational Field inside and outside a Homogeneous Planet Differentiated planets such as Earth, with an iron core, have higher mass densities near the core. However, in this problem, assume a spherical planet of radius R_{p} and total mass M_{p} that has uniform mass density throughout. Use Gauss's law for gravity to find expressions for the acceleration due to gravity for points at distance r from the centre of the planet, for (a) $r > R_{\text{p}}$ and (b) $r < R_{\text{p}}$.

APPROACH In this case, we have spherical symmetry, therefore, to find the acceleration due to gravity at different radial distances, r, we draw spherical Gaussian surfaces at distance r from the center of the planet. When we are outside the planet, the enclosed mass is simply the total mass of the planet. Inside the planet, we have to use the mass density to find the mass enclosed by the spherical Gaussian surface. In both cases, the symmetry of the situation requires that the direction of the acceleration due to gravity be radially inward.

SOLUTION (a) $r > R_{\text{p}}$: Here we have a situation similar to that of Figure 1.69 and draw a Gaussian sphere concentric with the planet, but larger:

$$\oint_{\text{surface}} \vec{g} \cdot d\vec{A} = -4\pi GM_{\text{enc}} \qquad (1.54)$$

Since $\vec{g}$ and $d\vec{A}$ are in opposite directions, therefore -

$$\vec{g}\cdot d\vec{A} = g\,dA\cos 180° = -g\,dA$$

From the symmetry of the situation, the gravitational field

strength is constant at any particular radial distance and can therefore be taken outside the integral:

$$\oint \vec{g} \cdot d\vec{A} = -4\pi\, GM_p$$

$$-\oint_{\text{surface}} g\,dA = -4\pi\, GM_p$$

$$-g\oint_{\text{surface}} dA = -4\pi\, GM_p$$

$$-4\pi r^2 g = -4\pi\, GM_p$$

$$\Rightarrow \qquad g = -\frac{GM_p}{r^2}$$

In vector form, since the gravitational field is directed radially inward,

$$\boxed{\vec{g} = -\frac{GM_p}{r^2}\hat{r}} \qquad (1.55)$$

(b) $r < R_p$: In this case, we draw a concentric spherical Gaussian sphere of radius r inside the planet (see Figure 1.69).

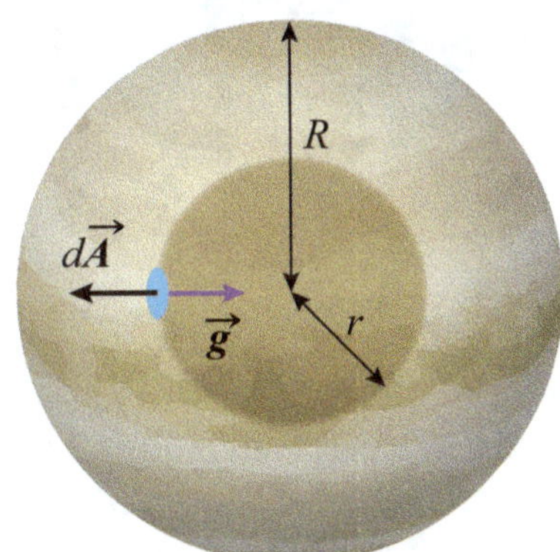

Figure 1.69: To calculate the gravitational field at a point inside the planet, we draw a concentric spherical Gaussian surface (the darker smaller sphere) of radius r less than the radius of the planet. Everywhere on this Gaussian surface, $\vec{g}$ will have a constant magnitude and will be directly opposite in direction to $d\vec{A}$.

First we calculate the mass density of the spherical planet:

$$\rho = \frac{M_p}{V} = \frac{M_p}{\frac{4}{3}\pi R^3} = \frac{3M_p}{4\pi R^3} \qquad (1.56)$$

The gravitational field $\vec{g}$ points radially inward and the differential area element $d\vec{A}$ radially outward:

$$\vec{g} \cdot d\vec{A} = g\,dA\cos 180° = -g\,dA \qquad (1.57)$$

We know from the symmetry of the situation that the gravitational field strength is constant at any particular radial distance. Therefore, we compute the gravitational flux using result (1.49). The integral over the differential area elements for the Gaussian surface simply yields the surface area of a sphere of radius r:

$$\Phi_g = \oint_{\text{sphere}} \vec{g} \cdot d\vec{A}$$

$$= -g\oint_{\text{sphere}} dA$$

$$\Rightarrow \qquad \Phi_g = -g\,4\pi r^2 \qquad (1.58)$$

For the Gaussian surface inside the planet, M_{enc} is not the entire mass but rather just the part of the mass inside the Gaussian surface. We can calculate it using the density we found earlier in Eq.(1.56):

$$M_{encl} = V\rho = \frac{4}{3}\pi r^3 \rho$$

$$= \frac{4}{3}\pi r^3 \left(\frac{3M_p}{4\pi R^3}\right)$$

$$\Rightarrow \qquad M_{encl} = M_p \frac{r^3}{R^3} \qquad (1.59)$$

Now, substituting the values obtained from Eq.1.58 and 1.59, in Eq.1.49, we get

$$\Phi_g = \oint_{\text{surface}} \vec{g} \cdot d\vec{A} = -4\pi GM_{encl}$$

or

$$-g4\pi r^2 = -4\pi GM_p \frac{r^3}{R^3}$$

or

$$g = \frac{GM_p r}{R^3}$$

Since the acceleration due to gravity is radially inward, we can write this in vector form, using the outward-pointing radial unit vector, as

$$\boxed{\vec{g} = -\frac{GMr}{R^3}\hat{r}} \qquad (1.60)$$

Note that The result we obtained in part (b) reduces to the result for gravitational acceleration at the surface of a planet when we set $r = R$. When we are inside the surface of the planet, we get the interesting result that (for this homogeneous-density planet) the strength of the gravitational field is directly proportional to the distance from the centre of the planet. Right at the centre, as expected, the relationship yields a value of zero.

Note: For the homogeneous planet, g verses r graph will be similar to Fig.1.41.

- A uniform shell of matter exerts no net gravitational force on a particle located inside it.

EXAMPLE 38. A thin spherical shell of radius R has a total mass M. Find the gravitational field inside and outside the shell.

APPROACH By symmetry, if any field exists inside the shell, it must be radial. For any point outside or on the surface, it behaves like a solid sphere. So, due to spherical symmetry, construct spherical Gaussian surfaces for both cases and apply Gauss's law:

$$\Phi_g = \oint \vec{g} \cdot d\vec{A} = -4\pi GM_{encl} \qquad (1.61)$$

Solve above equation for g in each case.

SOLUTION (i) **Inside the shell** $(r < R)$: Let us construct a spherical Gaussian surface of radius $r < R$ concentric with the shell (Fig.1.70). Since, there is no enclosed mass within the Gaussian surface, i.e., $M_{encl} = 0$, therefore, Eq.(1.61) gives-

$$\oint \vec{g} \cdot d\vec{A} = -g\left(4\pi r^2\right) = 0$$

$$\Rightarrow \qquad \boxed{g = 0} \qquad (1.62)$$

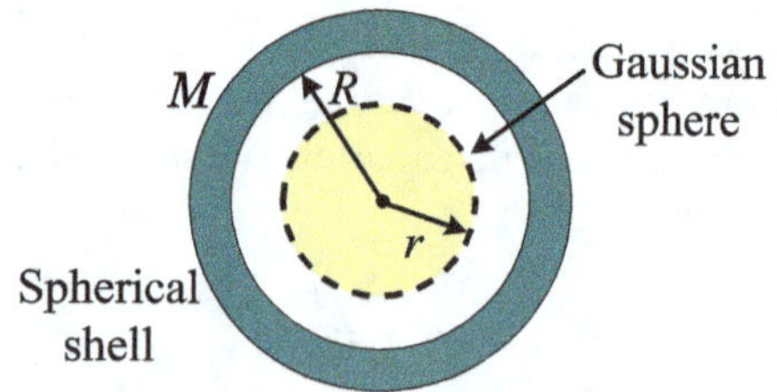

Figure 1.70: Cross sectional view of shell with Gaussian surface inside it.

So, we conclude that there is no gravitational field inside a spherical shell.

(ii) Outside the shell $(r \geq R)$**:** Outside the shell, we construct a spherical Gaussian surface of radius $r > R$ concentric with the charged shell as shown in Fig.1.71. Symmetry suggests that $g =$ constant on that surface and $\vec{g}$ is opposite to $d\vec{A}$, i.e., $\oint \vec{g} \cdot d\vec{A} = -g\left(4\pi r^2\right)$. Since the net mass M_{encl} inside the Gaussian surface is equal to the total mass M on the shell, the shell is equivalent to a point mass located at the center. Therefore, Eq.(1.61) gives-
$$-g\left(4\pi r^2\right) = -4\pi GM$$

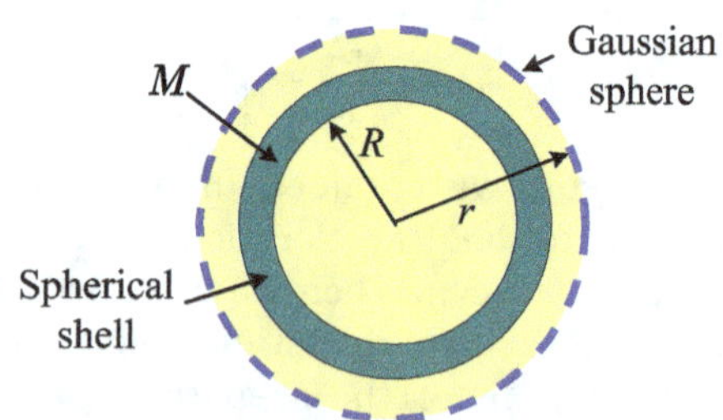

Figure 1.71: Cross sectional view of shell with Gaussian surface outside it.

$$\Rightarrow \qquad \boxed{g = \dfrac{GM}{r^2} \quad (r > R)} \qquad (1.63)$$

(iii) At the surface of the shell $(r = R)$**:** In this case, Eq.(1.63) takes the form,

$$\boxed{g = \dfrac{GM}{R^2} \quad (r = R)} \qquad (1.64)$$

For a shell of mass M, and radius R, gravitational field 'g' vs 'r' graph is for shell of mass M, is shown in Fig.1.72

☞ A uniform spherical shell of matter attracts a particle that is outside the shell as if all the shell's mass were concentrated at its center.

EXAMPLE 39. Two concentric shells of masses m_1 and m_2 are shown in Fig.1.73. Calculate the net gravitational force on m due to both shells at points P, Q and R.

APPROACH The magnitude of gravitation force on a mass m in a gravitational field $\vec{g}$ is mg. Therefore, to find net gravitational force on mass m at all given points, we have to know the gravitational fields at that points. For it, we draw Gaussian spherical surfaces through these points (these are shown by dashed circles in Fig.1.73) and then find g by

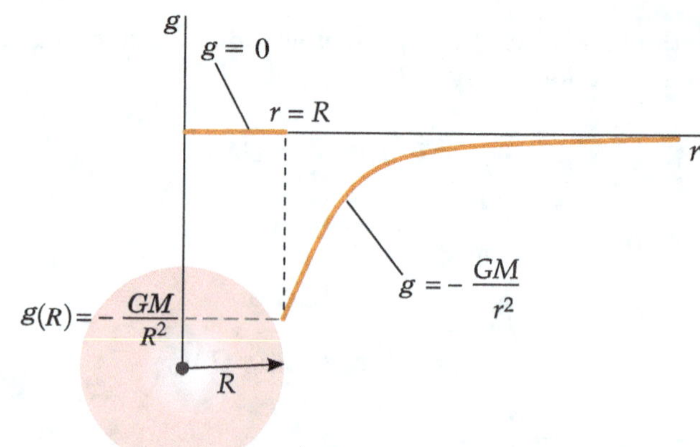

Figure 1.72: Gravitational field g vs r graph for spherical shell of radius R and mass M

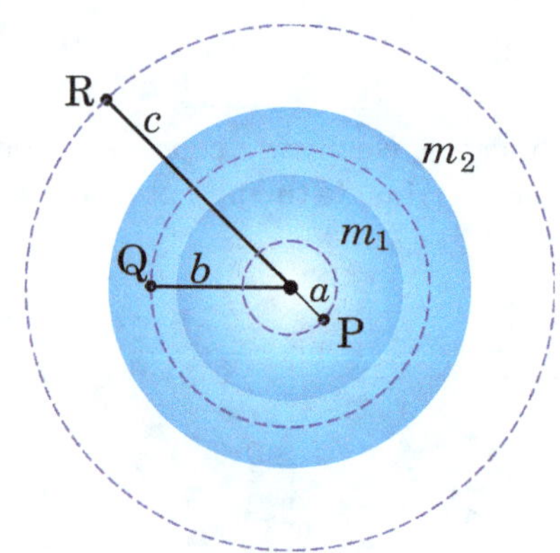

Figure 1.73

applying Gauss's theorem as given in Eq.1.49:
$$\oint_{\text{surface}} \vec{g} \cdot d\vec{A} = -4\pi GM_{\text{encl}} \qquad (1)$$
here, M_{encl} is the net mass enclosed within the Gaussian surface.

SOLUTION At P: Applying Gausses' theorem (Eq.(1)), for the spherical Gaussian surface passing through P.
$$\oint_{\text{surface}} \vec{g} \cdot d\vec{A} = -4\pi G(0) \qquad [\text{Since, } M_{\text{encl}} = 0]$$
$$\Rightarrow \qquad -g\left(4\pi a^2\right) = 0$$
$$\Rightarrow \qquad g = 0$$
Therefore, net gravitational force on mass m placed at P, is
$$F = mg = 0$$

At Q: Applying Gausses' theorem (Eq.(1)), for the spherical Gaussian surface passing through Q.
$$\oint_{\text{surface}} \vec{g} \cdot d\vec{A} = -4\pi Gm_1 \qquad [\text{Since, } M_{\text{encl}} = m_1]$$
$$\Rightarrow \qquad -g\left(4\pi b^2\right) = -4\pi Gm_1$$
$$\Rightarrow \qquad g = \dfrac{Gm_1}{b^2}$$
Therefore, net gravitational force on mass m placed at Q, is
$$F = mg = \dfrac{Gmm_1}{b^2}$$

At R: Applying Gausses' theorem (Eq.(1)), for the spherical Gaussian surface passing through Q.
$$\oint_{\text{surface}} \vec{g} \cdot d\vec{A} = -4\pi G\left(m_1 + m_2\right) \quad [\text{Since, } M_{\text{encl}} = m_1 + m_2]$$
$$\Rightarrow \qquad -g\left(4\pi c^2\right) = -4\pi G\left(m_1 + m_2\right)$$

$$\Rightarrow \qquad g = \frac{G(m_1 + m_2)}{c^2}$$

Therefore, net gravitational force on mass m placed at R, is

$$F = mg = \frac{Gm(m_1 + m_2)}{c^2}$$

EXAMPLE 40. Consider a particle at a point P anywhere inside a spherical shell of matter. Assume that the shell is of uniform thickness and density. Construct a narrow double cone with apex at P intercepting areas dA_1 and dA_2, on the shell (Fig. 1.74). (a) Show that the resultant gravitational force exerted on the particle at P by the intercepted mass elements is zero. (b) Show then that the resultant gravitational force of the entire shell on an internal particle is zero. (This method was devised by Newton.)

APPROACH From Fig.1.74, it is clear that $dA_1 \neq dA_2$, but

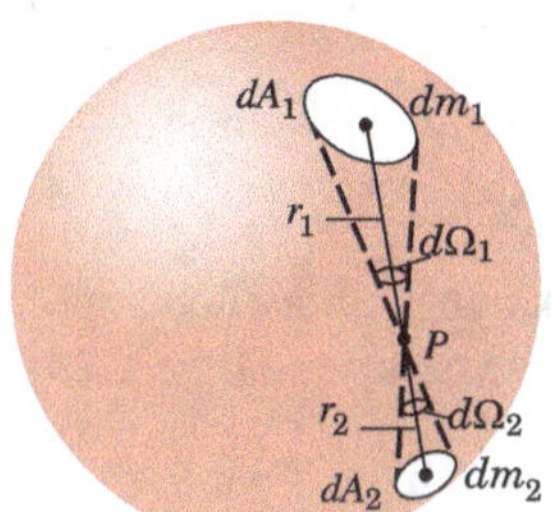

Figure 1.74

the solid angle made by dA_1 at point P is identical to the solid angle made by dA_2 at point P, i.e., $d\Omega_1 = d\Omega_2$, so by using $d\Omega = dA/r^2$, find a relation between dA_1, r_1, dA_2, and r_2 and then use this relation to find the gravitational forces of masses of areas dA_1 and dA_2 on a particle placed at point P.
SOLUTION Since, $d\Omega_1 = d\Omega_2$, therefore, we cn write
$$\Rightarrow \qquad dA_1/r_1^2 = dA_2/r_2^2$$
If σ is the mass per unit area of the shell, then multiplying both sides of above equation by σ gives-
$$\sigma dA_1/r_1^2 = \sigma dA_2/r_2^2$$
$$\Rightarrow \qquad dm_1/r_1^2 = dm_2/r_2^2$$
$$\text{(here, } \sigma dA = dm, \text{ mass of area } dA)$$
On multiplying both sides of above equation by Gm ($m =$ mass of the particle at P), we get

$$\frac{Gmdm_1}{r_1^2} = \frac{Gmdm_2}{r_2^2} \qquad\qquad (1.65)$$

Equation 1.65 shows that the force on an object at point P is balanced by both cones.

(b) Evaluate $\int d\Omega$ for the top and bottom halves of the sphere. Since every $d\Omega$ on the top is balanced by one on the bottom, the net force is zero.

☞ Students are advised to prove above results by using Gauss's theorem of Gravitation.

EXAMPLE 41. A solid sphere of radius R and mass M is spherically symmetric but not uniform. Its density ρ, defined as its mass per unit volume, is proportional to the distance r from the center for $r \leq R$. That is, $\rho = Cr$ for $r \leq R$, where C is a constant. (a) Find C. (b) Find $\vec{g}$ for all $r \geq R$. (c) Find $\vec{g}$ at

$r = R/2$.

APPROACH (a) You can find C by integrating the density over the volume of the sphere and setting the result equal to M. For an infinitely small volume element, take a spherical shell of radius r and thickness dr (Figure 1.75). Its volume is $4\pi r^2 dr$ and its mass is $dM = \rho dV = Cr(4\pi r^2 dr)$.
(b) If we draw a Gaussian surface of radius $r \geq R$, passing through the given point and concentric with the center of the sphere, it contains total mass of the sphere, i.e., $M_{\text{encl}} = M$. So, the field outside the sphere ($r \geq R$) is the same as if the total mass M were at the centre of the sphere.
(c) Now, if we draw a Gaussian surface of radius $r = R/2$, passing through the given point and concentric with the centre of the sphere, it contains only the portion of sphere of radius $r = R/2$. So, the field at $r = R/2$ is the same as if mass M_{encl} were at the center of the sphere, where M_{encl} is the amount of mass within the Gaussian sphere of radius $R/2$. The mass out side of the Gaussian surface, i.e., between $r = R/2$ and $r = R$ produces zero field at $r = R/2$.
SOLUTION (a) 1. Integrate $dM = \rho dV$ to relate C to the

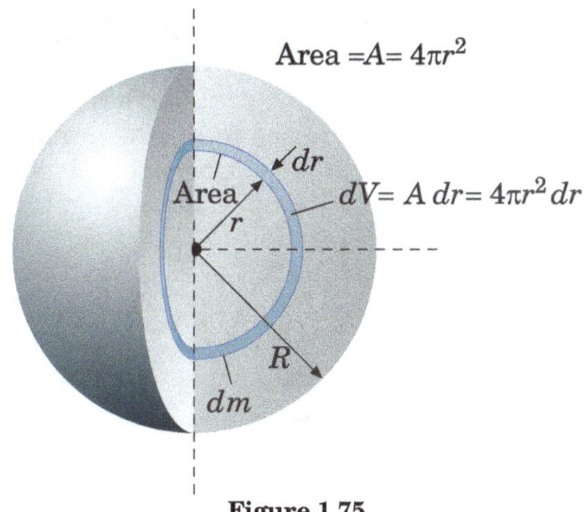

Figure 1.75

mass M, where $dV = 4\pi r^2\, dr$. ($4\pi r^2$) is the area of a sphere of radius r so $4\pi r^2 dr$ is the volume of a spherical shell of radius r and thickness dr):

$$M = \int dM = \int \rho dV$$
$$= \int_0^R Cr(4\pi r^2 dr) = C\pi R^4$$

2. Solve for C in terms of the given quantities M and R.
$$C = \frac{M}{\pi R^4}$$
(b) By Gauss's law of gravitation [Eq.1.44], for $r > R$, we have
$$\oint \vec{g}.d\vec{A} = -4\pi GM_{\text{encl}} \Rightarrow \qquad -g\left(4\pi r^2\right) =$$
$$-4\pi GM \qquad\qquad [\text{since, } M_{\text{encl}} = M]$$
$$\Rightarrow \qquad g = \frac{GM}{r^2}$$
In vector form, the above expression can be written as
$$\vec{g} = \frac{GM}{r^2}\hat{r}$$
Here, the unit vector $\hat{r}$ is in the direction of increasing r.
(c) 1. Compute the mass M_{encl} that is within the Gaussian sphere of radius $\frac{1}{2}R$ by integrating $dm = \rho dV$ from $r = 0$ to

$\frac{1}{2}R$ and use the value of C found in Part (a), step 2.

$$M_{\text{encl}} = \int \rho dV = \int_0^{R/2} Cr(4\pi r^2 dr) = C\pi R^4/16$$

or $\quad M_{\text{encl}} = \dfrac{M}{16}$

2. Apply Gauss's law for $r = R/2$,

By Gauss's law of gravitation, we have

$$\oint \vec{g}.d\vec{A} = -4\pi G M_{\text{encl}}$$

$$\Rightarrow \quad -4\pi g\left(\frac{R}{2}\right)^2 = -4\pi G \frac{M}{16}$$

$$\Rightarrow \quad g = \frac{GM}{4R^2}$$

In vector form,

$$g = -\frac{GM}{4R^2}\hat{r} \tag{1.66}$$

The units for C are kg/m^4, so the units for ρ are kg/m^3, which is mass per volume.

EXAMPLE 42. A solid sphere of radius R and mass M is spherically symmetric but not uniform. Its density ρ, defined as its mass per unit volume, is is, $\rho = \rho_0 r$ for $r \le R$, where ρ is a constant. (a) Find ρ_0 (b) Find $\vec{g}$ for all $r \ge R$. (c) Find $\vec{g}$ for all $r < R$.

APPROACH (a) You can find ρ_0 by integrating the density over the volume of the sphere and setting the result equal to M. For a volume element, take a spherical shell of radius r and thickness dr (Figure 1.75). Its volume is $4\pi r^2 dr$ and its mass is $dM = \rho dV = \rho_0 r^2(4\pi r^2 dr)$.

(b) If we draw a Gaussian surface of radius $r \ge R$, passing through the given point and concentric with the center of the sphere, it contains total mass of the sphere. So,the field outside the sphere ($r \ge R$) is the same as if the total mass M were at the centre of the sphere.

(c) Now, if we draw a Gaussian surface of radius $r(< R)$, passing through the given point and concentric with the center of the sphere, it contains only the portion of sphere of radius r. So, the field at r is the same as if mass M_{encl} were at the centre of the sphere, where M_{encl} is the amount of mass within the sphere of radius r. The mass out side of the Gaussian surface, i.e., between $r = r$ and $r = R$ produces zero field at the point under consideration.

SOLUTION (a) 1. Integrate $dM = \rho dV$ to relate ρ_0 to the mass M, where $dV = 4\pi r^2\, dr$. $(4\pi r^2)$ is the area of a sphere of radius r so $4\pi r^2 dr$ is the volume of a spherical shell of radius r and thickness dr):

$$M = \int dM = \int \rho dV$$
$$= \int_0^R \rho_0 r^2(4\pi r^2 dr) = \frac{4}{5}\rho_0 \pi R^5$$

2. Solve for ρ_0 in terms of the given quantities M and R.

$$\rho_0 = \frac{5M}{4\pi R^5}$$

(b) By Gauss's law [Eq.1.44], for $r > R$, we have

$$\oint \vec{g}.d\vec{A} = -4\pi G M_{\text{encl}}$$

$$\Rightarrow \quad -g(4\pi r^2) = -4\pi G M_{\text{encl}}$$

or $\quad g = \dfrac{GM}{r^2} \qquad$ [since, for $r \ge R$, $M_{\text{encl}} = M$]

In vector form, above equation can be written as

$$\vec{g} = -\frac{GM}{r^2}\hat{r} \qquad\qquad (r > R)$$

Here, the unit vector $\hat{r}$ is in the direction of increasing r.

(c) 1. Compute the mass M_{encl} that is within the radius $r(< R)$ by integrating $dm = \rho dV$ from $r = 0$ to r.

$$M_{\text{encl}} = \int \rho dV = \int_0^r \rho_0 r^2(4\pi r^2 dr) = 4\rho_0 \pi r^5/5$$

2. By Gauss's law [Eq.1.44], for $r < R$, we have

$$\oint \vec{g}.d\vec{A} = -4\pi G M_{\text{encl}}$$

$$\Rightarrow \quad -g(4\pi r^2) = -4\pi G M_{\text{encl}}$$

or $\quad g = \dfrac{GM_{\text{encl}}}{r^2}$

In vector form, $\vec{g} = -\dfrac{GM_{\text{encl}}}{r^2}\hat{r}$

On substituting the value of M_{encl}, obtained in step 1, in above expression, we get

$$\vec{g} = -\frac{G4\rho_0\pi r^5/5}{r^2}\hat{r} = -\frac{4\pi\rho_0 G r^3}{5}\hat{r}$$

On substituting the value of ρ_0 obtaind in part (a), we get

$$\vec{g} = -\frac{4\pi(5M/4\pi R^5)G r^3}{5}\hat{r} = -\frac{GMr^3}{R^5}\hat{r}$$

EXAMPLE 43. A non-homogeneous sphere of radius R has the following density variations [Fig.1.76]:

$$\begin{aligned} \rho &= \rho_0, &\quad r &\le R/3 \\ \rho &= \tfrac{1}{2}\rho_0, &\quad \tfrac{R}{3} &\le r \le \tfrac{3R}{4} \\ \rho &= \tfrac{1}{8}\rho_0, &\quad \tfrac{3R}{4} &\le r \le R \end{aligned}$$

What is the gravitational field due to the sphere at $r = R/4$, $R/2$, $5R/6$ and $2R$?

APPROACH To find gravitational field at any position, we

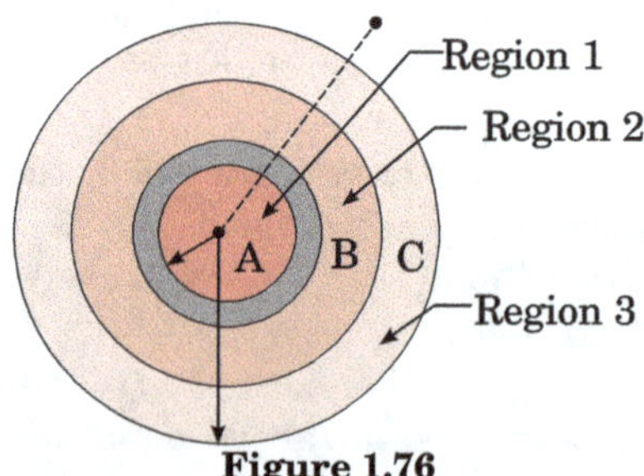

Figure 1.76

draw a Gaussian surface passing through that point and then consider the source mass equal to the mass enclosed within the Gaussian surface (M_{encl}). Now, we can apply the relation $g = GM_{\text{encl}}/r^2$.

SOLUTION (a) For $r = R/4$:

The gravitational field at a point due to a sphere of radius r is given by

$$g = \frac{GM_{\text{encl}}}{r^2} = G \cdot \frac{\frac{4\pi}{3}r^3\rho}{r^2} = \frac{4\pi}{3}r\rho G$$

Here $r = R/4$ and in this region ($r \le R/3$), $\rho = \rho_0$

$$\therefore \qquad g = \frac{4\pi}{3}\left(\frac{R}{4}\right)\rho_0 = 0.33\pi G\rho_0 R$$

(b) For $r = R/2$:

For the field at a point $r = R/2$, the mass (M) is the mass of the region 1 and the mass of the shaded portion where the density is $\rho = \tfrac{1}{2}\rho_0$

$$\therefore \quad M_{\text{encl}} = \frac{4\pi}{3}\left(\frac{R}{3}\right)^3 \rho_0 + \frac{4\pi}{3}\left[\left(\frac{R}{2}\right)^3 - \left(\frac{R}{3}\right)^3\right]\frac{1}{2}\rho_0$$

$$= \frac{4\pi}{3}\rho_0 R^3 \left[\frac{1}{27} + \frac{1}{16} - \frac{1}{54}\right]$$

The field at the point $r = R/2$ is

$$= \frac{GM_{\text{encl}}}{\left(\frac{R}{2}\right)^2} = \frac{4G}{R^2} \times \frac{4\pi}{3}\rho_0 R^3 \left[\frac{1}{27} + \frac{1}{16} - \frac{1}{54}\right]$$

$$= \frac{16\pi G}{3}\rho_0 R \left[\frac{1}{27} + \frac{1}{16}\right] = 0.4\rho G \rho_0 R$$

(c) For $r = 5R/6$:

Hence, r corresponds to $5R/6$, and M_{encl} corresponds to mass of the portion of the sphere of radius $5R/6$. Density of the portion of radius $R/3$ is ρ_0 and of the portion between $r = R/3$ to $r = 3R/4$ is $\rho_0/2$ and the portion between $r = \frac{3R}{4}$ to $\frac{5R}{6}$ is $\rho_0/8$

$$\therefore \quad M_{\text{encl}} = \frac{4\pi}{3}\left(\frac{R}{3}\right)^3 \rho_0 + \frac{4\pi}{3}\left[\left(\frac{R}{3}\right)^3 - \left(\frac{R}{3}\right)^3\right]\frac{\rho_0}{2}$$

$$+ \frac{4\pi}{3}\left[\left(\frac{5R}{6}\right)^3 - \left(\frac{3R}{4}\right)^3\right]\frac{\rho_0}{8}$$

$$= \frac{4\pi}{3}\rho_0 R^3 \left[\frac{1}{27} + \frac{27}{128} - \frac{1}{54} + \frac{125}{1728} - \frac{27}{512}\right]$$

$$= 0.332\pi R^3 \rho_0$$

The field at the point $r = 5R/6$

$$= \frac{GM_{\text{encl}}}{\left(\frac{5R}{6}\right)^2} = \frac{36G}{25R^2} \times 0.332\pi R^3 \rho_0 = 0.48\pi G\rho_0 R$$

(d) For $r = 2R$:

Here, r corresponds to $2R$ and M_{encl} corresponds to the total mass of the sphere of radius R.

$$\therefore \quad \text{Mass} = \frac{4\pi}{3}\left(\frac{R}{3}\right)^3 \rho_0 + \frac{4\pi}{3}\left[\left(\frac{3R}{4}\right)^3 - \left(\frac{R}{3}\right)^3\right]\frac{\rho_0}{2}$$

$$+ \frac{4\pi}{3}\left[R^3 - \left(\frac{3R}{4}\right)^3\right]\frac{\rho_0}{8}$$

$$= \frac{4\pi}{3}\rho_0 R^3 \left[\frac{1}{27} + \frac{27}{128} - \frac{1}{54} + \frac{1}{8} - \frac{27}{512}\right]$$

$$= 0.402\pi R^3 \rho_0$$

The field at the point $r = 5R/6$

$$= \frac{GM_{\text{encl}}}{(2R)^2} = \frac{G}{4R^2} \times 0.402\pi R^3 \rho_0 = 0.1\pi G\rho_0 R$$

Note: Table 1.2 shows gravitational field intensities at position r due to some geometrical shapes/objects. Distance r is measured from the centre of the object.

1.10.2 Check Point 2

1. •Draw g versus depth and g versus height graphs.
2. •The distance between earth and moon is about 3.8×10^5 km. At what point or points will the gravitational field strength of earth-moon system be zero? Given mass of earth is 81 times the moon's mass.
3. •Find the height above the surface of the earth where the acceleration due to gravity reduces by (a) 36% (b) 0.36% of its value on the surface of the earth. Radius of the earth $R = 6400$ km.
4. ••An astronaut landed on a planet and found that his weight at the pole of the planet was one third of his

Table 1.2: Gravitational field intensities at position r due to some geometrical shapes/objects

Geometrical Shape/Object	Gravitational Field Intensity
Point mass of mass M	$g = -\frac{GM}{r^2}\hat{r}$
On the axis of a uniform circular ring	$g = GM\dfrac{r}{(R^2+r^2)^{3/2}}$
At an axial point of rod of mass M and length L	$g = \frac{GM}{r^2}\left(\dfrac{1}{1+\frac{L}{r}}\right)$
At an equatorial point of a rod of mass M and length L	$g = \dfrac{2GM}{r\sqrt{L^2+4r^2}}$
On the axis of a uniform circular disc of mass M and radius R	$g = \frac{2GM}{R^2}\left[\dfrac{1}{r} - \dfrac{1}{\sqrt{r^2+R^2}}\right]$
Circular arc of mass M and radius R	$g = \frac{2\pi GM}{L^2}$
Uniform solid sphere of mass M and radius R	$g = \frac{GM}{r^2}$ (Outside sphere); $g \propto r$ (Inside sphere)
Spherical shell of mass M and radius R	$g = \frac{GM}{r^2}$ (Outside shell); $g = 0$ (Inside shell)
Long thread	$g = \frac{2G\lambda}{r}$

weight at the pole of the earth. He also found himself to be weightless at the equator of the planet. The planet is a homogeneous sphere of radius half that of the earth. Find the duration of a day on the planet. Given that the density of the earth is ρ_0 and the density of planet is ρ_P.

5. ••A gravity meter can detect change in acceleration due to gravity (g) of the order of 10^{-9}%. Calculate the smallest change in altitude near the surface of the earth that results in a detectable change in g. Radius of the earth $R = 6.4 \times 10^6$ m.

6. ••Suppose you leave the solar system and arrive at a planet that has the same mass-to-volume ratio as Earth but has 10 times Earth's radius. What would you weigh on this planet compared with what you weigh on Earth?

7. ••Suppose that Earth retained its present mass but was somehow compressed to half its present radius. What would be the value of g at the surface of this new, compact planet?

8. ••The earth is a homogeneous sphere of mass M and radius R. There is another spherical planet of mass M and radius R whose density changes with distance r from the centre as $\rho = \rho_0 r$. (a) Find the ratio of acceleration due to gravity on the surface of the earth and that on the surface of the planet, (b) Find ρ_0.

9. •••Angular speed of rotation of the earth is ω_e. A train is running along the equator at a speed v from west to east. A very sensitive balance inside the train shows the weight of an object as W_1. During the return journey when the train is running at same speed from east to west the balance shows the weight of the object to be W_2. Weight of the object at pole is W_0 by the balance. Calculate $W_2 - W_1$.

10. $\bullet\bullet\bullet$ If a planet rotates too fast, rocks from its surface will start flying off its surface. If density of a homogeneous planet is ρ and material is not flying off its surface then show that its time period of rotation must be greater than $\sqrt{\dfrac{3\pi}{G\rho}}$.

11. $\bullet\bullet$ (a) The angular speed of rotation of the earth is $\omega = 7.27 \times 10^{-5}$ rad s^{-1} and its radius is $R = 6.37 \times 10^{6}$m. Calculate the centripetal acceleration of a man standing at a place at $40°$ latitude. [$\cos 40° = 0.77$]

(b) If the earth suddenly stops rotating, find the effective value of g at latitude $40°$.

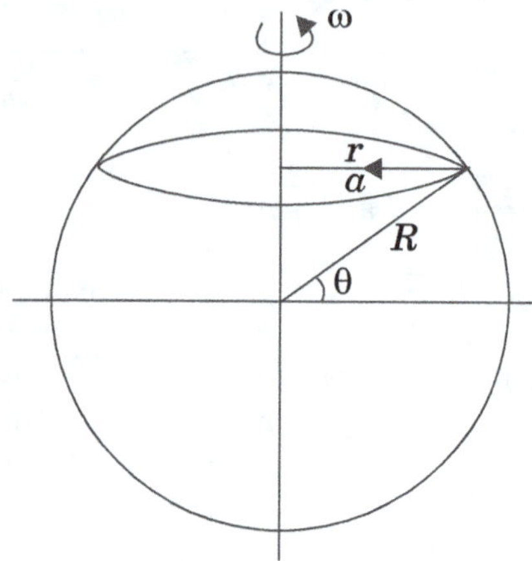

Figure 1.77

12. $\bullet\bullet$ A large non rotating star of mass M and radius R begins to collapse under its own gravity and ultimately becomes very small (nearly a point mass). Assume that the density remains uniform inside the sphere in any stage. Plot the variation of gravitational field intensity (you can also call it acceleration due to gravity) at a distance $R/2$ from the centre vs the radius (r) of the star.

13. $\bullet\bullet$ At a depth $h_1 = R/2$ from the surface of the earth acceleration due to gravity is g_1. It's value changes by Δg_1 when one moves down further by $1\ km$. At a height h_2 above the surface of the earth acceleration due to gravity is g_2. It's value changes by Δg_2 when one moves up further by 1km. If $\Delta g_1 = \Delta g_2$ find h_2. Assume the earth to be a uniform sphere of radius R.

14. $\bullet\bullet$ A non-uniform thin rod of length L lies on the x axis. One end of the rod is at the origin, and the other end is at $x = L$. The rod's mass per unit length λ varies as $\lambda = Cx$, where C is a constant. (Thus, an element of the rod has mass $dm = \lambda dx$.) (a) Determine the total mass of the rod. (b) Determine the gravitational field due to the rod on the x axis at $x = x_0$, where $x_0 > L$.

15. $\bullet\bullet\bullet$ A uniform thin rod of mass M and length L lies on the positive x axis with one end at the origin. Consider an element of the rod of length dx, and mass dm, at point x, where $0 < x < L$. (a) Show that this element produces a gravitational field at a point x_0 on the x axis in the region $x_0 > L$ is given by $dg_x = -\dfrac{GM}{L(x_0-x)^2}dx$. (b) Integrate this result over the length of the rod to find the total gravitational field at the point x_0 due to the rod. (c) Find the gravitational force on a point particle of mass m_0 at x_0.

(d) Show that for $x_0 \gg L$, the field of the rod approximates the field of a point particle of mass M at $x = 0$.

16. $\bullet\bullet$ Due to rotation of the earth the direction of vertical at a place is not along the radius of the earth and actually makes a small angle ϕ with the true vertical (i.e. with radius). At what latitude (θ) is this angle β maximum?

17. $\bullet\bullet$ A planet is a homogeneous ball of radius R having mass M. It is surrounded by a dense atmosphere having density $\rho = \sigma_\circ/r$ where $\sigma_\circ$ is a constant and r is distance from the centre of the planet. It is found that acceleration due to gravity is constant throughout the atmosphere of the planet. Find $\sigma_\circ$ in terms of M and R.

18. $\bullet\bullet$ A planet orbits a massive star. When the planet is at perihelion, it has a speed of 5.0×10^4 m/s and is 1.0×10^{15} m from the star. The orbital radius increases to 2.2×10^{15} m at aphelion. What is the planet's speed at aphelion?

19. $\bullet\bullet$ What is the magnitude of the gravitational field at the surface of a neutron star whose mass is 1.60 times the mass of the Sun and whose radius is 10.5 km? (Take mass of Sun, $M_S = 1.99 \times 10^{30}$kg)

20. $\bullet\bullet$ Find the force of attraction on a particle of mass m_0 placed at the centre of a quarter ring of mass m and radius R as shown in Figure.1.78

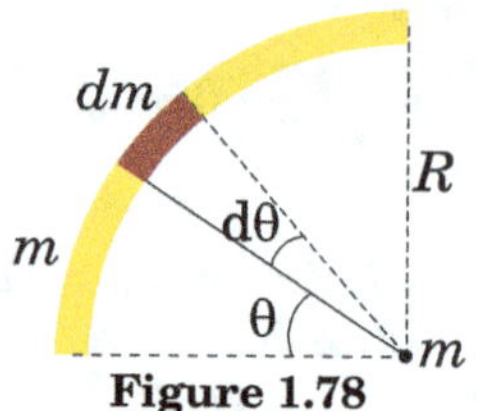

Figure 1.78

21. $\bullet\bullet$ Earth's radius is 6370 km and the moon's radius is 1738 km. The acceleration of gravity at the surface of the moon is 1.62 m/s^2. What is the ratio of the average density of the moon to that of Earth?

22. $\bullet\bullet$ The weight of a standard object defined as having a mass of exactly 1.00 kg is measured to be 9.81 N. In the same laboratory, a second object weighs 56.6 N. (a) What is the mass of the second object? (b) Is the mass you determined in Part (a) gravitational or inertial mass?

23. $\bullet\bullet$ A uniform thin spherical shell has a radius of 2.0 m and a mass of 300 kg. What is the gravitational field at the following distances from the centre of the shell: (a) 0.50 m, (b) 1.9 m, (c)2.5 m?

24. $\bullet\bullet$ A thick spherical shell of mass M and uniform density has an inner radius R_1 and an outer radius R_2. Find the gravitational field g_r as a function of r for $0 < r < \infty$. Sketch a graph of g_r versus r.

25. $\bullet\bullet$ (a) A thin uniform ring of mass M and radius R lies in the $x = 0$ plane and is centered at the origin. Sketch a plot of the gravitational field g_x versus x for all points on the x axis. (b) At what point, or points, on the axis is the magnitude of g_x a maximum? Find maximum value of g_x

Multiple Choice Questions

26. $\bullet\bullet$ Dependence of intensity of gravitational field (E) of earth with distance (r) from centre of earth is correctly represented by plot [Fig.1.79]

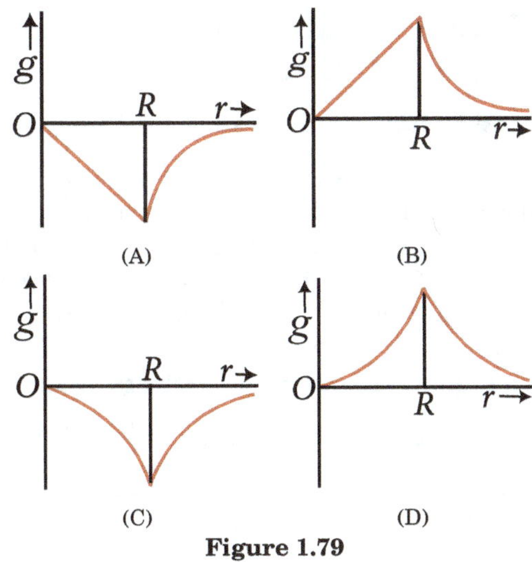

Figure 1.79

27. ••Which one of the following plots [Fig.1.80] represents the variation of gravitational field on a particle with distance r due to a thin spherical shell of radius R ? (r is measured from the centre of the spherical shell)

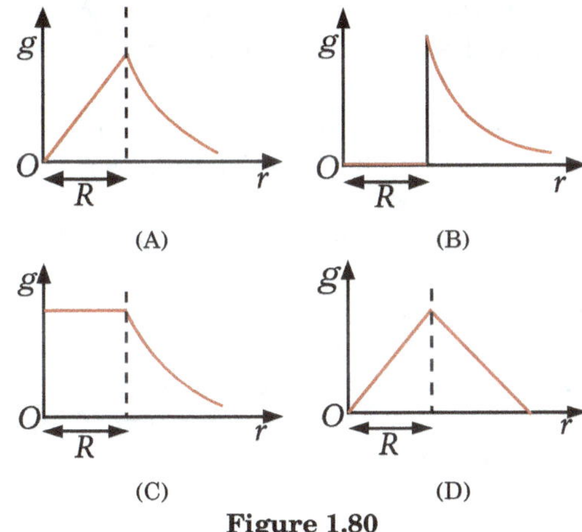

Figure 1.80

28. ••If the mass of the Sun were ten times smaller and the universal gravitational constant were ten times larger in magnitude, which of the following is not correct?
   (A) Raindrops will fall faster.
   (B) Walking on the ground would become more difficult.
   (C) Time period of a simple pendulum on the Earth would decrease.
   (D) g on the Earth will not change.

29. ••A spherical planet has a mass M_p and diameter D_p. A particle of mass m falling freely near the surface of this planet will experience an acceleration due to gravity, equal to
   (A) $\frac{4GM_p}{D_p^2}$
   (B) $\frac{GM_p m}{D_p^2}$
   (C) $\frac{GM_p}{D_p^2}$
   (D) $\frac{4GM_p m}{D_p^2}$

30. ••Imagine a new planet having the same density as that of earth but it is 3 times bigger than the earth in size. If the acceleration due to gravity on the surface of earth is g and that on the surface of the new planet is g', then

(A) $g' = g/9$
(B) $g' = 27g$
(C) $g' = 9g$
(D) $g' = 3g$

31. ••The density of a newly discovered planet is twice that of earth. The acceleration due to gravity at the surface of the planet is equal to that at the surface of the earth. If the radius of the earth is R, the radius of the planet would be
   (A) $2R$
   (B) $4R$
   (C) $\frac{1}{4}R$
   (D) $\frac{1}{2}R$

32. ••The acceleration due to gravity on the planet A is 9 times the acceleration due to gravity on planet B. A man jumps to a height of $2m$ on the surface of A. What is the height of jump by the same person on the planet B ?
   (A) $(2/9)m$
   (B) $18m$
   (C) $6m$
   (D) $(2/3)m$

33. ••What will be the formula of mass of the earth in terms of g, R and G ?
   (A) $G\frac{R}{g}$
   (B) $g\frac{R^2}{G}$
   (C) $g^2\frac{R}{G}$
   (D) $G\frac{g}{R}$

34. ••The acceleration due to gravity g and mean density of the earth ρ are related by which of the following relations? (where G is the gravitational constant and R is the radius of the earth.)
   (A) $\rho = \frac{3g}{4\pi GR}$
   (B) $\rho = \frac{3g}{4\pi GR^3}$
   (C) $\rho = \frac{4\pi gR^2}{3G}$
   (D) $\rho = \frac{4\pi gR^3}{3G}$

35. ••The radius of earth is about 6400 km and that of mars is 3200 km. The mass of the earth is about 10 times mass of mars. An object weighs 200 N on the surface of earth. Its weight on the surface of mars will be
   (A) $20N$
   (B) $8N$
   (C) $80N$
   (D) $40N$

36. ••A body weighs 72 N on the surface of the earth. What is the gravitational force on it, at a height equal to half the radius of the earth?
   (A) $48N$
   (B) $32N$
   (C) $30N$
   (D) $24N$

37. ••A body weighs 200 N on the surface of the earth. How much will it weigh half way down to the centre of the earth?
   (A) $100N$
   (B) $150N$
   (C) $200N$
   (D) $250N$

38. ••The acceleration due to gravity at a height $1km$ above the earth is the same as at a depth d below the surface of earth. Then
   (A) $d = 1km$
   (B) $d = \frac{3}{2}km$
   (C) $d = 2km$
   (D) $d = \frac{1}{2}km$

39. ••Starting from the centre of the earth having radius R, the variation of g (acceleration due to gravity) is shown by [Fig.1.81]

40. ••The height at which the weight of a body becomes $\left(\frac{1}{16}\right)^{th}$, its weight on the surface of earth (radius R), is
   (A) $5R$
   (B) $15R$
   (C) $3R$
   (D) $4R$

41. ••Consider an infinite plane sheet having surface mass density σ. The gravitational field intensity at a point P at perpendicular distance r from such a sheet is
   (A) Zero
   (B) $-\sigma G$
   (C) $2\pi\sigma G$
   (D) $-4\pi\sigma G$

1.11   Gravitational Potential Energy

In Chapter "Work, Energy and Power", we obtained the potential energy change due to gravity for a body that moves

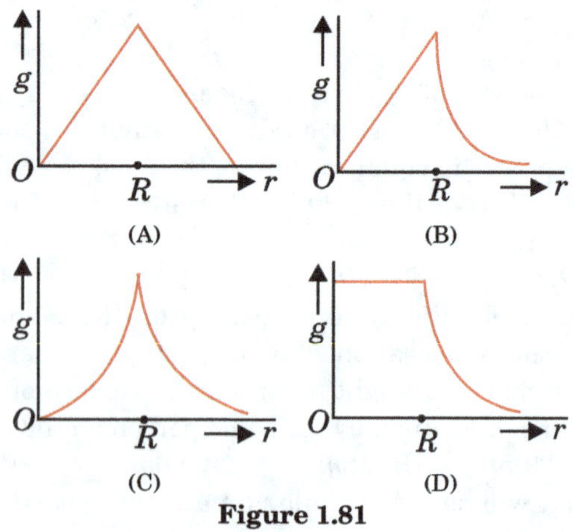

Figure 1.81

1.11.2 Gravitational Potential Energy From the Law of Gravity

1.11.2.1 One Dimensional Case

Let us consider two objects 1 and 2 of masses m_1 and m_2. Suppose, the mass of object 1 is much greater than the mass of object 2, $m_1 \gg m_2$, so that if we place object 1 at the origin, we can consider it being fixed there. Now let object 2 is moving along the positive direction of x axis from $x = x_i$ to $x = x_f$ [Fig.1.82].

The gravitational force exerted by object 1 on object 2, is

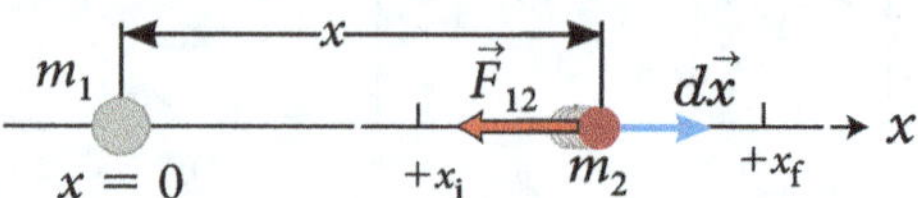

Figure 1.82: An object of mass m_2 moves toward an object of mass $m_1 \gg m_2$ held fixed at the origin.

not uniform, but decreases as $1/x^2$. The general definition of change in potential energy is-

Change in gravitational potential energy

$$= -\text{Work done by conservative system force}$$

Therefore, to find the change in gravitational potential energy, we first calculate the work done of gravitational force of object 1 on object 2.

Let at any arbitrary position x of object 2, the gravitational force applied by object 1 on object 2 is $\vec{F}_{12}$, then from Fig.1.82

$$\vec{F}_{12} = -F_{12}\hat{i} = -G\frac{m_1 m_2}{x^2}\hat{i} \tag{1.69}$$

Here, $\hat{i}$ is the unit vector along the positive direction of x axis. Negative sign in above expression shows that the gravitational force om m_2 is opposite to the direction of $\hat{i}$. If at position x, the elementary displacement of object 2 along the positive direction of x axis is $d\vec{x}$, then-

$$d\vec{x} = dx\hat{i} \tag{1.70}$$

The work done by gravitational force $\vec{F}_{12}$, in displacement of object 2 from $x = x_i$ to $x = x_f$ is given by-:

$$W = \int_{x_i}^{x_f} \vec{F}_{12}.d\vec{x} \tag{1.71}$$

Using Eq. 1.69 and Eq.1.70, in Eq.1.71, we obtain-

$$W = \int_{x_i}^{x_f} \left(-G\frac{m_1 m_2}{x^2}\hat{i}\right).\left(dx\hat{i}\right)$$

$$W = -Gm_1 m_2 \int_{x_i}^{x_f} \left(\frac{1}{x^2}\right)dx = Gm_1 m_2 \left[\frac{1}{x}\right]_{x_i}^{x_f}$$

or $$\boxed{W = Gm_1 m_2 \left(\frac{1}{x_f} - \frac{1}{x_i}\right)} \tag{1.72}$$

This is the work done by the gravitational force exerted by object 1 on the system consisting of object 2 only (Figure 1.83a). Because the force and the displacement point in the opposite direction, we know that the work must be negative. Indeed, the right side of Eq. 1.72 is negative because $x_f > x_i$.

As object 2 moves from x_i to x_f, it's velocity decreases, and so its kinetic energy decreases. Because no other energy associ-

---

through a height y near the Earth's surface: $\Delta U = mgy$. However, this applies only near the Earth's surface, where (for changes in height that are small compared with the distance from the center of the Earth) we can regard the gravitational force as approximately constant. When we are far from the surface of Earth, we must take into account the fact that the gravitational force exerted by Earth is not uniform, but decreases as $1/r^2$. In following sub section we obtain a general expressions for the gravitational potential energy from the law of gravity.

1.11.1 Calculation of Change in Potential Energy by Integration Method

The concept of potential energy has already been discussed in the chapter "Work, Energy and Power (WEP)". The word potential energy is defined only for a conservative force field. If a conservative force $\vec{F}$ displaces a particle from position $\vec{r}_i$ to position $\vec{r}_f$, and the potential energy of system changes from U_i to U_f, then change in potential energy will be equal to $-$ve of work done by conservative field force ($\vec{F}$), i.e.,

$$\int_{U_i}^{U_f} dU = -\int_{r_i}^{r_f} \vec{F}\cdot d\vec{r}$$

or $$\boxed{U_f - U_i = -\int_{r_i}^{r_f} \vec{F}\cdot d\vec{r}} \tag{1.67}$$

We generally choose the reference point at infinity and assume potential energy to be zero there, i.e., corresponding to $r_i = \infty$ (infinite) the gravitational potential energy $U_i = 0$. Substituting these values in Eq.1.67, we get

$$U = -\int_{\infty}^{r} \vec{F}\cdot d\vec{r} = -W \tag{1.68}$$

From Eq.1.68, it is clear that the potential energy of a body or system for given configuration is negative of work done by the conservative forces in bringing it from infinity to the present configuration.

ated with it, the changes, i.e. decrease in the kinetic energy must be equal to the work done on it by object $1, \Delta K = W$.

If, instead of object 2 by itself, we now consider the (closed) system of the two interacting objects as a whole, then the decrease in the kinetic energy of object 2 is due not to the work done by the external gravitational force exerted by object 1 on it but to a increase in the gravitational potential energy of the system (Figure 1.83b. This gravitational potential energy is a measure of the configuration of the system. The change in the gravitational potential energy of the system is thus $\Delta U_g = -\Delta K$. Note that the change in kinetic energy does not depend on the choice of system. Thus, we see that the change in potential energy of the two-object system is the negative of W in Eq. 1.72:

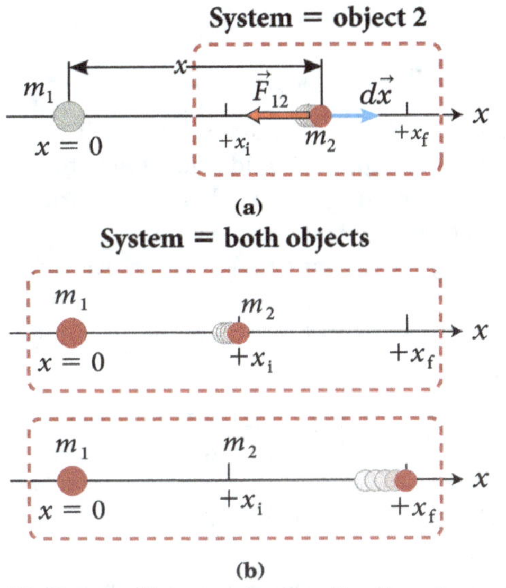

Figure 1.83: Energy diagrams for the situation shown in Figure

$$\Delta U_g = -\Delta K = -W = Gm_1 m_2 \left(\frac{1}{x_i} - \frac{1}{x_f}\right) \quad (1.73)$$

It is customary to choose the gravitational potential energy of a system to be zero at infinite separation ($x = \infty$), which is the separation distance at which the gravitational force is zero: $U_g(\infty) = 0$. So, if we let object 2 move in from $x = \infty$ to an arbitrary position x under the influence of the gravitational acceleration due to object 1, then $x_i = \infty$, $x_f = x$, so Eq.1.73 becomes

$$\Delta U_g = U_g(x) - U_g(\infty)$$

$$\text{or} \quad \Delta U_g = U_g(x) - 0 = Gm_1 m_2 \left(0 - \frac{1}{x}\right) \quad (1.74)$$

So, the gravitational potential energy for separation 'x' is

$$\boxed{U_g(x) = -G\frac{m_1 m_2}{x}} \quad (1.75)$$

which is zero at infinite x. As object 2 moves in from infinitely far away, its kinetic energy increases, and so the potential energy of the two-object system decreases. Therefore setting $U_g(\infty) = 0$ at $x = \infty$ causes the potential energy to be negative for all values of $x < \infty$.

1.11.2.2 Three Dimensional Case

Because the force of gravity is a central force, we can easily generalize Eq. 1.75 to more than one dimension.

Suppose, object 2 moves from P to Q in Figure 1.84a along part of an elliptical orbit. Let us first approximate the displacement of object 2 by two successive displacements: one along the radial direction from P to R and one along a circular arc from R to Q. The work done by the force of gravity on object 2 over the radial displacement is given by Eq. 1.72. Along the circular arc RQ, the magnitude of the force of gravity is constant but always oriented perpendicular to the displacement. Because the work is given by the scalar product of the force and the force displacement, the work done by the force of gravity on object 2 along the circular arc is zero:

$$W_{R \to Q} = \int_R^Q \vec{F}_{12} \cdot d\vec{r} = 0 \quad (1.76)$$

So the work done by object 1 on object 2 as it moves from P

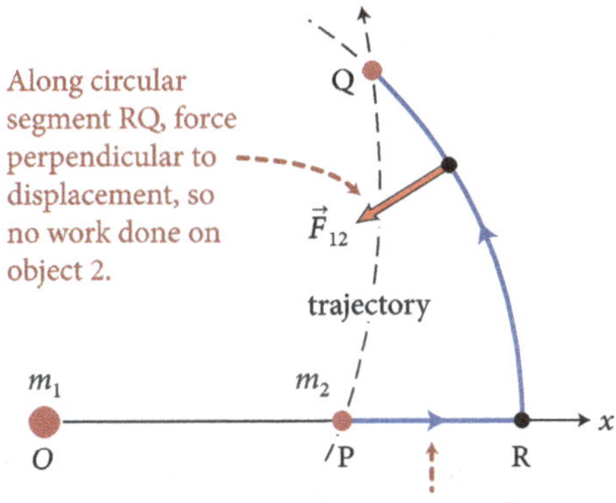

(a) Trajectory approximated by one radial segment PR and one circular segment RQ

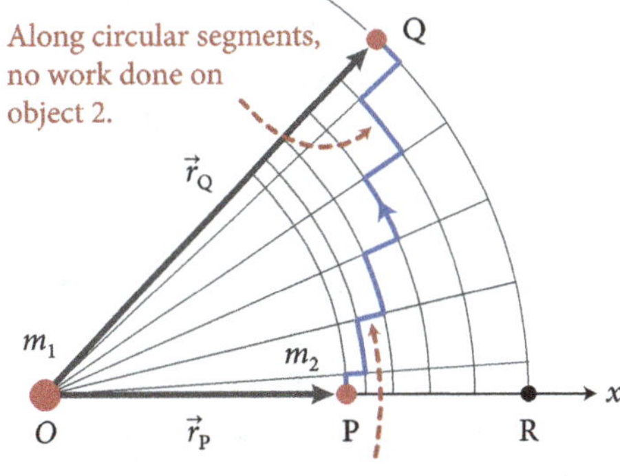

(b) Trajectory approximated by multiple radial and circular segments

Figure 1.84: To calculate the work done by object 1 on object 2 while object 2 moves along the dashed path from P to Q, we approximate the path by (a) one radial and one circular dis-placement or (b) a number of successive such displacements.

to Q along the path PRQ is equal to the work done along the radial displacement only.

A better approximation of the actual path of object 2 can be obtained by breaking down the motion into small segments and approximating each segment by the sum of radial and circular displacements, as shown in Figure 1.84b. Again, the work done on object 2 along each circular displacement is zero. Furthermore, all the radial displacements simply add up to the same radial displacement as in Figure 1.84b. So, only the change in radial distance matters and the gravitational work done by object 1 on object 2 while object 2 moves from P to Q is

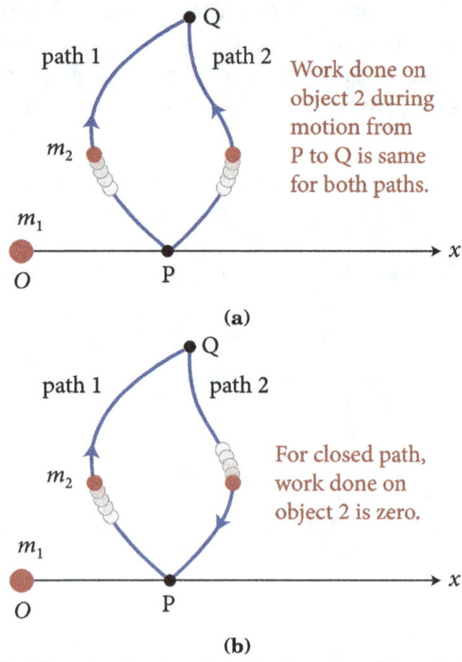

Figure 1.85: The gravitational work done on object 2 depends only on the endpoints.

$$W_{P \to Q} = \int_{r_P}^{r_Q} \left(-G\frac{m_1 m_2}{r^2}\right) dr = G m_1 m_2 \left(\frac{1}{r_Q} - \frac{1}{r_P}\right) \quad (1.77)$$

where r_P and r_Q are the magnitudes of the position vectors of object 2 at P and Q, respectively.

If we consider the initial position of mass m_2 at infinite distance from m_1, i.e., $r_P \to \infty$ and final position of m_2 at separation r from m_1, i.e., $r_Q \to r$, then work done by gravitational force:

$$W_{\infty \to r} = \int_{\infty}^{r} \left(-G\frac{m_1 m_2}{r^2}\right) dr = G\frac{m_1 m_2}{r} \quad (1.78)$$

Now, following the same line of reasoning that led from Eq. 1.72 to Eq. 1.75, we can write for the gravitational potential energy of the system

$$\boxed{U(r) = -W_{\infty \to r} = -G\frac{m_1 m_2}{r}} \quad \text{(zero at infinite r)} \quad (1.79)$$

where r is the distance between the two objects.

Note that the result we obtained in Eq.1.79 holds for any path that leads from P to Q. Any arbitrary path can be broken

down into small radial and circular segments, and as Figure 1.84b shows, the work done by the gravitational force depends only on the position of the endpoints relative to m_1, not on the path taken. Consider, for example, paths 1 and 2 from P to Q in Figure 1.85a. The work done is the same along both paths:

$$W_{P \to Q, \text{ path 1}} = W_{P \to Q, \text{ path 2}}. \quad (1.80)$$

This path independence is a characteristic of non-dissipative conservative interactions.

We also know that the work done along a path moving in one direction is the negative of the work done along that path moving in the opposite direction. This can readily be seen in Figure 1.85b: Reversing the direction of travel inverts the sign of the displacement (and therefore the sign of the work) along all radial components. So if we let object 2 move from Q to P along path 2 in Figure 1.85a, the work done on it is

$$W_{Q \to P, \text{ path 2}} = -W_{P \to Q, \text{ path 2}} \quad (1.81)$$

A consequence of Eq. 1.80 and 1.81 is that if an object moves from P to Q along path 1 and then back to P along path 2, as shown in Figure 1.85b, then the work done by the gravitational force on the object is zero:

To obtain the gravitational potential energy of Earth (mass $= M$) and a given point or spherical mass ($= m$) system, just replace m_1 by the mass of earth M and m_2 by given mass m in Eq.1.79, you get

$$\boxed{U(r) = -G\frac{Mm}{r}} \quad \text{(zero at infinite r)} \quad (1.82)$$

Here, r is the separation between spherical or point mass m and center of the earth.

If the spherical body is very near to the surface of earth, then we can replace r by the radius of Earth (R). In this case Eq.1.82 becomes

$$\boxed{U(r) = -G\frac{Mm}{R}} \quad (1.83)$$

EXAMPLE 44. Find an expression of gravitational potential energy of Earth (mass $= M$) - particle of (mass $= m$) system, having separation r from the center of Earth.

APPROACH In sub section 1.11.2.2, we have already derived an expression for gravitational potential energy of Earth-mass system (see Eq.1.83), however, we can also directly find an expression for gravitational potential energy in three dimensional space by applying the general definition:

$$dU = -\vec{F}_g \cdot d\vec{l} = -(-F_g \hat{r}) \cdot d\vec{l} = F_g \hat{r} \cdot d\vec{l} \quad (1.84)$$

where dU is the change in potential energy of Earth-particle system in general displacement $d\vec{l}$ and $\vec{F}$ [$= (-F_g \hat{r})$] is conservative gravitational force of earth on the particle. From Fig.1.86, $\hat{r} \cdot d\vec{l} = dl \cos\phi = dr$, therefore-

$$dU = F_g dr \quad (1.85)$$

Therefore, change in potential energy on moving the particle from initial position $r = r_i$ to $r = r_f$ is given by

$$U_f - U_i = \int_{r_i}^{r_f} F_g \, dr \tag{1.86}$$

Here, U_i and U_f are the gravitational potential energies of Earth-particle system for $r = r_i$ and $r = r_f$, respectively.

At general position r of the particle, from the center of Earth, the magnitude of gravitational force F_g of Earth on the particle is given by Eq.1.2:

$$F_g = G\frac{Mm}{r^2}$$

Substituting this value of F_g, in Eq.1.86 gives:

$$U_f - U_i = \int_{r_i}^{r_f} G\frac{Mm}{r^2} \, dr \tag{1.87}$$

Now, take reference point i.e., zero potential energy $(U_i = 0)$ where F_g is zero, i.e., at $r = \infty$ and simplify Eq.1.87 for $U_f = U$ at position $r_f = r$.

SOLUTION Integrating Equation 1.87, we get

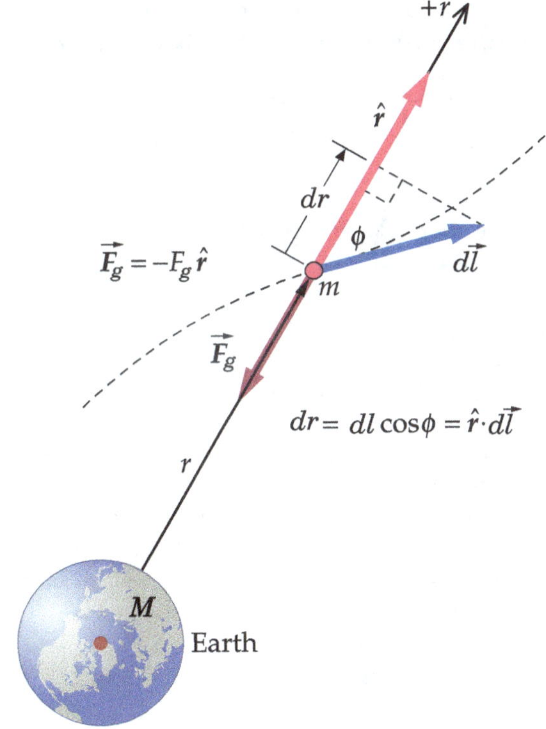

Figure 1.86

$$U_f - U_i = -\left[G\frac{Mm}{r}\right]_{r_i}^{r_f}$$

$$\text{or} \qquad U_f - U_i = -GMm\left[\frac{1}{r_f} - \frac{1}{r_i}\right] \tag{1.88}$$

$$\text{or} \qquad U_f = -GMm\left[\frac{1}{r_f} - \frac{1}{r_i}\right] + U_i$$

Considering zero potential energy point at $r_i = \infty$, i.e., $U_i = 0$ for $r_i = \infty$, gives

$$U_f = -GMm\left[\frac{1}{r_f}\right]$$

If we take $U_f = U$ for $r_f = r$, then

$$U(r) = -\frac{GMm}{r} \tag{1.89}$$

Special case: At the surface of Earth, $r = R$, therefore at the surface of earth Eq.1.89 becomes:

$$U(R) = -\frac{GMm}{R} \tag{1.90}$$

Thus, a choice of zero for $U_i = 0$ means that U approaches zero as r approaches infinite. At first this may seem like a strange choice, because for finite values of r all values of U are negative. This just means, however, that the potential energy is maximum when Earth and the particle are at infinite separation. Negative potential energy is nothing new. When we use the potential energy function $U = mgh$, where h is the height above some reference point on a tabletop, the potential energy is negative any time the particle of mass m is anywhere below the level of the tabletop. This reflects the fact that when the particle is below the level of the tabletop the potential energy is less than it is when the particle is at the level of the table-top.

☞ From Eq.1.88, we have
$$U_f - U_i = -GMm\left[\frac{1}{r_f} - \frac{1}{r_i}\right] = GMm\left[\frac{r_f - r_i}{r_i r_f}\right]$$
If a particle of mass m, is displaced from the surface of earth (i.e., $r_i = R$, the radius of earth) to height h from earth's surface (i.e., $r_f = R + h$), then change in gravitational potential energy
$$\Delta U = U_f - U_i$$
$$= -GMm\left[\frac{1}{R+h} - \frac{1}{R}\right] = GMm\left[\frac{h}{R(R+h)}\right]$$
$$= gR^2m\left[\frac{h}{R(R+h)}\right] \qquad \left[\because \quad g = \frac{GM}{R^2}\right]$$

$$\text{or} \qquad \Delta U = \frac{mgh}{1+\frac{h}{R}} \tag{1.91}$$

For $h << R$, $\qquad \Delta U \approx mgh$
Therefore, we can say that, mgh is the difference in potential energy (not the absolute potential energy), for $h << R$.

☞ Again from Eq.1.88, we have
$$U_f - U_i = -GMm\left[\frac{1}{r_f} - \frac{1}{r_i}\right] = GMm\left[\frac{r_f - r_i}{r_i r_f}\right]$$
If the body stays close to the earth, then in the denominator we may replace r_i and r_f by the earth's radius R, so
$$U_f - U_i = GMm\left[\frac{r_f - r_i}{R^2}\right] = \frac{GMm}{R^2}[r_f - r_i]$$
$$\because \qquad \frac{GM}{R^2} = g$$
$$\therefore \qquad U_f - U_i = mg[r_f - r_i]$$
If we replace the r's by y's, then, equation becomes-
$$U_f - U_i = mg[y_f - y_i]$$

If we assume, $y_f - y_i = y$, then above expression becomes:
$$U_f - U_i = mgy$$
We have already derived it in chapter "Work power and energy". So, we may consider this expression for gravitational potential energy to be a special case of the more general Eq.1.88

☞ From equation (1.89) it is clear that gravitational potential energy is always negative.

If we wanted, we could make $U = 0$ at the surface of the earth, where $r = R$ by simply adding the quantity GMm/R to Eq. (1.89). This would make U positive when $r > R$. We won't do this for two reasons: One, it would make the expression for U more complicated; and two, the added term would not affect the *difference* in potential energy between any two points, which is the only physically significant quantity.

Figure 1.87 shows a plot of $U(r)$ versus r for $r \geq R$. When the

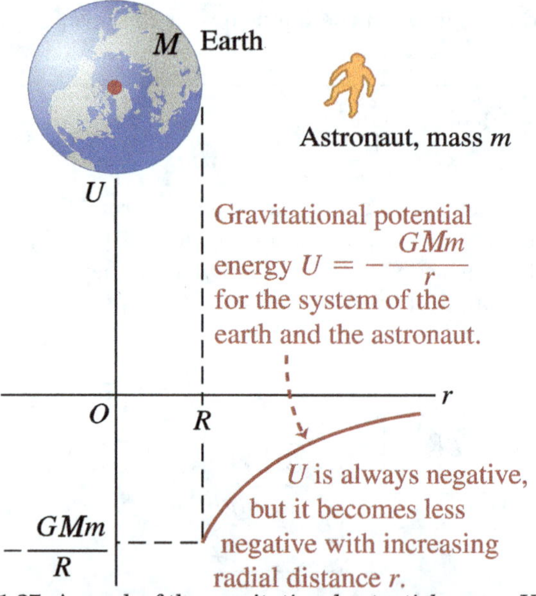

Figure 1.87: A graph of the gravitational potential energy U for the system of the earth (mass M and an astronaut (mass m) versus the astronaut's distance r from the centre of the earth.

particle or a spherically symmetric body moves away from the earth, r increases, the gravitational force does negative work, as a result of which U increases (i.e. becomes less negative). This is the reason why $U(r)$ vs r plot begins at the negative value $U = -GMm/R$ at Earth's surface and increases as r increases, approaching zero as r approaches infinity.

The slope of this curve at any general value of r (>= R), is given by
$$\frac{\partial U}{\partial r} = \frac{\partial}{\partial r}\left(-\frac{GMm}{r}\right) = \frac{GMm}{r^2} = mg'$$
It is equal to the magnitude of the gravitational force between Earth and the given particle.

For, $r = R$, the slope is $\frac{GMm}{R^2} = mg$. It is the magnitude of gravitational force of earth on the particle at the surface of the Earth.

When the body 'falls' toward earth, r decreases, the gravitational work is positive, and the potential energy decreases (i.e. becomes more negative).

EXAMPLE 45. If g is the acceleration due to gravity on the earth's surface, then find the gain in the potential energy of an object of mass m raised from the surface of the earth to a height equal to the radius R of the earth.

APPROACH Apply Eq.1.82 for $r = R$ and $r = 2R$ and then find the difference in gravitational potential energies. You can also directly apply Eq.1.88, with $r_i = R$ and $r_f = 2R$.

SOLUTION The acceleration due to gravity on the earth's surface is given by
$$g = GM/R^2,$$
where M is the mass of the earth and R is its radius. The potential energies of Earth-particle system for particle of mass 'm' on the earth's surface and at a height R above the earth's surface are
$$U_1 = -GMm/R$$
$$U_2 = -GMm/(R + R) = -GMm/(2R).$$
The increase in potential energy of mass m when it is raised from the earth's surface to a height R above the earth's surface is
$$\Delta U = U_2 - U_1 = -GMm/(2R) - (-GMm/R)$$
$$= GMm/(2R) = \left(GM/R^2\right)(mR/2)$$
$$= mgR/2$$
Note that the increase in the potential energy is not mgR, because g is not constant.

EXAMPLE 46. From the surface of earth, a particle of mass m is projected vertically upwards with a speed v. Find the maximum height, h, attained by the particle.

APPROACH Here, the force acting on the particle is gravity due to earth which is conservative, therefore, we can use the principle of conservation of mechanical energy, i.e., decrease in kinetic energy = increase in gravitational potential energy of Earth-particle system. Also note that gravitational acceleration changes with height from the earth's surface.

SOLUTION By conservation of energy,
$$\text{Decrease in KE} = \text{Increase in PE}.$$
$$\Rightarrow \quad \frac{1}{2}mv^2 = \Delta U$$
On substituting the value of ΔU obtained from Eq.1.91:
$\Delta U = \dfrac{mgh}{1 + \frac{h}{R}}$, in above relation, we get
$$\frac{1}{2}mv^2 = \frac{mgh}{1 + \frac{h}{R}}$$
On simplifying it for h, we get
$$h = \frac{v^2}{2g - \frac{v^2}{R}}$$
Note: In case of smaller values of v, $h \approx \frac{v^2}{2g}$.

EXAMPLE 47. Two particles of masses m_1 and m_2 are released from rest at a large separation distance. Find their speeds v_1 and v_2 when their separation distance is r. The initial separation distance is given as large, but large is a relative term. Relative to what distance is it large?

APPROACH The only force acting between the particles, is the gravitational mutual interaction which is conservative, so the principle of conservation of mechanical energy, is applicable. Now, if we consider both particles as a system then the external force on the system will be zero and hence the linear momentum of the system will also be conserved. Because the two-particle system has zero initial energy and zero initial linear momentum; we can use energy and momentum conservation to obtain simultaneous equations in the variables r, v_1 and v_2. We'll assume that initial separation distance of the particles and their final separation r is large compared to the size of the particles so that we can treat them as though they are point particles.

SOLUTION Use conservation of energy to relate the speeds of the particles when their separation distance is r :

$$E_i = E_f$$

or
$$0 = \frac{1}{2}m_1 v_1^2 + \frac{1}{2}m_2 v_2^2 - \frac{Gm_1 m_2}{r} \qquad (1)$$

Use conservation of linear momentum to obtain a second relationship between the speeds of the particles and their masses:

$$p_i = p_f$$

or
$$0 = m_1 v_1 + m_2 v_2 \qquad (2)$$

Solve equation (2) for v_1 and substitute in equation (1) to obtain:

$$v_2^2 \left(m_2 + \frac{m_2^2}{m_1} \right) = \frac{2Gm_1 m_2}{r} \qquad (3)$$

Solve equation (3) for v_2 :

$$v_2 = \sqrt{\frac{2Gm_1^2}{r(m_1 + m_2)}}$$

Solve equation (2) for v_1 and substitute for v_2 to obtain:

$$v_1 = \sqrt{\frac{2Gm_2^2}{r(m_1 + m_2)}}$$

EXAMPLE 48. Show that if a body be projected vertically upward from the surface of the earth so as to reach a height nR above the surface of earth

(i) the increase in its potential energy is $\dfrac{n}{n+1}mgR$,

(ii) the velocity with which it must be projected is $\sqrt{\dfrac{2ngR}{n+1}}$, where R is the radius of the earth and m the mass of body.

APPROACH In first part, apply Eq.1.91,

$$U = \left[\frac{mgh}{1 + \frac{h}{R}} \right] \qquad (1)$$

with $h = nR$ and in second part apply the principle of conservation of mechanical energy.

SOLUTION (i) From above Eq.(1), we have

$$\Delta U = \frac{mgh}{1 + \frac{h}{R}}$$

On substituting, $h = nR$, in Eq.(1), we get

$$\Delta U = \left(\frac{n}{n+1} \right) mgR$$

(ii) By conservation of mechanical energy, increase in potential energy = decrease in kinetic energy

$$\Rightarrow \quad \Delta U = \frac{1}{2}mv^2 - 0$$

or
$$\Delta U = \left(\frac{n}{n+1} \right) mgR = \frac{1}{2}mv^2$$

$$\therefore \qquad v = \sqrt{\frac{2ngR}{n+1}}$$

1.11.3 Gravitational Potential Energy and Binding Energy of a System of Particles

We now consider another interpretation for $U(r)$. Consider two objects, of masses m and M, separated by an infinitely large distance and at rest. We take one of the particles (m, for example) and move it slowly and at constant velocity toward the other, until the separation of the two particles is r. To move the particle at constant velocity, the net work done on the particle must be zero:

$$W_{\text{net}} = W_{\text{ext}} + W_{\text{grav}} = 0$$

where W_{ext} is the work done by our hand (i.e., by an external agent) and W_{grav} is the work done by the gravitational force. From Eq. 1.78, the work done by the gravitational force as the object moves from infinite separation to a separation r is $W_{\text{grav}} = W_{\infty \to r} = GMm/r$. Thus the work done by our hand is $W_{\text{ext}} = -W_{\text{grav}} = -GMm/r$. Noting that this is equal to $U(r)$ as given in Eq. 1.79, we can give this alternative view of the potential energy:

The potential energy of a system of particles is equal to the work done by an external agent to assemble the system, starting from the standard reference configuration.

Here, "standard reference configuration" means that the particles start out at rest at infinite separation. As we have seen, we also specify that the final assembled system is at rest in the same reference frame in which the particles are at rest in their initial state.

These considerations also hold for systems that contain more

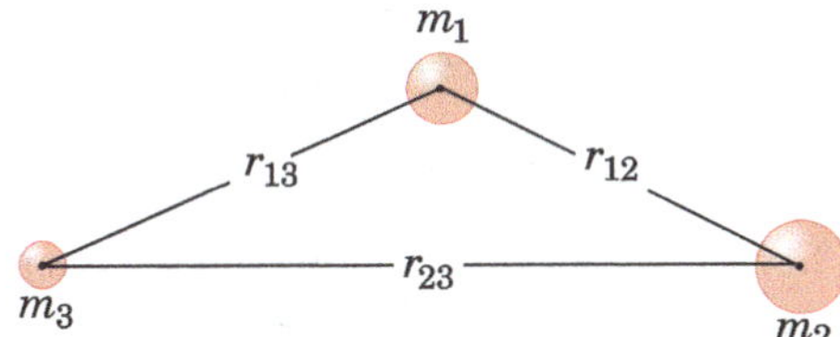

Figure 1.88: Three masses brought together from infinity and held in place by nongravitational forces.

than two particles. Consider three bodies of masses m_1, m_2, and m_3. Let them initially be at rest infinitely far from one another. The problem is to compute the work done by an external agent to bring them into the positions shown in Fig. 1.88. We first bring m_1 in from infinity to its final position and hold it in place. No work is done by gravity or the external agent because the separation between the three particles remains infinite. Let us then bring m_2 in toward m_1 from an infinite separation to the separation r_{12} and then hold it in place. The work done by the external agent in opposing the gravitational force exerted by m_1 on m_2 is $-Gm_1 m_2/r_{12}$. Now let us bring m_3 in from infinity to the separation r_{13} from m_1 and r_{23} from m_2. The work done by the external agent opposing the gravitational force exerted by m_1 on m_3

is $-Gm_1m_3/r_{13}$, and that opposing the gravitational force exerted by m_2 on m_3 is $-Gm_2m_3/r_{23}$. The total potential energy of this system is equal to the total work done by the external agent in assembling the system, or

$$U = -\left(\frac{Gm_1m_2}{r_{12}} + \frac{Gm_1m_3}{r_{13}} + \frac{Gm_2m_3}{r_{23}}\right) \qquad (1.92)$$

No matter how we assemble the system—that is, regardless of the order in which the particles are moved or the paths they take, we always find this same amount of work required to bring the bodies into the configuration of Fig. 1.88 from an initial infinite separation. The potential energy must therefore be associated with the system rather than with any one or two bodies. If we wanted to separate the system into three isolated masses once again, we would have to supply an amount of energy

$$E = +\left(\frac{Gm_1m_2}{r_{12}} + \frac{Gm_1m_3}{r_{13}} + \frac{Gm_2m_3}{r_{23}}\right) \qquad (1.93)$$

This energy is regarded as the **binding energy** holding the particles together in the configuration shown.

These concepts occur again in connection with forces of electric or magnetic origin, or, in fact, of nuclear origin. Their application is rather broad in physics. An advantage of the energy method over the dynamical method is that the energy method uses scalar quantities and scalar operations rather than vector quantities and vector operations. When the actual forces are not known, as is often the case in nuclear physics, the energy method is essential.

Generalization

Now, consider a system of N particles with masses m_1, m_2, ... m_n separated from each other by a distance r_{12}, r_{13} where r_{12} is the separation between m_1 and m_2 and so on. The total interaction energy of system is,

$$U = -\frac{1}{2}\sum_{i=1}^{N}\sum_{j=1}^{N}\frac{Gm_im_j}{r_{ij}} \qquad (1.94)$$

The factor 1/2 is taken because the interaction energy for each possible pair of system is taken twice during summation as for masses m_1 and m_3

$$U = -\frac{Gm_1m_3}{r_{13}} = -\frac{Gm_3m_1}{r_{31}} \quad (i \neq j)$$

EXAMPLE 49.   Three masses of 1 kg, 2 kg and 3 kg are placed at the vertices of an equilateral triangle of side 1 m [Fig.1.89]. How many interaction pairs are there? Find the gravitational potential energy of this system. Take $G = 6.67 \times 10^{-11}$ N.m^2/kg^2

APPROACH Since, total number of particles, $n = 3$, therefore, total number of interaction pairs are $^nC_2 = \frac{n(n-1)}{2} = \frac{3(3-1)}{2} = 3$.
Now, for three particles system, the gravitational potential energy is given by Eq.1.92:

$$U = -G\left(\frac{m_1m_2}{r_{12}} + \frac{m_1m_3}{r_{13}} + \frac{m_2m_3}{r_{23}}\right) \qquad \ldots (1)$$

Substitute the values of G, m_1, m_2, m_3, r_{12}, r_{13} and r_{23} in

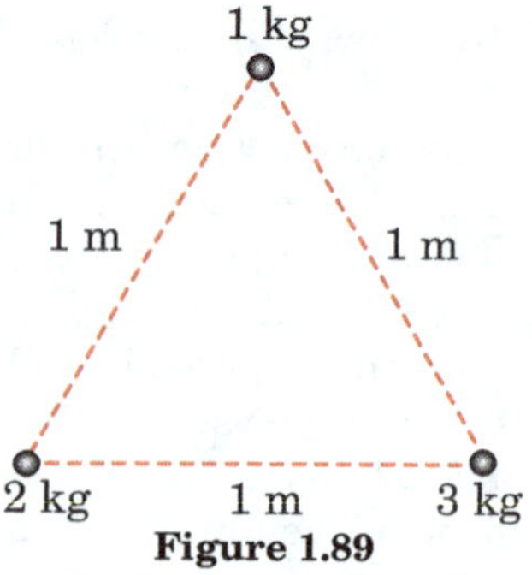

Figure 1.89

Eq.(1) and simplify for U.

SOLUTION Given that, $G = 6.67 \times 10^{-11}$ N.m^2/kg^2, $r_{12} = r_{13} = r_{23} = 1$ m, $m_1 = 1$ kg, $m_2 = 2$ kg, $m_3 = 3$ kg
On substituting these values in Eq.(1), we get

$$U = -\left(6.67 \times 10^{-11}\right)\left(\frac{1 \times 2}{1} + \frac{1 \times 3}{1} + \frac{2 \times 3}{1}\right) = -7.337 \times 10^{-10} \text{ J}$$

EXAMPLE 50. Eight particles of mass 'm' each are placed at the vertices of a cube of side 'a' [Fig.1.90]. Find gravitational potential energy of this system.

APPROACH Eight particle system with the distances

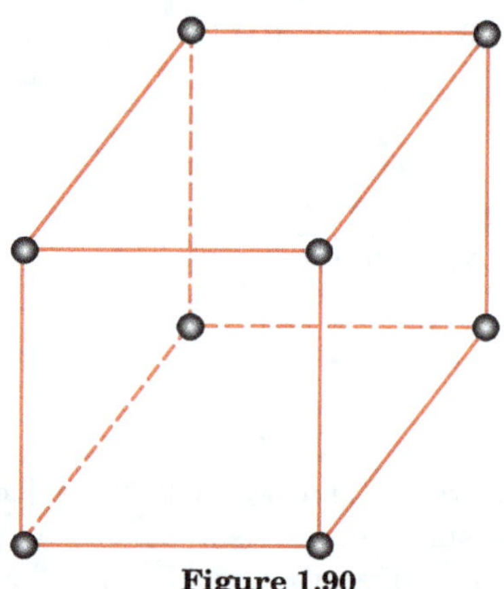
Figure 1.90

between particles is shown in Fig.1.91. Since, total number of particles, $n = 8$, therefore, total number of interaction pairs are $^nC_2$, i.e., $\frac{n(n-1)}{2} = \frac{8(8-1)}{2} = 28$.
Now, for interaction pair system, the gravitational potential energy is given by Eq.1.94. Find distances between interaction pairs and simplify for U.

SOLUTION All interaction pairs and the distances between them are given as:
(1) In 12 pairs (AB, BC, CD, DA, AH, BE, CF, DG, GH, HE, EF, FG), the distance between the particles is 'a'.
(2) In 12 pairs (AC, BD, CG, DF, FH, EG, BH, AE, DH, AG, CE, BF), the distance between particles, is $a\sqrt{2}$.
(3) in 4 pairs (AF, CH, BG, DE), the distance between particles, is $a\sqrt{3}$.

Therefore, the net gravitational potential energy of the system,

$$U = 12\left(-\frac{Gmm}{a}\right) + 12\left(-\frac{Gmm}{a\sqrt{2}}\right) + 4\left(-\frac{Gmm}{a\sqrt{3}}\right)$$

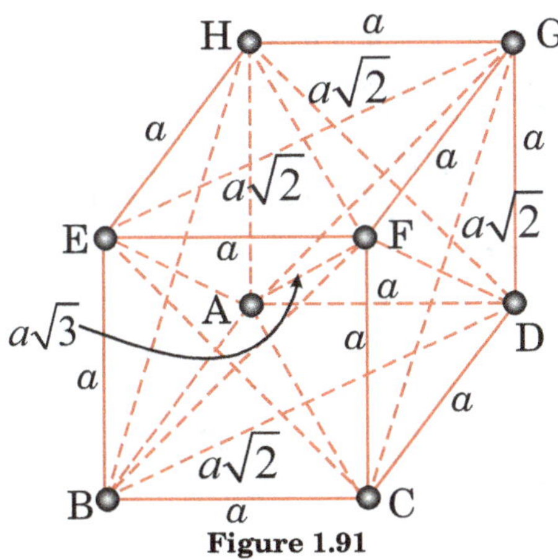

Figure 1.91

$$= -\frac{Gm^2}{a}\left[12 + 6\sqrt{2} + \frac{4}{\sqrt{3}}\right]$$

EXAMPLE 51. The gravitational potential energy between two point masses, m_1 and m_2 separated by distance r, is given by

$$U = -\frac{Km_1m_2}{r^n}$$

here, K is a positive constant. Find an expression for the time period of a satellite in an orbit of radius r around a planet of mass M.

APPROACH For the motion of a satellite (mass $= m$) in a circular orbit around a planet (mass $= M$), the required centripetal force is provided by gravitational force between them. The gravitational force and gravitational potential energies are related as-

$$F = -\frac{dU}{dr} \qquad \text{... (1)}$$

According to given gravitational potential energy formula, the gravitational potential energy between the planet and satellite system can be written as

$$U = -\frac{KMm}{r^n} \qquad \text{... (2)}$$

If ω is the angular speed of satellite about the planet, then, required centripetal force is given by

$$F_c = mr\omega^2 = mr\left(\frac{2\pi}{T}\right)^2 \qquad [\text{since, } \omega = 2\pi/T] \text{ ... (3)}$$

To find the gravitational force F, use Eq.(2) in Eq.(3). It gives-

$$F = -\frac{dU}{dr} = \frac{nKMm}{r^{n+1}}$$

Now this gravitational force will provide the required centripetal force for circular motion of satellite around the planet, i.e.,

$$F_c = F$$

or $$mr\left(\frac{2\pi}{T}\right)^2 = \frac{nKMm}{r^{n+1}}$$

or $$r\left(\frac{2\pi}{T}\right)^2 = \frac{nKM}{r^{n+1}} \qquad (4)$$

Now, simplify Eq.(4) for time period T.

SOLUTION From Eq.(4), we have

$$T^2 = \frac{4\pi^2}{nKM}r^{n+2}$$

or $$T = \frac{2\pi}{\sqrt{nKM}}r^{(n+2)/2}$$

This is the required expression of time period of the satellite From above it is clear that the time period of the satellite

$$T \propto r^{\frac{n}{2}+1}$$

For, $n = 1$, $T \propto r^{3/2}$, which is Kepler's third law.

1.12 Gravitational Potential

If we replace the object mass 'm' by unit mass in the definition of gravitational energy, then definition extended to gravitational potential at the position of unit mass. It is denoted by 'V'.

Definition 1. *"The gravitational potential at a point is equal to "negative" of work done by the gravitational force as a particle of unity mass is brought from infinity to it's position in the gravitational field.*

Definition 2. *The gravitational potential at a point is equal to the work done by the external force in bringing a particle of unit mass from infinity to it's position in gravitational field.*

Definition 3. *Gravitational potential at a given point is equal to the change in potential energy per unit mass, as the mass is brought from the infinity (reference point) to the given point.*

If we bring a mass m from infinity to the given point where the gravitational potential is required, and the change in potential energy is U, then gravitational potential at the given point will be

$$V = \frac{U}{m}$$

Gravitational potential is a point function. It's value depends only on source mass and the distance of the point under consideration.

If M is the source point mass producing gravitational potential V at distance r from it, then-

$$U = -G\frac{Mm}{r}, \text{ therefore,}$$

$$\boxed{V = \frac{U}{m} = -\frac{GM}{r}} \qquad (1.95)$$

Dimensional Formula of gravitational potential, is

$$[V] = \frac{[U]}{[m]} = \frac{[ML^2T^2]}{[M]} = [M^0L^2T^{-2}].$$

1.12.1 Gravitational Potential at Any Point Due to a Uniform Solid Sphere

Let us consider a solid sphere of mass M and radius R. Suppose, the density of the sphere is uniform and it is ρ, then

$$\rho = \frac{\text{mass of sphere}}{\text{volume}} = \frac{M}{\frac{4}{3}\pi R^3}$$

Suppose P, is any general point, at a distance r from the centre of the sphere, where gravitational potential due to this solid sphere, is to be determined.

Now, three cases are possible:

1. P is an external point ($r > R$)
2. P is on the surface of the sphere ($r = R$)
3. P is an internal point ($r < R$)

Case I: Potential at an external point $r > R$

In Fig. 1.92, P is an external point at distance $OP = r$. O is the centre of the sphere. Let us draw two spheres of radii x and $x + dx$ concentric with the given sphere. These two spheres enclose a thin spherical shell of volume $4\pi x^2 dx$
The mass of the shell is, $dm = \rho 4\pi x^2 dx$

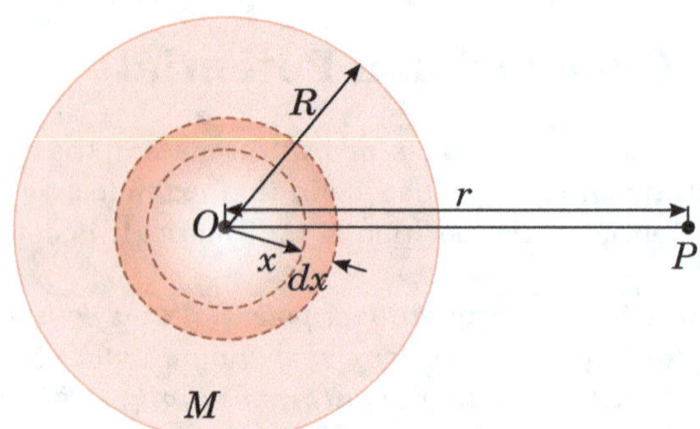

Figure 1.92

The gravitational potential due to this shell at the point P is
$$dV = -\frac{G\,dm}{r}$$
or $$dV = -\frac{G\,dm}{r} = -\frac{G\,\rho 4\pi x^2 dx}{r}$$
Thus, the potential due to the whole sphere is
$$V = -\frac{G4\pi\rho}{r}\int_0^R x^2 dx = -\frac{G4\pi\rho}{r}\left[\frac{R^3}{3}\right]$$
or $$V = -\frac{G\left[\frac{4}{3}\pi R^3 \rho\right]}{r} = -\frac{GM}{r}$$

$$\boxed{V = -\frac{GM}{r}}\qquad (1.96)$$

From Eq.1.96, it is clear that *the gravitational potential due to a uniform sphere at an external point is same as that due to a single particle of equal mass placed at its centre.*

Case II: Potential at the Surface of Sphere ($r = R$)
In this case, the approach will be same as above. Put $r = R$ in Eq. 1.96, you get-

$$\boxed{V = -\frac{GM}{R}}\qquad (1.97)$$

Case III: Potential at an internal point
Let P be an internal point to a sphere as shown in Fig. 1.93. Let us divide the sphere in two parts by imagining a concentric spherical surface passing through P. The net potential at P will be equal to sum of gravitational potentials produced at P due to inner and outer parts.
The inner part has a mass
$$M' = \frac{4}{3}\pi r^3 \rho$$
The point P is outer to this part, therefore, gravitational potential at P due to this part-
$$V_1 = -\frac{G\,M'}{r} \text{ for } x < r$$
Substituting the value of M', gives
$$V_1 = -\frac{G}{r}\left(\frac{4}{3}\pi r^3 \rho\right) = -\frac{4}{3}\pi Gr^2\rho \qquad (1.98)$$

Calculation of gravitational potential at P due to the outer part:

To get the gravitational potential at P due to the outer part of the sphere, we divide this part in concentric shells. The mass of the shell between radii x and $x + dx$ is $dm = \rho(4\pi x^2 dx)$. The gravitational potential at P due to this spherical shell of thickness dx and radius x, mass dm is given by
$$dV = -\frac{G\,dm}{x} \text{ for } x > r.\ \text{Here, } dm = \rho\left(4\pi x^2 dx\right)$$

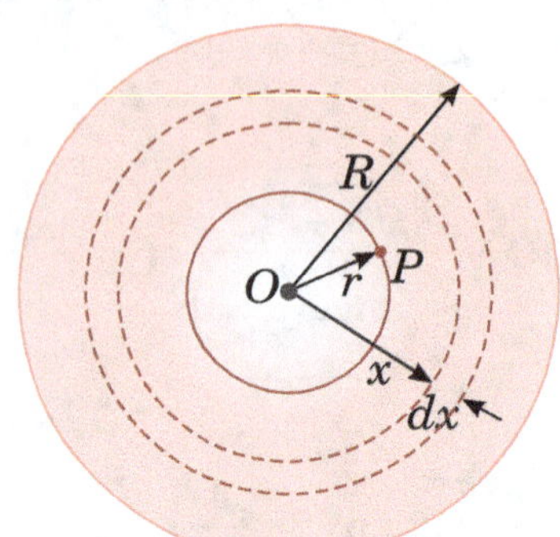

Figure 1.93

The potential at P due to this shell is,
$$dV = -\frac{Gdm}{x} = -\frac{G\rho\left(4\pi x^2 dx\right)}{x} = -4\pi G\rho x\,dx$$
So, the potential due to the outer part is
$$V_2 = \int_r^R (-4\pi G\rho x)\,dx = -4\pi G\rho\left[\frac{x^2}{2}\right]_r^R$$
or $$V_2 = -4\pi G\rho\left[\frac{R^2}{2} - \frac{r^2}{2}\right] \qquad (1.99)$$

From Eq.1.98 and Eq.1.99, the net gravitational potential at P
$$V = V_1 + V_2$$
$$V = -\frac{4}{3}\pi Gr^2\rho - 4\pi G\rho\left[\frac{R^2}{2} - \frac{r^2}{2}\right]$$
$$= -4\pi G\rho\left[\frac{R^2}{2} - \frac{r^2}{6}\right] = -2\pi G\rho\left[R^2 - \frac{r^2}{3}\right]$$

$$\Rightarrow \boxed{V = -\frac{2}{3}\pi G\rho\left[3R^2 - r^2\right]}\qquad (1.100)$$

In terms of mass of the sphere, the potential is given by
$$V = -\frac{2}{3}\pi G\left(\frac{M}{\frac{4}{3}\pi R^3}\right)\left[3R^2 - r^2\right]$$
$$\Rightarrow \boxed{V = -\frac{GM}{2R^3}\left[3R^2 - r^2\right]}\qquad (1.101)$$

At the centre of the sphere ($r = 0$) the potential is
$$V = -\frac{GM}{2R^3}\left[3R^2 - 0^2\right] = -\frac{3GM}{2R}$$
and at the surface of sphere $r = R$, the potential

$$\boxed{V = -\frac{GM}{R}}\qquad (1.102)$$

Thus, at the centre of the sphere the potential is (3/2) times the potential at the surface.
Figure 1.94 shows variation of potential with distance.
Note: All above results also hold for any planet/star having uniform mass density.

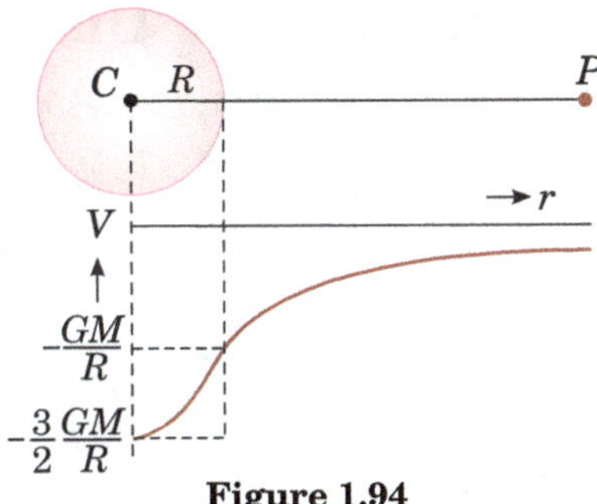

Figure 1.94

EXAMPLE 52. At a point above the surface of earth, the gravitational potential is -5.12×10^7 J/kg and the acceleration due to gravity is 6.4 m/s^2. Assuming the mean radius of the earth to be 6400 km, calculate the height of this point above the earth's surface.

APPROACH Write expressions for gravitational potential and field and then simplify for the position, r, of the given point.

SOLUTION If r be the distance of the given point from the centre of the earth. Then, gravitational potential

$$V = -\frac{GM}{r} = -5.12 \times 10^7 \qquad \dots (1)$$

and acceleration due to gravity,

$$g = \frac{GM}{r^2} = 6.4 \qquad \dots (2)$$

Dividing Eq.(1) by (2), we get

$$r = \frac{5.12 \times 10^7}{6.4} = 8 \times 10^6 m = 8000 \text{ km}$$

Therefore, the height of the point from earth's surface

$$h = r - R = 8000 \text{ km} - 6400 \text{ km} = 1600 \text{ km}$$

EXAMPLE 53. A solid sphere of uniform density and radius 4 units is located with its centre at the origin O of coordinates. Two spheres of equal radii 1 unit, with their centres at $A(-2,0,0)$ and $B(2,0,0)$ respectively, are taken out of the solid sphere leaving behind spherical cavities as shown in the figure. Then, which of the following options is/are correct?

(A) the gravitational field due to this object at the origin is

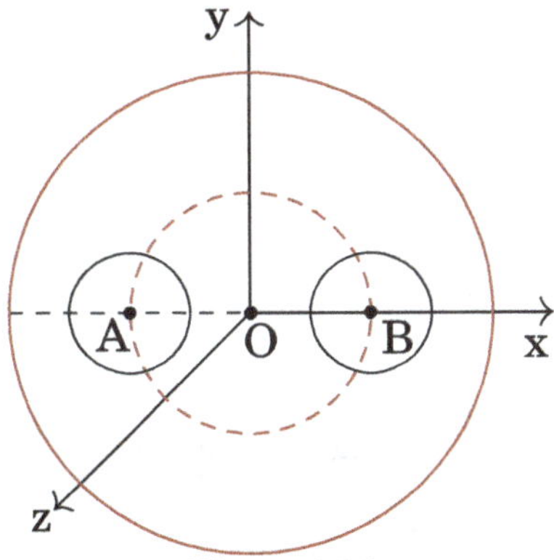

Figure 1.95

zero.

(B) the gravitational field at the point $B(2,0,0)$ is zero.

(C) the gravitational potential is the same at all points on cir-

cle $y^2 + z^2 = 36$.

(D) the gravitational potential is the same at all points on circle $y^2 + z^2 = 4$.

SOLUTION The mass of a solid sphere of radius $R = 4$ units is $M = \frac{4}{3}\pi(4)^3\rho = 64m_0$, where $m_0 = 4\pi\rho/3$. A and B are two identical spheres taken out from the larger solid sphere of radius 4 unit. Since, the density is uniform throughout the larger sphere and both spheres A and B have equal radii of 1 unit, therefore, their masses will also be equal. If m is the mass of each sphere A and B, then, $m = \frac{4}{3}\pi(1)^3\rho = m_0$. The solid sphere is a combination of three bodies, sphere A, sphere B, and remaining part C. The gravitational field due the solid sphere at its centre O, before A and B being taken out, is zero i.e.,

$$\vec{E}(O) = \vec{E}_A(O) + \vec{E}_B(O) + \vec{E}_C(O) = \vec{0}$$

Thus, gravitational field at O by the remaining part C is

$$\vec{E}_C(O) = \vec{E}(O) - \vec{E}_A(O) - \vec{E}_B(O)$$

$$= \vec{0} - \frac{Gm_0}{2^2}(-\hat{\imath}) - \frac{Gm_0}{2^2}\hat{\imath} = \vec{0}$$

Note the direction of gravitational field. The gravitational field by the solid sphere at $(2,0,0)$ is $\vec{E}(2,0,0) =$

$$-\frac{GMx}{R^3}\hat{\imath} = -\frac{G(64m_0)(2)}{4^3}\hat{\imath} = -2Gm_0\hat{\imath}$$

Thus, the gravitational field at $(2,0,0)$ by the remaining part C is

$$\vec{E}_C(2,0,0) = \vec{E}(2,0,0) - \vec{E}_A(2,0,0) - \vec{E}_B(2,0,0)$$

$$= -2Gm_0\hat{\imath} - \frac{Gm_0}{4^2}(-\hat{\imath}) - \vec{0}$$

$$= -\frac{31}{16}Gm_0\hat{\imath}$$

($\because$ Point $(2,0,0)$ is the centre for sphere at B, so $\vec{E}_B(2,2,0) = 0$)

The gravitational potential by the object C at the point

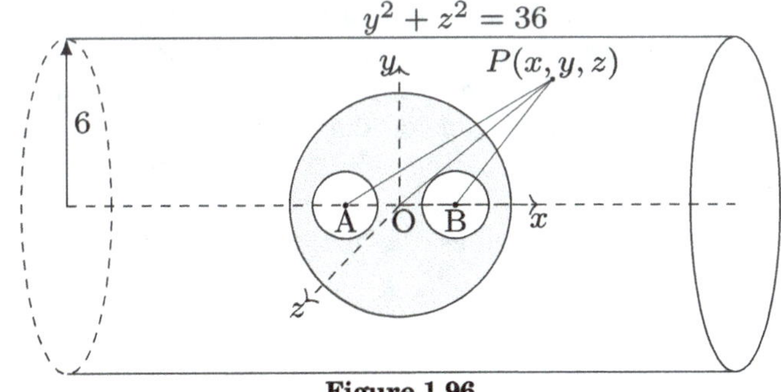

$P(x,y,z)$ is given by

$$V_C = V - V_A - V_B$$

$$= \frac{G(64m_0)}{\sqrt{x^2 + y^2 + z^2}} - \frac{Gm_0}{\sqrt{(x+2)^2 + y^2 + z^2}}$$

$$- \frac{Gm_0}{\sqrt{(x-2)^2 + y^2 + z^2}}.$$

The curve $y^2 + z^2 = 36$ is a cylinder of radius 6 units with its axis along x. The potential is not constant on this cylinder as x varies along its length although $y^2 + z^2$ is constant. However, potential is a constant on any circle on this cylinder (each circle is cross-section of a cylinder at fixed x) so any circle on this cylinder is an equipotential surface. Same argument is valid for $y^2 + z^2 = 4$

So, options, A, C, D are correct.

1.12.2 Potential due to a Uniform Thin Spherical Shell

Suppose, we have a spherical thin shell of mass M and radius R. We want to determine the gravitational potential at any point due to this shell. Now, three cases are possible:
1. Point is an external to the shell.
2. Point is an internal to the shell.
3. Point is at the surface of the shell.

1.12.2.1 Gravitational Potential at an External Point and on the Surface of the shell

For an external point, a uniform spherical shell can be considered like a point mass of same magnitude at its center. Thus, potential at a distance r is given by

$$V(r) = \begin{cases} -\dfrac{GM}{r}, & \text{if } r > R \\ -\dfrac{GM}{R}, & \text{if } r = R \end{cases} \qquad (1.103)$$

1.12.2.2 Gravitational Potential at an Internal Point

Since, inside the shell, gravitational field is zero, therefore by definition, there is no extra work required in bringing a test mass from surface of the shell to any point inside the shell.
So, the gravitational potential due to a uniform spherical shell is constant at any point inside the spherical shell and it is equal to the potential at the surface of the shell which is: $-\dfrac{GM}{R}$.
The, $V - r$ graph for a spherical shell is shown in Fig.1.97.

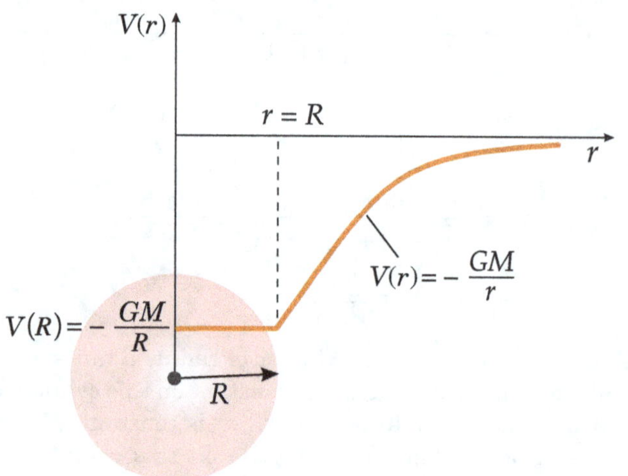

Figure 1.97

1.13 Potential due to a Uniform Ring at some Point on its Axis

The gravitational potential at a distance r from the centre on the axis of a ring of mass M and radius R is given by,

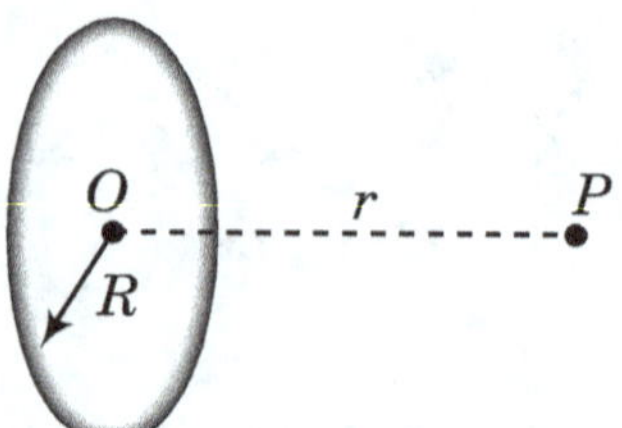

Figure 1.98

$$V = -\int_{\infty}^{r} g(r)\,dr$$

On replacing x by r in Eq.1.157, we get

$$g(r) = \frac{GM}{(R^2 + r^2)^{3/2}} r$$

Therefore, $\quad V = -\displaystyle\int_{\infty}^{r} \frac{GM}{(R^2 + r^2)^{3/2}} r\, dr = -\frac{GM}{\sqrt{R^2 + r^2}}$

$$\boxed{V(r) = -\frac{GM}{\sqrt{R^2 + r^2}} \qquad 0 \le r \le \infty} \qquad (1.104)$$

At $r = 0$, $V = -\frac{GM}{R}$, i.e. at the centre of the ring gravitational potential is $V = -\frac{GM}{R}$.
The $V - r$ graph is shown in Fig.1.99.

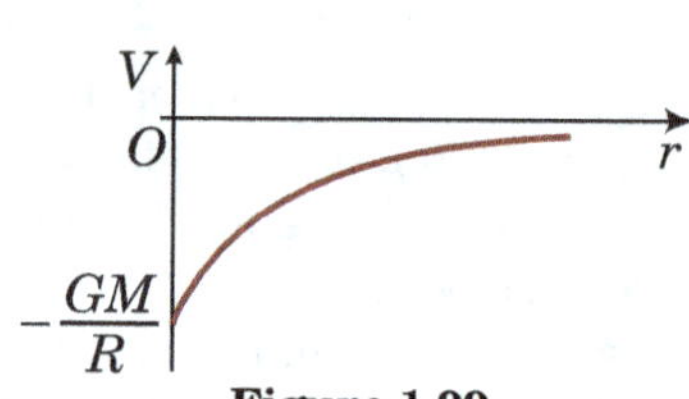

Figure 1.99

EXAMPLE 54. Three point masses 'm' each are kept at three vertices of a square of side 'a' as shown in Fig.1.100. Find gravitational potential and field strength at point O.

APPROACH We know that gravitational potential is a scalar

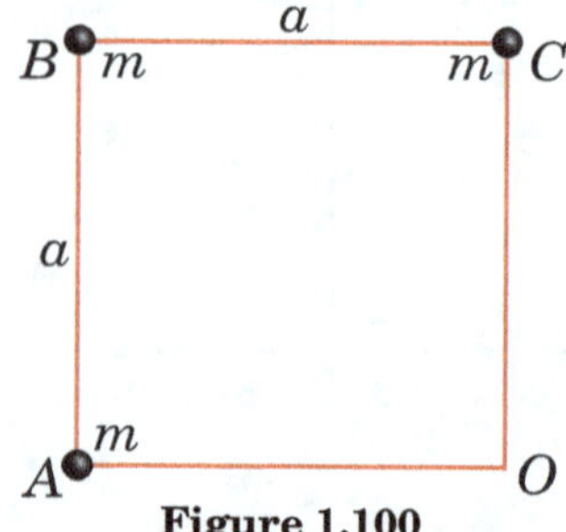

Figure 1.100

quantity whereas gravitational field is a vector. Therefore, we will be using scalar and vector sums for net gravitational potential and gravitational field respectively at vertex O.
SOLUTION Calculation of Gravitational Potential at Vertex O [Fig.1.101]
From Eq.1.95, the gravitational potential at distance r from a

point mass M, is given by

$$V = -\frac{GM}{r} \qquad (1)$$

By principle of super position of scalar field (like potential, temperature etc.), net potential at vertex O is

V_O = scalar sum of potentials due to point masses at A, B and C.

$$= -\frac{Gm}{a} - \frac{Gm}{\sqrt{2}a} - \frac{Gm}{a}$$

$$= -\frac{Gm}{a}\left[2 + \frac{1}{\sqrt{2}}\right]$$

Gravitational field strength is a vector quantity. By superposition principle of gravitational field, the net gravitational field strength at O will be equal to the vector sum of three field strengths produced by three point masses placed at A, B and C.

From Eq.1.25, the gravitational field strength at any position

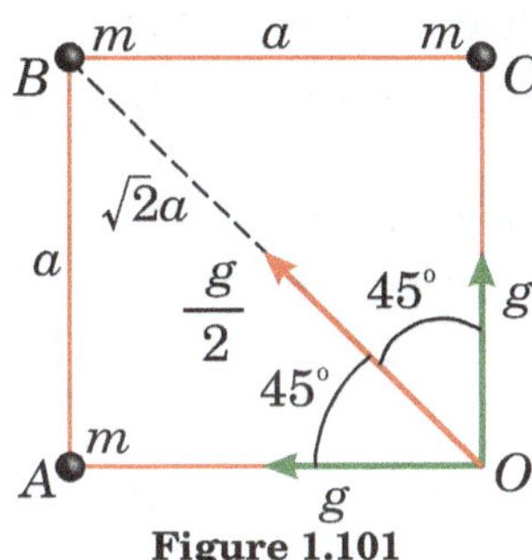
Figure 1.101

due to a point mass M, is given by

$$\vec{g} = -\frac{GM}{r^2}\hat{r} \qquad (2)$$

Since, points A and C are equal distant from O, therefore, at O, the magnitude of field strength produced by point masses of A and C will be equal, i.e.,

$$g_A = g_C = \frac{Gm}{a^2} = g \text{ (say)} \qquad (3)$$

here, g_A is the gravitational field strength at O due to point mass m at A and g_C is the the gravitational field strength at O due to point mass m at C.

Vector $\vec{g}_A$ is directed towards point A whereas, the vector $\vec{g}_C$ is directed towards the point C.

The resultant of theses two fields is

$$g_1 = \sqrt{g_A{}^2 + g_B{}^2} = \sqrt{g^2 + g^2} = g\sqrt{2} \qquad (4)$$

It is directed towards point B as shown in Fig.1.101

Gravitational field strength produced at point C due to point mass m placed at B

$$g_B = \frac{Gm}{(\sqrt{2}a)^2} = \frac{Gm}{2a^2} = \frac{g}{2} \text{ It is also directed towards}$$

point B.

The resultant of g_B and g_1 is given by:

$$g_{\text{net}} = g_1 + g_B = g\sqrt{2} + \frac{g}{2} = \left(\sqrt{2} + \frac{1}{2}\right)g$$

or $\qquad g_{\text{net}} = \left(\sqrt{2} + \frac{1}{2}\right)\frac{Gm}{a^2}$

This net gravitational field is also directed towards point B.

EXAMPLE 55. Four point masses each of mass 'm' are placed at four vertices A, B, C and D of a regular hexagon of side 'a' as shown in Fig.1.102. Find gravitational potential

and field strength at the centre O of the hexagon.

APPROACH Follow the same approach as given in last

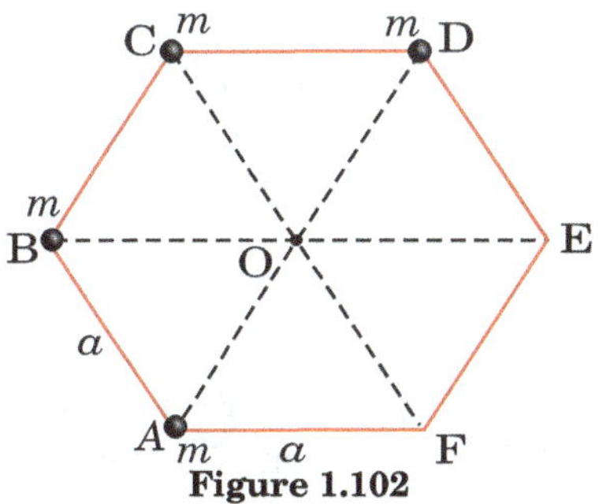
Figure 1.102

Example.

SOLUTION Gravitational potential is a scalar quantity. Therefore, net gravitational potential at O, V_O = scalar sum of gravitational potentials produced by four point masses at A, B, C and D.

$$= -\frac{Gm}{a} - \frac{Gm}{a} - \frac{Gm}{a} - \frac{Gm}{a}$$

$$= -\frac{4Gm}{a}$$

Gravitational field strength is a vector quantity. Therefore, the net gravitational field at O will be the vector sum of gravitational fields produced by point masses placed at vertices A, B, C and D respectively. Since, all these masses are at equal distance from O, therefore, their field magnitudes will be equal and it is given by:

$$g = \frac{Gm}{a^2} = g_A = g_B = g_C = g_D$$

g_A and g_D are oppositely directed, therefore they get can-

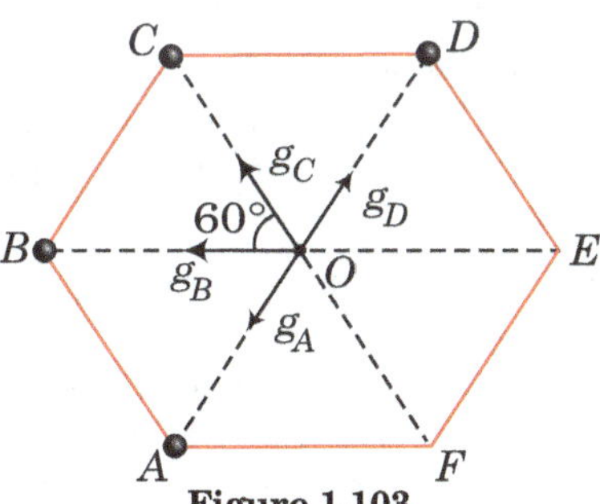
Figure 1.103

celled with each other. So, net field strength is a vector sum of g_B and g_C at angle 60°.

$$\therefore \qquad g_{\text{net}} = \sqrt{g^2 + g^2 + 2(g)(g)\cos 60°} = \sqrt{3}g$$

$$= \frac{\sqrt{3}Gm}{a^2}$$

If θ is the angle between g_{net} and g_B, then[5],

$$\tan\theta = \frac{g_C \sin 60°}{g_B + g_C \cos 60°} = \frac{g\sqrt{3}/2}{g + (g/2)}$$

---

[5]If $\vec{A}$ and $\vec{B}$ are two vectors acting at α with each other (Fig.1.104), then their resultant $\vec{R}$, is given by-

$$R = \sqrt{A^2 + B^2 + 2AB\cos\alpha}$$

The direction of resultant from A is given by

$$\tan\theta = \frac{B\sin\alpha}{A + B\cos\alpha} \quad \text{or} \quad \theta = \tan^{-1}\left(\frac{B\sin\alpha}{A + B\cos\alpha}\right)$$

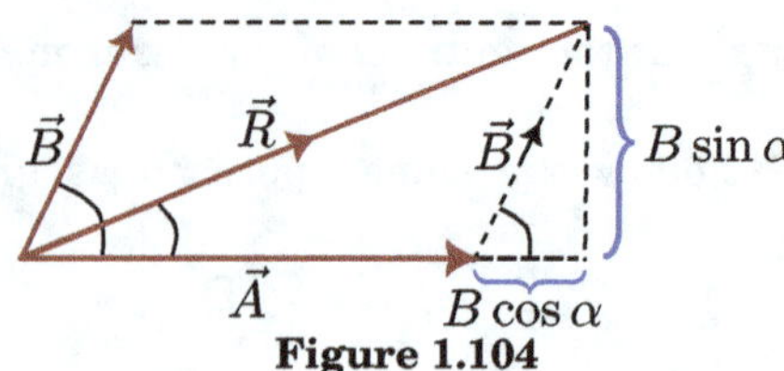

Figure 1.104

$$= \frac{1}{\sqrt{3}}$$

$$\Rightarrow \qquad \theta = 30°$$

It shows that, net gravitational field strength is along the bisector line of $\angle COB$, away from O, between g_C and g_B.

EXAMPLE 56. At what distance 'd' from the surface of a solid sphere of radius R,

(a) potential is same as at a distance $R/2$ from the centre?

(b) field strength is same as at a distance $R/4$ from centre.

Discussion: (a) From Eq.1.101, the value of gravitational potential at any point inside a solid sphere of mass M and radius R, is given by

$$V = -\frac{GM}{2R^3}\left[3R^2 - r^2\right] \qquad \text{... (1)}$$

At the center of the sphere, $r = 0$, therefore, above equation gives

$$V_{center} = -\frac{3GM}{2R}$$

At a point on the surface of sphere, $r = R$, therefore from Eq.(1), we get

$$V_{surface} = -\frac{GM}{R}$$

From Eq.1.96, the gravitational potential out side a sphere of mass M is

$$V = -\frac{GM}{r}$$

At infinity, $r \to \infty$, therefore $V \to 0$

So, it is clear that, from centre to surface potential varies from $-\frac{1.5GM}{R}$ to $\frac{-GM}{R}$ and then from surface to infinity potential varies from $\frac{-GM}{R}$ to zero.

Potential (V) vs (r) graph for a solid sphere is also shown in Fig.1.94.

Clearly, gravitational potential continuously increases from $-1.5GM/R$ to 0 as one moves from center of sphere to out side the sphere at infinity. So, there can not be any two points for which potential is same.

(b) From Eq.1.60, gravitational field at any point inside a solid sphere of radius R is given by:

$$\vec{g} = -\frac{GMr}{R^3}\hat{r}$$

Therefore, magnitude of gravitational field at distance $r(\leq R)$, inside the sphere of mass M and radius R is

$$\vec{g} = -\frac{GMr}{R^3}\hat{r} \qquad \text{... (1)}$$

Now, from Eq.1.55, the gravitational field outside a sphere of mass M and radius R, i.e., for $r > R$, is-

$$\vec{g} = -\frac{GM}{r^2}\hat{r}$$

It's magnitude

$$g = \frac{GM}{r^2} \qquad \text{... (2)}$$

SOLUTION (a) From above discussion, we can say that po-

tential inside a solid sphere can not be same as potential outside the solid sphere. So, such points will not exist.

(b) For, $g_{inside} = g_{outside}$, Eq. (1), and (2) give

$$\frac{GM}{R^3}(r_1) = \frac{GM}{r_2^2}$$

Here, r_1 and r_2 are the distances from centre. or

$$\frac{r_1}{R^3} = \frac{1}{r_2^2}$$

or

$$r_2 = \sqrt{\frac{R^3}{r_1}}$$

Corresponding to, $r_1 = \dfrac{R}{4}, r_2 = \sqrt{\dfrac{R^3}{R/4}} = 2R$

Therefore, distance of required point from the surface of the sphere,

$$= r_2 - R = 2R - R = R$$

1.14 Internal Potential Energy or Gravitational Self Energy

It is the energy possessed by a body due to the interaction forces between the constituent particles of the body. This can be defined as the work done in assembling all the particles of a body in a definite shape and size, i.e., it is the work done in creating a body.

In other words, the gravitational potential energy due to internal gravitational interactions of constituent particles of the body, is called the "gravitational self energy" or internal potential energy of the body.

1.14.1 Calculation of Gravitational Self energy of a Uniform Solid Sphere (or a Planet)

Suppose we have a sphere having uniform mass density ρ, total mass M and radius R.

The gravitational self energy of this sphere is the work done by external agent (against conservative gravitational interactions between the particles of the sphere) in creating this sphere in assembling all particles from infinite separation to form the sphere.

Suppose, at any instant, the radius of the sphere is r and it's

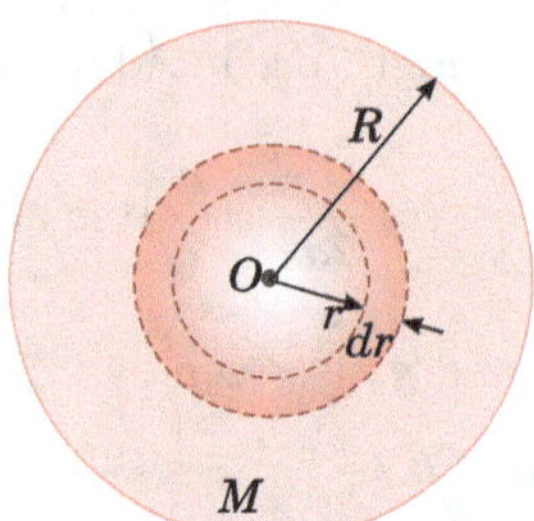

Figure 1.105

mass is m. The gravitational potential at each point on the surface of this sphere:

$$V = -G\frac{m}{r}$$

here, $m =$ volume $\times$ density $= \frac{4}{3}\pi r^3 \rho$ with $\rho = \frac{M}{\frac{4}{3}\pi R^3}$

Now, let us increase the particles over the surface of the sphere symmetrically such that its mass increases by dm and the radius of the sphere increases from r to $r + dr$. In doing so, the change in gravitational potential energy is given by-

$$dU_{\text{self}} = V.dm = -G\frac{m.dm}{r} \qquad (1.105)$$

From Fig.1.105, we have
$$dm = \left(4\pi r^2 dr\rho\right)$$
Substituting the values of m and dm in Eq.1.105, gives
$$dU_{\text{self}} = -G\frac{\left(\frac{4}{3}\pi r^3\rho\right)\left(4\pi^2 dr\rho\right)}{r}$$
$$= -\frac{1}{3}G\left(4\pi\rho\right)^2 r^4 dr$$
Therefore, the net self energy or internal potential energy of the sphere of radius R:

$$U_{\text{self}} = -\frac{1}{3}G\left(4\pi\rho\right)^2 \int_0^R r^4 dr = -\frac{1}{3}G\left(4\pi\rho\right)^2 \left[\frac{r^5}{5}\right]_0^R$$
$$= -\frac{3}{5}G\left(\frac{4\pi}{3}R^3\rho\right)^2 \frac{1}{R}$$

or $\qquad \boxed{U_{\text{self}} = -\frac{3}{5}\frac{GM^2}{R}} \qquad \left[\because \quad \rho = M/\frac{4}{3}\pi R^3\right] \qquad (1.106)$

Note that each body having continuous mass distribution possesses internal potential energy. Generally, when apply conservation of mechanical energy, we don't use this energy until the body changes it's shape or size. Because if body not change it's shape or size, it's internal potential energy remains unchanged.

1.14.2 Conversion of V Function into g Function

In chapter "Work, Energy and Power", we have seen that a gravitational force $\vec{F}_g$ on an object of test mass m in a gravitational field g, can be calculated from the gradient of potential energy as follows:

$$\vec{F}_g = -\vec{\nabla}U_g \qquad (1.107)$$

here, $\vec{\nabla}U$ is called the gradient of U and $\vec{\nabla}$ is defined as:
$$\vec{\nabla} = \frac{\partial}{\partial x}\hat{i} + \frac{\partial}{\partial y}\hat{j} + \frac{\partial}{\partial z}\hat{k}$$
Dividing both sides of Eq.1.107 by the mass (m) of object, yields:
$$\frac{\vec{F}_g}{m} = -\frac{\vec{\nabla}U_g}{m}$$
Since, $\vec{F}_g/m = \vec{g}$ (gravitational field at the position of the object of test mass m) and $\frac{U_g}{m} = V$, (Gravitational potential at the position of object of test mass m), therefore above equation can also be written as:
$$\vec{g} = -\vec{\nabla}V = -\text{gradient } V = -\text{ grad } V, \text{ i.e.,}$$

$$\vec{g} = -\vec{\nabla}V = -\left[\frac{\partial V}{\partial x}\hat{i} + \frac{\partial V}{\partial y}\hat{i} + \frac{\partial V}{\partial z}\hat{i}\right] \qquad (1.108)$$

$\Rightarrow \qquad \boxed{\vec{g} = g_x\hat{i} + g_y\hat{j} + g_z\hat{k}} \qquad (1.109)$

where, $g_x = -\frac{\partial V}{\partial x}$, $g_y = -\frac{\partial V}{\partial y}$, and $g_z = -\frac{\partial V}{\partial z}$.
Magnitude of the gravitational field,

$$g = \sqrt{(g_x)^2 + (g_y)^2 + (g_z)^2} = \sqrt{\left(-\frac{\partial V}{\partial x}\right)^2 + \left(-\frac{\partial V}{\partial y}\right)^2 + \left(-\frac{\partial V}{\partial z}\right)^2} \qquad (1.110)$$

For one dimensional case (say along x axis only), we can write:

$$\boxed{g_x = -\frac{dV}{dx}} \qquad (1.111)$$

Similar to Eq.1.111, if V is a function of distance from the source, i.e., r, then we have

$$\boxed{g = -\frac{dV}{dr}} \qquad (1.112)$$

1.14.3 Conversion of g Function into V Function

(i) When g is a function of only one variable, x (say):
From Eq.1.111, we have
$$g_x = -\frac{dV}{dx} \quad \text{or} \quad dV = -\vec{g}_x d\vec{x}$$
If the gravitational potential at position $x = x_i$ is V_i and at position $x = x_f$, it is V_f, then, integration of above equation with these limits, gives

$$\boxed{\int_{V_i}^{V_f} dV = -\int_{x_i}^{x_f} \vec{g}(x)\cdot d\vec{x}} \qquad (1.113a)$$

or $\qquad \boxed{V_f - V_i = -\int_{x_i}^{x_f} \vec{g}(x)\cdot d\vec{x}} \qquad (1.113b)$

Since, we have already considered zero gravitational potential at $x = \infty$, therefore if we take $x_i = \infty$ then $V_i = 0$ and let at $x_f = x$, $V_f = V$, then Eq.1.113b, can be written as
$$V - 0 = -\int_{\infty}^{x} \vec{g}(x)\cdot d\vec{x}$$

or $\qquad \boxed{V = -\int_{\infty}^{x} \vec{g}(x)\cdot d\vec{x}} \qquad (1.114)$

(ii) When g is a function of more than one variable, say, x, y and z : In this case, we replace, $\vec{x}$ with $\vec{r}$, in Eq.(1.113a), (1.113b) and (1.114); where $\vec{r} = x\hat{i} + y\hat{j} + z\hat{k}$. The results will be:

$$\boxed{\int_{V_i}^{V_f} dV = -\int_{r_i}^{r_f} \vec{g}\cdot d\vec{r}} \qquad (1.115a)$$

or $\qquad \boxed{V_f - V_i = -\int_{r_i}^{r_f} \vec{g}\cdot d\vec{r}} \qquad (1.115b)$

or $\qquad \boxed{V = -\int_{\infty}^{r} \vec{g}\cdot d\vec{r}} \qquad (1.116)$

with, $\vec{g} = g_x\hat{i} + g_y\hat{j} + g_z\hat{k}$ and $d\vec{r} = dx\hat{i} + dy\hat{j} + dz\hat{k}$
Note: Always use proper limits when calculate gravitational potential difference between two points by using gravitational

field function (see Eq.1.115). In case, you want to find potential at any point in gravitational field, set lower limit of position at infinity and at infinity, we consider gravitational potential value equal to zero. Now, assume that at required point, the gravitational potential is V, so put V for upper limit and simplify for V (see Eq.1.116)

EXAMPLE 57. Gravitational potential in x-y plane varies with x and y coordinates as $V = x^2 y + 2xy$. Find gravitational field strength g.

APPROACH Given that, $V = x^2 y + 2xy$, which is basically a function of two variables x and y, therefore, from Eq.1.108, the expression for gravitational field strength, will be:

$$\vec{g} = -\left[\frac{\partial V}{\partial x}\hat{i} + \frac{\partial V}{\partial y}\hat{j}\right] \qquad \ldots (1)$$

Now, calculate $\frac{\partial V}{\partial x}$ and $\frac{\partial V}{\partial y}$ and put these values in Eq.(1) and then simplify for g.

SOLUTION Since, $V = x^2 y + 2xy$, therefore
$$\frac{\partial V}{\partial x} = (2xy + 2y), \text{ and } \frac{\partial V}{\partial y} = x^2 + 2x$$
On substituting these values in Eq.(1), we get
$$\vec{g} = -\left[(2xy + 2y)\hat{i} + (x^2 + 2x)\hat{j}\right]$$
This is the required field strength at point (x, y) in the gravitational field.

EXAMPLE 58. If the gravitational potential at a distance 'r' from a point mass 'm' is $V = -\dfrac{Gm}{r}$, find gravitational field strength at that point.

APPROACH Given potential function is,
$$V = -\frac{Gm}{r} \qquad \ldots (1)$$
Since, V is a function of r, therefore to find g, we have to use Eq.1.112,
$$g = -\frac{dV}{dr} \qquad \ldots (2)$$
SOLUTION Substituting the value of V, from Eq.(1) in (2), we get
$$g = -\frac{d}{dr}\left(-\frac{Gm}{r}\right) = -\frac{Gm}{r^2}$$
Negative sign just implies that direction of $\vec{g}$ is towards the point mass.

Therefore, $\qquad |g| = \dfrac{Gm}{r^2}$

EXAMPLE 59. Gravitational potential varies along x-axis as shown in Fig.1.106.
(a) Plot E versus x graph corresponding to given V-x graph.
(b) A mass of 2 kg is kept at $x = 3$ m. Find gravitational force on it.

APPROACH Since, gravitational field is defined as the $-$ ve

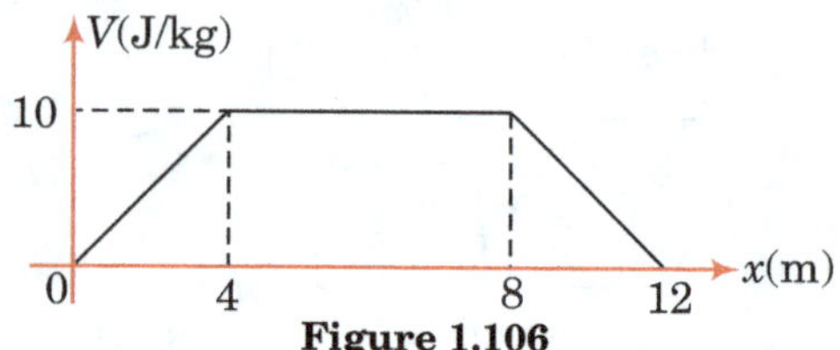

Figure 1.106

slope of V - x graph, therefore, find slopes of each segment of

the graph and trace it in g - x graph.

SOLUTION (a) $g = -\dfrac{dV}{dx} = -$ slope of V-x graph.
From $x = 0$ to $x = 4$ m
$$\text{slope} = +\frac{10}{4} = +2.5 \text{ N/kg}$$
$\therefore \qquad g = -2.5 \text{ N/kg}$
From $x = 4$ m to $x = 8$ m,
$$\text{slope} = 0$$
$\therefore \qquad g = 0$
From $x = 8$ m to $x = 12$ m
$$\text{slope} = -\frac{10}{4} = -2.5 \text{ N/kg}$$
$\therefore \qquad g = +2.5 \text{ N/kg}$
Therefore, g-x graph is as shown in Fig.1.107.
(b) At $x = 3$ m, $g = -2.5$ N/kg
$\therefore$ Gravitational force, $F = mg \quad \left(\text{as } g = \dfrac{F}{m}\right)$
$$= (2)(-2.5) = -5 \text{ N}$$
Here, negative sign implies that this force is acting towards

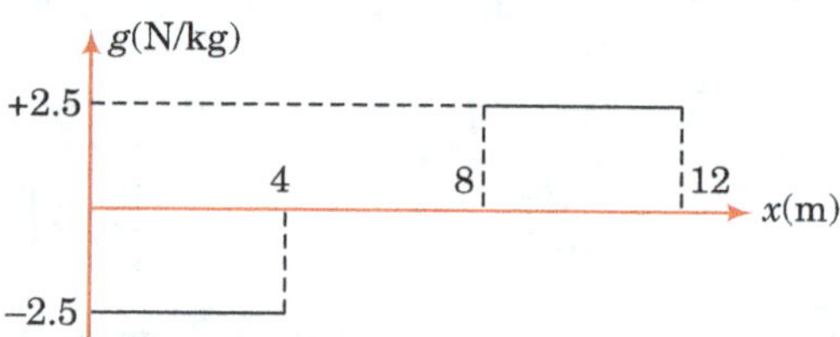

Figure 1.107: g - x graph.

negative x-direction.

EXAMPLE 60. Gravitational field in x-y plane is given as $\vec{g} = \left(2x\hat{i} + 3y^2\hat{j}\right)$ N / kg. Find difference in gravitation potential between two points A and B, where co-ordinates of A and B are (2 m, 4 m) and (6 m, 0).

APPROACH Here, we have to determine gravitational potential from gravitational field vector
$$\vec{g} = \left(2x\hat{i} + 3y^2\hat{j}\right) \text{ N/kg.} \qquad \ldots (1)$$
So, we have to apply Eq.1.115a:
$$\int_{V_i}^{V_f} dV = -\int_{r_i}^{r_f} \vec{g} \cdot d\vec{r} \qquad \ldots (2)$$
Since, in Eq. (1), g is a function of coordinates x and y only, therefore, we use, $d\vec{r} = dx\hat{i} + dy\hat{j}$

SOLUTION On substituting the values of $\vec{g}$ and $d\vec{r}$ in Eq. (2), we get
$$\int_{V_A}^{V_B} dV = -\int_{(2m,\,4m)}^{(6m,\,0)} \left(2x\hat{i} + 3y^2\hat{j}\right) \cdot (dx\hat{i} + dy\hat{j})$$
or
$$V_B - V_A = -2\int_{2m}^{6m} x\,dx - 3\int_{4m}^{0} y^2\,dy$$
or
$$= -\left[x^2\right]_{2m}^{6m} - \left[y^3\right]_{4m}^{0}$$
$$= -\left[(6)^2 - (2)^2\right] \text{ J/kg} - \left[0^3 - 4^3\right] \text{ J/kg}$$
$$= 32 \text{ J/kg}$$

EXAMPLE 61. The gravitational potential at a point $P(x, y, z)$, due to a mass distribution, is $V = 3x^2 y + y^3 z$. Find the gravitational field at this point.

APPROACH Given that, $V = 3x^2 y + y^3 z$, which is a function of three variables x, y and z; therefore, from Eq.1.108, the expression for gravitational field strength, will be:
$$\vec{g} = -\frac{\partial V}{\partial x}\hat{i} + \frac{\partial V}{\partial x}\hat{j} + \frac{\partial V}{\partial z}\hat{k} \qquad \ldots (1)$$
Now, substitute the values of $\frac{\partial V}{\partial x}$, $\frac{\partial V}{\partial y}$ and $\frac{\partial V}{\partial z}$ from given relation of potential and then substitute these values in Eq.(1)

and simplify for g.

SOLUTION Since, $V = 3x^2 y + y^3 z$, therefore
$$\frac{\partial V}{\partial x} = 6xy, \quad \frac{\partial V}{\partial y} = 3x^2 + 3y^2 z \text{ and } \frac{\partial V}{\partial z} = y^3$$
On substituting these values in Eq.(1), we get
$$\vec{g} = -\left[(6xy)\hat{i} + \left(3x^2 + 3y^2 z\right)\hat{j} + y^3 \hat{k}\right]$$
This is the required field strength at point (x, y, z) in the gravitational field.

EXAMPLE 62. Gravitational potential at $x = 2$ m is decreasing at a rate of 10 J/kg.m along the positive x-direction. It implies that the magnitude of gravitational field at $x = 2$ m is also 10 N/Kg. Is this statement true or false?

APPROACH The magnitude of gravitational field $\vec{g}$ and gravitational potential are related by Eq. 1.110:
$$g = \sqrt{\left(-\frac{\partial V}{\partial x}\right)^2 + \left(-\frac{\partial V}{\partial y}\right)^2 + \left(-\frac{\partial V}{\partial z}\right)^2} \quad \dots (1)$$
Since, each term within the square bracket is a complete square, so, no term can be negative.

SOLUTION Given that, $-\dfrac{\partial V}{\partial x} = 10$ J/kg.m, nothing is given about other terms. So, in Eq.(1), second and third terms within the square root are, either zero or positive.
Therefore, we can say that $g \geq 10$ N/kg.

EXAMPLE 63. The gravitational potential at point $P(x, y)$ is, $V = 20(x + y)$J/kg. Find the magnitude of the gravitational force on a particle of mass 0.5 kg placed at the origin.

APPROACH Given that, $V = 20(x + y)$ J/kg, which is a function of two variables x and y, therefore, from Eq.1.108, the expression for gravitational field strength, will be:
$$\vec{g} = -\left[\frac{\partial V}{\partial x}\hat{i} + \frac{\partial V}{\partial x}\hat{j}\right] \quad \dots (1)$$
Now, find the values of $\frac{\partial V}{\partial x}$ and $\frac{\partial V}{\partial x}$ from given potential function and then substitute these values in Eq.(1) and simplify for g.
The gravitational force on a particle of mass m, in the gravitational field $\vec{g}$, can be obtained by,
$$\vec{F} = m\vec{g} \quad \dots (2)$$
It's magnitude is given by
$$F = mg \quad \dots (3)$$

SOLUTION Since, $V = 20(x + y)$ J/kg, therefore
$$\frac{\partial V}{\partial x} = 20 \text{ J/kg.m, and } \frac{\partial V}{\partial y} = 20 \text{ J/kg.m}$$
On substituting these values in Eq.(1), we get
$$\vec{g} = -\left[20\hat{i} + 20\hat{j}\right] \text{ m/s}^2$$
This is the required field strength at point (x, y) in the gravitational field.
The mass of the particle placed in above gravitational field, $m = 0.5$ kg, therefore, from Eq.(2), the gravitational force on it is
$$\vec{F} = m\vec{g} = -0.5 \text{ kg}\left[20\hat{i} + 20\hat{j}\right] \text{ m/s}^2 = -\left[10\hat{i} + 10\hat{j}\right] \text{ N}$$
It's magnitude
$$F = \sqrt{\left[10^2 + 10^2\right]} \text{ N} = 10\sqrt{2} \text{ N}$$

1.15 Work Done in Displacing of a Body in Gravitational Field

Let us consider two points A and B in a gravitational field. Suppose, an external agent displaces a point mass m from point A to B along the path shown in Fig. 1.108. If the gravitational potential energies of mass m and source of gravitational field system for positions of test mass (m) at A and B are U_A and U_B respectively, then, the work done by external agent in displacing the test mass from point A to B against the conservative gravitational force, is
$$W_{\text{ext}} = U_B - U_A = (mV_{gB}) - (mV_{gA})$$

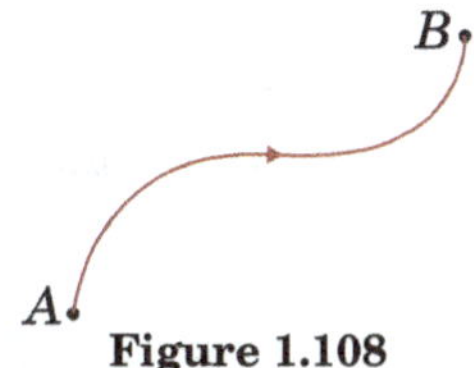

Figure 1.108

$$= m(V_{gB} - V_{gA})$$
here, V_{gA} and V_{gB} are gravitational potentials at points A and B respectively.
Therefore,
$$\boxed{W_{\text{ext}} = U_B - U_A = m(V_{gB} - V_{gA})} \quad (1.117)$$
Thus, Work done = mass of body × gravitational potential difference between the terminal points.

EXAMPLE 64. Find work done in shifting a body of mass m from a height h above the earth's surface to a height $2h$ above the earth's surface.

APPROACH Given Example is an application of Eq.1.117, so we write the gravitational potentials at heights $2h$ and h and then apply Eq.1.117:

SOLUTION From Eq.1.96, the gravitational potential at a height h and $2h$ above the earth surface is given as:
$$V_h = -\frac{GM_e}{R_e + h}, \quad V_{2h} = -\frac{GM_e}{R_e + 2h}$$
If a body of mass m is shifted from h to $2h$, then from Eq.1.117, the work done by external force against system gravitational force, is
$$W = m(V_{2h} - V_h) = m\left[\frac{GM_e}{R_e + h} - \frac{GM_e}{R_e + 2h}\right]$$

EXAMPLE 65. A body of mass m is placed on the surface of earth (take, mass of Earth $= M$, radius $= R$). Find work required to lift this body by a height (i) $h = R/1000$ (ii) $h = R$

APPROACH Given example is an application of Eq.1.117. If V_i and V_f are the gravitational potentials for initial and final positions of mass m, then from Eq.1.117, the work done by an external agent, in shifting the mass from initial position to final position against the system gravitational force is given by,
$$W = m(V_f - V_i) \quad \dots (1)$$

SOLUTION (i) Initially, the body was at the surface of earth, therefore, from Eq.1.97, gravitational potential for initial position,

$$V_i = -\frac{GM}{R}$$

Finally, the body is at height h, over the surface of earth, therefore, from Eq.1.96, the gravitational potential at height h, over the surface of earth,

$$V_f = -\frac{GM}{r} = -\frac{GM}{(R+h)} \qquad [\text{since, } r = R + h]$$

Since, $h = R/1000$, therefore

$$V_f = -\frac{1000GM}{(1001R)}$$

Therefore, from Eq.(1), we can write

$$W = m\left[-\frac{1000GM}{(1001R)} + \frac{GM}{R}\right]$$

or $\qquad W = \frac{GMm}{1001R}$

(ii) In this case too, initially, the body was at earth's surface, so initial gravitational potential is same as given in part (i), i.e.

$$V_i = -\frac{GM}{R}$$

Finally, the body is at height h, over the surface of earth, therefore, from Eq.1.96, the gravitational potential at height h over the surface of earth,

$$V_f = -\frac{GM}{r} = -\frac{GM}{(R+h)} \qquad [\text{since, } r = R + h]$$

Since, $h = R$, therefore

$$V_f = -\frac{GM}{r} = -\frac{GM}{2R}$$

So, from Eq.1.117, the work done by external agent against sytem gravitational force, in displacing the body from earth's surface to height $h = R$, is given by

$$W = U_f - U_i = m\left(V_f - V_i\right);$$
$$W = m\left[\left(-\frac{GM}{2R}\right) - \left(-\frac{GM}{R}\right)\right]; \quad \text{or} \quad W = -\frac{GM\,m}{2R}$$

EXAMPLE 66. The gravitational field in a region is given by $\vec{g} = (2\hat{i} + 3\hat{j})$N/kg.
Find the work done by the gravitational field when a particle of mass 1 kg is moved on the line $3y + 2x = 5$ from point A $(1\text{ m}, 1\text{ m})$ to B $(-2\text{ m}, 3\text{ m})$.

APPROACH Since, gravitational field is conservative in nature, therefore, work done by gravitational force is independent on the path followed by particle. So, find the displacement $\overrightarrow{AB}$, and then work done by applying the relation

$$W = \vec{F}.\overrightarrow{AB} = m\vec{g}.\overrightarrow{AB} \qquad \dots (1)$$

SOLUTION Particle is displaced from point A $(1\text{ m}, 1\text{ m})$ to B $(-2\text{ m}, 3\text{ m})$, therefore, displacement vector

$$\overrightarrow{AB} = \overrightarrow{OB} - \overrightarrow{OA}$$
$$= (-2-1)\hat{i} + (3-1)\hat{j} = -3\hat{i} + 2\hat{j} \text{ m}$$

Given that, gravitational force,

$$\vec{F} = m\vec{g} = m(2\hat{i} + 3\hat{j})$$

Here, m is the mass of particle.
Now, from Eq. (1), the required work done by gravitational force in displacement $\overrightarrow{AB}$ is,

$$W = m(2\hat{i} + 3\hat{j}).\left(-3\hat{i} + 2\hat{j}\right) = m[-6 + 6] = 0 \text{ J}$$

EXAMPLE 67. A thin uniform annular disc (Fig.1.109) of mass M has outer radius $4R$ and inner radius $3R$. Calculate, the work done in moving a unit mass from point P on its axis to infinity.

APPROACH If ΔU is the change in potential energy in moving the object from initial position to final position, then from Eq.1.117, the work done by external agent can be written as

$$W = \Delta U$$

SOLUTION We need to find gravitational potential energy of

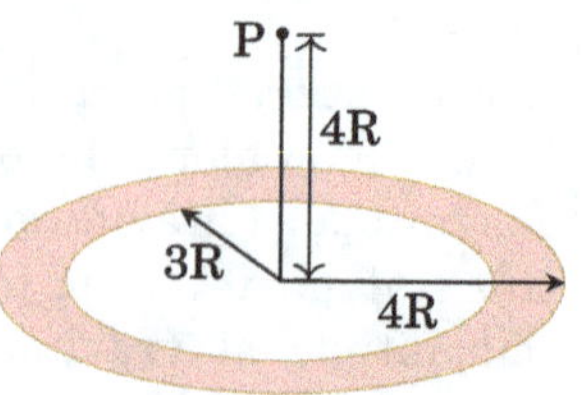

Figure 1.109

a unit mass placed at the point P.
The surface mass density of the annular disc is

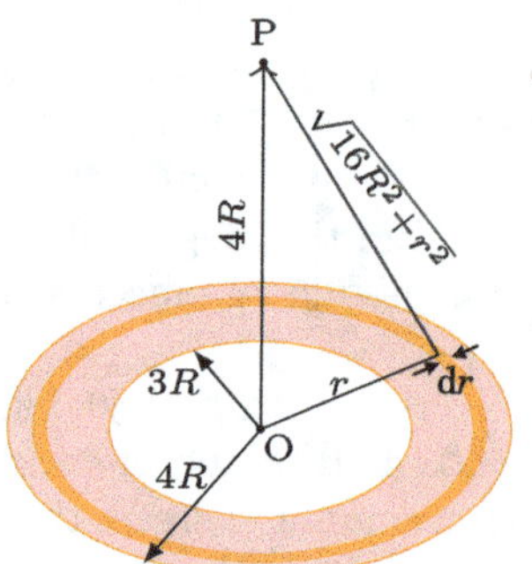

Figure 1.110

$$\sigma = \frac{\text{mass}}{\text{area}} = M/\left(16\pi R^2 - 9\pi R^2\right) = M/\left(7\pi R^2\right)$$

Consider a small ring of radius r and thickness dr [Fig.1.110]. The mass of the ring is $dm = 2\pi r\sigma dr$. As distance of any point of the ring from P is same, the potential at P due to the ring is

$$dU_{\text{ring}} = -\frac{G(2\pi r\sigma dr)}{\sqrt{16R^2 + r^2}}.$$

Integrate from $r = 3R$ to $r = 4R$ to get the potential energy of the unit mass placed at P

$$U_{\text{disc}} = -\int_{3R}^{4R} \frac{2\pi r\sigma G}{\sqrt{16R^2 + r^2}}\, dr = -\frac{2GM}{7R}\left(4\sqrt{2} - 5\right).$$

The work required to take the unit mass to infinity is

$$W = \Delta U = [0 - U_{\text{disc}}] = \frac{2GM}{7R}\left(4\sqrt{2} - 5\right).$$

EXAMPLE 68. In a region of only gravitational field a particle of mass 'M' is shifted from A to B via three different paths as shown in the Fig.1.111. If the work done in different paths are W_1, W_2, W_3 respectively, then find a relation between them.

SOLUTION The gravitational force is conservative and hence the work done by a gravitational force is independent of the path and depends only on the end points.
So, the required relation will be: $W_1 = W_2 = W_3$

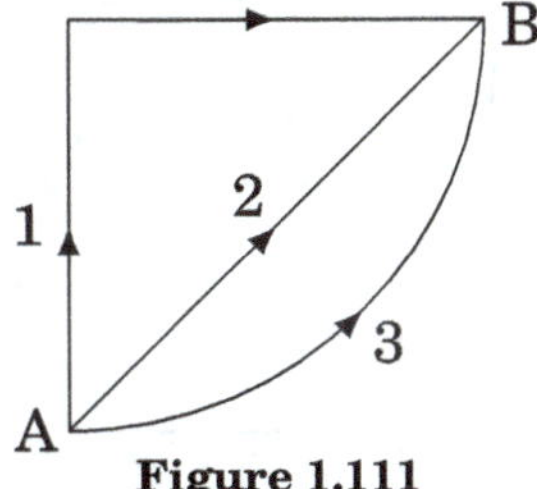

Figure 1.111

1.15.1 How to Draw g vs r and V vs r Graph for Point Masses?

Gravitational field and gravitational potential due to a point mass m at distance r are given as follows:

$$g = \frac{Gm}{r^2} \qquad \text{(towards the point mass)}$$

and $\qquad V = -\dfrac{Gm}{r}$

From above expressions, it is clear that-

1. At very large distance from a point charge, i.e., when $r \to \infty$, gravitational field $g \to 0$ and gravitational potential $V \to 0$.

2. Very near to point mass, i.e., for $r \to 0$, $g \to \infty$ and $V \to -\infty$

3. Since, g is a vector quantity, therefore, on both sides of a point mass the directions of $\vec{g}$ will be different. So, on one side very close to point mass, if the value of g is $+\infty$ (because, just over the mass, distance $r \to 0$), then on its other side very close to point mass, the value of g will be $-\infty$ [See Fig.1.112a].

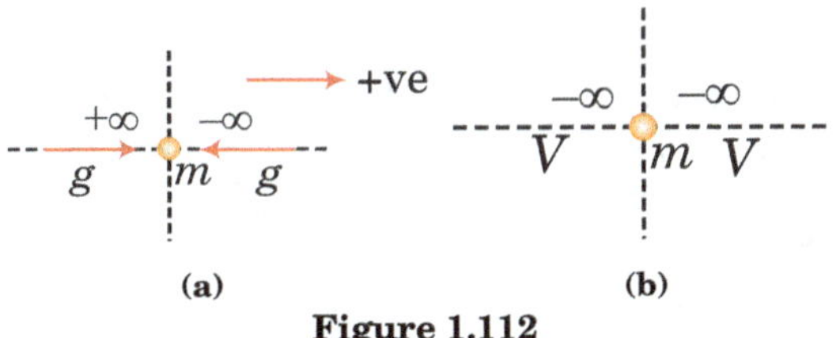

Figure 1.112

4. V is a scalar quantity. On both sides very close to the mass (i.e., for $r \to 0$) value will be $-\infty$ [See Fig.1.112a].

5. Between two zero values, we will always have an extremum.

EXAMPLE 69. Two point masses 'm' and $2m$ are place at certain separation as shown in Fig.1.114.

Draw g-r and V-r graphs along the line joining them corresponding to given mass system.

APPROACH Follow the points given in section 1.15.1.

SOLUTION $g - r$ **Graph:** In g - r graph, the separation between point masses, is represented on horizontal axis whereas gravitational field is shown on vertical axis.

The positive direction of horizontal axis is towards right and upward direction is considered as positive direction of gravitational field $\vec{g}$.

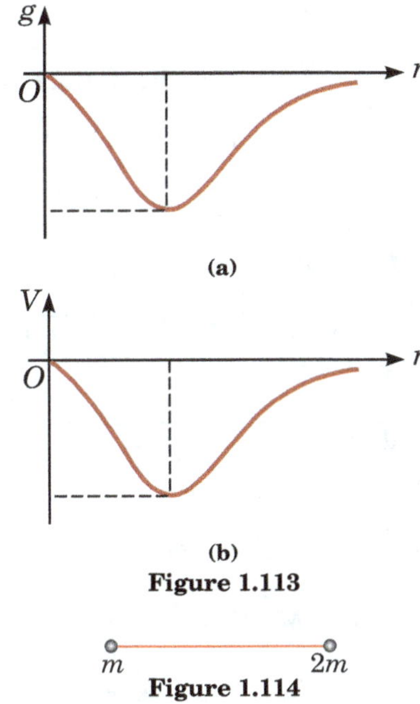

(a)

(b)

Figure 1.113

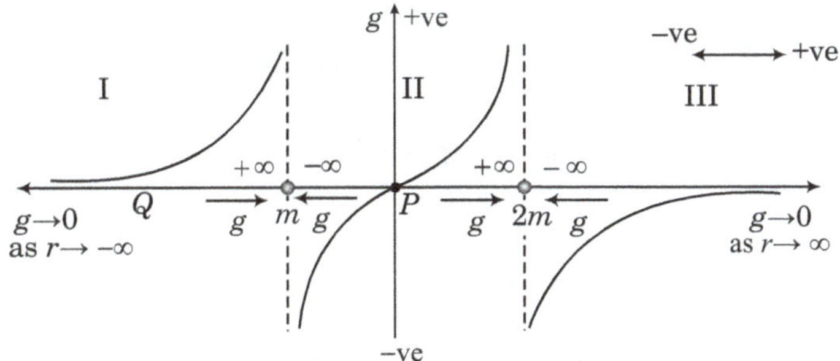

$m \qquad\qquad 2m$

Figure 1.114

If at any position, the net gravitational field is towards right, we consider it as positive and it is shown on vertical axis in upward direction. If gravitational field is towards left, it is considered as negative and it is shown in downward direction on vertical axis.

In Region I of horizontal axis (Fig.1.115), fields of m and $2m$ both are directed toward right, i.e., towards the positive direction of horizontal axis. Hence, on g-axis it will be in upward direction. Very close to point mass m, the field strength will be infinite and for $r \to \infty$, it is zero.

In region II of horizontal axis, gravitational field due to point

Figure 1.115: g-r Graph

mass m is toward left, i.e., it is negative and field of $2m$ is toward right, i.e., it is positive. At some point P (nearer to m) two fields are equal and opposite and net field is zero. Very close to m, the field is infinite and along the negative direction of horizontal axis. Very close to point mass $2m$, the field is again infinite but it is along positive direction of horizontal axis. In region III, fields of m and $2m$ both are negative. Hence, net field at any point in this region, is also negative. Very close to point mass $2m$, it is infinite and for $r \to \infty$, it is zero.

V - r **Graph:** Gravitational potential is always negative for all regions. At a position very close to any point mass it is also infinite as like gravitational field. It is zero for $r \to \infty$. Between

Table 1.3: Gravitational Potentials at position r due to some geometrical shapes/objects

Geometrical Shape/Object	Gravitational Potential (V)
Point mass of mass M	$V = \dfrac{-GM}{r}$
At a point on the axis of ring of mass M and radius R	$V = \dfrac{-GM}{\sqrt{R^2+r^2}} \quad 0 \le r \le \infty$
At an axial point of rod of length of mass M and length L	$V = -\dfrac{GM}{L} In\left(1+\dfrac{L}{r}\right)$
At an equatorial point of a rod of mass M and length L	$V = \dfrac{2GM}{r\sqrt{L^2+4r^2}}$
Circular arc of mass M and radius R	$V = \dfrac{2\pi GM}{L}$
Hollow sphere of mass M and radius R	$V(r) = \dfrac{-GM}{r} \quad (r \ge R)$ $V(r) = \dfrac{-GM}{R} \quad (r \le R)$
Solid sphere of mass M and radius R	$V(r) = \dfrac{-GM}{r} \quad (r \ge R)$ $V(r) = -\dfrac{GM}{2R^3}\left[3R^2 - r^2\right] \quad (r \le R)$
Long thread	$V = \infty$

two masses potential varies between $-\infty$ and $-\infty$. So, graph is as shown in Fig.1.116.

Note: Table 1.3 shows gravitational potentials at position r

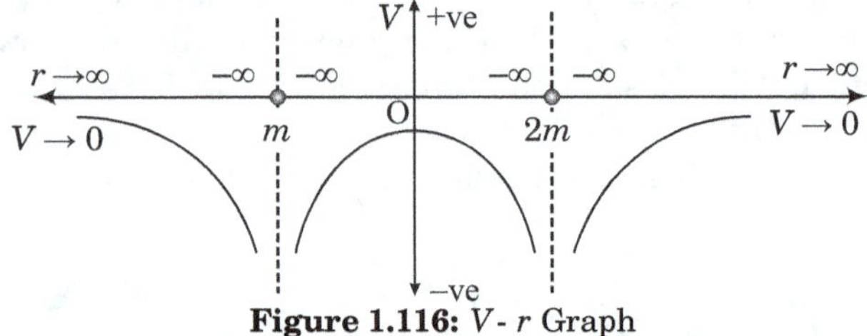

Figure 1.116: V-r Graph

due to some geometrical shapes/objects of mass M. Distance r is measured from the centre of the object. L is the length of object and R is it's radius.

1.15.2 Check Point 3

1. ••Two particles of masses m_1 and m_2, initially at rest at infinite distance from each other, move under the action of mutual gravitational pull. By applying, the principle of conservation of mechanical energy, show that at any instant their relative velocity of approach is $\sqrt{2G(m_1+m_2)/r}$, where r is their separation at that instant.

2. ••Figure 1.117 shows four particles, each of mass 20.0 g, that form a square with an edge length of $d = 0.600$ m. If d is reduced to 0.200 m, what is the change in the gravitational potential energy of the four-particle system?

3. ••An object is projected straight upward from the surface of Earth with an initial speed of 4.0 km/s. What is the maximum height it reaches?

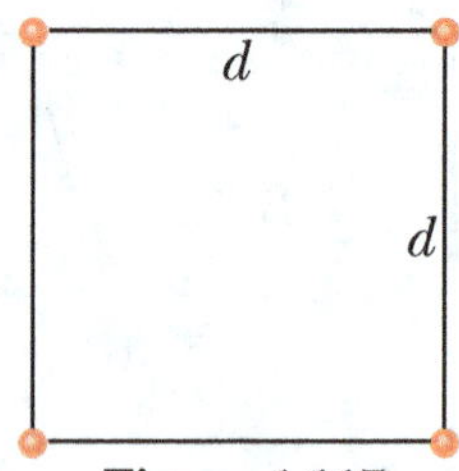

Figure 1.117

4. ••Zero, a hypothetical planet, has a mass of $5.0 \times 10^{23} kg$, a radius of $3.0 \times 10^6 m$, and no atmosphere. A $10 kg$ space probe is to be launched vertically from its surface.
(a) If the probe is launched with an initial energy of $5.0 \times 10^7 J$, what will be its kinetic energy when it is 4.0×10^6 m from the centre of Zero?
(b) If the probe is to achieve a maximum distance of 8.0×10^6 m from the centre of Zero, with what initial kinetic energy must it be launched from the surface of Zero?

5. ••The three spheres in Fig. 1.118, with masses $m_A = 80$ g, $m_B = 10$ g, and $m_C = 20$ g, have their centres on a common line, with $L = 12$ cm and $d = 4.0$ cm. You move sphere B along the line until its centre-to-centre separation from C is $d = 4.0$ cm. How much work is done on sphere B (a) by you and (b) by the net gravitational force on B due to spheres A and C?

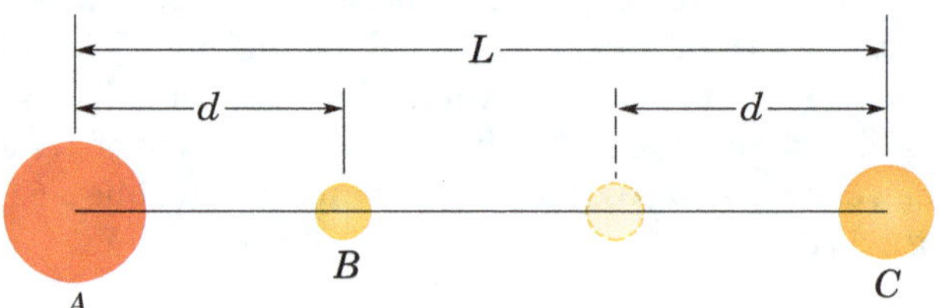

Figure 1.118

6. ••The force exerted by Earth (Radius $= R$, mass $= M$) on a particle of mass m a distance $r(r > R)$ from the center of Earth has the magnitude mgR^2/r^2, where $g = GM/R^2$. (a) Calculate the work you must do to move the particle from distance r_1 to distance r_2. (b) Show that when $r_1 = R$ and $r_2 = R + h$, the result can be written as $W = mgR^2\left[\frac{1}{R} - \frac{1}{R+h}\right]$. (c) Show that when $h \ll R$, the work is given approximately by $W = mgh$.

7. ••In deep space, sphere A of mass 20 kg is located at the origin of an x axis and sphere B of mass 10 kg is located on the axis at $x = 0.80$ m. Sphere B is released from rest while sphere A is held at the origin. (a) What is the gravitational potential energy of the two-sphere system just as B is released? (b) What is the kinetic energy of B when it has moved 0.20 m toward A?

8. ••Two neutron stars are separated by a distance of $1.0 \times 10^{10} m$. They each have a mass of $1.0 \times 10^{30} kg$ and a radius of $1.0 \times 10^5 m$. They are initially at rest with respect to each other. As measured from that rest frame, how fast are they moving when (a) their separation has decreased to one-half its initial value and (b) they are about to collide?

9. ••A rocket is fired vertically with a speed of 5 km/s from

the Earth's surface. How far from the Earth does the rocket go before returning to the Earth? Mass of the Earth $= 6.0 \times 10^{24}$ kg; mean radius of the Earth $= 6.4 \times 10^6 m$; $G = 6.67 \times 10^{-11}$ N - m^2/ kg^2.

10. ••An asteroid, headed directly toward Earth, has a speed of 12 km/s relative to a stationary space point when the asteroid is at 10 Earth radii from Earth's centre. Neglecting the effects of Earth's atmosphere on the asteroid, find the asteroid's speed v_f when it reaches Earth's surface.

11. ••Two stars each of one solar mass $(= 2 \times 10^{30}$ kg) are approaching each other for a head on collision. When they are at a distance of 10^9 km, their speeds are negligible. What is the speed with which they collide? The radius of each star is 10^4 km. Assume the stars to remain undistorted until they collide. (Use known value of G).

12. ••Two heavy spheres each of mass 100 kg and radius 0.10 m are placed 1.0 m apart on a horizontal plane. What is the gravitational field and potential at the mid point of the line joining the centres of the spheres? Is the object placed at that point in equilibrium? If so, is the equilibrium stable or unstable?

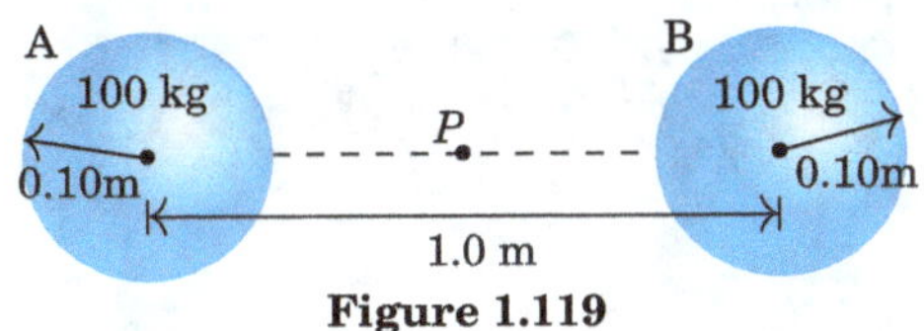

Figure 1.119

13. ••An asteroid is moving directly towards the centre of the earth. When at a distance of $10R$ (R is the radius of the earth) from the earth's centre, it has a speed of 12 km/s. Neglecting the effect of earth's atmosphere, what will be the speed of the asteroid when it hits the surface of the earth (escape velocity from the earth is 11.2 km/s)? Give your answer to the nearest integer (in km/s)

14. •••A tunnel is dug along a chord of non rotating earth at a distance $d = \frac{R}{2}$ [R = radius of the earth] from its centre [Fig.1.120]. A small block is released in the tunnel from the surface of the earth. The block comes to rest at the centre (C) of the tunnel. Assume that the friction coefficient between the block and the tunnel wall remains constant at μ.

(a) Calculate work done by the friction on the block.

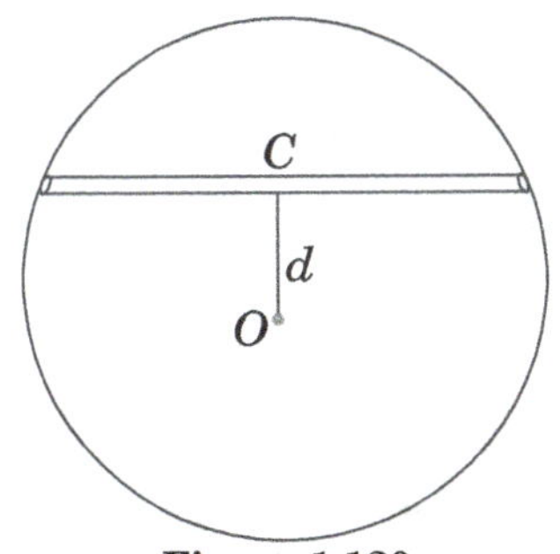

Figure 1.120

(b) Calculate μ.

Multiple Choice Questions

15. ••The work done to raise a mass m from the surface of the earth to a height h, which is equal to the radius of the earth, is

(A) $\frac{3}{2}mgR$ (B) mgR
(C) $2mgR$ (D) $\frac{1}{2}mgR$

16. ••At what height from the surface of earth the gravitation potential and the value of g are -5.4×10^7 J kg^{-1} and 6.0 m s^{-2} respectively? Take the radius of earth as 6400 km.

(A) 1400 km   (B) 2600 km
(C) 2000 km   (D) 1600 km

17. ••Infinite number of bodies, each of mass 2 kg are situated on x-axis at distances 1 m, 2 m, 4 m, 8 m,$\ldots$, respectively, from the origin. The resulting gravitational potential due to this system at the origin will be

(A) $-\frac{4}{3}G$ (B) $-4G$ (C) $-G$ (D) $-\frac{8}{3}G$

18. ••A body of mass 'm' is taken from the earth's surface to the height equal to twice the radius (R) of the earth. The change in potential energy of body will be

(A) $3mgR$ (B) $\frac{1}{3}mgR$ (C) $mg2R$ (D) $\frac{2}{3}mgR$

19. ••A particle of mass M is situated at the centre of a spherical shell of same mass and radius R. The magnitude of the gravitational potential at a point situated at $a/2$ distance from the centre, will be

(A) $\frac{GM}{R}$ (B) $\frac{2GM}{R}$ (C) $\frac{3GM}{R}$ (D) $\frac{4GM}{R}$

20. ••A body of mass m is placed on earth's surface which is taken from earth surface to a height of $h = 3R$, then change in gravitational potential energy is

(A) $\frac{mgR}{4}$ (B) $\frac{2}{3}mgR$ (C) $\frac{3}{4}mgR$ (D) $\frac{mgR}{2}$

1.16   Satellite and Planetary Motion

Figure 1.121 shows the orbits of an object launched multiple times from a fixed location A in the direction AB, tangent to the earth's surface. The distance of the point A from the centre of earth is r. The launch speed is increased for each successive launch. Because the initial gravitational potential energy U_g is always the same, the value of the mechanical energy E_{mech} is determined by the initial kinetic energy:

$$E_{\text{mech}} = U_g + K \qquad (1.118)$$

If the object is released from rest, $K = 0$ and so from Eq.1.118, $E_{\text{mech}} = U_g$. In this case, the object will fall in a straight line toward the surface of Earth. When the object is given a small initial speed, however, the orbit is an ellipse with Earth's centre at the focus farther from the launch point. For small v_i, the orbit intersects Earth and the object crashes. As v_i is increased, the ellipse becomes rounder and longer, and for sufficiently large v_i, the object makes it around the earth. If there is no retarding force, the object's speed when it returns to point A, is the same as its initial speed and it repeats its motion indefinitely. As we continue to increase v_i, we reach a point when v_i has just the right magnitude to make v_i^2/r equal to the centripetal acceleration caused by Earth's gravitational force, and the object goes in a circular orbit. For greater v_i, the orbit is again elliptical, but Earth's center now coincides with the nearer focus of the ellipse.

As we continue to increase v_i, we reach a point where the mechanical energy E_{mech} of the Earth-object system is no longer negative and the orbit is unbound. This happens when

$$E_{\text{mech}} = \frac{1}{2}mv_e^2 - \frac{GMm}{r} = 0 \qquad (1.119)$$

where v_e is the object's escape speed, m is the mass of the launched object, M is Earth's mass, and r is the distance from the launch point to Earth's center. An object launched at the escape speed and moving without any further propulsion travels out to infinity along a parabolic path. At launch speeds greater than v_e the orbit corresponds to one branch of a hyperbola.

In Fig.1.121, trajectories ① through ⑦ show the effect of increasing the initial speed. In trajectories ③ through ⑤ the projectile misses the Earth and becomes a satellite. Trajectories ① through ⑤ close on themselves and are called **closed orbits**. All closed orbits are ellipses or segments of ellipses; trajectory ④ is a circle, a special case of an ellipse. Trajectories ⑥ and ⑦ are **open orbits**. For these paths the object never returns to its starting point but travels ever farther away from the earth.

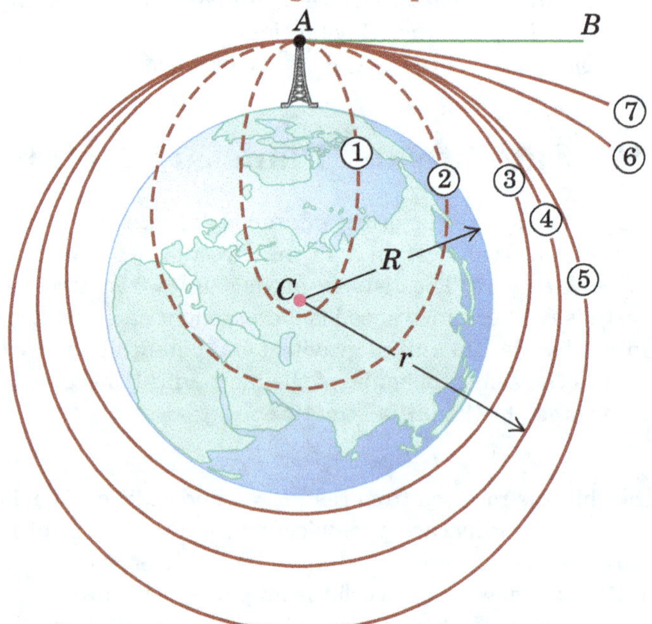

Figure 1.121: Trajectories of a projectile launched from a great height (ignoring air resistance). Orbits ① and ② would be completed as shown if the earth were a point mass at C.

1.16.1   Motion of a Satellite in a Circular Orbit

1.16.1.1   Orbital Speed

A circular orbit, like trajectory ④ in Fig. 1.121, is the simplest case. It is also an important case, since many artificial satellites have nearly circular orbits and the orbits of the planets around the sun are also fairly circular. The only force acting on a satellite in circular orbit around the earth is the earth's gravitational attraction, which is directed toward the center of the earth and hence toward the center of the orbit (Fig. 1.122a).

Since, the centripetal gravitational force is always perpendic-

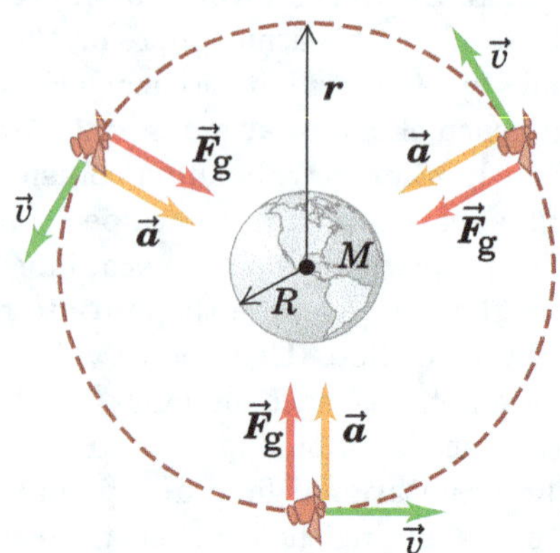

(a) The force $\vec{F_g}$ due to the earth's gravitational attraction provides the centripetal acceleration that keeps a satellite in orbit.

(b) These space shuttle astronauts are in a state of apparent weightlessness.

Figure 1.122

ular to the orbital velocity vector '$\vec{v}$', of the satellite, therefore,

$$\vec{F}_g.\vec{v} = 0$$

or $\qquad \vec{F}_g.\dfrac{d\vec{s}}{dt} = 0$

here, $d\vec{s}$ is the instantaneous displacement vector on circular path of orbit. Above equation can also be written as

or $\qquad \vec{F}_g.d\vec{s} = 0$

or $\qquad dW = 0$, here, dW is the work done by $\vec{F}_g$ in displacement $d\vec{s}$

or $\qquad dK = 0$ (by work energy theorem)

or $\qquad \dfrac{1}{2}mv_2^2 - \dfrac{1}{2}mv_1^2 = 0$

or $\qquad v^2 = $ constant

or $\qquad$ speed $=$ constant

So, the satellite is in uniform circular motion and its speed is constant. The satellite isn't falling toward the earth; rather, it's constantly falling around the earth. In a circular orbit the speed is just right to keep the distance from the satellite to the center of the earth constant.

Calculation of speed of satellite in circular orbit

Let the radius of the orbit is r, measured from the center of the earth; the acceleration of the satellite has magnitude $a_{rad} = v^2/r$ and is always directed toward the center of the circle. By the law of gravitation, the net force (gravitational force) on the satellite of mass m has magnitude $F_g = GMm/r^2$ and is in the same direction as the acceleration. So, by Newton's second law $\left(\sum \vec{F} = m\, \vec{a} \right)$, we can write

$$\frac{GMm}{r^2} = \frac{mv^2}{r}$$

Solving this for v, we find that, $v = \sqrt{\dfrac{GM}{r}}$

From now on, we denote this orbital speed by v_o, i.e.,

$$\boxed{v_o = \sqrt{\frac{GM}{r}} \quad \text{(orbital speed)}} \qquad (1.120)$$

Putting $GM/R^2 = g$ (acceleration due to gravity on the surface of Earth), in Eq.1.120, we obtain:

$$\boxed{v_o = \sqrt{\frac{gR^2}{r}}} \qquad (1.121)$$

If the altitude of the orbit is h, then $r = R + h$, so Eq. 1.121 can also be written as

$$v_o = \sqrt{\frac{gR^2}{R+h}} = \sqrt{\frac{gR^2}{R(1+h/R)}}$$

$$\boxed{v_o = \sqrt{\frac{gR}{(1+h/R)}}} \qquad (1.122)$$

If satellite is very near to the surface of Earth, i.e., $h << R$, then above Eq.1.122 gives

$$\boxed{v_o = \sqrt{gR} = \sqrt{\frac{GM}{R}}} \qquad (1.123)$$

$$[\text{since, } g = GM/R^2]$$

If we substitute the values of g and R in Eq.1.123, we get,

$$\boxed{v_0 = \sqrt{gR} \simeq 8 \text{ km/s}}.$$

So, for a near by satellite, the orbital speed $\boxed{\simeq 8 \text{ km/s}}$
For **Moon** $h = 380,000$ km, therefore, from Eq.1.122:

$$v_o = \sqrt{\frac{gR}{(1+h/R)}}$$

we get, $v_{om} \simeq 1.04$ km/sec.
We can also calculate the orbital speed of moon by applying the relation: $v_{om} = \dfrac{2\pi(R+h)}{T_m}$

Here, T_m is the time period of moon $= 27$ days , therefore

$$v_{om} = \frac{2\pi(R+h)}{T_m} = \frac{2\pi\left(386400 \times 10^3\right)}{27 \times 24 \times 60 \times 60} \simeq 1.04 \text{ kmsec.}$$

Relations 1.120 or 1.121 show that we can't choose the orbit radius r and the speed v_o independently; for a given radius r, the speed v_o for a circular orbit is determined by Eq.1.120. The satellite's mass m doesn't appear in Eq.1.120, which shows that the motion of a satellite does not depend on its

mass. An astronaut on board an orbiting space station is himself a satellite of the earth, held by the earth's gravity in the same orbit as the station. The astronaut has the same velocity and acceleration as the station, so nothing is pushing him against the station's floor or walls. He is in a state of apparent weightlessness, as in a freely falling elevator; (True weightlessness would occur only if the astronaut were infinitely far from any other masses, so that the gravitational force on his would be zero.) Indeed, every part of his body is apparently weightless; he feels nothing pushing his stomach against his intestines or his head against his shoulders (Fig.1.122b).
Apparent weightlessness is not just a feature of circular orbits; it occurs whenever gravity is the only force acting on a spacecraft. Hence it occurs for orbits of any shape, including open orbits such as trajectories ⑥ and ⑦ in Fig. 1.121.

1.16.1.2 Mathematical Explanation of Weightlessness in a Satellite

When the weight of a body (either true or apparent) becomes zero, the body is said to be in the state of weightlessness. If a body is in a satellite (which does not produce its own gravity) orbiting the Earth at a height h above its surface then

True weight, $= mg_h = \dfrac{mGM}{(R+h)^2} = \dfrac{mg}{\left(1+\frac{h}{R}\right)^2}$

and Apparent weight $= m\left(g_h - a\right)$

But $a = \dfrac{v_0^2}{r} = \dfrac{GM}{r^2} = \dfrac{GM}{(R+h)^2} = g_h$

$\Rightarrow$ Apparent weight $= m\left(g_h - g_h\right) = 0$

Note: The condition of weightlessness can be overcome by creating artificial gravity by rotating the satellite in addition to its revolution.

1.16.1.3 Angular Speed

The angular speed of a satellite orbiting in a circular orbit of radius r, is,

$$\omega = v_o/r$$

Substituting $v_o = \sqrt{\dfrac{GM}{r}}$, in above expression, we get

$$\boxed{\omega = \sqrt{\frac{GM}{r^3}} = \sqrt{\frac{GM}{(R+h)^3}}} \qquad (1.124)$$

1.16.1.4 Calculation of Time Period

Time period of the satellite, $T = \dfrac{\text{circumference of the orbit}}{\text{orbital speed}}$

$$= \frac{2\pi r}{v_o}$$

On substituting the value of v_o, from Eq.1.120 in above expression, we get-

$$T = \frac{2\pi r}{\sqrt{GM/r}}$$

$$\boxed{T = 2\pi \sqrt{\frac{r^3}{GM}} = 2\pi \sqrt{\frac{(R+h)^3}{GM}}} \qquad (1.125)$$

$$\boxed{T^2 = \frac{4\pi^2}{GM}r^3 \quad \Rightarrow \quad T^2 \propto r^3 \quad (r = R + h)} \qquad (1.126)$$

here, h is the height of satellite from the surface of Earth.

Note: You can also find time period by applying the relation $T = 2\pi/\omega$.

For a nearby satellite, $T_{NS} = 2\pi\sqrt{R/g} = 84$ minute

$$= 1\ \text{hour}\ 24\ \text{minute} = 1.4\,\text{hr} = 5063\ \text{s}$$

In terms of density, $T_{NS} = \dfrac{2\pi (R)^{1/2}}{(G \times 4/3\pi R \times \rho)^{1/2}} = \sqrt{\dfrac{3\pi}{G\rho}}$

Time period of near by satellite only depends upon density of planet.

Equations (1.120) and (1.125) show that larger orbits correspond to slower speeds and longer periods. As an example, the International Space Station (Fig.1.123) orbits 6800 km from the centre of the earth (400 km above the earth's surface) with an orbital speed of $7.7\,km/s$ and an orbital period of 93 min. The moon orbits the earth in a much larger orbit of radius $384,000$ km, and so has a much slower orbital speed (1.0 km/s) and a much longer orbital period (27.3 days).

Figure 1.123: With a mass of approximately 4.5×10^5 kg and a width of over 10^8 m, the International Space Station is the largest satellite ever placed in orbit.

EXAMPLE 70. Three particles, each of mass m, are situated at the vertices of an equilateral triangle of side length a. The only force acting on the particles are their mutual gravitational forces. It is desired that each particle moves in a circle while maintaining the original mutual separation a. Find the initial velocity that should be given to each particle and also the time period of the circular motion.

APPROACH For circular motion of each particle, the net force on each particle should be towards the centre of the circle and must provide required centripetal force for circular motion.

If v is the speed of the particle in circular path, and r is the radius of the circular path, then time period can be written as

$$T = \frac{2\pi r}{v}$$

SOLUTION In right angle triangle OBL [Fig.1.124], we have

$$\cos 30° = \frac{a/2}{r}$$

or $\qquad r = a/\sqrt{3}$

The gravitational forces on the particle at A due to the particles at B and C have magnitude

$$F| = Gm^2/a^2$$

Resolve $\vec{F}$ along and perpendicular to AO. The resultant force on the particle at A is towards the centre O and its magnitude

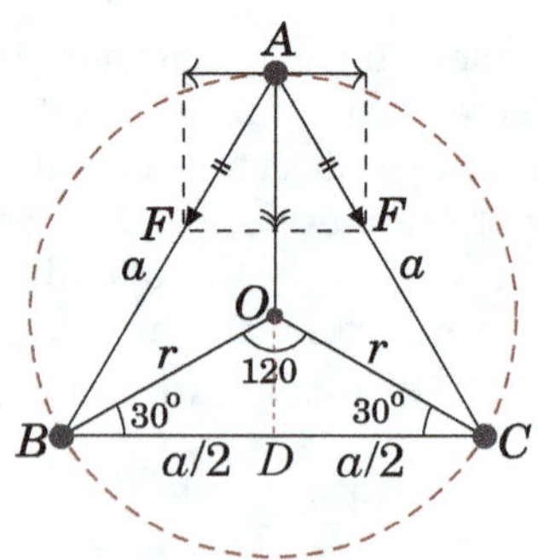

Figure 1.124

is

$$F_r = F\cos 30° + F\cos 30° = \sqrt{3}Gm^2/a^2$$

By symmetry, the resultant forces on the particles at B and C are also towards the centre O with the same magnitude F_r. The force F_r provides the necessary centripetal force for the particle to revolve in a circle of radius r with constant speed v i.e.,

$$\sqrt{3}Gm^2/a^2 = mv^2/r = \sqrt{3}mv^2/a$$

$$\Rightarrow \qquad v = \sqrt{GM/a}$$

The time period in a uniform circular motion is given by

$$T = \frac{2\pi r}{v} = \frac{2\pi a/\sqrt{3}}{\sqrt{GM/a}} = 2\pi\sqrt{\frac{a^3}{3Gm}}$$

EXAMPLE 71. A satellite moves eastwards very near to the surface of the Earth in equatorial plane with speed (v_0). Another satellite moves at the same height with the same speed in the equatorial plane but westwards. If R = radius of the Earth and ω be the angular speed of Earth about its own axis. Then find the approximate difference in the two time period as observed on the Earth.

APPROACH Time period of a satellite

$$T = \frac{\text{circumference}}{\text{absolute linear speed of satellite}}$$

Find the time periods of both satellites and then find their difference.

SOLUTION $T_{\text{west}} = \dfrac{2\pi R}{v_0 + R\omega}$ and $T_{\text{east}} = \dfrac{2\pi R}{v_0 - R\omega}$

$$\Rightarrow \qquad \Delta T = T_{\text{east}} - T_{\text{west}} = 2\pi R\left[\frac{2R\omega}{v_0^2 - R^2\omega^2}\right]$$

$$= \frac{4\pi\omega R^2}{v_0^2 - R^2\omega^2}$$

EXAMPLE 72. Imagine a light planet revolving around a very massive star in a circular orbit of radius R with a period of revolution T. If the gravitational force of attraction between the planet and the star is proportional to $R^{-5/2}$, then, show that $T^2 \propto R^{7/2}$.

APPROACH For the circular motion of a planet about a star, the required centripetal force is provided by gravitational force of star on the planet. By applying this condition, you can obtain a relation between linear velocity of the planet v and it's distance from the star. Now, time period can be obtained by using the relation

$$\text{Time period} = \frac{\text{circumference}}{\text{speed}}$$

SOLUTION Let the gravitational force on the planet by the star is

$$F = kR^{-5/2}$$

where k is some constant and R is the distance between the planet and the star. This force provides the centripetal acceleration to the planet of mass m moving with a velocity v in the circular orbit of radius R,

$$mv^2/R = kR^{-5/2}$$

which gives

$$v = \left(\frac{k}{mR^{3/2}}\right)^{1/2}$$

The time period of revolution is the time taken to complete one revolution i.e.,

$$T = \frac{2\pi R}{v} = \left(\frac{4\pi^2 mR^{7/2}}{k}\right)^{1/2}$$

which gives, $T^2 \propto R^{7/2}$.

Note: Now, if we generalize the result for $F = kR^{-n}$, we get $T^2 \propto R^{n+1}$.

EXAMPLE 73. A satellite that has a mass of 300 kg moves in a circular orbit 5.00×10^7 m above Earth's surface. (a) What is the gravitational force on the satellite? (b) What is the speed of the satellite? (c) What is the period of the satellite?

APPROACH We'll use the law of gravity to find the gravitational force acting on the satellite. The application of Newton's second law will lead us to the speed of the satellite and its period can be found from its definition.

SOLUTION (a) Let m represent the mass of the satellite and h its elevation. The gravitational force acting on it, is given by

$$F_g = \frac{GmM_E}{(R_E + h)^2}$$

Because $GM_E = gR_E^2$, therefore,

$$F_g = \frac{mR_E^2 g}{(R_E + h)^2}$$

Dividing the numerator and denominator of right hand side of this expression by R_E^2, we get

$$F_g = \frac{mg}{(1 + \frac{h}{R_E})^2}$$

On substituting the numerical values, we get

$$F_g = \frac{(300 \text{ kg})(9.81 \text{ N/kg})}{(1 + \frac{5.00 \times 10^7 \ m}{6.37 \times 10^6 \ m})^2}$$

$$= 37.58 \text{ N} = 37.6 \text{ N}$$

(b) Using Newton's second law, relate the gravitational force acting on the satellite to its centripetal acceleration:

$$F_g = m\frac{v^2}{r} \Rightarrow v = \sqrt{\frac{F_g r}{m}}$$

On substituting numerical values, we get

$$v = \sqrt{\frac{(3758 \ N)637 \times 10^6 \ m + 500 \times 10^7 \ m}{300 \ kg}}$$

$$= 2.66 km/s$$

(c) The period of the satellite is given by-

$$T = \frac{2\pi r}{v}$$

Substituting numerical values, we get

$$T = \frac{2\pi(6.37 \times 10^6 \text{ m} + 5.00 \times 10^7 \text{ m})}{2657 \times 10^3 \text{ m/s}}$$

$$= 1.333 \times 10^5 s \times \frac{1 \ h}{3600 \ s} = 37.0 \text{ h}$$

EXAMPLE 74. Suppose that the attractive interaction between a star of mass M and a planet of mass $m \ll M$ is of the form $F = KMm/r$, where K is the gravitational constant. What would be the relation between the radius of the planet's circular orbit and its period?

APPROACH We can use the law of gravity and Newton's second law to relate the force exerted on the planet by the star to its orbital speed and the definition of the period to relate it to the radius of the orbit.

SOLUTION The period of the planet is related to its orbital speed:

$$T = \frac{2\pi r}{v} \qquad \qquad \dots (1)$$

Using the law of gravity and Newton's second law, relate the force exerted on the planet by the star to its centripetal acceleration:

$$F_{net} = \frac{KMm}{r} = m\frac{v^2}{r} \quad \Rightarrow v \ = \sqrt{KM}$$

Substituting for v in equation (1), we get

$$T = \frac{2\pi}{\sqrt{KM}}r$$

EXAMPLE 75.   Lagrange Point. The mathematician Joseph-Louis Lagrange discovered five special points in the vicinity of the Earth's orbit about the Sun where a small satellite (mass m) can orbit the Sun with the same period T as Earth's (= 1 year). One of these "Lagrange Points," called L1, lies between the Earth (mass M_E) and Sun (mass M_S), on the line connecting them (Fig. 1.125). That is, the Earth and the satellite are always separated by a distance d. If the Earth's orbital radius is R_{ES}, then the satellite's orbital radius is $(R_{ES} - d)$. Determine d.

APPROACH For motion of earth and the satellite in a

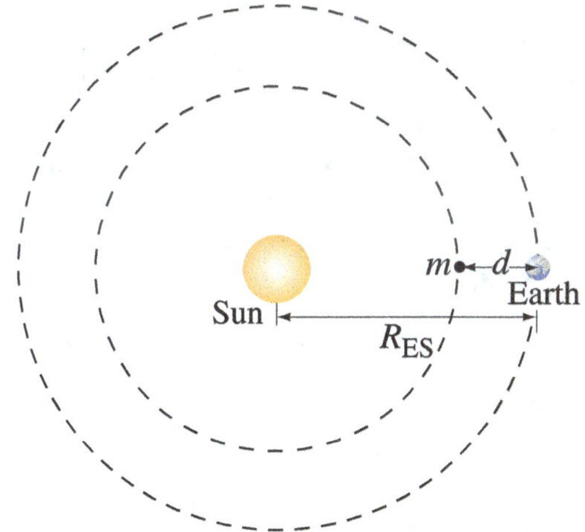

Figure 1.125: Finding the position of the Lagrange Point $L1$ for a satellite that can remain along the revolving line between the Sun and Earth, at distance d from the Earth. Thus a mass m at $L1$ has the same period around the Sun as the Earth has. (Not to scale.)

circular orbit around the Sun, the required centripetal force is provided by net gravitational force on these objects. Since, mass of satellite is very small as compared to earth, therefore it's gravitational force on earth does not produce sufficient acceleration to effect the circular motion of earth around the Sun. So, for circular motion of earth, we consider only gravitational force of Sun on it and equate it to required centripetal force. But for circular motion of the satellite, we consider the force applied by Sun and earth both because it's mass is very small as compared to earth and so force applied by earth and sun both produce a larger acceleration. The net gravitational force on it provides the required centripetal force for circular motion. So, we equate the net gravitational force acting on it with the net centripetal force.

But how could an object with a smaller orbit than Earth's have the same period as Earth? Kepler's third law clearly tells us a smaller orbit around the Sun results in a smaller period. But that law depends on only the Sun's gravitational attraction. Our mass m is pulled by both the Sun and the Earth.

SOLUTION Because the satellite is assumed to have negligible mass in comparison to the masses of the Earth and Sun, to an excellent approximation the Earth's orbit will be determined solely by the Sun. Applying Newton's second law to the Earth gives

$$\frac{GM_E M_S}{R_{ES}^2} = M_E \frac{v^2}{R_{ES}} = \frac{M_E}{R_{ES}} \frac{(2\pi R_{ES})^2}{T^2}$$

or

$$\frac{GM_S}{R_{ES}^2} = \frac{4\pi^2 R_{ES}}{T^2} \qquad \ldots (1)$$

Next we apply Newton's second law to the satellite m (which has the same period T as Earth), including the pull of both Sun and Earth.

$$\frac{GM_S m}{(R_{ES}-d)^2} - \frac{GM_E m}{d^2} = \frac{mv^2}{R_{ES}-d}$$

or

$$\frac{GM_S}{(R_{ES}-d)^2} - \frac{GM_E}{d^2} = \frac{\left(\frac{2\pi(R_{ES}-d)}{T}\right)^2}{R_{ES}-d}$$

or

$$\frac{GM_S}{(R_{ES}-d)^2} - \frac{GM_E}{d^2} = \frac{4\pi^2(R_{ES}-d)}{T^2}$$

or

$$\frac{GM_S}{R_{ES}^2}\left(1-\frac{d}{R_{ES}}\right)^{-2} - \frac{GM_E}{d^2} = \frac{4\pi^2 R_{ES}}{T^2}\left(1-\frac{d}{R_{ES}}\right)$$

We now use the binomial expansion $(1+x)^n \approx 1+nx$, if $x \ll 1$. Setting $x = d/R_{ES}$ and assuming $d \ll R_{ES}$, we have

$$\frac{GM_S}{R_{ES}^2}\left(1+2\frac{d}{R_{ES}}\right) - \frac{GM_E}{d^2} = \frac{4\pi^2 R_{ES}}{T^2}\left(1-\frac{d}{R_{ES}}\right) \qquad \ldots (2)$$

Substituting GM_S/R_{ES}^2 from Eq. (1) into Eq. (2) we find

$$\frac{GM_S}{R_{ES}^2}\left(1+2\frac{d}{R_{ES}}\right) - \frac{GM_E}{d^2} = \frac{GM_S}{R_{ES}^2}\left(1-\frac{d}{R_{ES}}\right).$$

$$\Rightarrow \quad \frac{GM_S}{R_{ES}^2}\left(3\frac{d}{R_{ES}}\right) = \frac{GM_E}{d^2}$$

On simplifying for d, we get

$$d = \left(\frac{M_E}{3M_S}\right)^{\frac{1}{3}} R_{ES}$$

If we substitute the values $M_E = 5.97219 \times 10^{24}$ kg, $M_S = 1.9891 \times 10^{30}$ kg, and $R_{ES} = 1.5 \times 10^{11}$ m, we get:

$$d = 1.0 \times 10^{-2} R_{ES} = 1.5 \times 10^9 \text{ m} = 1.5 \times 10^6 \text{ km}$$

NOTE 1. Since $d/R_{ES} = 10^{-2}$, we were justified in using the binomial expansion.

2. Placing a satellite at L1 has two advantages: the satellite's view of the Sun is never eclipsed by the Earth, and it is always close enough to Earth to transmit data easily. The L1 point of the Earth-Sun system is currently home to the Solar and Heliospheric Observatory (SOHO) satellite.

EXAMPLE 76. A science-fiction tale describes an artificial "planet" in the form of a band completely encircling a sun (Fig.1.126).The inhabitants live on the inside surface (where it is always noon).Imagine that this sun is exactly like our own,

that the distance to the band is the same as the Earth–Sun distance (to make the climate temperate), and that the ring rotates quickly enough to produce an apparent gravity of gas on Earth. What will be the period of revolution, this planet's year, in Earth days?

APPROACH Net force on the object should be towards the

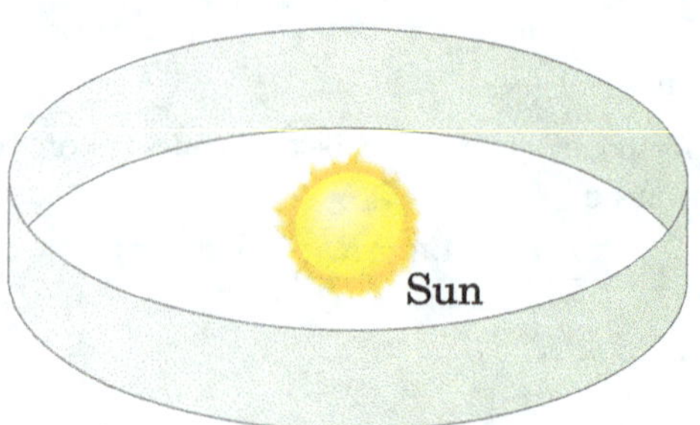

Figure 1.126

centre of the sun and must provide the centripetal force for circular motion. Convert the speed of ring in terms of time period of ring and then simplify for required time period.

SOLUTION If the ring is to produce an apparent gravity equivalent to that of Earth, then the normal force of the ring on objects must be given by $F_N = mg$. The Sun will also exert a force on objects on the ring. See the free-body diagram [Fig.1.127]. Write Newton's second law for the object, with the fact that the acceleration is centripetal.

$$\sum F = F_R = F_{Sun} + F_N = mv^2/r$$

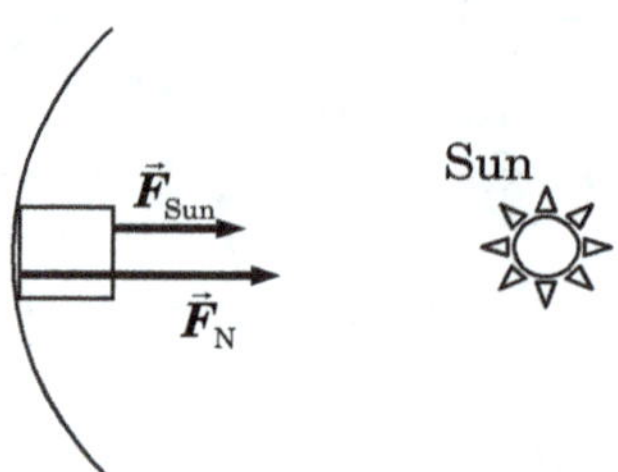

Figure 1.127

Substitute in the relationships that $v = 2\pi r/T, F_N = mg$, and $F_{Sun} = G\frac{M_{Sun}m}{r^2}$, and solve for the period of the rotation.

$$F_{Sum} + F_N = mv^2/r$$

$$\Rightarrow \quad G\frac{M_{Sum}m}{r^2} + mg = \frac{4\pi^2 mr}{T^2}$$

$$\Rightarrow \quad G\frac{M_{Sun}}{r^2} + g = \frac{4\pi^2 r}{T^2}$$

$$T = \sqrt{\frac{4\pi^2 r}{G\frac{M_{Sun}}{r^2} + g}}$$

$$= \sqrt{\frac{4\pi^2\left(1.50 \times 10^{11}m\right)}{\left(6.67 \times 10^{-11}N\cdot m^2/kg^2\right)\frac{(1.99\times 10^{30}kg)}{(1.50\times 10^{11}m)^2} + 9.80m/s^2}}$$

$$= 7.77 \times 10^5 s = 8.99 \text{ days}$$

The force of the Sun is only about 1/1600 the size of the normal force. The force of the Sun could have been ignored in the calculation with no significant change in the result given above.

EXAMPLE 77. When an object is in a circular orbit of radius r around the earth (mass M), the period of the orbit is T, given by Eq.1.125, and the orbital speed is v, given by Eq. 1.120. Show that when the object is moved into a circular orbit of slightly larger radius $r + \Delta r$, where $\Delta r \ll r$, its new period is $T + \Delta T$ and its new orbital speed is $v - \Delta v$, where $\Delta r, \Delta T$, and Δv are all positive quantities and

$$\Delta T = \frac{3\pi \Delta r}{v} \quad \text{and} \quad \Delta v = \frac{\pi \Delta r}{T}$$

APPROACH In Eqs. 1.125 and 1.120 replace T by $T + \Delta T$ and r by $r + \Delta r$. Use the expression $(1+x)^n \approx 1 + nx$, valid for $|x| << 1$.

SOLUTION The earth has $M = 5.97 \times 10^{24} kg$ and $R = 6.38 \times 10^6 m$. $r = h + R$, where h is the altitude above the surface of the earth.

$$T = \frac{2\pi r^{3/2}}{\sqrt{GM}} \text{ therefore}$$

$$T + \Delta T = \frac{2\pi}{\sqrt{GM}}(r + \Delta r)^{3/2} = \frac{2\pi r^{3/2}}{\sqrt{GM}}\left(1 + \frac{\Delta r}{r}\right)^{3/2}$$

$$\approx \frac{2\pi r^{3/2}}{\sqrt{GM}}\left(1 + \frac{3\Delta r}{2r}\right) = T + \frac{3\pi r^{1/2}\Delta r}{\sqrt{GM}}$$

Since $v = \sqrt{\frac{GM}{r}} = \left(\sqrt{GM}\right)r^{-1/2}$, $\Delta T = \frac{3\pi \Delta r}{v}$. $v = \sqrt{GM}r^{-1/2}$, and therefore

$$v - \Delta v = \sqrt{GM}(r + \Delta r)^{-1/2}$$

$$= \sqrt{GM}r^{-1/2}\left(1 + \frac{\Delta r}{r}\right)^{-1/2} \text{ and}$$

$$\approx \sqrt{GM}r^{-1/2}\left(1 - \frac{\Delta r}{2r}\right) = v - \frac{\sqrt{GM}}{2r^{3/2}}\Delta r.$$

$$\Rightarrow \quad \Delta v = \frac{\sqrt{GM}}{2r^{3/2}}\Delta r$$

Since, $T = \frac{2\pi r^{3/2}}{\sqrt{GM}}$, therefore, $\Delta v = \frac{\pi \Delta r}{T}$.

Key Points

1. **Essential Condition's for Satellite Motion**

   (a) Centre of satellite's orbit should coincide with centre of the Earth.

   (b) Plane of orbit of satellite should pass through center of the Earth.

2. There is only one speed that a satellite can have if the satellite is to remain in an orbit with a fixed radius.

3. For a given orbit, a satellite with a large mass has exactly the same orbital speed as a satellite with a small mass.

4. Time period of a satellite very close to earth's surface

   $(r \approx R)$ is, $\qquad T = 2\pi \sqrt{\dfrac{R}{g}} = 84.6 \text{ min}$

1.17 Energy Considerations in Planetary and Satellite Motion

Consider an object of mass m moving with a speed v in the vicinity of a massive object of mass M, where $M >> m$. The system might be a planet moving around the Sun, a satellite in orbit around the Earth, or a comet making a one-time flyby of the Sun. If we assume the object of mass M is at rest in an inertial reference frame, the total mechanical energy E of the two-object system when the objects are separated by a distance r is the sum of the kinetic energy of the object of mass m and the potential energy of the system:

$$E = K + U, \text{ i.e.,}$$

$$E = \frac{1}{2}mv^2 - \frac{GMm}{r} \tag{1.127}$$

Equation 1.127 shows that E may be positive, negative, or zero, depending on the value of v. For a bound system such as the Earth-Sun system, however, E is necessarily less than zero because we have chosen the convention that $U \to 0$ as r → ∞.

We can easily establish that $E < 0$ for the planet-satellite system:

Kinetic Energy of a Satellite

The kinetic energy of a satellite in it's orbit is given by

$$K = \frac{1}{2}mv_0^2, \text{ where } v_o = \sqrt{\frac{GM}{r}} \text{ (from Eq.1.120),}$$

$$\text{or} \qquad \boxed{K = \frac{GMm}{2r}} \tag{1.128}$$

Potential Energy

The Gravitational potential energy of the planet satellite system is

$$\boxed{U = -\frac{GMm}{r}} \tag{1.129}$$

Here, $-ve$ sign indicates that the satellite is subjected to planet's gravitational field.

Total Energy of a Planet-Satellite System

The total mechanical energy of the satellite is equal to the sum of its kinetic and potential energies.

Total energy: $E = K + U$

$$\text{or} \qquad E = \left(\frac{GMm}{2r}\right) + \left(-\frac{GMm}{r}\right)$$

$$\text{or} \qquad \boxed{E = -\frac{GMm}{2r}} \tag{1.130}$$

The total energy is negative because the satellite is bound by the planet's gravitational field.

On comparing Eq.(1.128), 1.129 and 1.130, we can write:

$$\boxed{E = -K = \frac{U}{2}} \tag{1.131}$$

So, from Eq.(1.130), and (1.131) we see that, the total mechanical energy E in a circular orbit is negative and equal to one-half the potential energy. Therefore, increasing the orbit radius r means decreasing the magnitude of negative mechanical energy, i.e., increasing the mechanical energy (or, making E less negative). If the satellite is in a relatively low orbit that encounters the outer fringes of earth's atmosphere, mechanical energy decreases due to negative work done by the force of air resistance; as a result, the orbit radius decreases until the satellite hits the ground or burns up in the atmosphere. Fig.1.128 shows energy vs r graph for a satellite orbiting around a planet in a circular orbit.

We have talked mostly about earth satellites, but we can

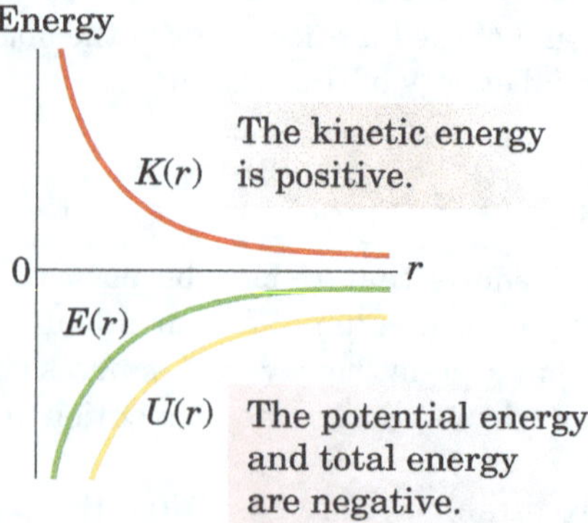

Figure 1.128: The variation of kinetic energy K, potential energy U, and total energy E with radius r for a satellite in a circular orbit. For any value of r, the values of U and E are negative, the value of K is positive, and $E = -K$. As $r \to \infty$, all three energy curves approach a value of zero.

apply the same analysis to the circular motion of any object under its gravitational attraction to a stationary object.

EXAMPLE 78. An artificial satellite moving in a circular orbit around the earth has a total (kinetic + potential) energy E_0. Find, it's potential energy.

APPROACH From Eq.1.131, the gravitational potential energy and total energy of a satellite, are related as:
$$E_0 = U/2.$$
SOLUTION Total energy, $E_0 = U/2 \Rightarrow U = 2E_0$
Since, E_0 is $-ve$ for bounded satellite, therefore, $U < E_0$.

EXAMPLE 79. In Fig. 1.129, two satellites, A and B, both of mass $m = 125$ kg, move in the same circular orbit of radius $r = 7.87 \times 10^6 m$ around Earth but in opposite senses of rotation and therefore on a collision course. (a) Find the total mechanical energy $E_A + E_B$ of the two satellites + Earth system before the collision. (b) If the collision is completely inelastic so that the wreckage remains as one piece of tangled material (mass $= 2m$), find the total mechanical energy immediately after the collision. (c) Just after the collision, is the wreckage falling directly toward Earth's center or orbiting around Earth?

(a) APPROACH For total energy of a satellite use Eq. 1.130:

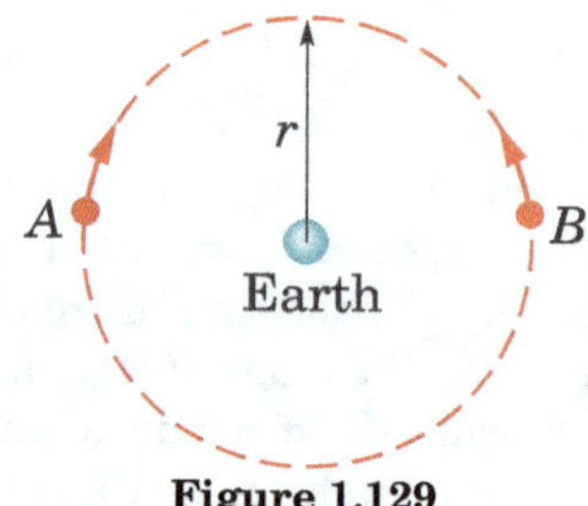

Figure 1.129

$$E = -\frac{GMm}{2r}$$
Net energy of two satellites before collision will be equal to the sum of their individual energies.

SOLUTION (a) From Eq. 1.130, we see that the energy of

each satellite is $-GMm/2r$. The total energy of the two satellites is twice that result:
$$E = E_A + E_B = -\frac{GMm}{r}$$
$$= -\frac{\left(6.67 \times 10^{-11} m^3/kg \cdot s^2\right)\left(5.98 \times 10^{24} kg\right)(125 kg)}{7.87 \times 10^6 m}$$
$$= -6.33 \times 10^9 J$$

(b) APPROACH Since both satellites (each of mass m) are revolving in opposite sense around the earth and their collision is completely inelastic, therefore, there is a complete loss of kinetic energy into thermal energy. So, net energy of the wreckage will be equal to the sum of their gravitational potential energies only.

SOLUTION Since, the collision is completely inelastic, therefore, the speed of the wreckage will be zero (just after the collision), so it has no kinetic energy at that moment. Replacing m with $2m$ in the potential energy expression (Eq.1.129), we therefore find the total energy of the wreckage at that instant is

$$E = U = -\frac{GM(2m)}{2r}$$
$$= -\frac{\left(6.67 \times 10^{-11} m^3/kg \cdot s^2\right)\left(5.98 \times 10^{24} kg\right) 2(125 kg)}{2\left(7.87 \times 10^6 m\right)}$$
$$= -6.33 \times 10^9 J$$

(c) An object with zero speed at that distance from Earth will simply fall toward the Earth, its trajectory being toward the center of the planet.

1.17.0.1　Binding Energy of Earth-Satellite System

It is defined as the minimum energy required to be given to a satellite to escape out from the gravitational influence of the Earth. Since the total energy of a satellite is given by Eq:1.130 $E = -\frac{GMm}{2r}$, in order to free the satellite from the orbit, it should be given an energy of $+\frac{GMm}{2r}$ so that the total energy becomes zero and the satellite is free from attractive force of the Earth. Thus,

Binding energy, $E_B = -$ Total energy of the system

or
$$\boxed{E_B = -\left[-\frac{GMm}{2r}\right] = \frac{GMm}{2r}}$$

$$(1.132)$$

Eq.1.132 is required expression of binding energy of Earth-satellite system.

1.17.0.2　Angular Momentum of Satellite

The angular momentum of an object, about a fixed point O, moving with velocity $\vec{v}$ having position vector $\vec{r}$ with respect to O is given by
$$\vec{L} = m\left(\vec{r} \times \vec{v}\right)$$
It's magnitude is given by:
$$L = mrv \sin\theta$$
here, $\vec{\theta}$ is the angle between $\vec{r}$ and $\vec{v}$.

For a satellite, moving in a circular orbit around the earth, $\theta = \pi/2$

Therefore, the angular momentum of the satellite, about an axis passing through the centre of earth, is given by

$$L = mv_o r \qquad (1.133)$$

here, $v_0 = \sqrt{\dfrac{GM}{r}}$ is the orbital speed of the satellite. Substituting this value in Eq.1.133, we get-

$$\boxed{L = m(\sqrt{GMr})} \qquad (1.134)$$

1.17.1 Escape Speed

It is the minimum speed required to escape a body from the gravitational field of a planet.

When an object is projected, from a planet, with the escape speed, v_e, it reaches to infinite distance ($r = \infty$) away from the planet with zero speed; so, the net mechanical energy of the object-planet system at infinite separation will be zero, i.e., $E_f = K + U = 0$.

Note that ∞, of course, is not a place, so a statement like this means that we want the object's speed to approach $v = 0$ asymptotically as $r \to \infty$.

Initial energy of the object-Earth system, when the object was at the surface of Earth:

$$E_i = \frac{1}{2}mv_e^2 - \frac{GMm}{R}$$

Here, M, m are the masses of spherical planet and the object respectively; R is the radius of the spherical planet.

Using mechanical energy conservation,

$$E_f = E_i$$
$$0 = \frac{1}{2}mv_e^2 - \frac{GMm}{R}$$

or $\qquad\boxed{v_e = \sqrt{\dfrac{2GM}{R}}} \qquad (1.135)$

If the planet is Earth, then $GM = gR^2$, therefore, Eq.1.135 can be written as

or $\qquad\boxed{v_e = \sqrt{2gR}} \qquad (1.136)$

Now, substitution of $R = 6.37 \times 10^6$ m and $M = 5.97 \times 10^{24}$ kg in above equation, yields

$$v_e = 11,200 \; m/s = 11.2 \; km/s \approx 25,000 \; \text{miles/h}.$$

Note that v_e does not depend on the direction in which a projectile is fired from the Earth. However, attaining that speed is easier if the projectile is fired in the direction of Earth's tangential surface speed i.e. eastward. For example, in India, rockets are launched eastward at Sriharikota to take the advantage of this "free" contribution of Sriharikota's eastward speed of 1500 km/h because of Earth's rotation.

On comparing Eq.1.123 to the escape speed obtained in Eq.1.136, we see that the escape speed from a spherical object with radius R is $\sqrt{2}$ times greater than the speed of a satellite in a circular orbit at that radius. If our spacecraft is in circular orbit around any planet, we have to multiply our speed by a factor of $\sqrt{2}$ to escape to infinity, regardless of the planet's mass.

Table 1.4: Escape Speeds from the Surfaces of the Planets, Moon, and Sun

Planet	v_e(km/s)
Mercury	4.3
Venus	10.3
Earth	11.2
Mars	5.0
Jupiter	60
Saturn	36
Uranus	22
Neptune	24
Moon	2.3
Sun	618

From Eq.1.135, it is clear that, the escape speed depends on:
(i) mass (M) and radius (R) of the planet,
(ii) the position from where the particle is projected.
and the escape speed does not depend on:
(i) Mass of the body (m) which is projected.
(ii) Angle of projection.

Table 1.5 shows the conditions for a projectile moving in bounded or unbounded orbits:

☞ If the velocity of a particle is less than v_e, then total mechanical energy of particle-earth system is negative and it cannot escape to infinity.

☞ If the velocity of particle is more than v_e then total mechanical energy will be positive. Even at infinity the particle possess some kinetic energy and speed. Although its potential energy becomes zero.

In derivation of expression for escape speed (Eq.1.135 or 1.136), we have considered the planet as an isolated system which is generally not true. If we consider the effect of gravitational field of other stars or planets, the result would be different (see Example90, 89).

EXAMPLE 80. Calculate the escape speed for an object launched from the Moon and Sun. Given that, mass of Moon $M_m = 7.34 \times 10^{22}$ kg and radius of moon $R_m = 1.74 \times 10^6$ m; mass of Sun $M_S = 1.99 \times 10^{30}$ kg and radius of Sun $R_S = 1.50 \times 10^{11}$ m.

APPROACH Similar to Eq.1.135, the escape speed for any planet of mass M_P, radius R_P, is given by

$$v_e = \sqrt{\frac{2GM_P}{R_P}} \qquad \text{... (1)}$$

SOLUTION From Eq. (1), the escape speed at the surface of moon is given by

$$v_{em} = \sqrt{\frac{2GM_m}{R_m}}$$
$$= \sqrt{\frac{2(6.67 \times 10^{11} N.m^2/kg^2)(7.34 \times 10^{22} kg)}{1.74 \times 10^6 m}}$$
$$= 2370 \; m/s = 2.37 \; km/s$$

Similarly, the escape speed at the surface of Sun is given by

$$v_{es} = \sqrt{\frac{2GM_S}{R_S}}$$

$$= \sqrt{\frac{2(6.67 \times 10^{11} N.m^2/kg^2)(1.99 \times 10^{30} kg)}{1.50 \times 10^{11} m}}$$

$$= 42000 \ m/s = 42 \ km/s$$

Discussion of Result: The relatively low escape speed of the Moon means that it is much easier to launch a rocket into space from the Moon than from the Earth.

The Moon's low escape speed is the reason the Moon has no atmosphere. Even if you could magically supply the Moon with an atmosphere, it would soon evaporate into space because the individual molecules in the air move with rms speeds great enough to escape. On Earth, however, where the escape speed is much higher, gravity can prevent the rapidly moving molecules from moving off into space. Even so, light molecules, like hydrogen and helium, move faster at a given temperature than the heavier molecules like nitrogen and oxygen. For this reason, the Earth's atmosphere contains virtually no hydrogen or helium.

EXAMPLE 81. A body is projected vertically upwards from the surface of earth with a velocity sufficient to carry it to infinity. Calculate the time taken by it to reach height h.

APPROACH First of all obtain the expression for speed of the body at any general distance r from the centre of the earth by applying the principle of conservation of energy. Now replace speed at distance r by dr/dt and integrate for $r = R$ to $r = R + h$ and simplify dor required time t.

SOLUTION If at a distance r from the centre of the earth the body has velocity v, by conservation of mechanical energy,

$$\frac{1}{2}mv^2 + \left(-\frac{GMm}{r}\right) = \frac{1}{2}mv_e^2 + \left(-\frac{GMm}{R}\right)$$

or $\qquad v^2 = v_e^2 + \frac{2GM}{R}\left[\frac{R}{r} - 1\right]$

But as $v_e = \sqrt{2gR}$ and $g = (GM/R^2)$

$$v^2 = 2gR + 2gR\left[(R/r) - 1\right]$$

or $\qquad v = \sqrt{\frac{2gR^2}{r}}$, i.e., $\frac{dr}{dt} = R\frac{\sqrt{2g}}{\sqrt{r}}$

$$\Rightarrow \qquad \int_0^t dt = \frac{1}{R\sqrt{2g}}\int_R^{R+h} r^{1/2} dr$$

$$\Rightarrow \qquad t = \frac{2}{3}\frac{1}{R\sqrt{2g}}\left[(R+h)^{3/2} - R^{3/2}\right]$$

$$\Rightarrow \qquad t = \frac{1}{3}\sqrt{\frac{2R}{g}}\left[\left(1 + \frac{h}{R}\right)^{3/2} - 1\right]$$

EXAMPLE 82. If velocity given to an object from the surface of the Earth is n times the escape velocity then what will be the residual velocity at infinity.

APPROACH Since, there is only a gravitational conservative force on the object-Earth system and assuming no dissipative force is acting on the object, the total mechanical energy of the system will be conserved. Therefore, to get the residual speed at infinity, apply the mechanical energy conservation principle.

SOLUTION Let residual velocity be v then from energy conservation, we gave

Net energy of the object-Earth system for object at Earth = Net energy of the object-Earth system for object at infinity.

$$\frac{1}{2}m(nv_e)^2 - \frac{GMm}{R} = \frac{1}{2}mv^2 + 0$$

$$\Rightarrow v^2 = n^2 v_e^2 - \frac{2GM}{R} = n^2 v_e^2 - v_e^2 = (n^2 - 1)v_e^2$$

$$\Rightarrow \qquad v = \left(\sqrt{n^2 - 1}\right)v_e$$

EXAMPLE 83. A very small groove is made in the Earth (mass = M, radius = R), and a particle of mass m_0 is placed at $R/2$ distance from the centre[Fig.1.130a]. Find the escape speed of the particle from that place.

APPROACH When a particle has such kinetic energy that it just reaches the infinite with zero kinetic energy, then the projection speed of the particle is said to be escape speed. By applying the principle of conservation of mechanical energy between initial and final positions, you can determine the required escape speed at distance $R/2$ from the centre of the earth.

SOLUTION Suppose we project the particle with speed v_e, so that it just reaches at $(r \rightarrow \infty)$.

Applying the principle of conservation of mechanical energy,

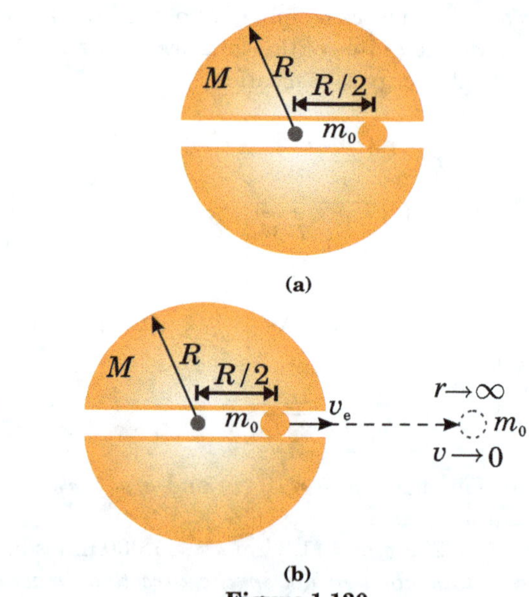

Figure 1.130

we get

$$K_i + U_i = K_f + U_f$$

$$\Rightarrow \frac{1}{2} m_0 v_e^2 + m_0\left[-\frac{GM}{2R^3}\left\{3R^2 - \left(\frac{R}{2}\right)^2\right\}\right] = 0$$

$$\Rightarrow \qquad v_e = \sqrt{\frac{11GM}{4R}}$$

EXAMPLE 84. Kinetic energy of a particle on the surface of earth, is E_0 and potential energy is $-E_0/2$.
(a) Will the particle escape to infinity?
(b) At some height its kinetic energy becomes $0.6E_0$. What is potential energy at this height ?
(c) If the particle escapes to infinity, what is kinetic energy of the particle at infinity?

APPROACH (a) Escaping of a particle from earth's surface means the minimum kinetic energy provided to it is just sufficient to send it to infinity. In this case, it reaches to infinity with zero velocity, i.e., with zero kinetic energy remaining.

We have already assumed that, net gravitational potential energy of a particle-earth system, is always zero at infinite. So, if a particle reaches to infinite with zero kinetic energy, then it's total mechanical energy (which is the sum of it's kinetic and potential energy) at infinite will be zero.

Thus, by conservation of mechanical energy, we can say that to escape a particle from earth's surface, the minimum required mechanical energy at the earth's surface must be zero.

(b) Since, the force acting on the particle is gravitational force, which is conservative in nature, therefore its total mechanical energy ($KE + PE$), will always be conserved. So, by applying the conservation of mechanical energy principle, you can find the required potential energy at the given height.

(c) By applying the principle of conservation of mechanical energy, you can find the kinetic energy of the particle at infinite, corresponding to total mechanical energy at earth's surface.

SOLUTION Total mechanical energy of the particle at earth's surface,

$$E = K + U = E_0 - \frac{E_0}{2} = 0.5E_0$$

Since, E is positive, therefore, the particle will escape to infinity.

(b) By conservation of mechanical energy:

Net mechanical energy of the earth-particle system, at the given height = net mechanical energy of the earth-particle system, when it was on the earth's surface.

i.e., $\quad 0.6E_0 + U = 0.5E_0$

$\Rightarrow \quad U = 0.5E_0 - 0.6E_0 = -0.1E_0$

So, the potential energy at the given height, $U = -0.1E_0$

(c) At infinity, the gravitational potential energy of earth-particle system is zero, i.e., $U_\infty = 0$,

By conservation of mechanical energy, we have

$$[E_{\text{total}}]_\infty = [E_{\text{total}}]_{\text{on earth's surface}}$$

or $\quad K + 0 = E_0 - \frac{E_0}{2}$

or $\quad K = 0.5\,E_0$

EXAMPLE 85. Kinetic energy of a particle on the surface of earth is E_0 and the potential energy is $-2E_0$.

(a) Will the particle escape to infinity?

(b) What is the value of potential energy where speed of the particle becomes zero?

APPROACH follow the approach given in "Example 84"

SOLUTION (a) Total mechanical energy of the earth-particle system, when the particle is on the surface of earth,

$$= E_0 - 2E_0 = -E_0$$

Since, it is negative, it will not escape to infinity.

(b) By conservation of energy, we have

$$E_i = E_f \Rightarrow E_0 - 2E_0 = 0 + U \Rightarrow U = -E_0$$

EXAMPLE 86. A spaceship is launched into a circular orbit close to the earth's surface. What additional speed has now to be imparted to the spaceship in the orbit to overcome the gravitational pull. Radius of earth $= 6400$ km, $g = 9.8$ m/s^2.

APPROACH If v_o is the orbital speed of spaceship near the earth's surface and v_e is the escape speed, then additional speed imparted to the spaceship in order to pull it from gravitational field of earth will be equal to the difference between escape speed and orbital speed, i.e.,

$$\Delta v = v_e - v_o$$

SOLUTION The orbital speed of the spaceship close to the earth's surface is given by,

$$v_o = \sqrt{gR}$$

and it's escape speed is

$$v_e = \sqrt{2gR}$$

So, the additional speed required to escape

$$v_e - v_o = \sqrt{2gR} - \sqrt{gR} = (\sqrt{2} - 1)\sqrt{gR}$$

On substituting the values of g and R in above relation, we get

$$v_e - v_o = 3.278 \times 10^3 \text{ m/s}$$

EXAMPLE 87. A planet of mass m moves along an ellipse around the Sun of mass M, so that its maximum and minimum distances from Sun are a and b respectively. Prove that the angular momentum L of this planet relative to the centre of the Sun is $L = m\sqrt{\frac{2GMab}{(a+b)}}$.

APPROACH Since, the gravitational force of Sun on the planet acts towards the centre of Sun, therefore the torque of gravitational force on the planet about the centre of the planet is zero. Thus, the angular momentum of planet about the centre of Sun will always be conserved.

As aphelion and perihelion distances are also given so you can make a relation between the magnitudes of velocities at these points by applying the principle of conservation of angular momentum.

The gravitational force is conservative in nature, therefore you can also make another relation between the magnitudes of velocities at aphelion and perihelion points by applying the principle of conservation of mechanical energy.

By simultaneously simplifying these relations, you can obtain the magnitudes of both velocities and hence the angular momentum.

SOLUTION Angular momentum at maximum distance from Sun = Angular momentum at minimum distance from Sun. i.e.,

$$mv_1 a = mv_2 b \Rightarrow v_1 = \frac{v_2 b}{a}$$

Now, on applying the mechanical energy conservation principle, we get

$$\frac{1}{2}mv_1^2 - \frac{GMm}{a} = \frac{1}{2}mv_2^2 - \frac{GMm}{b}$$

From above equations $v_1 = \sqrt{\frac{2GMb}{a(a+b)}}$

Angular momentum of planet $L = mv_1 a = ma\sqrt{\frac{2GMb}{(a+b)a}}$

EXAMPLE 88. A satellite is moving with a constant speed v in a circular orbit about earth. An object of mass m is ejected from the satellite such that it just escapes from the gravitational pull of the earth. Find the kinetic energy of the object at the time of it's ejection.

APPROACH A particle escapes from the gravitational pull if its total energy (E), i.e., sum of kinetic energy (K) and potential energy (U), is greater than or equal to zero. The condition for just escape is $E = K + U = 0$, i.e., $K = -U$

From Eq.1.129: $U = -\dfrac{GMm}{r}$

Therefore, $K = -U = \dfrac{GMm}{r}$ ... (1)

In a circular orbit of radius r, gravitational pull provides the required centripetal force, i.e., $mv^2/r = GMm/r^2$, which gives

$$r = GM/v^2 \qquad \ldots (2)$$

Now, substitute this value of r from Eq.(2) to Eq.(1), and simplify for required kinetic energy.

SOLUTION On substituting this value in Eq.(1), we get

$$K = \frac{GMm}{GM/v^2} = mv^2$$

EXAMPLE 89. A rocket is launched normal to the surface of the earth, away from the Sun, along the line joining the Sun and the earth. The Sun is 3×10^5 times heavier than the earth and is at a distance 2.5×10^4 times larger than the radius of the earth. The escape velocity from the earth's gravitational field is $v_e = 11.2$ km/s. Find the minimum approximate value of initial velocity (v_s) required for the rocket to be able to leave the Sun-earth system. [Ignore the rotation and revolution of the earth and the presence of any other planet.]

APPROACH The rocket (mass = m, say) is launched from the point P along the line joining the Sun and the earth as shown in Fig.1.131. The rocket escapes from the Sun-Earth system if its total energy (E) i.e., sum of kinetic energy (K) and potential energy (U) of the Sun-earth-rocket system, is greater than or equal to zero. The condition for just escape is $E = K + U = 0$ i.e.,

$$K = -U \qquad \ldots (1)$$

If v_s is the escape speed of rocket from point P, then

$$K = \frac{1}{2}mv_s^2$$

On substituting this values in Eq.(1), we get

$$\frac{1}{2}mv_s^2 = -U$$

or $$v_s = \sqrt{\frac{2(-U)}{m}} \qquad \ldots (2)$$

Now, determine U and substitute it in Eq.(2) and then simplify it for v_s.

SOLUTION Let M_e be the mass of the earth, $M_s = 3 \times 10^5 M_e$ be the mass of the sun, R_e be the radius of the earth and $r = 2.5 \times 10^4 R_e$ be the distance of the sun from the earth.

The potential energy of the rocket at the point P (due to the

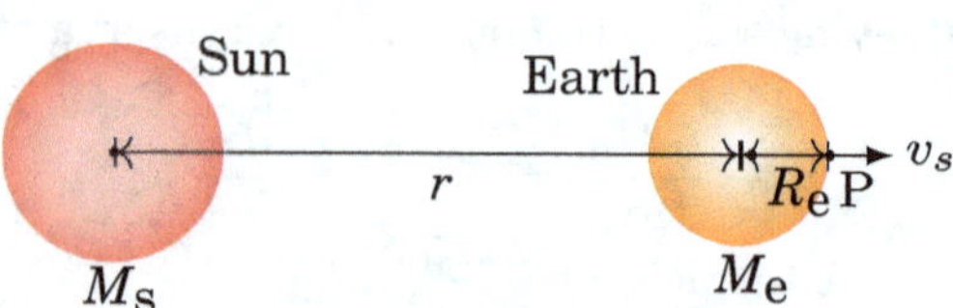

Figure 1.131

earth and the sun) is

$$U = -\frac{GM_e m}{R_e} - \frac{GM_s m}{r + R_e}$$

On substituting it in Eq.(2), we get

$$v_s = \sqrt{\frac{2GM_e}{R_e}\left(1 + \frac{\frac{M_s}{M_e}}{\frac{r}{R_e} + 1}\right)}$$

$$\approx 11.2\sqrt{1 + \frac{3 \times 10^5}{2.5 \times 10^4}} = 40.38 \text{ km/s}$$

where, we have, used $(r/R_e + 1) \approx r/R_e$ (since $r \gg R_e$) and $v_e = \sqrt{2GM_e/R_e} = 11.2$ km/s (escape velocity from the earth's gravitational field).

EXAMPLE 90. When we calculate escape speeds, we usually do so with the assumption that the object from which we are calculating escape speed is isolated. This is, of course, generally not true in the solar system. Show that the escape speed at a point near a system that consists of two stationary massive spherical objects is equal to the square root of the sum of the squares of the escape speeds from each of the two objects considered individually.

APPROACH The pictorial representation shows the two massive objects from which the object (whose mass is m), located at point P, is to escape. This object will have escaped the gravitational fields of the two massive objects provided, when its gravitational potential energy has become zero, its kinetic energy will also be zero.

SOLUTION Express the total energy of the system consist-

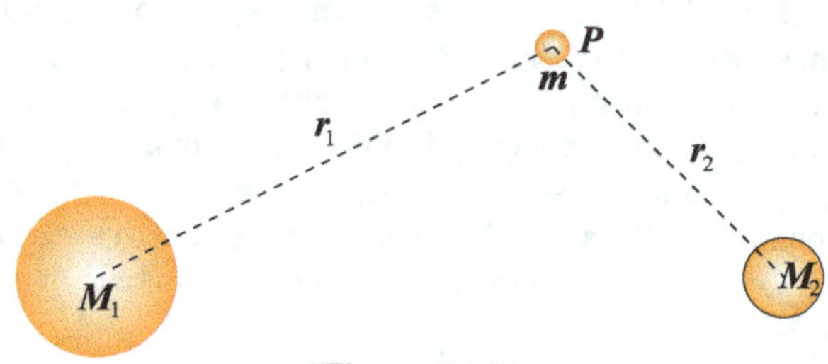

Figure 1.132

ing of the two massive objects and the object whose mass is m :

$$E = \frac{1}{2}mv^2 - \frac{GM_1 m}{r_1} - \frac{GM_2 m}{r_2} \qquad \ldots (1)$$

When the object whose mass is m has escaped, $E = 0$, therefore Eq.(1), gives

$$0 = \frac{1}{2}mv_e^2 - \frac{GM_1 m}{r_1} - \frac{GM_2 m}{r_2}$$

Solving for v_e yields:

$$v_e^2 = \frac{2GM_1}{r_1} + \frac{2GM_2}{r_2}$$

The terms on the right-hand side of the equation are the squares of the escape speeds from the objects whose masses are M_1 and M_2. Hence;

$$v_e^2 = v_{e,1}^2 + v_{e,2}^2$$

EXAMPLE 91. Two bodies, each of mass M, are kept fixed with a separation $2L$. A particle of mass m is projected from the midpoint of the line joining their centres, perpendicular to the line. Find the minimal initial velocity of the mass m to escape the gravitational field of the two bodies.

APPROACH Since, there are only gravitational forces on the particle therefore, mechanical energy of the system will be conserved. Also corresponding to minimum speed needed to escape the particle means it reaches at infinite with zero velocity. Keeping this point in mind, apply the law of conservation of mechanical energy and simplify for required speed of projection

SOLUTION Suppose, A and B are two points separated by distance $2L$. It's mid point is O. P is a point at a perpendicular distance x from O. Now, we consider the given bodies each of mass M at A and B respectively. Let us project a particle of mass m from point O with initial velocity u and suppose it reaches to P with velocity v [1.133].

The gravitational potential energy of the system and kinetic energy of the particle for the positions O and P are given by:

$$U_O = -\frac{2GMm}{L}, \qquad U_P = -\frac{2GMm}{\sqrt{L^2 + x^2}}.$$

$$K_O = \frac{1}{2}mu^2, \qquad K_P = \frac{1}{2}mv^2.$$

Gravitational force is conservative so, total mechanical en-

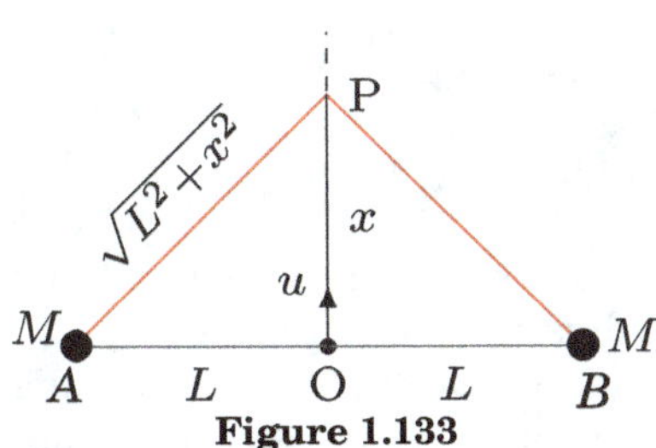

Figure 1.133

ergy of the system is conserved, i.e.,

$$K_O + U_O = K_P + U_P \qquad (1)$$

Substituting the values of all terms in above equation, we get-

$$\frac{1}{2}mu^2 - \frac{2GMm}{L} = \frac{1}{2}mv^2 - \frac{2GMm}{\sqrt{L^2 + x^2}} \qquad (2)$$

Now, if point P is at infinity, i.e., $x \to \infty$, then second term of RHS:

$$\lim_{x \to \infty} U_P = \frac{2GMm}{\sqrt{L^2 + x^2}} = 0$$

In this case, Eq. (2) gives:

$$\frac{1}{2}mu^2 - \frac{2GMm}{L} = \frac{1}{2}mv^2 \qquad (3)$$

If the projection speed ($u = u_{\min}$) of the particle at O, is just sufficient to reach it at infinite, then it's speed at infinity, will be zero, i.e., $v = 0$. So, to get minimum or escape speed at O, put $v = 0$ in Eq. (3), you get

$$\frac{1}{2}mu_{\min}^2 - \frac{2GMm}{L} = 0$$

Simplifying for $u_{\min}$ gives:

$$u_{\min} = 2\sqrt{\frac{GM}{L}}$$

Aliter: The gravitational binding energy of the particle at O is BE $= -U_O$. The particle will escape if $K_O \geq$ BE i.e., $K_O \geq -U_O$

EXAMPLE 92. A bullet is fired vertically upwards with velocity v from the surface of a spherical planet. When it reaches its maximum height, its acceleration due to the planet's gravity is $1/4^{\text{th}}$ of its value at the surface of the planet. Find the escape speed from the surface of planet. [ignore energy loss due to atmosphere]

APPROACH The escape speed from the surface of a planet of mass M and radius R is given by Eq.1.135:

$$v_e = \sqrt{\frac{2GM}{R}} \qquad \dots (1)$$

Here, right hand side of above equation is unknown, so we have to find it in terms of the given velocity (v) of the bullet. It is given that if a bullet is fired at velocity v in vertically upward direction, it reaches to a height where gravitational acceleration of the planet becomes $1/4^{\text{th}}$ of its value at the surface of the planet.

So by applying Eq.1.12:

$$g' = g\left(1 + \frac{h}{R}\right)^{-2}$$

or

$$\frac{g}{4} = g\left(1 + \frac{h}{R}\right)^{-2}$$

or

$$\left(1 + \frac{h}{R}\right)^2 = 4$$

or

$$\left(1 + \frac{h}{R}\right) = 2$$

or

$$h = R$$

Now, use the mechanical energy conservation between initial and final points (at final point, $h = R$) of bullet and then simplify it for, GM/R in terms of v.

Finally, substitute it in Eq.(1) to get v_e

SOLUTION At the maximum height, kinetic energy of the bullet becomes zero. Apply conservation of energy between the initial and the final (maximum height) positions of the bullet to get

$$-GMm/R + \frac{1}{2}mv^2 = -GMm/(R + h) + 0$$

Substituting $h = R$ in above expression gives

$$GM/R = v^2.$$

On substituting it in Eq.(1), we get

$$v_e = \sqrt{2(GM/R)} = \sqrt{2v^2} = v\sqrt{2}$$

EXAMPLE 93. Gravitational acceleration on the surface of a planet is $g\sqrt{6}/11$, where g is the gravitational acceleration on the surface of the earth. The average mass density of the planet is 2/3 times that of the earth. If the escape speed on the surface of the earth is taken to be 11 km/s, find the escape speed on the surface of the planet.

APPROACH Similar to Eq.1.136, the escape speed from the surface of a planet of mass M_p, radius R_p and gravity g_p, at it's surface, is given by

$$v_p = \sqrt{2g_p R_p} \qquad \dots (1)$$

Similar to Eq.1.6, the gravitational acceleration at the surface of the planet is given by

$$g_p = \frac{GM_p}{R_p^2} \qquad \dots (2)$$

The mass is related to average mass density of planet, ρ_p, by

$$M_p = (4/3)\pi R_p^3 \rho_p \qquad \dots (3)$$

On substituting it in equation (2) and simplifying for R_p, we get

$$R_p = 3g_p/(4\pi G \rho_p) \qquad \dots (4)$$

On substituting this value of R_p from Eq.(4) in Eq.(1), we get

$$v_p = \sqrt{2g_p 3g_p/(4\pi G \rho_p)}$$

$$v_p = g_p\sqrt{3/(2\pi G \rho_p)} \qquad \dots (5)$$

Similarly, for earth, we can write

$$v_e = g\sqrt{3/(2\pi G \rho_e)} \qquad \dots (6)$$

Divide Eq.(5), by Eq.(6) and substitute the given values of all the variables and then simplify for v_p.

SOLUTION On dividing Eq.(5) by Eq.(6), we get

$$\frac{v_p}{v_e} = \frac{g_p\sqrt{3/(2\pi G \rho_p)}}{g\sqrt{3/(2\pi G \rho_e)}}$$

or

$$\frac{v_p}{v_e} = \frac{g_p}{g}\sqrt{\frac{\rho_e}{\rho_p}} \qquad \dots (6)$$

It is also given that the acceleration at the surface of the planet

$$g_p = g\sqrt{6}/11, \quad \rho_p = 2\rho_e/3$$

On substituting these values in Eq.(6), we get

$$\frac{v_p}{v_e} = \frac{\sqrt{6}}{11}\sqrt{\frac{3}{2}} = \frac{3}{11} \qquad \ldots (7)$$

Given that $v_e = 11$ km/s

Therefore, Eq.(7), gives: $v_p = 11$ km/s

EXAMPLE 94. Two spherical planets P and Q have the same uniform density ρ, masses M_P and M_Q, and surface areas A and $4A$, respectively. A spherical planet R has uniform density ρ and its mass is $(M_P + M_Q)$. The escape speeds from the planets P, Q and R, are v_P, v_Q and v_R, respectively. Then establish a relation between them,

APPROACH Similar to Eq.1.135, the escape speed on a planet can be written as:

$$v_p = \sqrt{\frac{2GM_p}{R_p}} \qquad \ldots (1)$$

here, M_p, and R_p are the mass and radius respectively of the planet. G is universal gravitational constant.

So, in order to find escape speed, we need mass and radius of the planet. For it, we use the fact that all the three spherical planets have same uniform mass density (ρ), areas of planets P and Q are A and $4A$ respectively and also the mass of planet R is equal to sum of mass of planet P and that of Q.

SOLUTION Let r_P, r_Q, and r_R be the radii of the planet P, Q, and R, respectively. The fact that the area of Q is four times the area of P and densities of all three are same, gives

$$A_Q = 4A_P \quad \Rightarrow \quad 4\pi R_Q^2 = 4\left(4\pi R_P^2\right)$$

or $\qquad R_Q = 2R_P$

Now, the mass of planet P: $M_P = \rho\left(\frac{4}{3}\pi R_P^3\right)$

and the mass of planet Q: $M_Q = \rho\left(\frac{4}{3}\pi R_Q^3\right) = \rho\left(\frac{4}{3}\pi (2R_P)^3\right)$

$$= 8M_P \qquad (\because \quad R_Q = 2R_P)$$

Mass of planet R: $M_R = M_P + M_Q = M_P + 8M_P = 9M_P$

Again, $M_R = 9M_P \quad \Rightarrow \quad \frac{4}{3}\pi R_R^3 = 9\left(\frac{4}{3}\pi R_P^3\right) \quad \Rightarrow \quad R_R = \sqrt[3]{9}R_P$

To get the escape speeds from planets P, Q and R, we substitute these values in Eq.(1)

Escape speed from planet P:

$$v_P = \sqrt{\frac{2GM_p}{R_P}}$$

Escape speed from planet Q:

$$v_Q = \sqrt{\frac{2GM_q}{R_Q}} = \sqrt{\frac{2G(8M_P)}{2R_P}} = 2v_P$$

Escape speed from planet R:

$$v_R = \sqrt{\frac{2GM_R}{R_R}} = \sqrt{\frac{2G(9M_P)}{\sqrt[3]{9}R_P}} = \sqrt[3]{9}\,v_P$$

EXAMPLE 95. A particle is projected vertically upwards from the surface of earth (radius $= R$) with a kinetic energy equal to half of the minimum value needed for it to escape. Find the height to which it rises above the surface of earth.

APPROACH Apply conservation of mechanical energy and use the given condition

SOLUTION The expressions for escape velocity and minimum kinetic energy for a particle to escape from a planet of radius R and mass M are

$$v_e = \sqrt{2GM/R}, \quad K_{\min} = \frac{1}{2}mv_e^2 = GMm/R$$

Initial potential energy and kinetic energy of the projectile are

$$U_i = -\frac{GMm}{R}, \quad K_i = \frac{1}{2}K_{\min} = \frac{GMm}{2R}$$

At the maximum height h, the velocity of the projectile is zero. The potential and kinetic energy of the projectile at the maximum height are

$$U_f = -\frac{GMm}{R+h}, \quad K_f = 0$$

By conservation of energy, we have

$$U_i + K_i = U_f + K_f$$

$$-\frac{GMm}{R} + \frac{GMm}{2R} = -\frac{GMm}{R+h} + 0$$

or $\quad -\dfrac{GMm}{2R} = -\dfrac{GMm}{R+h} + 0 \quad \Rightarrow \quad h = R$

EXAMPLE 96. There is a crater of depth $R/100$ on the surface of moon (radius R)[Fig.1.134]. A projectile is fired vertically upward from the crater with velocity equal to the escape velocity v from the surface of the moon. Find the maximum height attained by the projectile.

APPROACH Apply the conservation of mechanical energy between initial and final position and simplify for required height.

SOLUTION Similar to Eq.1.101, the gravitational potential at a point inside a homogeneous spherical solid sphere of radius R, at a distance r $(<R)$:

$$V = -\frac{GM}{2R^3}\left[3R^2 - r^2\right] \qquad \ldots (1)$$

Therefore, the potential energy of the particle-sphere system when the particle of mass m placed at the position where the gravitational potential is V:

$$U = Vm = -\frac{GMm}{2R^3}\left(3R^2 - r^2\right)$$

From Eq.1.135, the escape speed from the surface is

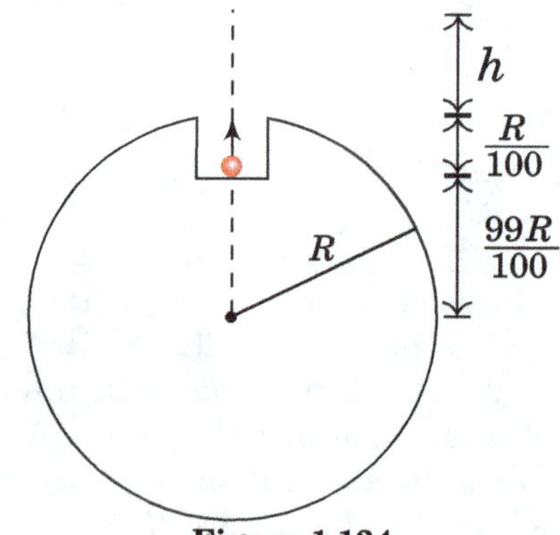

Figure 1.134

$$v_e = \sqrt{\frac{2GM}{R}}$$

and kinetic energy of the projectile of mass m at the point of projection is

$$K_i = \frac{1}{2}mv_e^2 = GMm/R$$

Therefore, the total energy of projectile at the point of projection is

$$E_i = U_i + K_i$$

$$= -\frac{GMm}{2R^3}\left[3R^2 - (99R/100)^2\right] + \frac{GMm}{R}.$$

At the maximum height h, kinetic energy, potential energy, and total energy of the projectile are

$$K_f = 0, \quad U_f = -\frac{GMm}{R+h}, \quad E_f = -\frac{GMm}{R+h} \text{ respectively.}$$

By conservation of mechanical energy, we have
$$E_i = E_f$$

$$-\frac{GMm}{2R^3}\left[3R^2 - (99R/100)^2\right] + \frac{GMm}{R} = -\frac{GMm}{R+h} \Rightarrow h = 99.5R$$

EXAMPLE 97. Distance between the centres of two stars is $10\,a$. The masses of the stars are M and $16\,M$ and their radii a and $2\,a$ respectively. A body of mass m is fired straight from the surface of the larger star towards the surface of the smaller star. What should be its minimum initial speed to reach the surface of the smaller star? Obtain the expression in terms of G, M and a.

APPROACH Since, both planets apply their gravitational force on the projected body, so it is not possible for the body to reach the smaller star with zero speed. In other words, it hits the smaller star with a definite speed. But here is a problem that we don't know the speed of projectile either on the surface of star of mass $16M$ or on the surface of star of mass M. So, the application of principle of conservation of mechanical energy between these two stars, only gives a relation between the unknown speeds of the body at the surfaces of stars of mass $16M$ and M. From this relation you cannot find the minimum speed of mass m on the surface of star of mass $16M$ so that it it reaches to the surface of star of mass M. So, it won't give any fruitful result. Now, we change our approach of thinking. We find the position of a point P, where the gravitational pull due to both stars is equal i.e., a zero gravitational field point. If we project the mass m with a velocity (from the surface of star of mass $16M$), such that it just reaches to P, with a very very small velocity ($v \to 0$) then beyond P, it is automatically attracted by the mass M. So, we apply the conservation of mechanical energy principle between the surface of star of mass $16M$ and the point P, where the the gravitational field due to both the stars becomes zero.

SOLUTION Let the gravitational attraction by the two stars is equal and opposite at the point P [Fig.1.135]. Suppose, point P is at a distance of r_1 from the centre of star of star of mass M and it is at distance of r_2 from the centre of star of mass $16M$.

The mass m should be projected with a velocity such that it

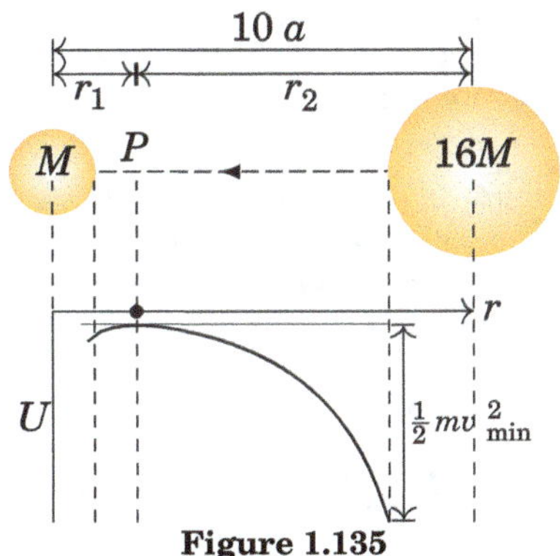

Figure 1.135

just reaches P, beyond which it is automatically attracted by the mass M. At P,
$$\frac{GMm}{r_1^2} = \frac{G(16M)m}{r_2^2}$$

On simplifying it, we get-
$$r_2 = 4r_1 \qquad \qquad \dots (1)$$
It is also given that-
$$r_1 + r_2 = 10a \qquad \qquad \dots (2)$$
On substituting Eq.(1) in Eq.(2), we get
$$r_1 = 2a \text{ and } r_2 = 8a$$
Let $v_{\min}$ be the minimum velocity required for mass m to reach P. The conservation of energy gives
$$\frac{1}{2}mv_{\min}^2 - \frac{16GMm}{2a} - \frac{GMm}{10a - 2a} = -\frac{16GMm}{r_2} - \frac{GMm}{r_1}$$
On substituting $r_1 = 2a$ and $r_2 = 8a$, we get
$$v_{\min} = \frac{3}{2}\sqrt{5GM/a}$$

EXAMPLE 98. An artificial satellite is moving in a circular orbit around the earth with a speed equal to half the magnitude of escape velocity from the earth.

(a) Determine the height of the satellite above the earth's surface.

(b) If the satellite is stopped suddenly in its orbit and allowed to fall freely onto the earth, find the speed with which it hits the surface of the earth.

APPROACH For both parts apply the law of conservation of mechanical energy

SOLUTION Consider a satellite of mass m in a circular orbit of radius $R + h$. In a circular orbit, the required centripetal force is provided by gravitational force, i.e.,
$$\frac{GMm}{(R+h)^2} = \frac{mv_o^2}{R+h} \qquad \qquad \dots (1)$$
The equation (1) gives the orbital speed of the satellite as
$$v_o = \sqrt{GM/(R+h)} \qquad \qquad \dots (2)$$
The escape velocity from the surface of earth is given by
$$v_e = \sqrt{2GM/R} \qquad \qquad \dots (3)$$
Given $v_o = \frac{1}{2}v_e$. Substitute v_o from equation (2) and v_e from equation 3 to get $h = R = 6400$ km. When satellite is stopped in the orbit, its kinetic and potential energy become
$$K_i = 0, \quad U_i = -\frac{GMm}{2R}$$
On reaching the earth's surface, kinetic and potential energies are
$$K_f = \frac{1}{2}mv^2, \quad U_f = -\frac{GMm}{R}$$
Applying the law of conservation of mechanical energy, we get
$$U_i + K_i = U_f + K_f$$
or $v = \sqrt{GM/R} = \sqrt{gR} = \sqrt{(9.8)(6400 \times 10^3)} = 7.9 \times 10^3$ m/s

1.17.1.1 Key Points

- Kinetic energy of a satellite, in it's orbit is-
$$K = \frac{1}{2}mv_0^2 = \frac{GMm}{2r} = \frac{L^2}{2mr^2}$$
here, $L = mv_0r$ is the angular momentum of satellite about the center of Earth.

- Potential energy the satellite,
$$U = -\frac{GMm}{r} = -mv_0^2 = -\frac{L^2}{mr^2}$$

- Total mechanical energy,

$$E = U + K = -\frac{mv_0^2}{2} = -\frac{GMm}{2r} = -\frac{L^2}{2mr^2}$$

- Binding energy of satellite,
$$BE = -E = \frac{mv_0^2}{2} = \frac{GMm}{2r} = \frac{L^2}{2mr^2}$$

- $v_e = \sqrt{\frac{2GM}{R}}$ (In terms of mass and radius). If $M =$ constant, then $v_e \propto \frac{1}{\sqrt{R}}$

- $v_e = \sqrt{2gR}$ (In terms of g and R). If $g =$ constant, then $v_e \propto \sqrt{R}$

- $v_e = R\sqrt{\frac{8\pi G \cdot \rho}{3}}$ (In form of density). If $\rho =$ constant $v_e \propto R$

- $v_e \propto m^0$ and $v_e \propto \theta$

- Escape speed at Earth's surface $v_e = 11.2$ km/s Moon surface $v_e = 2.31$ km/s

- Atmosphere on Moon is absent because root mean square velocity of gas particle is greater then escape velocity. $v_{\text{rms}} > v_e$

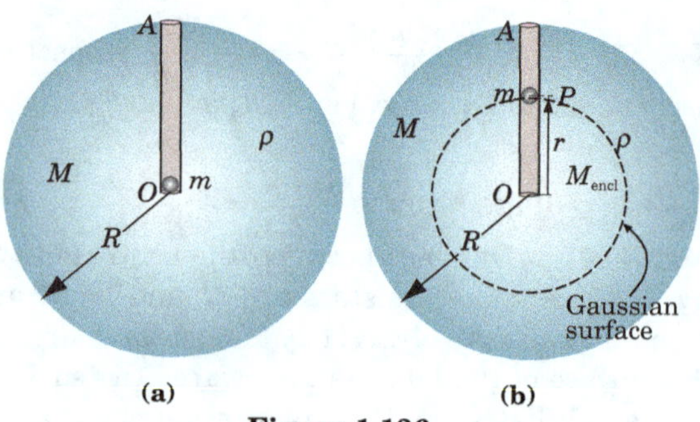

Figure 1.136

Why did the ISRO choose Sriharikota?
To take the advantage of eastward tangential surface velocity of earth, the launch site must be as close to the equator as possible. Now, in India-

1. Sriharikota has the advantage of having the sea to it's east which is also the direction of launch to take the advantage of the tangential surface velocity of Earth.

2. It has wide azimuth angle angle, so that we can launch spacecrafts to both polar and Geo orbits without danger of spent rockets stages impacting foreign soil.

3. 44000 acres space for doing all sort of dangerous activities like propellent handling, launcing.

EXAMPLE 99. A hole is drilled from the surface of Earth to its centre, as in Figure **??**. Ignore Earth's rotation and any effects due to air resistance, and model Earth as a uniform sphere. (a) How much work is required to lift a particle of mass m from the centre of Earth to Earth's surface? (b) If the particle is dropped from rest at the surface of Earth, what is its speed when it reaches the centre of Earth? (c) What is the escape speed for a particle projected from the center of Earth? Express your answers in terms of m, g, and R.

APPROACH Let r represent the separation of the particle from the centre of Earth and assume a uniform density for Earth[Fig.1.136b]. Since, we don't know the speed of projection of the particle from the centre, therefore, it is not advised to apply the law of conservation of mechanical energy or work-energy theorem. In this case, we use the definition of work in terms of force and displacement. The work required to lift the particle from the centre of Earth to its surface is the line integral of the gravitational force function. This function can be found from the law of gravity and by relating the mass of Earth enclosed within the spherical Gaussian surface concentric with the centre of earth and passing through the position of the particle, to Earth's mass. We can use the work-kinetic energy theorem to find the speed with which the particle, when released from the surface of Earth, will strike the centre of Earth. Finally, the energy required for the particle to escape Earth from the centre of Earth is the sum of the energy required to get it to the surface of Earth and the kinetic energy it must have to escape from the surface of Earth.

SOLUTION (a) Express the work required to lift the particle from the center of Earth to Earth's surface:

$$W = \int_0^R \vec{F} \cdot d\vec{r} = -\int_0^R \vec{F}_g \cdot d\vec{r} = \int_0^{R_E} F_g dr \quad \ldots (1)$$

where F_g is the gravitational force acting on the particle. Using the law of gravity, express the force acting on the particle as a function of its distance from the center of Earth:

$$F_g = \frac{GmM_{\text{encl}}}{r^2} \quad \ldots (2)$$

where M_{encl} is the mass of a sphere whose radius is r [Fig.1.136b].
Express the ratio of M_{encl} to M and simplify to obtain:

$$\frac{M_{\text{encl}}}{M} = \frac{\rho\left(\frac{4}{3}\pi r^3\right)}{\rho\left(\frac{4}{3}\pi R^3\right)} = \frac{r^3}{R^3} \Rightarrow M_{\text{encl}} = M\frac{r^3}{R^3}$$

Substitute for M_{encl} in equation (2) to obtain:

$$F_g = \frac{GmM}{R^3}r = \frac{mgR^2}{R^3}r = \frac{mg}{R}r$$

Substitute for F_g in equation (1) and evaluate the integral:

$$W = \frac{mg}{R}\int_0^R r dr = \frac{1}{2}gmR$$

(b) Use the work-kinetic energy theorem to relate the kinetic energy of the particle as it reaches the center of Earth to the work done on it in moving it to the surface of Earth:

$$W = \Delta K = \frac{1}{2}mv^2$$

Substituting for W yields:

$$\frac{1}{2}gmR = \frac{1}{2}mv^2 \Rightarrow v = \sqrt{gR}$$

(c) Express the total energy required for the particle to escape when projected from the centre of Earth:

$$E_{\text{esc}} = W + \frac{1}{2}mv_e^2 = \frac{1}{2}mv_e'^2$$

where v_e is the escape speed from the surface of Earth.
Substituting for W yields:

$$\frac{1}{2}gmR_E + \frac{1}{2}mv_e^2 = \frac{1}{2}mv_e'^2$$

or, simplifying,

$$gR + v_e^2 = v_e'^2$$

Because $v_e^2 = \dfrac{2GM}{R}$

Apply Newton's second law to an object of mass m at the surface of Earth to obtain:

$$mg = \frac{GMm}{R^2} \Rightarrow \frac{GM}{R} = gR$$

Substitute for GM/R in equation (3) to obtain:

$$gR + 2gR = v_e'^2 \Rightarrow v_e' = \sqrt{3gR}$$

Remarks: This escape speed is approximately 122% of the escape speed from the surface of Earth.

EXAMPLE 100.   Assume that there is a tunnel in the shape of a circular arc through the earth. Wall of the tunnel is smooth. A ball of mass m is projected into the tunnel at A with speed v. The ball comes out of the tunnel at B and escapes out of the gravity of the earth. Mass and radius of the earth are M and R respectively and radius of the circle shaped tunnel is also R.

(a) Find minimum possible value of v (call it v_0)

(b) If the ball is projected into the tunnel with speed v_0, calculate the normal force applied by the tunnel wall on the ball when it is closest to the centre of the earth. It is given that the closest distance between the ball and the centre of the earth is $R/2$.

APPROACH Since, non-conservative normal reaction N of the tunnel always acts perpendicular to the velocity of the ball, therefore, it's work done for each elementary displacement will always be zero and hence the law of conservation of mechanical energy is still valid. So, for part (a), apply conservation of energy between positions A and B. Since, the ball just escapes from B, therefore by mechanical energy conservation, it's total energy at every point is always zero.

For part (b) we first find speed of the ball at the centre of the tunnel by using the principle of conservation of energy and then by applying Newton's law of motion we can find the normal reaction of the wall.

SOLUTION (a) From energy conservation, it is easy to see that

$$v_0 = v_{\text{escape}} = \sqrt{\frac{2GM}{R}}$$

(b) From Eq.1.101, the gravitational potential at a distance $r = R/2$

$$V = -\frac{GM}{2R^3}\left[3R^2 - r^2\right]$$

$$= -\frac{GM}{2R^3}\left[3R^2 - \left(\frac{R}{2}\right)^2\right]$$

$$= -\frac{GMR^2}{2R^3}\left[3 - \frac{1}{4}\right] = -\frac{11}{8}\frac{GM}{R}$$

Speed of the ball (u, say) when it is at distance $R/2$ from the centre[Fig.1.137] can be obtained by applying the law of conservation of mechanical energy.

Net energy of the ball at C (Fig.1.137) = net energy required to escape the ball from B, i.e., $E_C = 0$

$$\Rightarrow \qquad \frac{1}{2}mu^2 - \frac{11}{8}\frac{GMm}{R} = 0$$

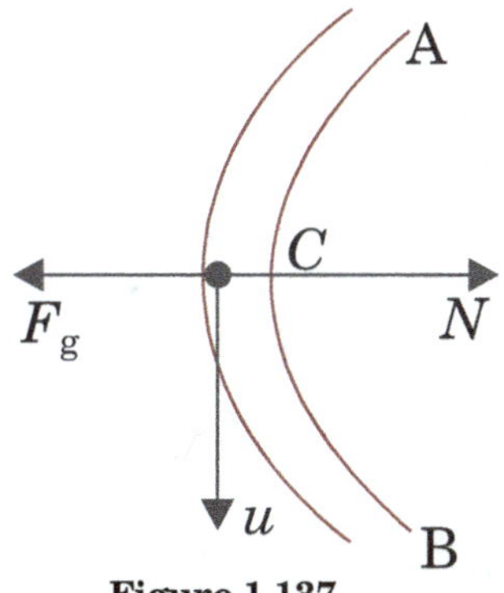

Figure 1.137

$$\Rightarrow \qquad u = \sqrt{\frac{11}{4}\frac{GM}{R}}$$

$$N - F_g = \frac{mu^2}{R}$$

or $\qquad N = \dfrac{mu^2}{R} + F_g = \dfrac{11}{4}\dfrac{GMm}{R^2} + \dfrac{GMm}{(R/2)^2}$

$$\Rightarrow \qquad N = \frac{27}{4}\frac{GMm}{R^2}$$

EXAMPLE 101. Imagine a smooth tunnel along a chord of nonrotating earth at a distance $\frac{R}{2}$ from the centre. R is the radius of the earth. A projectile is fired along the tunnel from the centre of the tunnel at a speed $v_0 = \sqrt{gR}$[g is acceleration due to gravity at the surface of the earth].

(a) Is the angular momentum [about the centre of the earth] of the projectile conserved as it moves along the tunnel?

(b) Calculate the maximum distance of the projectile from the centre of the earth during its course of motion.

APPROACH In part (a), observe whether the torque of all the forces acting on the ball about the centre of the earth is zero or not. For conservation of angular momentum about the centre of the earth, the net torque of all the forces acting on the projectile about the centre of the earth should be zero. In part (b), apply conservation of mechanical energy between points C and A and determine the speed of the projectile at A (Fig.1.138). When projectile comes out at A and reaches the out side space, the only force acting it is the gravitational and it is directed towards the centre of the earth. Hence torque of gravitational force about the centre of earth is zero and hence the angular momentum of the projectile when it comes out of A about the centre of earth is conserved. At maximum separation from earth, the projectile becomes parallel to earth and hence it's velocity vector becomes perpendicular to radius vector. By applying the conservation of angular momentum and mechanical energy conservation between this farthest point and A, you can find the farthest distance and the speed at farthest point.

SOLUTION (a) No. The tunnel wall applies a normal force which produces a torque about the centre.

(b) Let the speed of the projectile be v when it comes out of the tunnel [Fig.1.138].

 Since, work done by a non conservative normal force N, when the projectile was inside the tunnel, is zero (because $\vec{N}$ is perpendicular to displacement), therefore, net mechanical energy of earth projectile system will remain conserved. So, mechan-

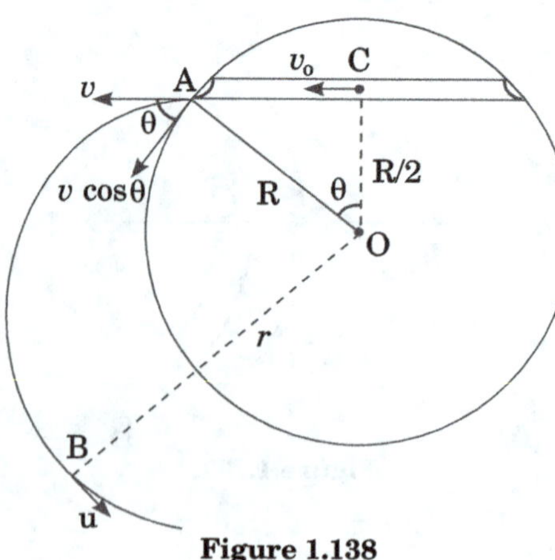

Figure 1.138

ical energy conservation can be applied between point C and A.

From Eq.1.101, the gravitational potential at any point at distance r, from the centre of earth is-

$$V = -\frac{GM}{2R^3}\left[3R^2 - r^2\right]$$

At point C, $r = R/2$, therefore

$$V_C = -\frac{GM}{2R^3}\left[3R^2 - \left(\frac{R}{2}\right)^2\right]$$

$$= -\frac{GM}{2R^3}R^2\left[\frac{11}{4}\right] = -\frac{11GM}{8R}$$

Therefore, gravitational potential energy of earth-projectile system for projectile's position at C, is-

$$U_C = V_C m = -\frac{11GMm}{8R}$$

By conservation of mechanical energy, we have

$$\therefore \quad K_A + U_A = K_C + U_C$$

$$\frac{1}{2}mv^2 - \frac{GMm}{R} = \frac{1}{2}m(\sqrt{gR})^2 - \frac{11GMm}{8R}$$

Simplifying after substituting $g = \frac{GM}{R^2}$, we get

$$v^2 = \frac{GM}{4R} \Rightarrow v = \frac{1}{2}\sqrt{\frac{GM}{R}}$$

When the projectile is at farthest distance its velocity u is perpendicular to its position vector r relative to the centre of the earth. Applying conservation of angular momentum between A and B we get-

$$mur = m(v\cos\theta)R \quad \Rightarrow \quad ur = \frac{vR}{2}$$

$$\Rightarrow \quad ur = \frac{1}{4}\sqrt{GMR} \qquad \qquad \dots (i)$$

Mechanical Energy conservation between A and B, gives

$$\frac{1}{2}mu^2 - \frac{GMm}{r} = \frac{1}{2}mv^2 - \frac{GMm}{R}$$

$$\Rightarrow \frac{1}{2}m\left(\frac{\sqrt{GMm}}{4r}\right)^2 - \frac{GMm}{r} = \frac{1}{2}m\frac{1}{4}\frac{GM}{R} - \frac{GMm}{R}$$

or

$$\frac{R}{32r^2} - \frac{1}{r} = -\frac{7}{8R}$$

$$\Rightarrow \quad 28r^2 - (32R)r + R^2 = 0$$

$$\therefore \quad r = \frac{32R \pm \sqrt{(32R)^2 - 28 \times 4R^2}}{56}$$

$$= \left(\frac{8 \pm \sqrt{57}}{14}\right)R$$

$$\therefore \quad r_{\max} = \left(\frac{8 + \sqrt{57}}{14}\right)R$$

The other solution is less than R and not acceptable.

EXAMPLE 102. A satellite is launched tangentially from point B which is at a height h above the surface of earth [Fig.1.139]-

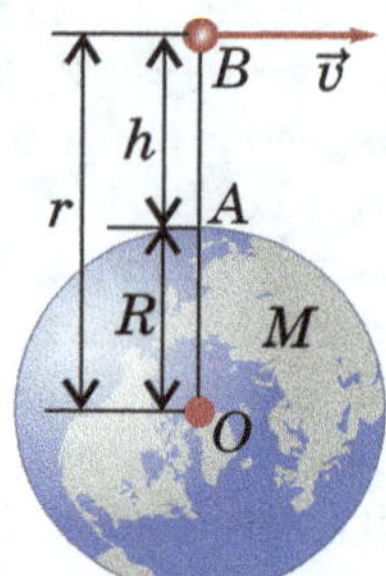

Figure 1.139

1. Find $v_{\min}$ so that it just touches the earth's surface
2. If $h = R$ and satellite is launched tangentially with speed $= \sqrt{\frac{3GM}{5R}}$, find the maximum distance of satellite from earth's centre.
3. If $h = R$ and satellite is launched tangentially with a speed $u = \sqrt{\frac{GM}{7R}}$. Find the angle with respect to vertical at which the satellite will crash on earth's surface.

APPROACH The problem is based on conservation of mechanical energy and conservation of angular momentum.

SOLUTION (I) The situation is shown in Fig.1.140. For minimum speed of projection the satellite will touch the earth at perihelion of the elliptical orbit at C. There is only a gravitational force on the satellite,. This force is directed towards centre of the Earth. So, the net torque on the satellite about the centre of earth will be zero. Hence, between points B and C (Fig.1.140), we can apply the principle of conservation of angular momentum about the centre of mass of Earth as:

$$mv(R + h) = mv'R$$

here, v' is the speed of satellite at point C.

Energy conservation between $r = R + h$ and $r = R$, gives:

$$\frac{-GMm}{R + h} + \frac{1}{2}mv^2 = -\frac{GMm}{R} + \frac{1}{2}mv'^2$$

On simplifying for v, we get,

$$v = \sqrt{\frac{2GMR}{(2R + h)(R + h)}}$$

(II) In this case, the height of point B from point A, directly below the surface of earth, is R. Now, point C changes it's position such that it's new distance from the centre of earth is $r_{\max}$. By conservation of angular momentum about the centre O of earth, we have-

$$mu(2R) = mvr_{\max}$$

By conservation of mechanical energy,

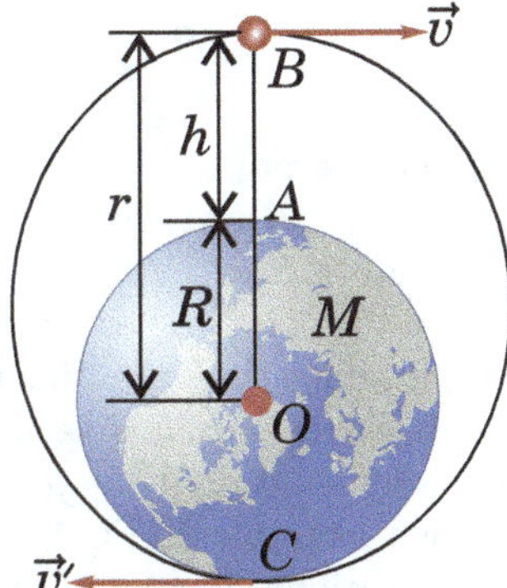

Figure 1.140

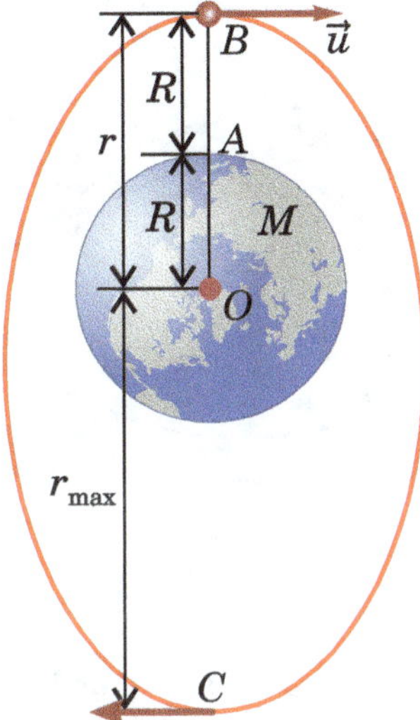

Figure 1.141

$$\frac{-GMm}{2R} + \frac{1}{2}mu^2 = \frac{-GMm}{r_{\max}} + \frac{1}{2}mv^2$$

$$\Rightarrow GMm\left(\frac{1}{2R} - \frac{1}{r_{\max}}\right) = \frac{1}{2}mu^2\left(1 - \left(\frac{2R}{r_{\max}}\right)^2\right)$$

$$\Rightarrow GM\left(\frac{1}{2R} - \frac{1}{r_{\max}}\right) = \frac{3GM}{10R}\left(1 - \frac{4R^2}{r_{\max}^2}\right)$$

$$\Rightarrow 2r_{\max}^2 - 10Rr_{\max} + 12R^2 = 0$$

$$\Rightarrow (r_{\max} - 2R)(r_{\max} - 3R) = 0$$

$$\Rightarrow r_{\max} = 3R$$

(III) Suppose, the satellite crashes at angle θ with vertical at point P(Fig.1.142). By conservation of mechanical energy between points B and P, we have-

$$\frac{1}{2}mu^2 - \frac{GMm}{2R} = \frac{1}{2}mv^2 - \frac{GMm}{R}$$

$$\Rightarrow \frac{1}{2}m(v^2 - u^2) = \frac{GMm}{2R}$$

$$v^2 = u^2 + \frac{GM}{R} \Rightarrow v = \sqrt{\frac{8GM}{7R}}$$

By conservation of angular momentum about the centre of earth, we have-

$$mu(2R) = mvR\sin\theta$$

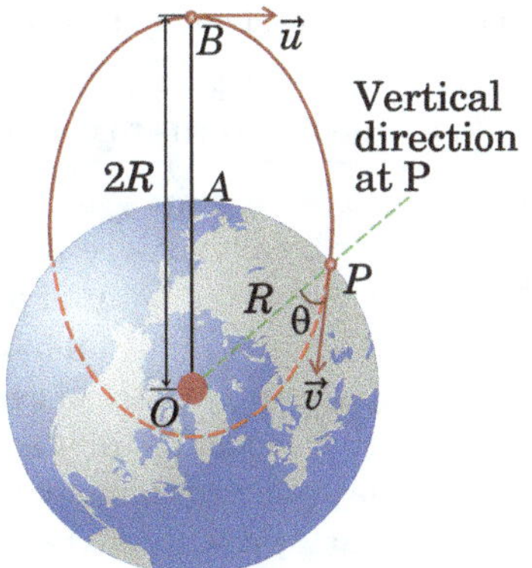

Figure 1.142

$$\Rightarrow \qquad \sin\theta = \frac{2u}{v}$$
$$\Rightarrow \qquad \theta = 45^o$$

From above analysis, we can say that the trajectory of a projectile above Earths surface depends on the velocity of projection $\vec{v}$. Table1.5 shows the trajectories of projectile for different velocities of projection [also see Fig.1.143]

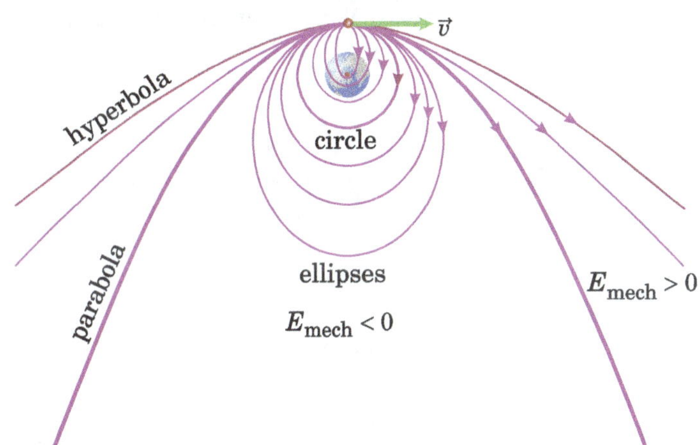

Figure 1.143: Orbits of an object launched at different speeds from a fixed location that is a fixed distance $r = R + h$ from Earth's centre.

1.18 Solid Angle

The solid angle is the extension of the concept of angle from two to three dimension. So let's start from $2D$: consider a circle(Fig.1.144) and pick two rays OA and OC starting from the centre O. They will divide the circumference in two parts ABC and ADC, called arcs. The length of each arc divided by the length of the radius will be the measure of the angle subtended by the arc itself, i.e.,

$$\text{angle} = \frac{\text{arc}}{\text{radius}}$$

From Fig.1.144, we have

$$\text{or} \qquad \boxed{\Delta\theta = \frac{\text{arcABC}}{AC} = \frac{\Delta s}{r}} \qquad (1.137)$$

Table 1.5: Conditions for bounded and unbounded trajectories of a projectile

1.	$v < \sqrt{\dfrac{GM}{r}\left(\dfrac{2R}{r+R}\right)}$	Body returns to earth
2.	$\sqrt{\dfrac{GM}{r}} > v > \sqrt{\dfrac{GM}{r}\left(\dfrac{2R}{r+R}\right)}$	Body acquires an elliptical orbit with earth as the far-focus with respect to the point of projection.
3.	$v < v_o$	The projectile follows an elliptical path with centre of earth as the farther focus. In this case, if projectile is projected from near the surface of earth, it will hit the earth's surface without completing the orbit.
4.	$v = v_o$	The projectile follows a circular orbit with the centre of earth as the centre of orbit.
5.	$v_o < v < v_e$	The projectile follows an elliptical orbit with the centre of earth as the focus nearer to the point of projection.
6.	$v = v_e$	The projectile escapes from the field of earth along a parabolic trajectory.
7.	$v > v_e$	The projectile escapes the field of earth along a hyperbolic trajectory.

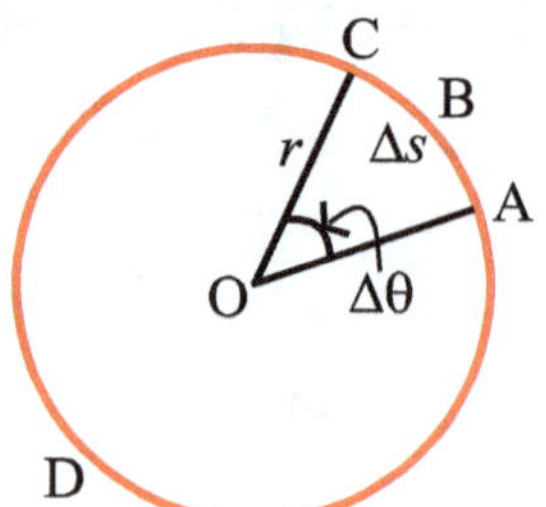

Figure 1.144: Plane angle

Here, Δs is the arc length and r is the radius of the circle. The SI unit of plane angle is "radian (rad)".

Now, extend this idea to three dimensions. Instead of a circle take a sphere, and instead of picking two rays pick a cone centred in the centre of the sphere [Fig.1.145]. The cone will cross the surface of the sphere: and now to define the *solid angle* measure the area (ΔA) of the surface bounded by the cone, divided by the square of the length of the radius (R^2) (so that we have an area divided by an area).

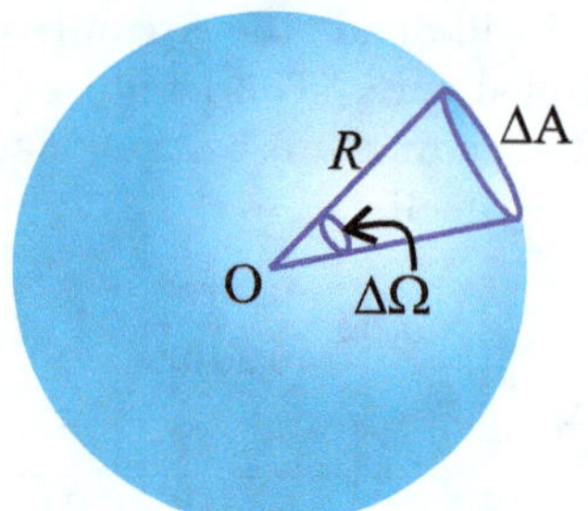

Figure 1.145: A solid angle measured on a sphere.

i.e., solid angle, $$\Delta\Omega = \frac{\Delta A}{R^2} \tag{1.138}$$

The unit of solid angle is steradian (sr).

Here, it is important to note that the line joining center of sphere O to ΔA is normal to ΔA.

Since, plane angles and the solid angles are the ratios of same physical quantities, therefore they are dimensionless quantities. Note that a small surface area as seen from a short distance can cover the same solid angle as a large area as seen from a long distance. For example, in Fig.1.146 different areas A and $A'(A' > A)$, at positions R, R' covers same solid angle at O.

Definition: *The solid angle is a three dimensional angle sub-*

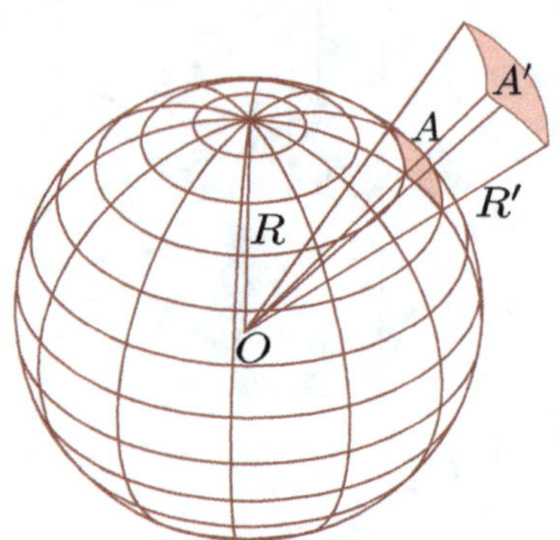

Figure 1.146

tended by an object (two dimensional or three dimensional) at a certain point in the space. It only depends on the relative distance of the object and its configuration with respect to the given point in the space. Solid angle subtended by a straight line or a point is always zero.

In Fig.1.147, we see that the projection of the area element

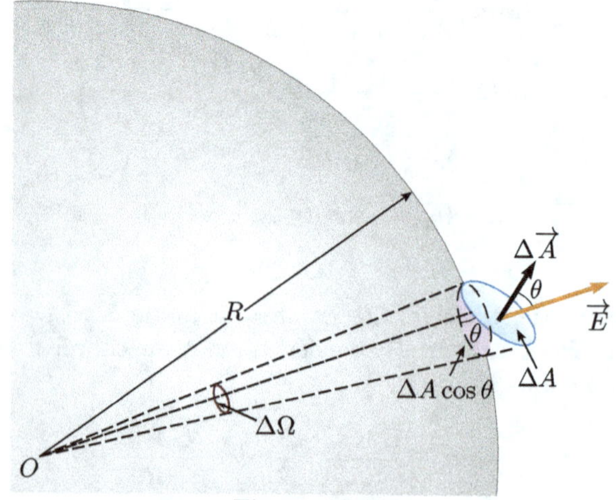

Figure 1.147

perpendicular to the radius vector is $\Delta A \cos\theta$. Thus, the quantity $\Delta A \cos\theta/R^2$ is equal to the solid angle $\Delta\Omega$ that the surface element ΔA subtends at the *origin O*. We also see that $\Delta\Omega$ is equal to the solid angle subtended by the area element of a spherical surface of radius R.

Note: *If the line joining O to ΔA makes an angle θ with the* *normal to* ΔA (Fig.1.147), we should write

$$\Delta\Omega = \frac{\Delta A \cos\theta}{R^2}$$

A complet circle subtends an angle -

$$\theta = \frac{\Delta s}{r} = \frac{2\pi r}{r} = 2\pi \text{ rad}$$

at the centre. In fact, any closed curve subtends an angle 2π at any of the internal points. Similarly, a complete sphere subtends a solid angle-

$$\Omega = \frac{A}{R^2} = \frac{4\pi R^2}{R^2} = 4\pi \text{ sr}$$

at the centre. Also, any closed surface subtends a solid angle of 4π sr at any internal point.

So for, we have defined the solid angles at an internal point. Now, we find the plane and solid angles at an external point.

From Fig.1.148, it is clear that on gradually closing the

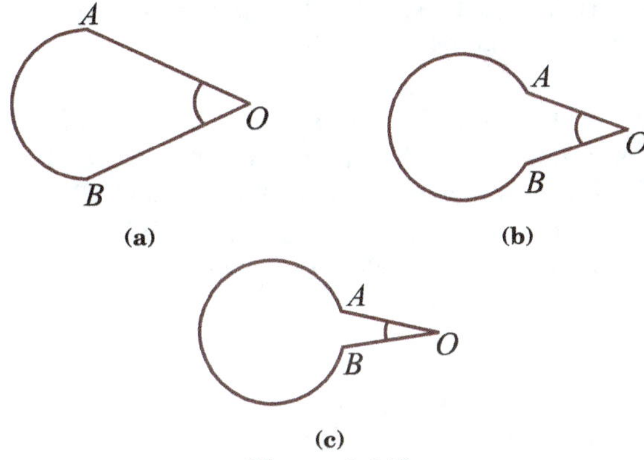

Figure 1.148

curve, the angle subtended by AB at an external point O, finally diminishes to zero. So, we can say that a closed curve subtends zero angle at an external point. Similarly, a closed surface also subtends zero solid angle at an external point.
Calculation of Solid Angle at the Centre of a Sphere:
Mathematically, the solid angle can be defined as-

$$\Omega = \frac{A}{R^2}$$

The definition of a solid angle Ω. is analogous to the definition of a plane angle. Just as the arc length s is everywhere perpendicular to the radius r, the area A must be everywhere perpendicular to the radius. The SI unit of solid plane angle is "steradian (sr)".

Fig.1.149 shows a sphere of radius R. Let us consider a conical section AOF having vertex at O. Suppose semi vertex angle of this cone is $\angle AOC = \theta$ and radius of its base is r. Now, consider a strip $ABEF$ of radius r on this sphere. If $\angle AOB = d\theta$, then thickness of the strip $AB = Rd\theta$ and radius of strip, $r = R\sin\theta$.
Curved area of this strip is given by -

$$dA = 2\pi r \times (\text{thickness } AB)$$

$\because \quad r = R\sin\theta$, therefore -

$$dA = 2\pi R \sin\theta \, R \, d\theta$$

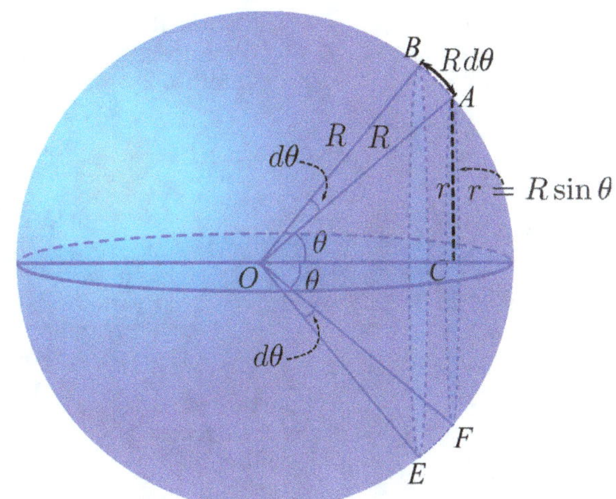

Figure 1.149

$$\Rightarrow \quad A = 2\pi R^2 \int_0^\theta \sin\theta \, d\theta$$

$$\Rightarrow \quad A = 2\pi R^2 (1 - \cos\theta)$$

$$\therefore \quad \Omega = \frac{A}{R^2} = \frac{2\pi R^2 (1 - \cos\theta)}{R^2} = 2\pi (1 - \cos\theta)$$

$$\Rightarrow \quad \boxed{\Omega = 2\pi (1 - \cos\theta) = 4\pi \sin^2 \frac{\theta}{2}} \qquad (1.139)$$

Eq.(1.139) gives the relation between semi vertex plane angle and the solid angle formed at the vertex of cone having its vertex at the centre of the sphere of radius R.

At $\quad \theta = 0°, \quad \Omega = 2\pi (1 - \cos 0°) = 0 \text{ sr}$

For a hemispherical surface, at it's center we have-

$$\theta = 90°, \quad \Omega = 2\pi (1 - \cos 90°) = 2\pi \text{ sr}$$

For a complete spherical surface, at it's center, we have-

$$\theta = 180°, \quad \Omega = 2\pi (1 - \cos 180°) = 4\pi \text{ sr}$$

Not only a spherical surface, all closed surfaces subtend a solid angle of $= 4\pi$ sr at their centres.

1.18.1 Communication Satellites

A communications satellite is a type of artificial satellite that is placed in Earth's orbit for the purpose of sending and receiving communication data between a source and destination. It is used to provide data communication and relaying services for televisions, radio, telecommunication, weather and Internet services.

Figure-1.150 shows as to how using satellites an information from an earth station, located at a point on earth's surface can be sent throughout the world.

First the information is sent to the nearest satellite in the range of earth station by means of electromagnetic waves then that satellite broadcasts the signal to the region of earth exposed to this satellite and also send the same signal to other satellite for broadcasting in other parts of the globe. Communications satellites are vital for remote areas that do

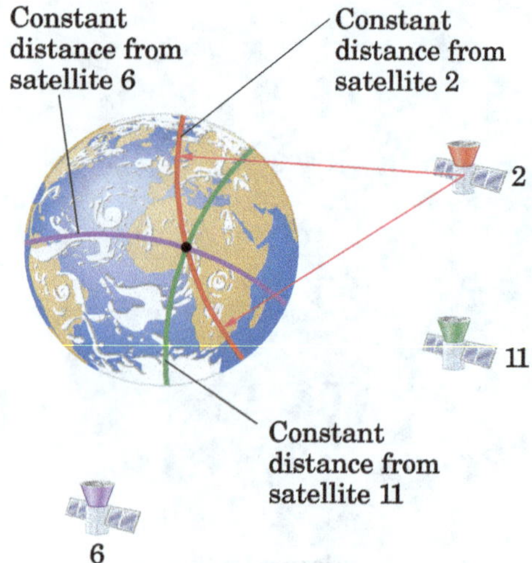

Figure 1.150: The Global Positioning System A system of 24 satellites in orbit about the Earth makes it possible to determine a person's location with great accuracy. Measuring the distance of a person from satellite 2 places the person somewhere on the red circle. Similar measurements using satellite 11 place the person's position somewhere on the green circle, and further measurements can pinpoint the person's location.

not have access to traditional landlines for telephone or Internet services.

1.18.2 Geostationary and geosynchronous Satellite and Parking Orbit

A Geostationary *satellite* , is one that orbits above the equator with a period equal to one day (24 hr). From the Earth, such a satellite appears to be in the same location in the sky at all times, making it particularly useful for applications such as communications and weather forecasting. Such satellite must be orbiting in an orbit of specific radius. This orbit is called parking orbit. If a Geostationary satellite is in an orbit of radius r at a height h above the earth's surface then its orbiting speed is given as

$$v_o = \sqrt{\frac{GM}{r}} \qquad \text{[here, } r = R + h\text{]}$$

The time period of its revolution can be given by-

$$T = \frac{2\pi r}{v_o} = \frac{2\pi r}{\sqrt{\frac{GM}{r}}} = \frac{2\pi}{\sqrt{GM}} r^{3/2}$$

or
$$\boxed{T^2 = \frac{4\pi^2}{GM} r^3} \qquad \text{[Kepler's III Law]} \quad (1.140)$$

or
$$T^2 = \frac{4\pi^2}{gR^2} r^3$$

or
$$T^2 = \frac{4\pi^2}{gR^2} (R+h)^3$$

or
$$h = \left[\frac{gR^2}{4\pi^2} T^2 \right]^{1/3} - R$$

or
$$h = \left[\frac{9.8 \times 6.4 \times 10^6 \times 86400^2}{4 \times 3.14^2} \right]^{1/3} \text{m} - 6.4 \times 10^6 \text{ m}$$
$$= 35954.6 \text{ km} \approx 36000 \text{ km} \approx 6R$$

or
$$\boxed{h \approx 6R} \qquad (1.141a)$$

therefore, the radial distance of satellite from the center of Earth,

$$\boxed{r = R + h \approx 7R} \qquad (1.141b)$$

Thus, when a satellite is launched in an orbit at a height of about 36000 km ($\approx 6R$) above the equator then it will appear to be at rest with respect to a point on Earth surface. A Geostationary satellite must have its orbit in equatorial plane due to the geographic limitation arose because of irregular geometry of earth (ellipsoidal shape). Figure 1.151 shows a cross sec-

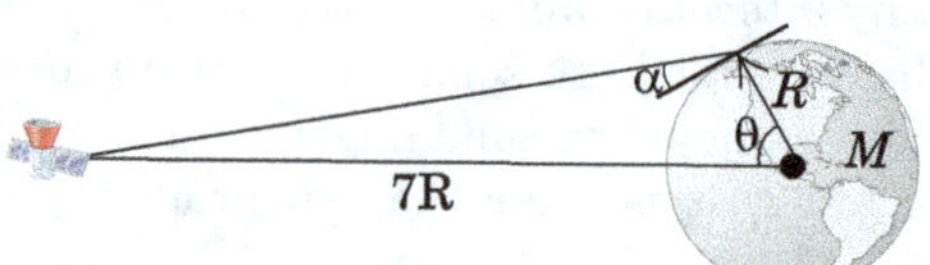

Figure 1.151: Angle of a satellite dish, α, relative to the local horizontal as a function of the angle of latitude, θ.

tion through Earth and the location of a geostationary satellite. This figure shows the angle α relative to the horizontal at which to orient a satellite dish for the best TV reception. Because any geostationary satellite is located in the plane of the Equator, a dish in the Northern Hemisphere should point in a southerly direction.

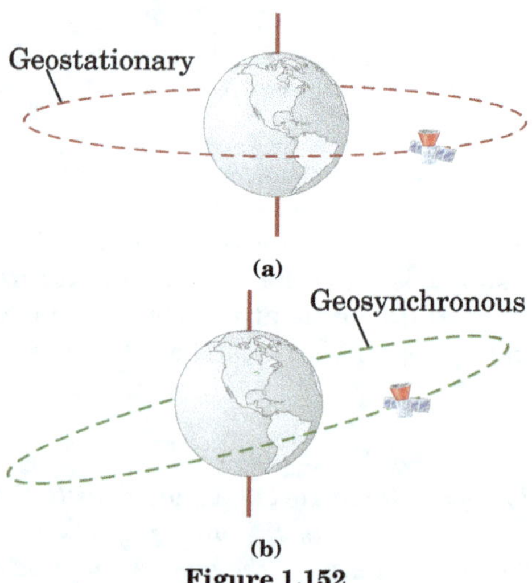

Figure 1.152

There are also geosynchronous satellites in orbit around the Earth. A geosynchronous satellite also has an orbital period of 1 day but does not need to remain at the same point in the sky as viewed from the surface of Earth. A geostationary orbit is a special case of a geosynchronous orbit.

1.18.3 Polar Satellite

Polar satellites travel around the earth in an orbit that travels around the earth over the poles [Fig.1.154]. The earth rotates on its axis as the satellite goes around the earth. Thus, over a period of many orbits it looks down on every part of the earth.

The angle of inclination of a polar orbit with equatorial plane of Earth is always, 90°. Polar satellites are Sun-synchronous satellites.

The polar satellites are used for getting the cloud images, atmospheric data, ozone layer in the atmosphere and to detect the ozone hole over Antarctica.

☞ A satellite must always rotate about the center of earth.

Fig.1.153 shows the orbits of some artificial satellites of earth with impossible orbit.

EXAMPLE 103. A geostationary satellite orbits around

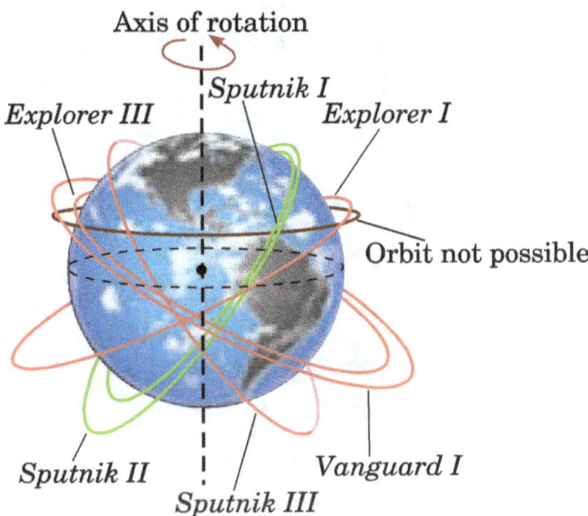

Figure 1.153: Orbits of the first artificial Earth satellites. All of these satellite orbits are ellipses.

the earth in a circular orbit of radius 36000 km. Find the approximate value of time period of a spy satellite orbiting a few hundred kilometers above the earth surface (radius of earth, $R = 6400$ km)

APPROACH From Eq.1.140

$$T^2 = \frac{4\pi^2}{GM} r^3$$

i.e.,
$$T \propto r^{3/2}$$

Therefore, if T_1 and T_2 are time periods of a satellite in orbits of radii r_1 and r_2 respectively, then

$$\frac{T_1}{T_2} = \left(\frac{r_1}{r_2}\right)^{3/2}$$

or
$$T_1 = T_2 \left(\frac{r_1}{r_2}\right)^{3/2} \qquad \text{... (1)}$$

SOLUTION If r_2 and T_2 are, respectively, the radius and time period of a geostationary satellite, then $r_2 = 36000$ km and $T_2 = 24$ h. Now, for $r_1 = 6400$ km, the time period, $T_1 = ?$. Substituting these values in Eq.(1), we get-

$$T_1 = 24 \times (6400/36000)^{3/2} \approx 1.8 \text{ h}$$

Since time period increases with r, time period of a satellite moving few hundred kilometers above the earth surface is ≈ 2 h.

EXAMPLE 104. A satellite is revolving around the earth in a circular orbit in a plane containing earth's axis of rotation. If the angular speed of satellite is equal to that of earth, find the time it takes to move from a point above north pole to a

point above the equator.

SOLUTION A satellite which rotates with angular speed equal to earth's rotation has an orbit radius $7R$ and the angular speed of revolution is

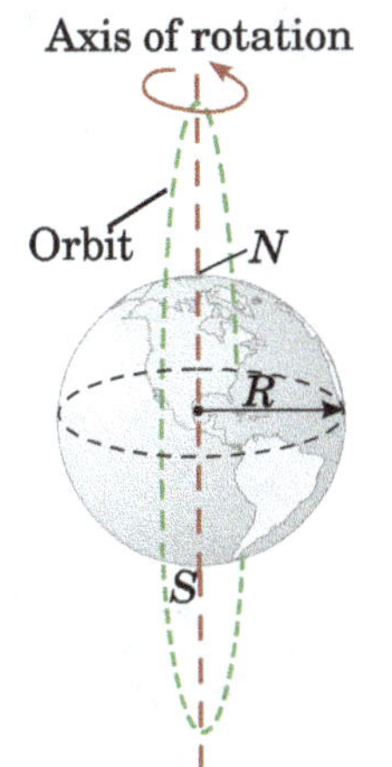

Figure 1.154: Polar orbit

$$\omega = \frac{2\pi}{T} = \frac{2\pi}{86400} = 7.27 \times 10^{-5} \text{rad/s}$$

When satellite moves from a point above north pole to a point above equator, it traverses an angle $\pi/2$. Time taken in it is given by:

$$t = \frac{\pi/2}{\omega} = 21600 \, s = 6\text{hr}$$

1.19 Orbital maneuvers

Let us consider a spacecraft in a circular orbit, and suppose we want to move it to a lower circular orbit. For it we use rockets to decrease the speed—that is, we fire the rockets from spacecraft that point in the forward direction so that their thrust is opposite to the direction of motion of spacecraft. The result of firing the decelerating rockets at a given point A in the original orbit of spacecraft is shown in Fig.1.155a. Notice that the new orbit is not a circle, as desired, but rather an ellipse. To produce a circular orbit we can simply fire the decelerating rockets once again at point B, on the opposite side of the Earth from point A. The net result of these two firings is that the spacecraft now moves in a circular orbit of smaller radius. Similarly, to move to a larger orbit, we must fire our accelerating rockets from spacecraft twice. The first firing puts spacecraft into an elliptical orbit that moves farther from the Earth, as Fig. 1.155a shows. After the second firing the spacecraft is again in a circular orbit. This simplest type of orbital transfer, requiring just two rocket burns, is referred to as a *Hohmann transfer*. The Hohmann transfer is the basic maneuver used to send spacecraft such as the Mars lander from Earth's orbit about the Sun to the orbit of Mars.

1.19.1 Broadcasting Region of a Satellite

In Eq.1.141a, we have seen that, the height of a geostationary satellite is approximately $6R$ from the surface of Earth. From

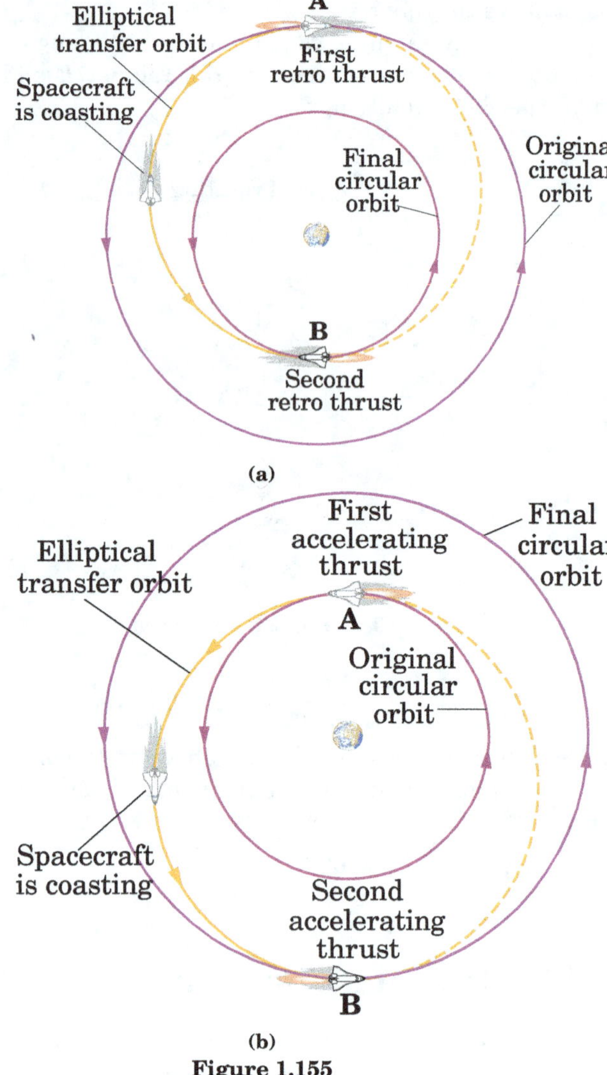

(a)

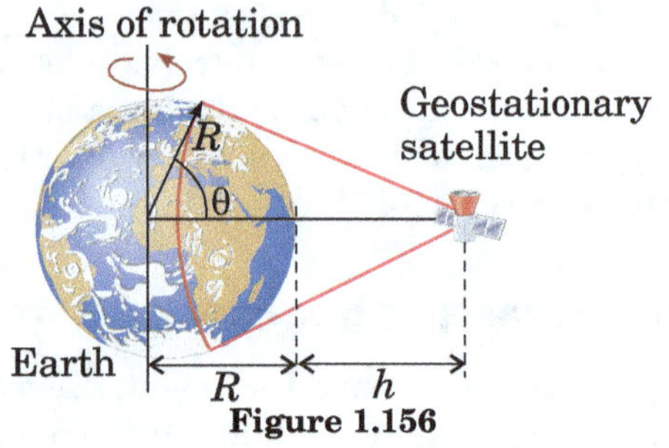

(b)

Figure 1.155

this information, we can easily find the area of earth exposed to the satellite or area of the region in which the communication can be made using this satellite.

Figure 1.156, shows Earth and its exposed area (in right side of Fig.1.156) to a geostationary satellite. Here the angle θ can be given as

$$\cos\theta = \frac{R}{R+h}$$

or

$$\theta = \cos^{-1}\left(\frac{R}{R+h}\right)$$

If the exposed area makes latitude angle θ [Fig.1.156], then

Figure 1.156

from Eq.1.139, the solid angle (Ω) formed at the center of the sphere will be given by-

$$\boxed{\Omega = 2\pi(1-\cos\theta)}$$

$$= 2\pi\left(1 - \frac{R}{R+h}\right) = \frac{2\pi h}{R+h}$$

Thus, the area of Earth's surface exposed to geostationary satellite is

$$S = \Omega R^2 = \frac{2\pi h R^2}{R+h} \qquad \left[\text{As } \Omega = \frac{\text{area}}{R^2} = \frac{S}{R^2}\right]$$

Lets take some examples to understand the concept in detail.

EXAMPLE 105. Find the minimum co-latitude which can directly receive a signal from a geostationary satellite.

APPROACH The farthest point on Earth, which can receive signals from the parking orbit, is the point where a line drawn from the satellite touches it tangentially 1.157.

To determine, the co-latitude λ of point P in Figure 1.157, we

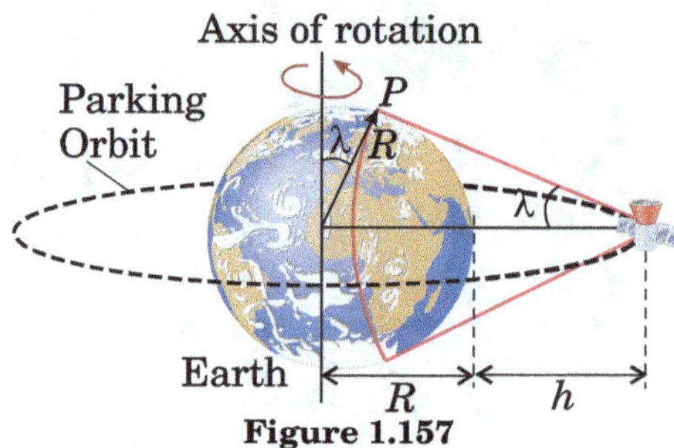

Figure 1.157

use simple trigonometry.

SOLUTION From Fig.1.157, we have

$$\sin\lambda = \frac{R}{R+h}$$

As for a parking orbit

$$h \simeq 6R$$

$$\therefore \qquad \sin\lambda = \frac{R}{R+h} \simeq \frac{1}{7}$$

or

$$\lambda = \sin^{-1}\left(\frac{1}{7}\right)$$

EXAMPLE 106. A satellite is orbiting around the earth in an orbit in equatorial plane of radius $2R$, where R is the radius of earth. Find the area on earth, this satellite covers for communication purpose in its complete revolution.

SOLUTION As shown in Figure 1.158, when satellite S

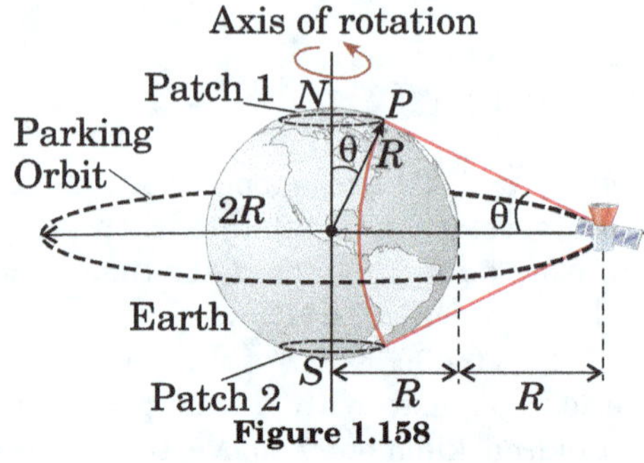

Figure 1.158

revolves, it covers a complete circular belt on earth's surface for communication. If the co-latitude of the farthest point on surface up to which signals can be received (point P) is θ then we have

$$\sin\theta = \frac{R}{2R} = \frac{1}{2} \Rightarrow \theta = \frac{\pi}{6}$$

During revolution satellite leaves two spherical patches 1 and 2 on earth surface at north and south poles where no signals can be transmitted due to curvature (see Fig.1.158). The areas of these patches can be obtained by using the relation between solid angle and surface area [Eq.1.138].

From Eq.1.139, the solid angle subtended by a patch on earth's centre is [Eq.1.139]

$$\Omega = 2\pi(1-\cos\theta) = 2\pi\left(1-\cos\frac{\pi}{6}\right) = \pi\left(2-\sqrt{3}\right)\text{ sr}$$

Area of patch 1 and 2 is

$$A_P = \Omega R^2 = \pi\left(2-\sqrt{3}\right)R^2$$

Thus total area on earths surface to which communication can be made is

$$A_C = 4\pi R^2 - 2A_P = 4\pi R^2 - 2\pi\left(2-\sqrt{3}\right)R^2 = 2\sqrt{3}R^2$$

EXAMPLE 107. Assuming that you live in a location at latitude 42.75° N and that the satellite TV company has a satellite aligned with your longitude, in what direction should you point the satellite dish?

APPROACH Satellite TV companies use geostationary satel-

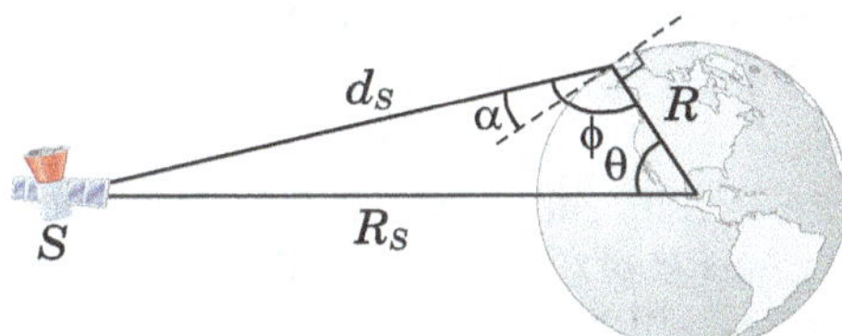

Figure 1.159: Geometry of a geostationary satellite in orbit around the Earth

lites to broadcast the signals. Thus, we know you need to point the satellite dish southward, toward the Equator, but you also need to know the angle of inclination of the satellite dish with respect to the horizontal. In Figure 1.159, this is the angle α. To determine α, we can use the law of cosines, incorporating the distance of a satellite in geostationary orbit, the radius of the Earth, and the latitude of the location of the dish.

SOLUTION Figure 1.159 is a sketch of the geometry of the location of the geostationary satellite and the point on the surface of the Earth where the dish is being set up. In this sketch, R is the radius of the Earth, R_S is the distance of the satellite from the center of the Earth, d_S is the distance from the satellite to the point on the Earth's surface where the dish is located, θ is the angle of the latitude of the surface of that location, and ϕ is the angle between d_S and R. Applying, law of cosines, we have

$$d_S^2 = R_S^2 + R^2 - 2\,R_S R\,\cos\theta \qquad (1.142a)$$

$$\text{and} \qquad R_S^2 = d_S^2 + R^2 - 2d_S R\,\cos\phi \qquad (1.142b)$$

SOLUTION We know that $R_S = 7R$ for geostationary satellites. The angle θ corresponds to the latitude, $\theta = 42.75$ °. We can substitute these quantities into Eq. 1.142a:

$$d_S^2 = (7R)^2 + R^2 - 2(7R)R\cos\theta$$

$$\text{or} \qquad d_S^2 = R^2\left[7^2 + 1^2 - 14\cos 42.75°\right]$$

$$\text{or} \qquad d_S^2 = R^2[50 - 14(0.7343)]$$
$$= R^2[50 - 10.2802] = 39.7198R^2$$

$$\therefore \qquad d_S = 6.3\,R$$

Now, we can solve Eq.1.142b for ϕ

$$\cos\phi = \frac{d_S^2 + R^2 - R_S^2}{2d_S R} = \frac{39.7198R^2 + R^2 - (7R)^2}{2(6.3\,R)R}$$

$$= \cos^{-1}\left(\frac{39.7198R^2 + R^2 - (7R)^2}{2(6.3\,R)R}\right) = 130.66°$$

The angle at which you need to aim the satellite dish with respect to the horizontal is then $= 130.66° - 90° = 40.66° \approx 40.7°$

Expressing our result with three significant figures gives $= 40.7°$

1.19.2 Check Point 4

1. ••(a) If we take the potential energy of a 100 kg object and Earth as zero when the two are separated by an infinite distance, what is the potential energy when the object is at the surface of Earth? (b) Find the potential energy of the same object at a height above Earth's surface equal to Earth's radius. (c) Find the escape speed for a body projected from this height.

2. ••A 100 kg spacecraft is in a circular orbit about Earth at a height $h = 2R$. (a) What is the orbital period of the spacecraft? (b) What is the spacecraft's kinetic energy? (c) Express the angular momentum L of the spacecraft about the centre of Earth in terms of the kinetic energy K and find the numerical value of L.

3. •••Many satellites orbit Earth at maximum altitudes above Earth's surface of 1000 km or less. Geosynchronous satellites, however, orbit at an altitude of 35790 km above Earth's surface. How much more energy is required to launch a 500 kg satellite into a geosynchronous orbit than into an orbit 1000 km above the surface of Earth?

4. ••Knowing that the acceleration of gravity on the moon is 0.166 times that on Earth and that the moon's radius is $0.273R_e$ find the escape speed for a projectile leaving the surface of the moon.

5. ••What initial speed would a particle need to be given at the surface of Earth if it is to have a final speed that is equal to its escape speed when it is very far from Earth? Neglect any effects due to air resistance.

6. ••An object is dropped from rest from a height of 4.0×10^6 m above the surface of Earth. If there is no air resistance, what is its speed when it strikes Earth?

7. ••A particle is projected from the surface of Earth with a speed twice the escape speed. When it is very far from Earth, what is its speed?

8. •••Calculate the minimum necessary speed, relative to Earth, for a projectile launched from the surface of Earth to escape the solar system [Fig.1.160]. The answer will depend on the direction of launch. Explain the choice of direction you would make for the direction of the launch in order to minimize the necessary launch speed relative to Earth. Neglect Earth's rotational motion and effects due to air resistance.

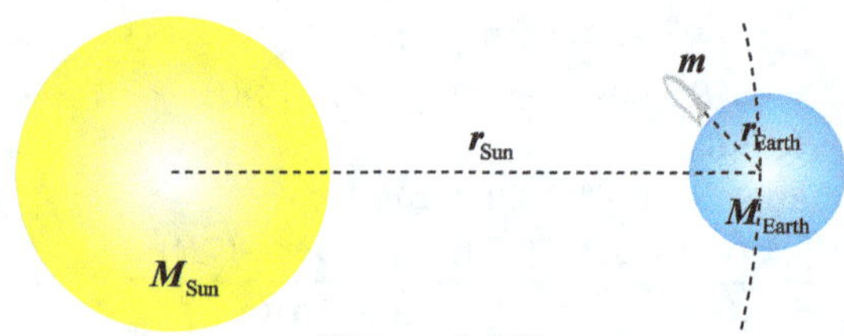

Figure 1.160

9. •••A satellite is launched in the equatorial plane in such a way that it can transmit signals upto 60° latitude on the earth [Fig.1.161]. The orbital velocity of the satellite is found to be $\sqrt{\frac{GM}{\alpha R}}$. Find the value of α.

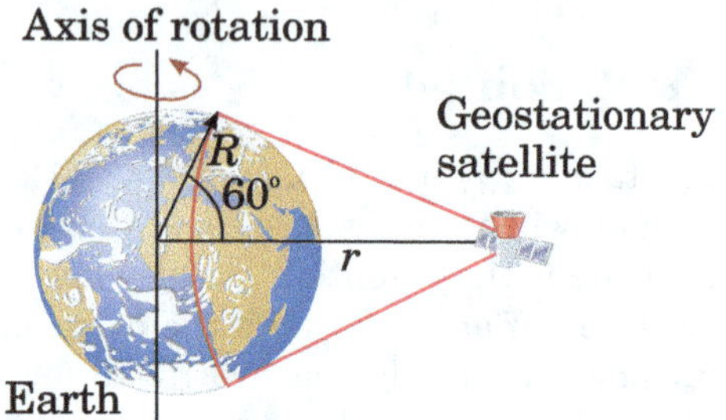

Figure 1.161

Multiple Choice Questions

10. ••The ratio of escape velocity at earth (v_e) to the escape velocity at a planet (v_p) whose radius and mean density are twice as that of earth is
    (A) $1:4$ (B) $1:\sqrt{2}$ (C) $1:2$ (D) $1:2\sqrt{2}$

11. ••The radius of a planet is twice the radius of earth. Both have almost equal average mass-densities. v_p and v_e are escape velocities of the planet and the earth, respectively, then
    (A) $v_p = 1.5 v_e$ (B) $v_e = 3 v_p$
    (C) $v_p = 2 v_e$ (D) $v_e = 1.5 v_p$

12. ••A particle of mass 'm' is kept at rest at a height '$3R$' from the surface of earth, where 'R' is radius of earth and 'M' is mass of earth. The minimum speed with which it should be projected, so that it does not return back, is (g is acceleration due to gravity on the surface of earth)//
    (A) $\left(\frac{GM}{2R}\right)^{1/2}$ (B) $\left(\frac{gR}{4}\right)^{1/2}$
    (C) $\left(\frac{2g}{R}\right)^{1/2}$ (D) $\left(\frac{GM}{R}\right)^{1/2}$

13. ••A particle of mass m is thrown upwards from the surface of the earth, with a velocity u. The mass and the radius of the earth are, respectively, M and R. G is gravitational constant and g is acceleration due to gravity on the surface of the earth. The minimum value of u so that the particle does not return back to earth, is
    (A) $\sqrt{\frac{2GM}{R^2}}$ (B) $\sqrt{\frac{2GM}{R}}$
    (C) $\sqrt{\frac{2gM}{R^2}}$ (D) $\sqrt{2gR^2}$

14. ••The earth is assumed to be a sphere of radius R. A platform is arranged at a height R from the surface of the earth. The escape velocity of a body from this platform is fv, where v is its escape velocity from the surface of the Earth. The value of f is
    (A) $1/2$ (B) $\sqrt{2}$ (C) $1/\sqrt{2}$ (D) $1/3$

15. ••With what velocity should a particle be projected so that its height becomes equal to radius of earth?
    (A) $\left(\frac{GM}{R}\right)^{1/2}$ (B) $\left(\frac{8GM}{R}\right)^{1/2}$
    (C) $\left(\frac{2GM}{R}\right)^{1/2}$ (D) $\left(\frac{4GM}{R}\right)^{1/2}$

16. ••For a planet having mass equal to mass of the earth but radius is one fourth of radius of the earth. The escape velocity for this planet will be
    (A) 11.2 km/s   (B) 5.6 km/s
    (C) 22.4 km/s   (D) 44.8 km/s

17. ••The escape velocity of a body on the surface of the earth is 11.2 km/s. If the earth's mass increases to twice its present value and radius of the earth becomes half, the escape velocity becomes
    (A) 22.4 km/s   (B) 5.6 km/s
    (C) 44.8 km/s   (D) 11.2 km/s

18. ••The escape velocity from earth is 11.2 km/s. If a body is to be projected in a direction making an angle 45° to the vertical, then the escape velocity is
    (A) 11.2×2 km/s   (B) 11.2 km/s
    (C) $11.2/\sqrt{2}$ km/s   (D) $11.2\sqrt{2}$ km/s

19. ••For a satellite escape velocity is 11 km/s. If the satellite is launched at an angle of 60° with the vertical, then escape velocity will be
    (A) 11 km/s   (B) $11\sqrt{3}$ km/s
    (C) $\frac{11}{\sqrt{3}}$ km/s   (D) 33 km/s

20. ••The time period of a geostationary satellite is 24 h, at a height $6R$ (R is radius of earth) from surface of earth. The time period of another satellite whose height is $2.5R$ from surface will be,
    (A) $6\sqrt{2}\, h$ (B) $12\sqrt{2}\, h$
    (C) $\frac{24}{2.5}\, h$ (D) $\frac{12}{2.5}\, h$

21. ••A remote-sensing satellite of earth revolves in a circular orbit at a height of 0.25×10^6 m above the surface of earth. If earth's radius is 6.38×10^6 m and $g = 9.8$ m s^{-2}, then the orbital speed of the satellite is
    (A) 9.13 km s^{-1} (B) 7.76 km s^{-1}
    (C) 6.67 km s^{-1} (D) 8.56 km s^{-1}

22. ••A satellite S is moving in an elliptical orbit around the earth. The mass of the satellite is very small compared to the mass of the earth. Then,
    (A) the linear momentum of S remains constant in magnitude
    (B) the acceleration of S is always directed towards the centre of the earth
    (C) the angular momentum of S about the centre of the earth changes in direction, but its magnitude remains constant
    (D) the total mechanical energy of S varies periodically with time

23. ••If v_e is escape velocity and v_o is orbital velocity of a satellite for orbit close to the earth's suface, then these are related by
    (A) $v_o = \sqrt{2} v_e$ (B) $v_e = \sqrt{2} v_o$
    (C) $v_o = v_e$ (D) $v_e = \sqrt{2} v_o$

24. ••The radii of circular orbits of two satellites A and B of the earth, are $4R$ and R, respectively. If the speed of satellite A is $3v$, then the speed of satellite B will be

(A) $\frac{3v}{4}$ (B) $6V$ (C) $12\,v$ (D) $\frac{3v}{2}$

25. ••A ball is dropped from a spacecraft revolving around the earth at a height of 120 km. What will happen to the ball?

   (A) it will fall down to the earth gradually

   (B) it will go very far in the space

   (C) it will continue to move with the same speed along the original orbit of spacecraft

   (D) it will move with the same speed, tangentially to the spacecraft.

26. ••A satellite A of mass m is at a distance of r from the centre of the earth. Another satellite B of mass $2m$ is at a distance of $2r$ from the earth's centre. Their time periods are in the ratio of

   (A) $1:2$ (B) $1:16$

   (C) $1:32$ (D) $1:2\sqrt{2}$

27. ••A satellite of mass m is orbiting the earth (of radius R) at a height h from its surface. The total energy of the satellite in terms of g, the value of acceleration due to gravity at the earth's surface, is

   (A) $\frac{mgR^2}{2(R+h)}$ (B) $-\frac{mgR^2}{2(R+h)}$

   (C) $\frac{2mgR^2}{R+h}$ (D) $-\frac{2mgR^2}{R+h}$

28. ••The additional kinetic energy to be provided to a satellite of mass m revolving around a planet of mass M, to transfer it from a circular orbit of radius R_1 to another of radius $R_2\,(R_2 > R_1)$ is

   (A) $GmM\left(\frac{1}{R_1^2} - \frac{1}{R_2^2}\right)$ (B) $GmM\left(\frac{1}{R_1} - \frac{1}{R_2}\right)$

   (C) $2GmM\left(\frac{1}{R_1} - \frac{1}{R_2}\right)$ (D) $\frac{1}{2}GmM\left(\frac{1}{R_1} - \frac{1}{R_2}\right)$

29. ••Two satellites of earth, S_1 and S_2 are moving in the same orbit. The mass of S_1 is four times the mass of S_2. Which one of the following statements is true?

   (A) The potential energies of earth and satellite in the two cases are equal.

   (B) S_1 and S_2 are moving with the same speed.

   (C) The kinetic energies of the two satellites are equal.

   (D) The time period of S_1 is four times that of S_2.

30. ••For a satellite moving in an orbit around the earth, the ratio of kinetic energy to potential energy is

   (A) $1/2$ (B) $1/\sqrt{2}$ (C) 2 (D) $\sqrt{2}$

31. ••The satellite of mass m is orbiting around the earth in a circular orbit with a velocity v. What will be its total energy?

   (A) $(3/4)mv^2$ (B) $(1/2)mv^2$

   (C) mv^2 (D) $-(1/2)mv^2$

32. ••The mean radius of earth is R, its angular speed on its own axis is ω and the acceleration due to gravity at earth's surface is g. What will be the radius of the orbit of a geostationary satellite?

   (A) $\left(R^2 g/\omega^2\right)^{1/3}$ (B) $\left(Rg/\omega^2\right)^{1/3}$

   (C) $\left(R^2\omega^2/g\right)^{1/3}$ (D) $\left(R^2 g/\omega\right)^{1/3}$

1.20 Kepler's Laws (Derivation)

1.20.1 Kepler's First Law (Law of orbit)

Each planet moves in an elliptical orbit, with the sun at one focus of the ellipse. This is called law of orbits.

Fig.1.162a shows that an ellipse has two foci (plural *of focus*), and the sun occupies one of these. The long axis of the ellipse is the *major axis*, and half the length of this axis is called the *semimajor-axis length (denoted by a)*. S and S' are the *foci*. The sun (mass M) is at S and the planet (mass m) is at P; we think of them both as point masses because the size of each is very small as compared to the distance between them. The sum of the distances from S to P and from S' to P is the same for all points on the curve. There is nothing at the other focus S'.

The distance of each focus from the centre of the ellipse is ea,

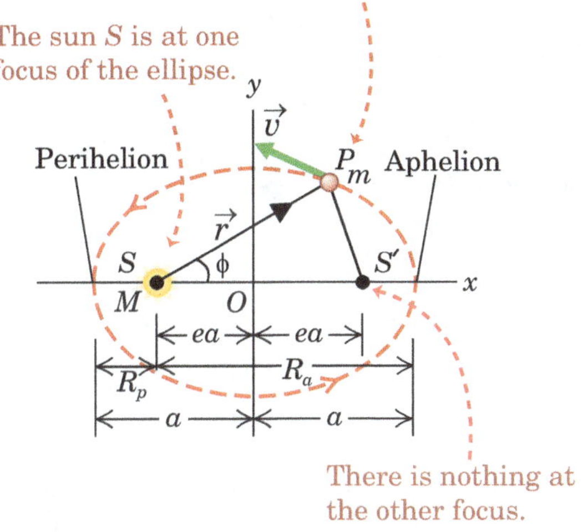

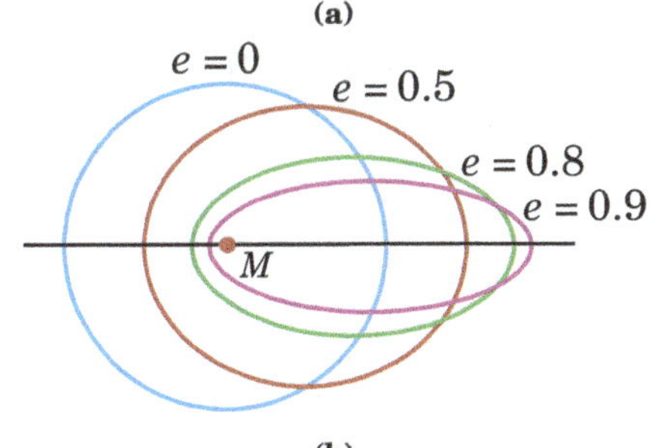

Figure 1.162: (a) A planet of mass m moving in an elliptical orbit around the Sun. The Sun of mass M, is at one focus S of the ellipse. The other focus is S' which is located in empty space. The semi major axis a of the ellipse, the perihelion (nearest the Sun)] distance R_p and the aphelion (farthest from the Sun)] distance R_a are also shown. (b) Four orbits with different eccentricities e about an object of mass M. All four orbits have the same semimajor axis and thus correspond to the same total mechanical energy E.

where e is a dimensionless number between 0 and 1 called the eccentricity[6]. If $e = 0$, the ellipse is a circle. The actual orbits of

---

[6]eccentricity (e) of a conic section is a non-negative real number that uniquely characterizes it's shape. The eccentricity of a conic section tells how much it deviates from being circular. For a circle, $e = 0$, for an ellipse, $0 < e < 1$, for parabola $e = 1$ and for a hyperbola $e > 1$

the planets are fairly circular; their eccentricities range from 0.007 for Venus to 0.206 for Mercury. (The earth's orbit has $e = 0.017$) The point in the planet's orbit closest to the sun is the *perihelion* (peri-, "around, near"; helios, "Sun"), and the point most distant from the sun is, the *aphelion* (api-, "away from"). The aphelion distance is also called "apogee distance" whereas perihelion distance is called "perigee distance".

1.20.1.1 Calculation of Eccentricity of the Orbit of a Planet

In Figure 1.162a, the maximum distance of a satellite from the Sun (*aphelion* distance):
$$R_a = a + ea = (1 + e)a \qquad \dots (1)$$
Minimum distance of a satellite from the Sun (*perihelion* distance):
$$R_p = a - ea = (1 - e)a \qquad (2)$$
For circular orbit $e = 0$ and for earth $e = 0.017$
On adding Eq.(1) and (2), we get-
$$R_a + R_p = 2a$$
or length of semi-major axis,

$$\boxed{a = (R_a + R_p)/2} \qquad (1.143)$$

On dividing Eq.(1) by (2), we get
$$\frac{1+e}{1-e} = \frac{R_a}{R_P}$$
or $\quad (1 + e)R_p = (1 - e)R_a$
or $\quad e\left(R_a + R_p\right) = R_a - R_p$

$$\Rightarrow \qquad \boxed{e = \frac{R_a - R_p}{R_a + R_p}} \qquad (1.144)$$

So, if we know, the perihelion and aphelion distances of an orbit, we can determine it's eccentricity by using Eq.1.144.

EXAMPLE 108. If a planet moving in an elliptical orbit around the Sun. The work done on the planet by the gravitational force of the Sun is zero for each segment of the orbit. Is this statement correct?

SOLUTION The statement is incorrect.

Explanation: For elliptical orbits, only at aphelion and perihelion positions the velocities and hence displacements of the planet are perpendicular to the gravitational force applied by Sun. So, only at these two positions, the work done by gravitational force is zero.

Since, gravitational force is conservative, therefore, the work done by it in a complete cycle is always zero.

1.20.2 Kepler's Second Law (Law of Area)

A line joining any planet to the Sun (i.e. radius vector of the planet form sun) sweeps equal areas in equal interval of time Fig.1.163a shows Kepler's second law. In a small time interval dt, the line from the sun S to the planet P turns through an angle $d\theta$. The area swept out is the coloured triangle with height r, base length $r\,d\theta$ and area $dA = \frac{1}{2}r^2\,d\theta$ (Fig.1.163b). The rate at which area is swept out, dA/dt is called the *sector velocity*:
$$\frac{dA}{dt} = \frac{1}{2}r^2\frac{d\theta}{dt} \qquad \dots (3)$$

Kepler's second law tells that the sector velocity has the same value at all points in the orbit. When the planet is close to the sun, r is small and $d\theta/dt$ is large; when the planet is far from the sun, r is large and $\frac{d\theta}{dt}$ is small.

Proof: Let v is the velocity of of planet at point P,

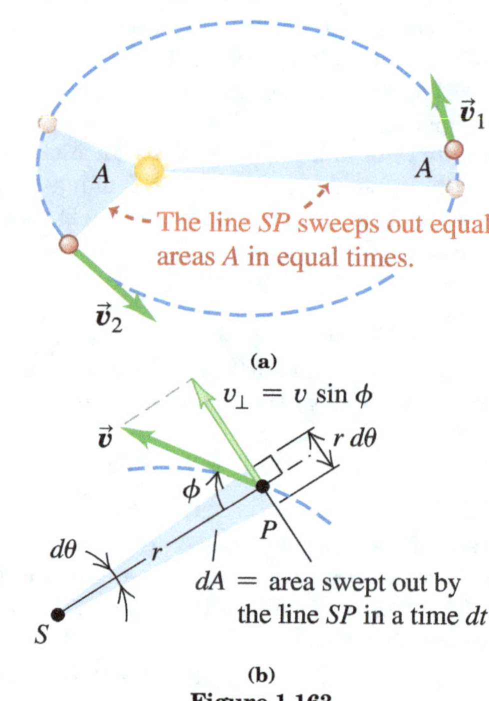

Figure 1.163

making angle ϕ with $\vec{r}$. It's component perpendicular to $\vec{r}$ is $v_\perp = v\sin\phi$. Transverse displacement of planet in time dt is $R\,d\theta$. Therefore,
Transverse velocity of of the planet
$$v_\perp = v\sin\phi = r\frac{d\theta}{dt}$$
From Eq.(3), the instantaneous rate at which area is being swept out is-

$$\frac{dA}{dt} = \frac{1}{2}r^2\frac{d\theta}{dt} = \frac{1}{2}r\left(r\frac{d\theta}{dt}\right) = \frac{1}{2}r\,(v_\perp) \qquad (1.145)$$

The angular momentum of the planet about Sun,
$$L = mrv_\perp$$
$$\text{or} \qquad v_\perp = \frac{L}{mr} \qquad (1.146)$$

Using 1.146 in 1.145, gives-

$$\frac{dA}{dt} = \frac{1}{2}r\left(\frac{L}{mr}\right) = \frac{L}{2m}$$

As net torque about the sun is zero, therefore $L = $ constant
Thus,

$$\boxed{\frac{dA}{dt} = \frac{L}{2m} = \text{constant.}} \qquad (1.147)$$

Note: By Newton's second law, acceleration of planet is towards the centre of the Sun. The torque on P about the centre of the sun is $\vec{\tau} = \vec{r} \times \vec{F} = \vec{0}$ because $\vec{r}$ and $\vec{F}$ are anti-parallel. Thus, the angular momentum, $\vec{L} = m\vec{r} \times \vec{v}$, of P

is conserved i.e., both magnitude and direction of $\vec{L}$ remain constant. The angular momentum is coming out of the paper for the orbital motion shown in the Fig.1.163a. Gravitational force is conservative so net mechanical energy of the Sun planet system, is also conserved. The magnitude of linear momentum of P remains constant in the circular orbit but not in the elliptical orbit. It is maximum in the elliptical orbit when P is closest to the Sun.

☞ In a planetary motion (in elliptical orbits), the physical quantities- angular momentum about centre of sun and mechanical energy, always remain conserved.

EXAMPLE 109. The ratio of earth's orbital angular momentum (about the sun) to its mass is 4.4×10^{15} m²/s. What is the area enclosed by earth's orbit?

SOLUTION The areal velocity of a planet of mass m and angular momentum L in an elliptical orbit are related as
$$dA/dt = L/(2m)$$
On integrating from $t = 0$ to $t = T$, we get
$$A = \int dA = \frac{L}{2m}\int_0^T dt = \frac{LT}{2m}$$
$$= \frac{L}{m}\frac{T}{2} = \left(4.4 \times 10^{15}\right)\left(365 \times 24 \times 60 \times 60/2\right)$$
$$\approx 6.94 \times 10^{22} \text{ m}^2$$
$$\left[\text{Since, in a circular orbit of radius } r,\right.$$
$$\left.\frac{L}{2m} = \frac{mvr}{2m} = \frac{vr}{2} = \frac{2\pi r}{T}\frac{r}{2} = \frac{\pi r^2}{T} = \frac{A}{T}.\right]$$

1.20.3 Third law (Law of period)

The square of the time periods of the planets are proportional to the cube of semi-major axis of ellipse.
$$T^2 \propto a^3$$

$$\text{or} \qquad \boxed{T^2 = ka^3} \qquad (1.148)$$

here, k is a proportionality constant.

Note: If we take the distance between Sun and earth in astronomical unit, and time period of earth around the Sun, in years i.e., $a = 1$AU, and time period of earth around the Sun, $T = 1$ yr, then by Kepler's third law, we have-
$$T^2 = ka^3, \quad \text{i.e.,} \quad (1 \text{ yr})^2 = k(1 \text{ AU})^3$$
$$\Rightarrow \qquad k = 1(\text{ yr})^2/(\text{AU})^3$$
So, for any planet orbiting the Sun with 'T' in years, and 'a' in astronomical units (1 AU = $1.495978707 \times 10^{11}$ m), $k = 1$ yr²/AU³, therefore, in astronomical units, Kepler's third law becomes

$$\boxed{T^2 = a^3} \quad \text{(when T is in years and a is in AU)} \qquad (1.149)$$

Proof: Since the mathematics of ellipses is difficult, we consider the orbit of planet as circular with radius r (equivalent to the length of semimajor axis of the ellipse) Fig.1.164a. Now, applying Newton's second law ($F = ma$) to the orbiting planet in Fig.1.164a yields
$$\frac{GMm}{r^2} = mr\omega^2$$
$$\because \qquad \omega = 2\pi/T$$

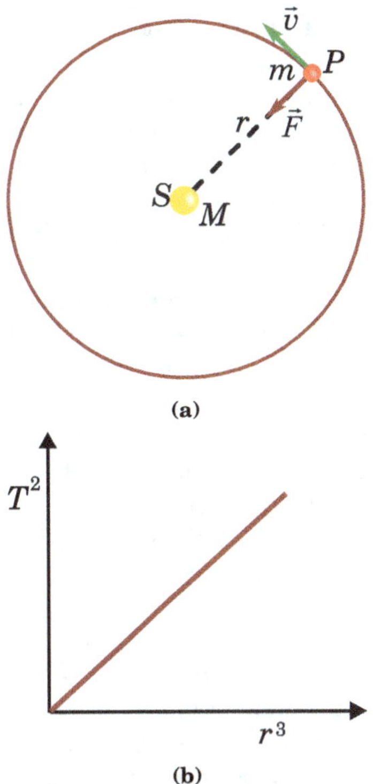

(a)

(b)

Figure 1.164

$$\therefore \qquad \frac{GMm}{r^2} = mr\left(\frac{2\pi}{T}\right)^2$$
$$\text{or} \qquad T^2 = \frac{4\pi^2}{GM}r^3$$

In other words, the square of the period is proportional to the cube of the radius. This is Kepler's third law. You can see that Kepler's third law is a direct consequence of Newton's law of gravity

Fig. 1.164b shows a graph between r^3 and T^2. It is a straight line.

Since, radius of circular path is equal to length of semi-major axis, therefore, we can also write, $r = a$ in above equation. In this case, we get

$$\boxed{T^2 = \frac{4\pi^2}{GM}a^3} \qquad (1.150a)$$

$$\text{or} \qquad \boxed{T = \left[\frac{2\pi}{\sqrt{GM}}\right]a^{3/2}} \qquad (1.150b)$$

From Eq.1.143, $a = (R_a + R_p)/2$, therefore Eq.1.150 can also be written as

$$\boxed{T^2 = \frac{\pi^2}{2GM}(R_a + R_p)^3} \qquad (1.151a)$$

$$\text{or} \qquad T = \left[\frac{\pi}{\sqrt{2GM}}\right](R_a + R_p)^{3/2} \qquad (1.151b)$$

On the basis of Kepler's laws, Newton concluded the following:
1. A force acting on the planet due to Sun is the centripetal force which is directed towards the Sun.
2. The force acting on the planet must be inversely proportional to the square of the distance from the Sun.

3. The force acting on the planet must be directly proportional to the product of the masses of the planet and Sun.

4. If e > 1 and total energy (KE + PE) > 0, the path of the satellite is hyperbolic and it escapes from its orbit.

5. If e < 1 and total energy is negative it moves in elliptical path.

6. If e = 0 and total energy is negative it moves in circular path.

7. If e = 0 and total energy is zero it will take parabolic path.

8. The path of the projectiles thrown to lower heights is parabolic and thrown to greater heights is elliptical.

9. Kepler's laws may be applied to natural and artificial satellites as well.

10. Gravitational force does not depend upon medium so no medium can shield it or block it.

11. The escape velocity and the orbital velocity are independent of the mass of the body being escaped or put into the orbit.

EXAMPLE 110. Two satellites S_1 and S_2 are revolving around a planet in coplanar and concentric circular orbit of radii R_1 and R_2 in the same direction respectively. Their respective periods of revolution are 1h and 8h respectively. The radius of the orbit of satellite S_1 is equal to 10^4 km. Find the relative speed in kmph when they are closest.

SOLUTION By Kepler's 3rd law, we have

$$\frac{T^2}{R^3} = \text{constant}$$

$$\therefore \quad \frac{T_1^2}{R_1^3} = \frac{T_2^2}{R_2^3} \quad \text{or} \quad \frac{1}{(10^4)^3} = \frac{64}{R_2^3}$$

or $\qquad R_2 = 4 \times 10^4$ km

Distances travelled by satellites S_1 and S_2, in one revolution, are $s_1 = 2\pi R_1 = 2\pi \times 10^4$ and $s_2 = 2\pi R_2 = 2\pi \times 4 \times 10^4$ respectively. If v_1 and v_2 are their respective speeds, then

$$v_1 = \frac{S_1}{t_1} = \frac{2\pi \times 10^4}{1} = 2\pi \times 10^4 \text{ km/h}$$

and $v_2 = \dfrac{S_2}{t_2} = \dfrac{2\pi \times 4 \times 10^4}{8} = \pi \times 10^4$ km/h

$\therefore$ Relative velocity $= v_1 - v_2 = 2\pi \times 10^4 - \pi \times 10^4 = \pi \times 10^4$ km/h

EXAMPLE 111. A satellite is revolving around the earth in an orbit of radius double that of the parking orbit and revolving in same sense. Find the periodic time duration between two instants when this satellite is closest to a geostationary satellite.

APPROACH Apply Kepler's third law of planetary of motion.

SOLUTION From Eq.1.150a, the time period of revolution of a satellite is given as

$$T^2 = \frac{4\pi^2}{GM} r^3 \qquad \text{[Kepler's III law]}$$

For satellite given in problem and for a geostationary satellite we have

$$\frac{T_1}{T_2} = \left(\frac{r_1}{r_2}\right)^{3/2} \Rightarrow T_1 = \left(\frac{r_1}{r_2}\right)^{3/2} \times T_1 = 2^{3/2} \times 24 = 67.9 \,\text{hr}$$

Suppose, at any instant both satellites are closest to each other, i.e., they are on the same radial line. Now, for next time they will be again on the same same radial line when the angular displacement covered by second satellite is exactly 2π radian greater than that of geostationary satellite.

If Δt be the time between two successive instants when the satellites are closest to each other, then

$$\Delta t = \frac{\theta}{\omega_1} = \frac{2\pi + \theta}{\omega_2} = \frac{2\pi}{\omega_2 - \omega_1}$$

Where ω_1 and ω_2 are the angular speeds of the two planets.

EXAMPLE 112. A body is orbiting around the earth at mean radius 9 times as great as the orbit of a geostationary satellite. In how many days it will complete one revolution around the earth and what is its angular velocity?

APPROACH To get required time, we apply Kepler's third law. The angular speed and time period are related as $T = 2\pi/\omega$

SOLUTION $(T_1^2/T_2^2) = (R_1^3/R_2^3)$

$$\Rightarrow T_2^2 = \left(\frac{R_2}{R_1}\right)^3 T_1^2 \Rightarrow T_2 = \left(\frac{R_2}{R_1}\right)^{3/2} T_1$$

For the geostationary satellite, $T_1 = 1$ day (24 hrs.)

$\therefore \qquad T_2 = (9R^3/R^3)^{1/2} = 27$ days

So it will complete one revolution in 27 days

$\omega = 2\pi/T_2 = \dfrac{2\pi}{27 \times 24 \times 3600}$ rad/s $= 2.693 \times 10^6$ rad/sec.

EXAMPLE 113. Two satellites S_1 and S_2 revolve around a planet in coplanar circular orbits in the same sense. Their periods of revolution are 1 h and 8 h respectively. The radius of the orbit of S_1 is 10^4 km. When S_2 is closest to S_1, find (a) the speed of S_2 relative to S_1 and (b) the angular speed of S_2 as observed by an astronaut in S_1.

APPROACH By Kepler's third law, the square of time period of a satellite is directly proportional to the cube of distance of it from the planet. From this principle, you can determine the radius of orbit of S_2 and then determine their linear speeds and hence relative speed of S_2 with respect to S_1.

If v_1 and v_2 are the linear speeds of satellites S_1 and S_2 respectively and their radii are R_1 and R_2 respectively then relative angular speed of S_2 with respect to S_1, when S_2 is closest to S_1, is-

$$\omega = \frac{|v_2 - v_1|}{R_2 - R_1} \qquad \qquad \dots (1)$$

SOLUTION Let, M, m_1 and m_2 are the masses of planet O, satellite S_1 and satellite S_2 respectively. If the radii of orbits S_1 and S_2 are R_1 and R_2 respectively, then it is given that, $R_1 \left(= 10^4 \text{ km}\right)$.

Let v_1 and v_2 be the linear speeds of S_1 and S_2 with respect to the planet [see Figure 1.165].

From Eq.1.148, the square of the time period is proportional

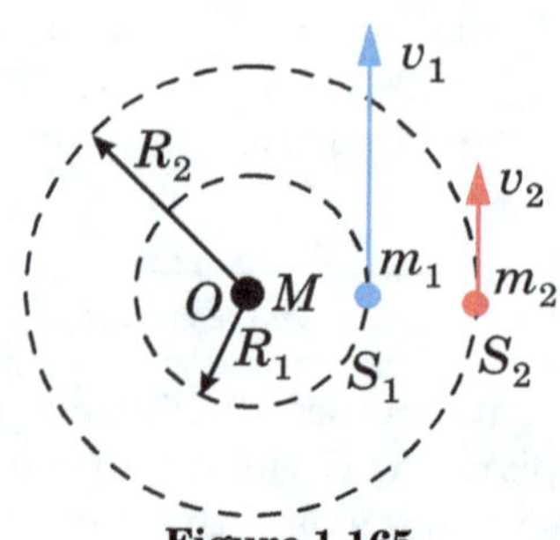

Figure 1.165

to the cube of the radius, therefore,

$$\left(\frac{R_2}{R_1}\right)^3 = \left(\frac{T_2}{T_1}\right)^2 = \left(\frac{8 \text{ h}}{1 \text{ h}}\right)^2 = 64$$

or, $\qquad \dfrac{R_2}{R_1} = 4$

or, $\qquad R_2 = 4R_1 = 4 \times 10^4$ km.

Since, the time period of S_1 is 1 h. So,

$$\frac{2\pi R_1}{v_1} = 1 \text{ h}$$

or, $v_1 = \dfrac{2\pi R_1}{1 \text{ h}} = 2\pi \times 10^4$ km h^{-1}

$$v_2 = \frac{2\pi R_2}{8 \text{ h}} = \pi \times 10^4 \text{ km h}^{-1}$$

(a) Note that, at the closest separation, they are moving in the same direction. Hence the linear speed of S_2 with respect to S_1 is

$$|v_2 - v_1| = \pi \times 10^4 \text{ km h}^{-1}.$$

(b) As seen from S_1, the satellite S_2 is at a distance $R_2 - R_1 = 3 \times 10^4$ km at the closest separation. Also, it is moving at $\pi \times 10^4$ km h^{-1} in a direction perpendicular to the line joining them. Thus, from Eq. (1), the angular speed of S_2 as observed by S_1 is

$$\omega = \frac{|v_2 - v_1|}{R_2 - R_1} = \frac{\pi \times 10^4 \text{ km h}^{-1}}{3 \times 10^4 \text{ km}} = \frac{\pi}{3} \text{rad h}^{-1}.$$

1.20.4   Binary Star System

It consists of two stars orbiting around their common centre of mass. The brighter star is classified as primary star and the other as companion star. There is only a gravitational force between the stars. The necessary centripetal force for circular motion about their common centre of mass is provided by the gravitational attraction between them. On each star, this gravitational force is directed towards the common centre of mass of both stars. So, the torque of gravitational force acting on stars about their common centre of mass is zero. Since toque is equal to rate of change of angular momentum, i.e., $\tau = dL/dt$, therefore, corresponding to zero toque, $dL/dt = 0$, i.e., $L =$ constant. So, the angular momentum of each star revolving around the common centre of mass, always remains conserved.

Therefore, we can apply the conservation of angular momentum about their common centre of mass.

If two stars of masses m_1 and m_2 are at distances r_1 and r_2 respectively from their common centre of mass, then by definition of centre of mass, we can write

$$m_1 r_1 = m_2 r_2$$
$$\frac{r_1}{r_2} = \frac{m_2}{m_1}, \quad r_1 + r_2 = r$$
$$r_1 = \frac{m_2}{m_1 + m_2} r, \quad r_2 = \frac{m_1}{m_1 + m_2} r$$

Since, for circular motion of stars, the required centripetal force is provided by gravitational force acting between them, therefore, for star of mass m_1, we have-

$$m_1 r_1 \omega_1^2 = \frac{Gm_1 m_2}{r^2}$$

If T is the time period of satellite, then

$$m_1 r_1 \left(\frac{2\pi}{T}\right)^2 = \frac{Gm_1 m_2}{r^2} \qquad [\text{since, } \omega_1 = 2\pi/T]$$

$$T^2 = \frac{4\pi^2 r^2 r_1}{Gm_2} = \frac{4\pi^2 r^2}{Gm_2} \frac{m_2}{(m_1 + m_2)} r = \frac{4\pi^2 r^3}{G(m_1 + m_2)} = \frac{4\pi^2 r^3}{GM}$$

here, $M = m_1 + m_2$ is the total mass of the system

If one of the bodies is massive as compared to other then CM of system shifts almost to the position of the massive body and

Star ②'s orbit around the center of mass

Centre of mass of the binary star system

v_1

Star ②   cm   Star ①

Star ①'s orbit

v_2

The star ① is more massive than the star ② and so orbits closer to the center of mass.

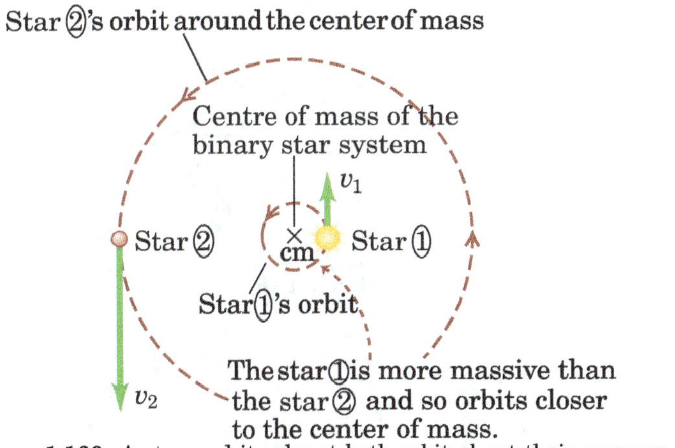

Figure 1.166: A star and its planet both orbit about their common centre of mass.

light body moves in circular orbit having its centre at the position of massive body.

Two bodies (double star) problem can also be solved with the help of reduce mass (μ). $\qquad \mu = \dfrac{m_1 m_2}{m_1 + m_2}$

Two bodies are replaced by a single body, whose mass is equal to reduced mass.

This single body revolves in a circular orbit, whose radius is equal to separation distance between two bodies and centripetal force of circular motion is equal to force of interaction between two bodies for actual separation.

MI of system about CM: $I = \mu r^2$

Angular momentum of system about CM: $L = \mu r^2 \omega = I\omega$

KE of system: $K = \frac{1}{2}\mu r^2 \omega^2 = \frac{1}{2}I\omega^2$

EXAMPLE 114.   In a double star, two stars (one of mass m and the other of $2\,m$) distant d apart rotate about their common center of mass [Fig.1.167]. Deduce an expression of the period of revolution. Show that the ratio of their angular momenta about the centre of mass is the same as the ratio of their kinetic energies.

SOLUTION The centre of mass C will be at distances $d/3$ and $2\,d/3$ from the masses $2\,m$ and m respectively. Both the stars rotate around C in their respective orbits with the same angular velocity ω. The gravitational force acting on each star due to the other supplies the necessary centripetal force. For rotation of the smaller star, the centripetal force $\left[m\left(\frac{2\,d}{3}\right)\omega^2\right]$ is provided by gravitational force.

$$\therefore \quad \frac{G(2\,m)m}{d^2} = m\left(\frac{2\,d}{3}\right)\omega^2 \quad \text{or} \quad \omega = \sqrt{\left(\frac{3Gm}{d^3}\right)}$$

Therefore, the period of revolution is given by

$$T = \frac{2\pi}{\omega} = 2\pi\sqrt{\left(\frac{d^3}{3Gm}\right)}$$

The ratio of the angular momenta is

$$\frac{(I\omega)_{\text{big}}}{(I\omega)_{\text{small}}} = \frac{I_{\text{big}}}{I_{\text{small}}} = \frac{(2\,m)\left(\frac{d}{3}\right)^2}{m\left(\frac{2\,d}{3}\right)^2} = \frac{1}{2}$$

here, I_{big} and I_{small} are the moment of inertia of bigger and smaller star respectively.

The ratio of their kinetic energies is

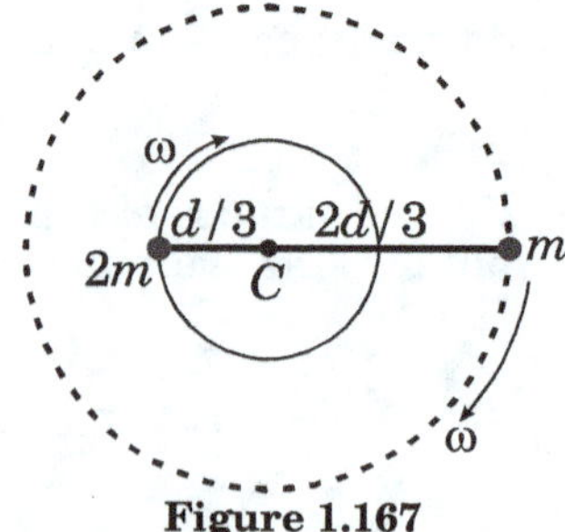

Figure 1.167

$$\frac{\left(\frac{1}{2}I\omega^2\right)_{\text{big}}}{\left(\frac{1}{2}I\omega^2\right)_{\text{small}}} = \frac{I_{\text{big}}}{I_{\text{small}}} = \frac{1}{2},$$

which is the same as the ratio of their angular momenta.

EXAMPLE 115. A double star system consists of two stars A and B which have time periods T_A and T_B, radius R_A and R_B and mass M_A and M_B. Find the relation between T_A and T_B.

SOLUTION Let the distance between A and B be r and C be the centre of mass of the double star system [Fig.1.168]. By definition of the centre of mass, the distances of A and B from C are given by

$$r_A = \frac{M_B r}{M_A + M_B}, \quad r_B = \frac{M_A r}{M_A + M_B}$$

The stars A and B move in a plane containing C with orbits

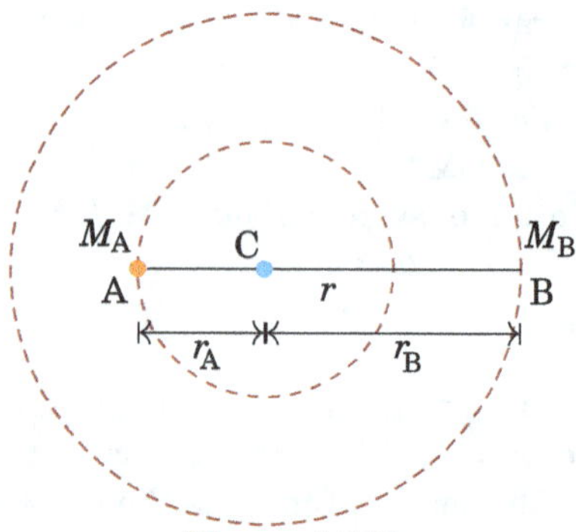

Figure 1.168

defined by the initial conditions. For simplicity, let orbits of A and B be circular. The gravitational force provides the centripetal acceleration i.e.,

$$GM_A M_B/r^2 = M_A v_A^2/r_A$$

The time period of A is given by

$$T_A = \frac{2\pi r_A}{v_A} = \frac{2\pi}{\sqrt{G(M_A + M_B)}}r^{3/2}$$

Similarly, $\quad T_B = \dfrac{2\pi r_B}{v_B} = \dfrac{2\pi}{\sqrt{G(M_A + M_B)}}r^{3/2}$

EXAMPLE 116. A planet of mass m moves along an ellipse around the Sun so that its maximum and minimum distances from the sun are equal to R and r respectively. Find the angular momentum of this planet relative to the centre of the Sun.

SOLUTION According to Kepler's Second Law the angular momentum of the planet is constant, i.e.,

$$mv_1 R = mv_2 r, \qquad v_1 R = v_2 r$$

If the mass of the Sun is M conserving total mechanical energy of the system at two given positions in Fig.1.169, gives,

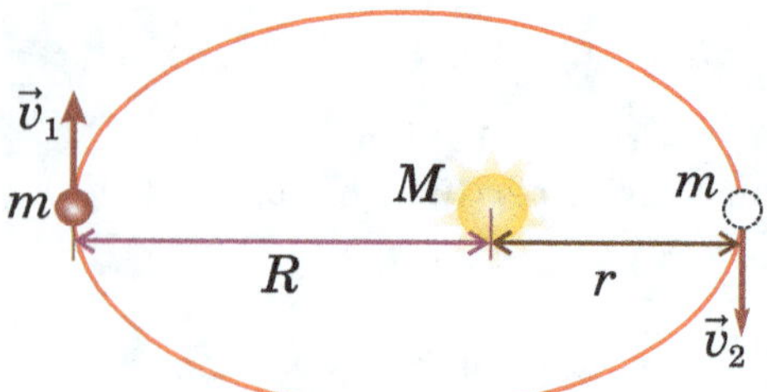

Figure 1.169

$$-\frac{GMm}{R} + \frac{1}{2}mv_1^2 = -\frac{GMm}{r} + \frac{1}{2}mv_2^2$$

$$\therefore \qquad GM\left[\frac{1}{R} - \frac{1}{r}\right] = \frac{v_2^2}{2} - \frac{v_1^2}{2}$$

or $\quad GM\left[\dfrac{r-R}{Rr}\right] = \dfrac{v_1^2 R^2}{2r^2} - \dfrac{v_1^2}{2}$

or $\quad GM\left[\dfrac{r-R}{Rr}\right] = \dfrac{v_1^2}{2}\left(\dfrac{R^2}{r^2} - 1\right) = \dfrac{v_1^2}{2}\left(\dfrac{R^2 - r^2}{r^2}\right)$

$$\therefore \quad v_1^2 = \frac{2GM(R-r)r^2}{Rr(R^2 - r^2)} = \frac{2GMr}{R(R+r)}$$

or $\quad v_1 = \sqrt{\dfrac{2GMr}{R(R+r)}}$

Now, angular momentum $mv_1 R = m\sqrt{\dfrac{2GMr}{R(R+r)}}$

EXAMPLE 117. If the distance between the earth and the sun were half of its present value, then what would be the number of days in a year.

APPROACH According to Kepler's third law, square of the time period is proportional to the cube of the distance (semi-major axis for the elliptical orbit) i.e., $T^2 \propto r^3$. If T_1 and T_2 are time periods corresponding to average distances r_1 and r_2 of a planet from sun, then

$$\frac{T_2^2}{T_1^2} = \frac{r_2^3}{r_1^3}$$

or $\qquad T_2 = T_1(r_2/r_1)^{3/2}$ ... (1)

Now, substitute the given values in Eq.(1), and simplify for T_2.

SOLUTION Let corresponding to distance r_1 between the earth and the sun, the time period of the orbital motion of the earth is $T_1 = 365$ days. When distance is halved, $r_2 = r_1/2$, then from Eq.(1), the time period becomes

$$T_2 = T_1(r_2/r_1)^{3/2} = 365(1/2)^{3/2} = 129 \text{ days}$$

EXAMPLE 118. A geostationary satellite is orbiting the earth at a height of $6R$ above the surface of earth where R is the radius of the earth. Find the time period of another satellite at a height of $2.5R$ from the surface of earth.

APPROACH According to Kepler's third law, square of the time period is proportional to the cube of the distance (semimajor axis for the elliptical orbit) i.e., $T^2 \propto a^3$. If T_1 and T_2 are time periods corresponding to average distances r_1 and r_2 of a satellites from the earth, then

$$\frac{T_2^2}{T_1^2} = \frac{r_2^3}{r_1^3}$$

or $\qquad T_2 = T_1(r_2/r_1)^{3/2}$ ... (1)

Now, substitute the given values in Eq.(1), and simplify for T_2.

SOLUTION For the geostationary satellite, distance from

the centre of earth is $r_1 = R + 6R = 7R$ and the time period is $T_1 = 24$ h. The distance of another satellite from the centre of the earth is $r_2 = R + 2.5R = 3.5R$. On applying these values in Eq.(1), we get

$$T_2 = (r_2/r_1)^{3/2} T_1 = (3.5/7)^{3/2} 24 = 8.48 \text{ h}$$

1.21 Black Hole

As we can see from Eq.1.135: $v_e = \sqrt{\frac{2GM}{R}}$, the escape speed of an object increases with increasing mass and decreasing radius. Thus, for example, if a massive star were to collapse to a relatively small size, its escape speed would become very large. According to Einstein's theory of general relativity, the escape speed of a compressed, massive star could even exceed the speed of light. In this case nothing—not even light—could escape from the star. For this reason, such objects are referred to as *black holes*.

1. **Black holes** were first predicted by J. Robert Oppenheimer and Hartland Snyder in 1939.
2. According to the general theory of relativity, if the density of an object such as a star is great enough, its gravitational attraction will be so great that once inside a critical radius, nothing can escape, not even light or other electromagnetic radiation. (The effect of a black hole on objects outside the critical radius is the same as that of any other mass.)
3. In Newtonian mechanics, the speed needed for a particle to escape from the surface of a planet or a star of mass M and radius R is given by Eq.1.135:

$$v_e = \sqrt{\frac{2GM}{R}}$$

4. If we set the escape speed equal to the speed of light and solve for the radius R_S, we obtain the critical radius called the **Schwarzschild radius:**

$$\boxed{R_S = \frac{2GM}{c^2}} \text{ [Black hole radius]} \qquad (1.152)$$

For an object that has a mass equal to five times that of our Sun (theoretically the minimum mass for a black hole) to be a black hole, its radius would have to be approximately 15 km. Because no radiation is emitted from a black hole and its radius is expected to be small, the detection of a black hole is not easy. The best chance of detection occurs in a case in which a black hole, together with a normal star, forms a binary system. In such a binary star system, both stars revolve around their centre of mass and the gravitational field of the black hole will pull gas from the normal star into the black hole. However, to conserve angular momentum, the gas does not go straight into the black hole. Instead, the gas orbits around the black hole in a disk, called an accretion disk, while slowly being pulled closer to the black hole. The gas in this disk emits X-rays because due to frictional heating, it's temperature reaches to greater than $1.0 \times 10^6 K$.

EXAMPLE 119. Schwarzschild Radius of a Stellar BI-ack Hole: One of the stellar black holes detected by LIGO in 2015, had a mass of about 36 times the mass of the Sun. What is the corresponding Schwarzschild radius?

APPROACH We first need to express the black hole mass in SI units:

$M = 36 \times 1.989 \times 10^{30}$ kg $= 7.16 \times 10^{31}$ kg

Now we can directly apply Equation 1.152 to find the Schwarzschild radius:

$$R_s = \frac{2GM}{c^2}$$

SOLUTION Substituting the given values in above equation, we get

$$R_s = \frac{2 \times 6.674 \times 10^{-11} \text{N} \cdot \text{m}^2 \cdot \text{kg}^{-2} \times 7.16 \times 10^{31} \text{kg}}{(2.998 \times 10^8 \text{m/s})^2}$$
$$= 1.06 \times 10^5 \text{m}$$

The input mass is only given to two significant digits, so we should express the final answer as $1.1 \times 10^5 m$ or 110 km.

Making sense of the result: Stellar black hole sizes are substantial, although this is still much smaller than the radius of our Sun, for example, which is about 7.0×10^5 km. The Schwarzschild radius is directly proportional to the black hole mass. Although a proper treatment of the size of black holes requires general relativity, it turns out that for a non-rotating spherically symmetric black hole, the result given by Eq.1.152, is same as given by general theory of relativity.

**EXAMPLE 120.   ** A black hole is an object whose gravitational field is so strong that even light cannot escape from it. To what approximate radius would earth (mass $= 5.98 \times 10^{24}$ kg) have to be compressed to be a black hole?

(A) 10^{-9} m   (B) 10^{-2} m   (C) 10^{-6} m   (D) 100 m

Sol (C) : Light cannot escape from a black hole,

$$v_e = c \Rightarrow \sqrt{\frac{2GM}{R}} = c \text{ or } R = \frac{2GM}{c^2}$$
$$R = \frac{2 \times 6.67 \times 10^{-11} \text{ N m}^2 \text{ kg}^{-2} \times 5.98 \times 10^{24} \text{ kg}}{\left(3 \times 10^8 \text{ m s}^{-1}\right)^2}$$
$$= 8.86 \times 10^{-3} \text{ m} \propto 10^{-2} \text{ m}$$

1.22 The Principle of Equivalence

Newton's law of gravity depends on the concept of mass. The concept of mass was introduced in Chapter "Newton's laws of Motion", by considering the relationship between force and acceleration. The inertial mass of an object, which is the mass that appears in Newton's second law, is found by measuring the object's acceleration a in response to force F:

$$\boxed{m_{\text{inert}} = \text{inertial mass} = \frac{F}{a}} \qquad (1.153)$$

Gravity plays no role in this definition of mass.

The quantities m_1 and m_2 in Newton's law of gravity [Eq.1.1]:$F = Gm_1 m_2/r^2$, are being used in a very different way. Masses m_1 and m_2 govern the strength of the gravitational attraction between two objects. The mass used in Newton's law of gravity is called the *gravitational mass*. The gravitational mass of an object can be determined by measuring the attractive force exerted on it by another mass M a distance r away:

$$m_{\text{grav}} = \text{gravitational mass} = \frac{r^2 F_{\text{M on m}}}{GM} \qquad (1.154)$$

Acceleration does not enter into the definition of the gravitational mass. These are two very different concepts of mass. Yet Newton, in his theory of gravity, asserts that the inertial mass in his second law is the very same mass that governs the strength of the gravitational attraction between two objects. The assertion that $m_{\text{grav}} = m_{\text{inert}}$ is called the principle of equivalence. It says that *inertial mass is equivalent to gravitational mass.*

Experiments comparing gravitational and inertial mass have improved steadily over the years. Their equivalence is now established to about 1 part in 5×10^{13}. Thus, the equivalence of gravitational and inertial mass is one of the best established of all physical laws. Principle of equivalence is the foundation of Einstein's general theory of relativity.

1.22.1   Check Point 5

1. ••A satellite, moving in an elliptical orbit, is 360 km above Earth's surface at its farthest point and 180 km above at its closest point. Calculate (a) the semimajor axis and (b) the eccentricity of the orbit.

2. ••The Sun's centre is at one focus of Earth's orbit. How far from this focus is the other focus, (a) in meters and (b) in terms of the solar radius, $6.96 \times 10^8 m$? The eccentricity is 0.0167, and the semimajor axis is 1.50×10^{11}m.

3. ••A 20 kg satellite has a circular orbit with a period of 2.4 h and a radius of 8.0×10^6 m around a planet of unknown mass. If the magnitude of the gravitational acceleration on the surface of the planet is 8.0 m/s^2, what is the radius of the planet?

4. ••The mean distance of Pluto from the Sun is 39.5 AU. Calculate the period of Pluto's orbital motion.

5. ••Observations of the light from a certain star indicate that it is part of a binary(two-star) system. This visible star has orbital speed $v = 270$ km/s, orbital period $T = 1.70$ days, and approximate mass $m_1 = 6M_s$, where M_s is the Sun's mass, 1.99×10^{30} kg. Assume that the visible star and its companion star, which is dark and unseen, are both in circular orbits (Fig. 1.170). What integer multiple of M_s gives the approximate mass m_2 of the dark star (black hole)?

6. ••The Jupiter's period of revolution around the Sun is 12 times that of the Earth. Assuming the planetary orbits to be circular, find
(a) how many times the distance between the Jupiter and the Sun exceeds that between the Earth and the Sun;
(b) the velocity and the acceleration of Jupiter in the he-

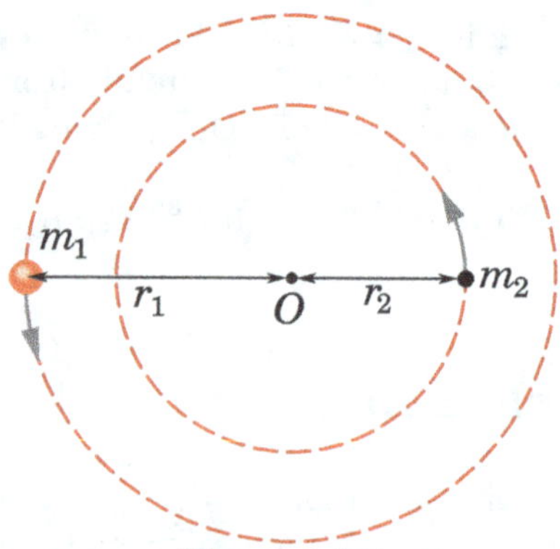

Figure 1.170

liocentric reference frame.

7. ••A planet of mass m moves around the Sun (mass $= M$), along an ellipse so that its minimum distance from the Sun is equal to r and the maximum distance to R. Making use of Kepler's laws, find its period of revolution around the Sun.

8. ••A small body starts falling onto the Sun from a distance equal to the radius of the Earth's orbit. The initial velocity of the body is equal to zero in the heliocentric reference frame. Making use of Kepler's laws, find how long the body will be falling.

9. ••Suppose, we have made a model of the Solar system scaled down in the ratio η but of materials of the same mean density as the actual materials of the planets and the Sun. How will the orbital periods of revolution of planetary models change in this case?

10. ••According to Kepler's second law, the radius vector to a planet from the sun sweeps out equal areas in equal intervals of time. This law is a consequence of the conservation of ...

Multiple Choice Problems

11. ••The kinetic energies of a planet in an elliptical orbit about the Sun, at positions A, B and C [Fig.1.171] are K_A, K_B and K_C, respectively. AC is the major axis and SB is perpendicular to AC at the position of the Sun S as shown in the figure. Then
(A) $K_A < K_B < K_C$ (B) $K_A > K_B > K_C$
(C) $K_B < K_A < K_C$ (D) $K_B > K_A > K_C$

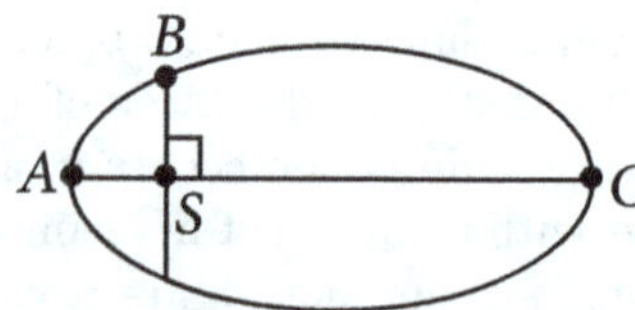

Figure 1.171

12. ••A planet moving along an elliptical orbit is closest to the sun at a distance r_1 and farthest away at a distance of r_2. If v_1 and v_2 are the linear velocities at these points respectively, then the ratio $\frac{v_1}{v_2}$ is
(A) $(r_1/r_2)^2$ (B) $(r_2/r_1)^2$ (C) r_2/r_1 (D) r_1/r_2

13. ••Fig.1.172 shows elliptical orbit of a planet m about the sun S. The shaded area SCD is twice of the shaded area

SAB. If t_1 is the time for the planet to move from C to D and t_2 is the time to move from A to B then

   (A) $t_1 = 4t_2$ (B) $t_1 = 2t_2$ (C) $t_1 = t_2$ (D) $t_1 > t_2$

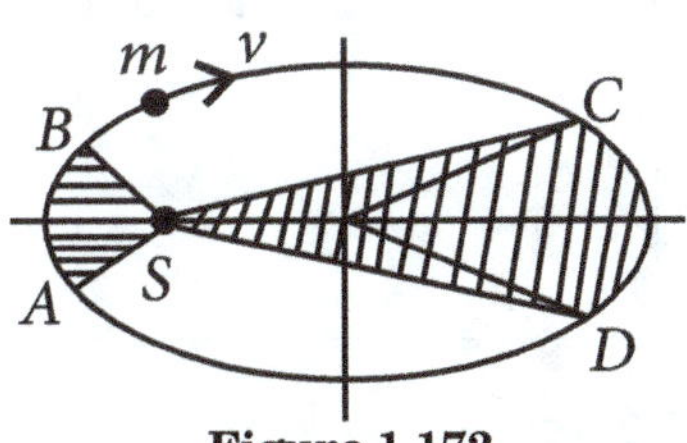

Figure 1.172

14. ••The period of revolution of planet A around the sun is 8 times that of B. The distance of A from the sun is how many times greater than that of B from the sun?

   (A) 4   (B) 5   (C) 2   (D) 3

15. ••The distance of two planets from the sun are 10^{13} m and 10^{12} m respectively. The ratio of time periods of the planets is

   (A) $\sqrt{10}$ (B) $10\sqrt{10}$ (C) 10   (D) $1/\sqrt{10}$

16. ••A planet is moving in an elliptical orbit around the sun. If T, V, E and L stand respectively for its kinetic energy, gravitational potential energy, total energy and magnitude of angular momentum about the centre of force, which of the following is correct?

   (A) T is conserved.

   (B) V is always positive.

   (C) E is always negative.

   (D) L is conserved but direction of vector L changes continuously.

17. ••The largest and the shortest distance of the earth from the sun are r_1 and r_2. Its distance from the sun when it is at perpendicular to the major-axis of the orbit drawn from the sun is

   (A) $\frac{r_1+r_2}{4}$ (B) $\frac{r_1+r_2}{r_1-r_2}$

   (C) $\frac{2r_1r_2}{r_1+r_2}$ (D) $\frac{r_1+r_2}{3}$

18. ••A planet of mass m moves around the sun of mass M in elliptical orbit. The maximum and minimum distances of the planet from the sun are r_1 and r_2 respectively. The time period of the planet is proportional to

   (A) $(r_1 - r_2)^{3/2}$ (B) $(r_1 + r_2)^{3/2}$

   (C) $r_1^{3/2}$ (D) $r_1^{2/5}$

19. ••A geostationary satellite is orbiting the earth at a height of $5R$ above the surface of the earth, R being the radius of the earth. The time period of another satellite in hours at a height of $2R$ from the surface of the earth is

   (A) 5   (B) 10   (C) $6\sqrt{2}$ (D) $\frac{6}{\sqrt{2}}$

20. ••Kepler's third law states that square of period of revolution (T) of a planet around the sun, is proportional to third power of average distance r between sun and planet i.e. $T^2 = kr^3$ here k is constant. If the masses of sun and planet are M and m respectively then as per Newton's law of gravitation force of attraction between them is $F = \frac{GMm}{r^2}$, here G is gravitational constant. The relation between G and K is described as

   (A) $K = G$ (B) $GK = 4\pi^2$

   (C) $K = \frac{1}{G}$ (D) $GMK = 4\pi^2$

1.23   Solved Problems

EXAMPLE 1. A uniform thin spherical shell has a radius of 2.00 m and a mass of 300 kg and its centre is located at the origin of a coordinate system. Another uniform thin spherical shell with a radius of 1.00 m and a mass of 150 kg is inside the larger shell, with its centre at 0.600 m on the x axis. What is the gravitational force of attraction between the two shells?

SOLUTION The gravitational force is zero. The gravitational field inside the 2.00 m shell due to that shell is zero; therefore, it exerts no force on the 1.00 m-shell, and, by Newton's third law, that shell exerts no force on the larger shell.

EXAMPLE 2. Two concentric uniform thin spherical shells have masses M_1 and M_2 and radii a and $2a$, as in Figure1.173. What is the magnitude of the gravitational force on a point particle of mass m (not shown) located (a) a distance $3a$ from the centre of the shells? (b) a distance $1.9a$ from the centre of the shells? (c) a distance $0.9a$ from the centre of the shells?

APPROACH The magnitude of the gravitational force is

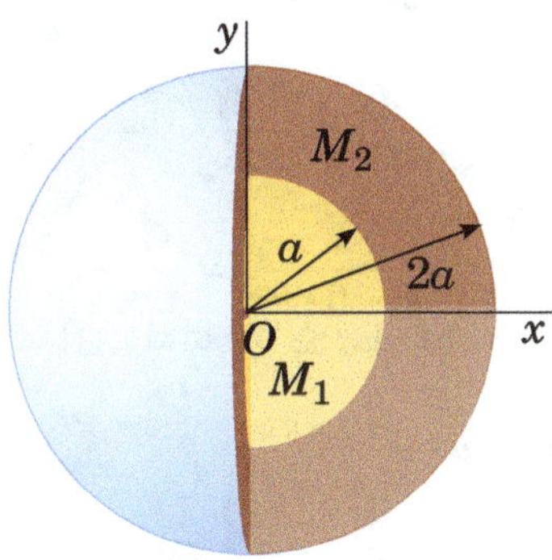

Figure 1.173

$F_g = mg$ where g inside a spherical shell is zero and outside is given by $g = GM/r^2$.

(a) The gravitational force on a particle of mass m is given by:

$$F_g = mg$$

At $r = 3a$, the masses of both spheres contribute to g :

$$F_g(3a) = m\frac{G(M_1 + M_2)}{(3a)^2}$$
$$= \frac{Gm(M_1 + M_2)}{9a^2}$$

(b) At $r = 1.9a$, g due to M_2 is zero and:

$$F_g(1.9a) = m\frac{GM_1}{(1.9a)^2} = \frac{GmM_1}{3.61a^2}$$

(c) At $r = 0.9a$, $g = 0$, therefore

$$F_g(0.9a) = 0$$

EXAMPLE 3. The inner spherical shell in **Example 2** is shifted so that its centre is now on the x axis at $x = 0.8a$. What is the magnitude of the gravitational force on a particle of point mass m located on the x axis at (a) (a) $x = 3a$? (b) $x = 1.9a$? (c) $x = 0.9a$?

APPROACH The configuration is shown in Fig.1.174. The centers of the spheres are indicated by the centerlines. The x coordinates of the mass m for Parts (a),(b), and (c) are indicated along the x axis. The magnitude of the gravitational force is $F_g = mg$ where g inside a spherical shell is zero and

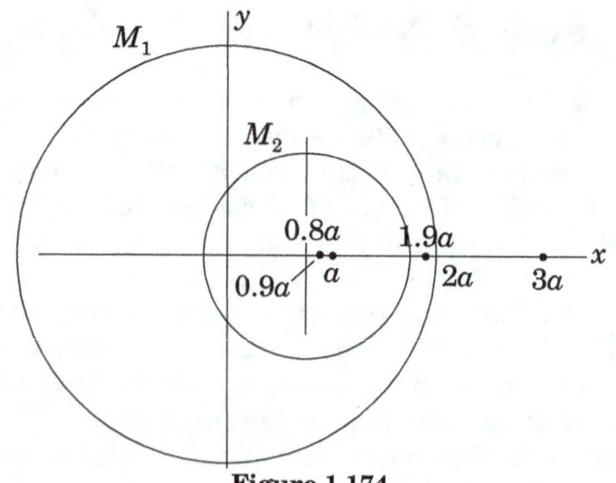

Figure 1.174

in Figure 1.175. The mass of a solid uniform lead sphere of radius R is M. Find the force of attraction on a point particle of mass m located a distance d from the center of the lead sphere as shown.

APPROACH The force of attraction of the small sphere of

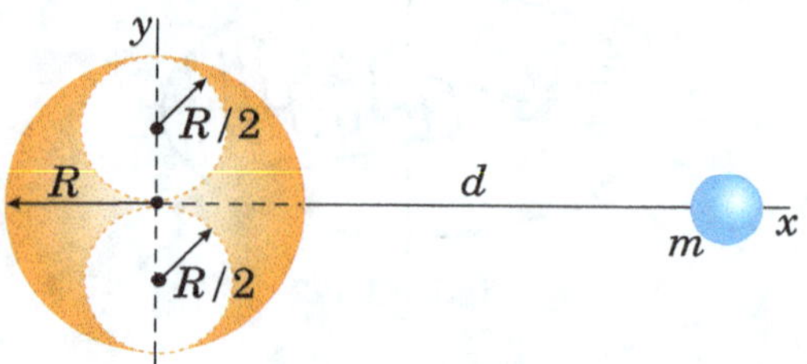

Figure 1.175

outside is given by $g = \frac{GM}{r^2}$.

SOLUTION (a) The gravitational force acting on the object whose mass is m is

$$F_g = m(g_{1x} + g_{2x}) \qquad \ldots (1)$$

Now, g_{1x} at $x = 3a$ is

$$g_{1x}(3a) = \frac{GM_1}{(3a)^2} = \frac{GM_1}{9a^2}$$

and g_{2x} at $x = 3a$ is

$$g_{2x}(3a) = \frac{GM_2}{(3a - 0.8a)^2} = \frac{GM_2}{4.84a^2}$$

Substituting the values of $g_{1x}(3a)$ and $g_{2x}(3a)$ in equation (1) and simplify we get

$$F(3a) = m\left(\frac{GM_1}{9a^2} + \frac{GM_2}{4.84a^2}\right)$$
$$= \frac{Gm}{a^2}\left(\frac{M_1}{9} + \frac{M_2}{4.84}\right)$$

(b) Now, g_{2x} at $x = 1.9a$:

$$g_{2x}(1.9a) = \frac{GM_2}{(1.9a - 0.8a)^2} = \frac{GM_2}{1.21a^2}$$

and g_{1x} at $x = 1.9a$: $g_{1x}(1.9a) = 0$

Substituting $g_{1x}(1.9a)$ and $g_{2x}(1.9a)$ in Eq.(1), we get

$$F(1.9a) = mg = \frac{GmM_2}{1.21a^2}$$

(c) At $x = 0.9a$, $g_{1x} = g_{2x} = 0$, therefore $F(0.9a) = 0$

EXAMPLE 4. Two widely separated solid spheres, S_1 and S_2, each have radius R and mass M. Sphere S_1 is uniform, whereas the density of S_2 is given by $\rho(r) = C/r$, where r is the distance from its centre. If the gravitational field strength at the surface of S_1 is g_1, what is the gravitational field strength at the surface of S_2?

APPROACH The gravitational field strength at the surface of a sphere is given by $g = GM/R^2$, where R is the radius of the sphere and M is its mass.

SOLUTION Express the gravitational field strength on the surface of S_1 :

$$g_1 = \frac{GM}{R^2}$$

Express the gravitational field strength on the surface of S_2 :

$$g_2 = \frac{GM}{R^2}$$

Divide the second of these equations by the first and simplify to obtain:

$$\frac{g_2}{g_1} = \frac{\frac{GM}{R^2}}{\frac{GM}{R^2}} = 1 \Rightarrow g_1 = g_2$$

EXAMPLE 5. Two identical spherical cavities are made in a lead sphere of radius R. The cavities each have a radius $R/2$. They touch the outside surface of the sphere and its center as

mass m to the lead sphere of mass M is the sum of the forces due to the solid sphere $\left(\vec{F}_S\right)$ and the cavities $\left(\vec{F}_C\right)$ of negative mass.

SOLUTION The net gravitational force of attraction:

$$\vec{F} = \vec{F}_S + \vec{F}_C \qquad \ldots (1)$$

The gravitational force due to the solid sphere:

$$\vec{F}_S = -\frac{GMm}{d^2}\hat{\mathbf{i}}$$

The magnitude of the force acting on the small sphere due to one cavity:

$$F_C = \frac{GM'm}{d^2 + \left(\frac{R}{2}\right)^2}$$

where M' is the negative mass of a cavity.

Relate the negative mass of a cavity to the mass of the sphere before hollowing:

$$M' = -\rho V = -\rho\left[\frac{4}{3}\pi\left(\frac{R}{2}\right)^3\right]$$
$$= -\frac{1}{8}\left(\frac{4}{3}\pi\rho R^3\right) = -\frac{1}{8}M$$

Let θ be the angle between the x axis and the line joining the centre of the small sphere to the center of either cavity, use the law of gravity to express the force due to the two cavities:

$$\vec{F}_C = 2\frac{GMm}{8\left(d^2 + \frac{R^2}{4}\right)}\cos\theta\hat{\mathbf{i}}$$

because, by symmetry, the y components add to zero.

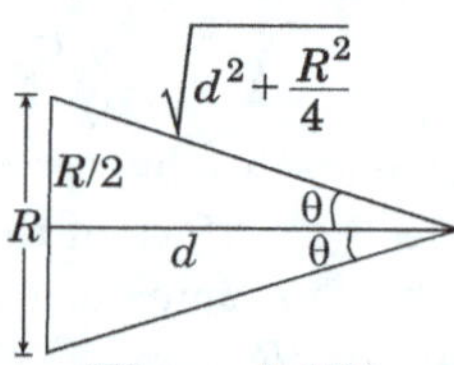

Figure 1.176

Use Fig.1.176, to express $\cos\theta$:

$$\cos\theta = \frac{d}{\sqrt{d^2 + \frac{R^2}{4}}}$$

On substituting the value of $\cos\theta$, we get

$$\vec{F}_C = \frac{GMm}{4\left(d^2 + \frac{R^2}{4}\right)}\frac{d}{\sqrt{d^2 + \frac{R^2}{4}}}\hat{i}$$
$$= \frac{GMmd}{4\left(d^2 + \frac{R^2}{4}\right)^{3/2}}\hat{i}$$

where M' is the negative mass of a cavity.

$$M' = -\rho V = -\rho \left[\frac{4}{3} \pi \left(\frac{R}{2} \right)^3 \right]$$
$$= -\frac{1}{8} \left(\frac{4}{3} \pi \rho R^3 \right) = -\frac{1}{8} M$$
$$\vec{F}_C = 2 \frac{GMm}{8\left(d^2 + \frac{R^2}{4}\right)} \cos\theta \,\hat{\mathbf{i}}$$

because, by symmetry, the y components add to zero.

$$\cos\theta = \frac{d}{\sqrt{d^2 + \frac{R^2}{4}}}$$
$$\vec{F}_C = \frac{GMm}{4\left(d^2 + \frac{R^2}{4}\right)} \frac{d}{\sqrt{d^2 + \frac{R^2}{4}}} \hat{i}$$
$$= \frac{GMmd}{4\left(d^2 + \frac{R^2}{4}\right)^{3/2}} \hat{i}$$

Substituting in equation (1), we get

$$\vec{F} = -\frac{GMm}{d^2} \hat{i} + \frac{GMmd}{4\left(d^2 + \frac{R^2}{4}\right)^{3/2}} \hat{i}$$
$$= -\frac{GMm}{d^2} \left[1 - \frac{d^3}{\left\{d^2 + \frac{R^2}{4}\right\}^{3/2}} \right] \hat{i}$$

EXAMPLE 6. Two widely separated uniform solid spheres, S_1 and S_2, have equal masses, but different radii, R_1 and R_2. If the gravitational field strength on the surface of S_1 is g_1, what is the gravitational field strength on the surface of S_2?

APPROACH The gravitational field strength at the surface of a sphere is given by $g = GM/R^2$, where R is the radius of the sphere and M is its mass.

Express the gravitational field strength on the surface of S_1:
$$g_1 = \frac{GM}{R_1^2}$$

Express the gravitational field strength on the surface of S_2:
$$g_2 = \frac{GM}{R_2^2}$$

Divide the second of these equations by the first and simplify to obtain:
$$\frac{g_2}{g_1} = \frac{\frac{GM}{R_2^2}}{\frac{GM}{R_1^2}} = \frac{R_1^2}{R_2^2} \Rightarrow g_2 = \frac{R_1^2}{R_2^2} g_1$$

Remarks: The gravitational field strengths depend only on the masses and radii because the points of interest are outside spherically symmetric distributions of mass.

EXAMPLE 7. A solid sphere of radius R has its centre at the origin. It has a uniform mass density ρ_0, except that the sphere has a spherical cavity in it of radius $r = \frac{1}{2}R$ centred at $x = \frac{1}{2}R$, as in Figure 1.177. Find the gravitational field at points on the x axis for $|x| > R$. Hint: The cavity may be thought of as a sphere of mass $m = (4/3)\pi r^3 \rho_0$ plus a sphere of "negative" mass $-m$.

APPROACH We can use the hint to find the gravitational field along the x axis.

SOLUTION Using the hint, express $g(x)$:
$$g(x) = g_{\text{solid sphere}} + g_{\text{hollow sphere}}$$

On substituting the values of $g_{\text{solid sphere}}$ and $g_{\text{hollow sphere}}$ in above expression, we get-
$$g(x) = \frac{GM_{\text{solid sphere}}}{x^2} + \frac{GM_{\text{hollow sphere}}}{\left(x - \frac{1}{2}R\right)^2}$$

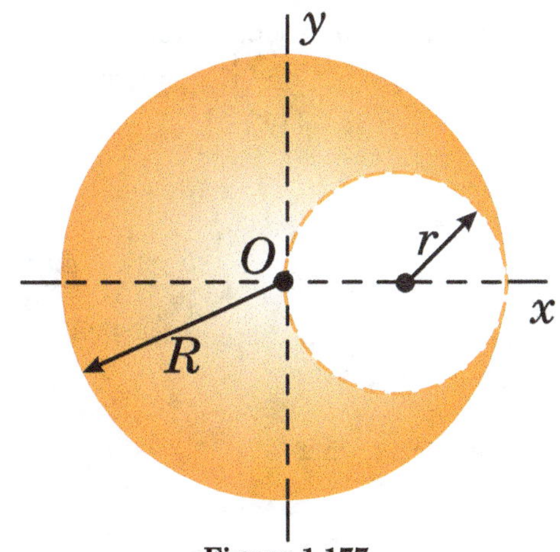

Figure 1.177

$$= \frac{G\rho_0\left(\frac{4}{3}\pi R^3\right)}{x^2} + \frac{G\rho_0\left[-\frac{4}{3}\pi\left(\frac{1}{2}R\right)^3\right]}{\left(x - \frac{1}{2}R\right)^2}$$
$$= G\left(\frac{4\pi\rho_0 R^3}{3}\right)\left[\frac{1}{x^2} - \frac{1}{8\left(x - \frac{1}{2}R\right)^2} \right]$$

EXAMPLE 8. For the sphere with the cavity in **Example 7**, show that the gravitational field is uniform throughout the cavity, and find its magnitude and direction there.

APPROACH Fig.1.178 shows the portion of the solid sphere in which the hollow sphere is embedded. $\vec{g}_1$ is the field due to the solid sphere of radius R and density ρ_0 and $\vec{g}_2$ is the field due to the sphere of radius $\frac{1}{2}R$ and negative density ρ_0 centred at $\frac{1}{2}R$. We can find the resultant field by adding the x and y components of $\vec{g}_1$ and $\vec{g}_2$.

Use it's definition to express $\left|\vec{g}_1\right|$:

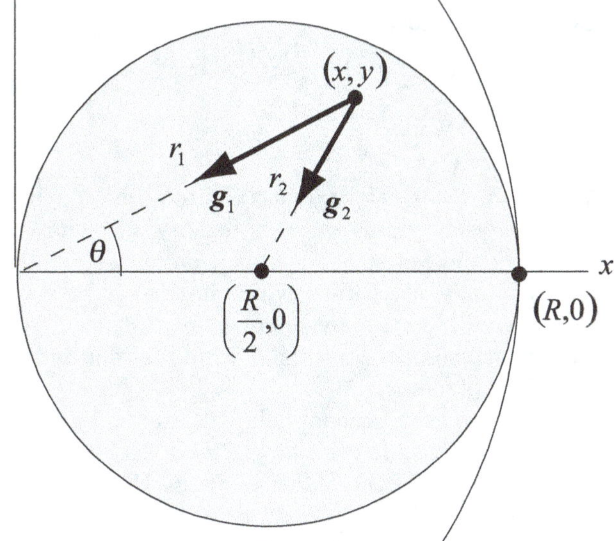

Figure 1.178

$$\left|\vec{g}_1\right| = \frac{F_g}{m}$$

Substitute for the gravitational force to obtain:
$$\left|\vec{g}_1\right| = \frac{\frac{GMm}{r_1^2}}{m} = \frac{GM}{r_1^2}$$

Substituting the product of its density and volume for the

mass of the sphere and simplifying yields:

$$|\vec{g}_1| = \frac{G\rho_0 V}{r_1^2} = \frac{4\pi\rho_0 r_1^3 G}{3r_1^2} = \frac{4\pi\rho_0 r_1 G}{3}$$

Find the x and y components of $\vec{g}_1$:

$$g_{1x} = -g_1\cos\theta = -g_1\left(\frac{x}{r_1}\right) = -\frac{4\pi\rho_0 Gx}{3}$$

and

$$g_{1y} = -g_1\sin\theta = -g_1\left(\frac{y}{r_1}\right) = -\frac{4\pi\rho_0 Gy}{3}$$

where the negative signs indicate that the field points inward. Proceed similarly to express $|\vec{g}_2|$:

Express the x and y components

$$|\vec{g}_2| = \frac{4\pi\rho_0 r_2 G}{3}$$

and

$$g_{2x} = g_2\left(\frac{x - \frac{1}{2}R}{r_2}\right) = \frac{4\pi\rho_0 G\left(x - \frac{1}{2}R\right)}{3} \quad // \text{ Add the } x$$

components and simplify

$$g_{2y} = g_2\left(\frac{y}{r_2}\right) = \frac{4\pi\rho_0 Gy}{3}$$

$$g_x = g_{1x} + g_{2x}$$

$$= -\frac{4\pi\rho_0 Gx}{3} + \frac{4\pi\rho_0 G\left(x - \frac{1}{2}R\right)}{3}$$

$$= -\frac{2\pi\rho_0 GR}{3}$$

where the negative sign indicates that the field points inward. Add the y components and simplify to obtain the y component of the resultant field:

$$g_y = g_{1y} + g_{2y}$$

$$= -\frac{4\pi\rho_0 Gy}{3} + \frac{4\pi\rho_0 Gy}{3} = 0$$

Express $\vec{g}$ in vector form:

$$\vec{g} = g_x\hat{\mathbf{i}} + g_y\hat{\mathbf{j}} = \left(-\frac{2\pi\rho_0 GR}{3}\right)\hat{\mathbf{i}} + 0\hat{\mathbf{j}}$$

The magnitude of $\vec{g}$ is:

$$|\vec{g}| = \sqrt{g_x^2 + g_y^2} = \frac{2\pi\rho_0 GR}{3}$$

EXAMPLE 9.   A straight, smooth tunnel is dug through a uniform spherical planet of mass density ρ_0. The tunnel passes through the centre of the planet and is perpendicular to the planet's axis of rotation, which is fixed in space. The planet rotates with a constant angular speed ω, so objects in the tunnel have no apparent weight. Find the required angular speed of the planet ω.

APPROACH The gravitational field will exert an inward radial force on the objects in the tunnel. We can relate this force to the angular speed of the planet by using Newton's second law of motion.

SOLUTION Let r be the distance from the objects to the centre of the planet, use Newton's second law to relate the gravitational force acting on the objects to their angular speed:

$$F_{\text{net}} = F_g = mr\omega^2$$

or

$$mg = mr\omega^2 \Rightarrow \omega = \sqrt{\frac{g}{r}} \qquad \ldots (1)$$

Now, gravitational acceleration can be written as,

$$g = \frac{F_g}{m} = \frac{\frac{GMm}{r^2}}{m} = \frac{GM}{r^2}$$

Since, $M = \rho_0 V$, therefore

$$g = \frac{G\rho_0 V}{r^2} = \frac{4\pi\rho_0 r^3 G}{3r^2} = \frac{4\pi\rho_0 rG}{3}$$

Substituting this value of g in equation (1), we get

$$\omega = \sqrt{\frac{\frac{\pi\rho_0 rG}{3}}{r}} = \sqrt{\frac{4\pi\rho_0 G}{3}}$$

EXAMPLE 10.   The density of a sphere is given by $\rho(r) = C/r$. The sphere has a radius of 5.0 m and a mass of 1.0×10^{11} kg. (a) Determine the constant C.   (b) Obtain expressions for the gravitational field for the regions (1) $r > 5.0$ m, and (2) $r < 5.0$ m.

APPROACH Because we're given the mass of the sphere, we can find C by expressing the mass of the sphere in terms of C. We can use its definition to find the gravitational field of the sphere both inside and outside its surface.

SOLUTION (a) The mass of a differential element of the sphere: $dm = \rho dV = \rho\left(4\pi r^2 dr\right)$ Therefore, the total mass of the sphere in terms of C:

$$M = 4\pi C \int_0^{5.0 \text{ m}} r\,dr = \left(50 \text{ m}^2\right)\pi C$$

Solving for C yields:

$$C = \frac{M}{\left(50 \text{ m}^2\right)\pi}$$

$$= \frac{1.0 \times 10^{11} \text{ kg}}{\left(50 \text{ m}^2\right)\pi} = 6.37 \times 10^8 \text{ kg/m}^2$$

$$= 6.4 \times 10^8 \text{ kg/m}^2$$

(b) The gravitational field of the sphere at a distance from its centre greater than its radius:

$$g = \frac{GM}{r^2}$$

(1) For $r > 5.0$ m :

$$g(r > 5.0 \text{ m}) = \frac{\left(6.673 \times 10^{-11} \text{ N} \cdot \text{m}^2/\text{kg}^2\right)\left(1.0 \times 10^{11} \text{ kg}\right)}{r^2}$$

$$= \frac{6.7 \text{ N} \cdot \text{m}^2/\text{kg}}{r^2}$$

The gravitational field of the sphere at a distance from its center less than its radius: $\quad g = \dfrac{GM_{\text{encl}}}{r^2} =$

$$G\frac{\int_0^r 4\pi r^2 \rho\, dr}{r^2} = G\frac{\int_0^r 4\pi r^2 \frac{C}{r}\, dr}{r^2}$$

$$= G\frac{4\pi C \int_0^r r\, dr}{r^2} = 2\pi GC$$

(2) For $r < 5.0$ m, we have-

$$g = 2\pi\left(6.673 \times 10^{-11} \text{ N} \cdot \text{m}^2/\text{kg}^2\right)\left(6.37 \times 10^8 \text{ kg/m}^2\right)$$

$$= 0.27 \text{ N/kg}$$

EXAMPLE 11.   The density of a sphere is given by $\rho(r) = C/r$. The sphere has a radius of 5.0 m and a mass of 1.0×10^{11} kg. A small-diameter hole is drilled into the sphere, toward the centre of the sphere to a depth of 2.0 m below the sphere's surface. A small mass is dropped from the surface into the hole. Determine the speed of the small mass when it strikes the bottom of the hole.

APPROACH We can use conservation of energy to relate the work done by the gravitational field to the speed of the small object as it strikes the bottom of the hole. Because we're given the mass of the sphere, we can find C by expressing the mass of the sphere in terms of C. We can then use the definition of the gravitational field to find the gravitational field of the sphere inside its surface. The work done by the field equals the negative of the change in the potential energy of the system as the small object falls in the hole.

SOLUTION Use conservation of energy to relate the work done by the gravitational field to the speed of the small object as it strikes the bottom of the hole:

$$U_f + K_f = U_i + K_i$$
$$\Rightarrow \quad K_f - K_i = -\left(U_f - U_i\right) = -\Delta U = W$$

Since, $K_i = 0$ therefore,

$$\frac{1}{2}mv^2 = W \quad \Rightarrow \quad v = \sqrt{\frac{2W}{m}} \qquad \ldots (1)$$

where v is the speed with which the object strikes the bottom of the hole and W is the work done by the gravitational field. The mass of an element of the sphere:

$$dm = \rho dV = \rho\left(4\pi r^2 dr\right)$$

Integrate to express the total mass of the sphere in terms of C:

$$M = 4\pi C \int_0^{5.0\text{ m}} r dr = \left(50\text{ m}^2\right)\pi C$$

Solving for C yields:

$$C = \frac{M}{\left(50\text{ m}^2\right)\pi}$$
$$= \frac{1.0 \times 10^{11} kg}{\left(50 m^2\right)\pi} = 6.37 \times 10^8 kg/m^2$$

The gravitational field of the sphere at a distance from its center less than its radius:

$$g = \frac{F_g}{m} = \frac{GM}{r^2} = G\frac{\int_0^r 4\pi r^2 \rho dr}{r^2}$$
$$= G\frac{\int_0^r 4\pi r^2 \frac{C}{r} dr}{r^2} = G\frac{4\pi C \int_0^r r dr}{r^2} = 2\pi GC$$

The work done on the small object by the gravitational force acting on it:

$$W = -\int_{5.0\text{ m}}^{3.0\text{ m}} mgdr = (2\text{ m})mg$$

Substituting this value of W, in equation (1), we get

$$v = \sqrt{\frac{2(2.0\text{ m})m(2\pi GC)}{m}} = \sqrt{(8.0\text{ m})\pi GC}$$

Substituting numerical values and evaluate v, we get :

$$v = \left[(8.0\text{ m})\pi\left(6.673 \times 10^{-11}\text{ N}\cdot\text{m}^2/\text{kg}^2\right)\right.$$
$$\left.\left(6.37 \times 10^8\text{ kg/m}^2\right)\right]^{1/2} = 1.0\text{ m/s}$$

EXAMPLE 12. As a geologist for a mining company, you are working on a method for determining possible locations of underground ore deposits. Assume that where the company owns land the crust of Earth is 40.0 km thick and has a density of about 3000 kg/m³. Suppose a spherical deposit of heavy metals with a density of 8000 kg/m³ and radius of 1000 m is centered 2000 m below the surface. You propose to detect it by determining its effect on the local surface value of g. Find $\Delta g/g$ at the surface directly above this deposit, where Δg is the increase in the gravitational field due to the deposit.

APPROACH The spherical deposit of heavy metals will increase the gravitational field at the surface of Earth. We can express this increase in terms of the difference in densities of the deposit and Earth and then form the quotient $\Delta g/g$

SOLUTION Δg due to the spherical deposit:

$$\Delta g = \frac{G\Delta M}{r^2} \qquad (1)$$

The mass of the spherical deposit:

$$M = \Delta\rho V = \Delta\rho\left(\frac{4}{3}\pi R^3\right) = \frac{4}{3}\pi\Delta\rho R^3$$

Substituting it in Eq.(1), we get

$$\Delta g = \frac{\frac{4}{3}G\pi\Delta\rho R^3}{r^2}$$

Therefore,

$$\frac{\Delta g}{g} = \frac{\frac{\frac{4}{3}G\pi\Delta\rho R^3}{r^2}}{g} = \frac{\frac{4}{3}G\pi\Delta\rho R^3}{gr^2}$$

On substituting numerical values, we get

$$\frac{\Delta g}{g} = \frac{\frac{4}{3}\pi\left(6.673 \times 10^{-11}\text{ N}\cdot\text{m}^2/\text{kg}^2\right)\left(5000\text{ kg/m}^3\right)\left(1000\text{ m}\right)^3}{(9.81\text{ N/kg})(2000\text{ m})^2}$$
$$= 3.56 \times 10^{-5}$$

EXAMPLE 13. A satellite is circling the moon (radius 1700 km) close to the surface at a speed v. A projectile is launched vertically up from the moon's surface at the same initial speed v How high will the projectile rise?

APPROACH We can use conservation of mechanical energy to establish a relationship between the height h to which the projectile will rise and its initial speed. The application of Newton's second law will relate the orbital speed, which is equal to the initial speed of the projectile, to the mass and radius of the moon.

SOLUTION Use conservation of mechanical energy to relate the initial energies of the projectile to its final energy:

$$K_f - K_i + U_f - U_i = 0$$

or, because $K_f = 0$, $-\frac{1}{2}mv^2 - \frac{GM_m m}{R_m + h} + \frac{GM_m m}{R_m} = 0$

Here, R_m and M_m are respectively, the radius and mass of the moon.

Solving for h yields:

$$h = R_m\left(\frac{1}{1 - \frac{v^2 R_m}{2GM_m}} - 1\right)$$

Use Newton's second law to relate the orbital speed of the satellite to the gravitational force acting on it:

$$\sum F_{\text{radial}} = \frac{GM_m m}{R_m^2} = \frac{mv^2}{R_m}$$

Solve for v^2 to obtain:

$$v^2 = \frac{GM_m}{R_m}$$

Substitute for v^2 in equation (1) and simplify to obtain:

$$h = R_m\left(\frac{1}{1 - \frac{1}{2}} - 1\right) = R_m = 1.74 \times 10^6\text{ m}$$

EXAMPLE 14. In a binary star system, two stars follow circular orbits about their common center of mass. If the stars have masses m_1 and m_2 and are separated by a distance r, show that the period of rotation is related to r by $T^2 = 4\pi^2 r^3/[G(m_1 + m_2)]$.

APPROACH Let the origin of our coordinate system be at the center of mass of the binary star system and let the distances of the stars from their centre of mass be r_1 and r_2. The period of rotation is related to the angular speed of the star system and we can use Newton's second law of motion to relate this speed to the separation of the stars.

SOLUTION Relate the square of the period of the motion of the stars to their angular speed:

$$T^2 = \frac{4\pi^2}{\omega^2} \qquad \ldots (1)$$

Using Newton's second law, relate the gravitational force acting on the star whose mass is m_2 to the angular speed of the system:

$$\sum F_{\text{radial}} = \frac{Gm_1m_2}{(r_1+r_2)^2} = m_2r_2\omega^2$$

Solving for ω^2 yields:

$$\omega^2 = \frac{Gm_1}{r_2(r_1+r)^2} \qquad \ldots(2)$$

From the definition of the center of mass we have:

$$m_1r_1 = m_2r_2 \qquad (3)$$

where
$$r = r_1 + r_2 \qquad \ldots\,(4)$$

Eliminate r_1 from equations (3) and (4) and solve for r_2 to obtain:

$$r_2 = \frac{rm_1}{m_1+m_2}$$

Eliminate r_2 from equations (3) and (4) and solve for r_1 to obtain:

$$r_1 = \frac{rm_2}{m_1+m_2}$$

Substituting for r_1 and r_2 in equation (2) yields:

$$\omega^2 = \frac{G(m_1+m_2)}{r^3}$$

Finally, substitute for ω^2 in equation (1) and simplify:

$$T^2 = \frac{4\pi^2}{\frac{G(m_1+m_2)}{r^3}} = \frac{4\pi^2r^3}{G(m_1+m_2)}$$

EXAMPLE 15. It is believed that there is a "supermassive" black hole at the centre of our galaxy. One datum that leads to this conclusion is the important recent observation of stellar motion in the vicinity of the galactic center. If one such star moves in an elliptical orbit with a period of 15.2 years and has a semimajor axis of 5.5 light-days (the distance light travels in 5.5 days), what is the mass around which the star moves in its Keplerian orbit?

APPROACH We can apply Kepler's third law to the orbital motion of the star to find the effective mass around which it is moving.

SOLUTION Using Kepler's third law, relate the orbital period of the star to the semi-major axis of its orbit:

$$T^2 = \frac{4\pi^2}{GM}a^3 \quad \Rightarrow \quad M = \frac{4\pi^2a^3}{GT^2}$$

where M is the mass around which the star moves in its Keplerian orbit.

Substitute numerical values and evaluate M:

$$M = \frac{4\pi^2\left(5.5\,d \times \frac{86400s}{d} \times 2.998\times10^8 m/s\right)^3}{\left(6.673\times10^{-11}N\cdot m^2/kg^2\right)\left(15.2\,y \times \frac{3.156\times10^7 s}{y}\right)^2}$$

$$= 7.434\times10^{36}kg = 7.434\times10^{36}kg \times \frac{1M_{\text{Sun}}}{1.99\times10^{30}kg}$$

$$= 3.7\times10^6 M_{\text{Sun}}$$

EXAMPLE 16. Four identical planets are arranged in a square, as shown in Figure 1.179. If the mass of each planet is M and the edge length of the square is a, what must their speed be if they are to orbit their common centre under the influence of their mutual attraction?

APPROACH Note that, due to the symmetrical arrangement of the planets, each experiences the same centripetal force [see Fig.1.180]. We can apply Newton's second law to any of the four planets to relate its orbital speed to this net (centripetal) force acting on it.

Applying $\sum \vec{F}_{\text{radial}} = m\vec{a}_{\text{radial}}$ to one of the planets gives:

$$F_c = 2F_1\cos\theta + F_2$$

Substituting for F_1, F_2, and θ and simplifying yields:

$$F_c = \frac{2GM^2}{a^2}\cos45° + \frac{GM^2}{(\sqrt{2}a)^2}$$

$$= \frac{2GM^2}{a^2}\frac{1}{\sqrt{2}} + \frac{GM^2}{2a^2}$$

$$= \frac{GM^2}{a^2}\left(\sqrt{2}+\frac{1}{2}\right)$$

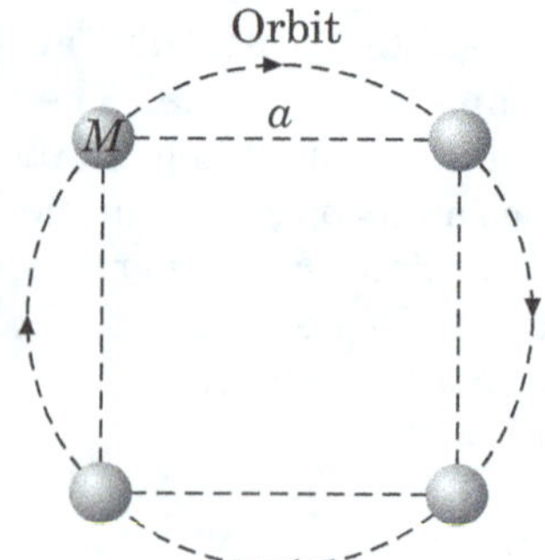

Figure 1.179

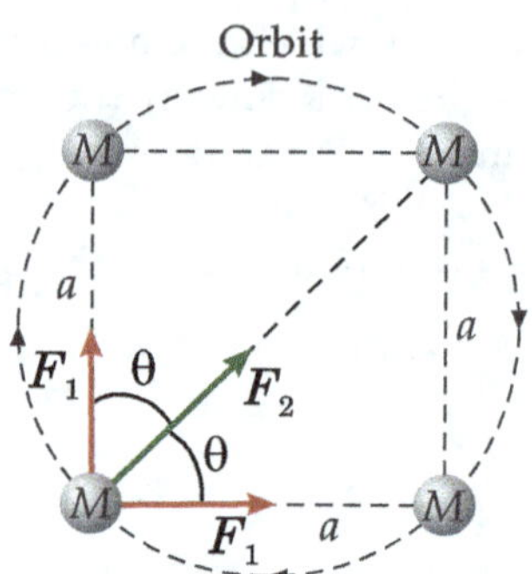

Figure 1.180

Because $F_c = \frac{Mv^2}{a/\sqrt{2}} = \frac{\sqrt{2}Mv^2}{a}$, therefore,

$$\frac{\sqrt{2}Mv^2}{a} = \frac{GM^2}{a^2}\left(\sqrt{2}+\frac{1}{2}\right)$$

Solve for v to obtain:

$$v = \sqrt{\frac{GM}{a}\left(1+\frac{1}{2\sqrt{2}}\right)} = 1.16\sqrt{\frac{GM}{a}}$$

EXAMPLE 17. The average density of the moon is $\rho = 3340$ kg/m^3. Find the minimum possible period T of a spacecraft orbiting the moon.

APPROACH Let m represent the mass of the spacecraft. From Kepler's third law we know that its period will be a minimum when it is in orbit just above the surface of the moon. We'll use Newton's second law to relate the angular speed of the spacecraft to the gravitational force acting on it.

SOLUTION Relate the period of the spacecraft to its angular speed:

$$T = \frac{2\pi}{\omega} \qquad \ldots\,(1)$$

Using Newton's second law, relate the gravitational force acting on the spacecraft when it is in orbit at the surface of the moon to the angular speed of the spacecraft:

$$\sum F_{\text{radial}} = \frac{GM_M m}{R_M^2} = mR_M\omega^2$$

Solving for ω and simplifying yields:

$$\omega = \sqrt{\frac{GM_M}{R_M^3}} = \sqrt{\frac{G\left(\frac{4}{3}\pi\rho R_M^3\right)}{R_M^3}}$$

$$= \sqrt{\frac{4}{3}G\pi\rho}$$

Substituting this value of ω in equation (1), we get

$$T_{\text{min}} = \frac{2\pi}{\sqrt{\frac{4}{3}G\pi\rho}} = \sqrt{\frac{3\pi}{\rho G}}$$

Substituting numerical values in above equation, we get

$$T_{\min} = \sqrt{\dfrac{3\pi}{\left(6.673 \times 10^{-11} \text{ N} \cdot \text{m}^2/\text{kg}^2\right)\left(3340 \text{ kg/m}^3\right)}}$$
$$= 6503 \text{ s} = 1 \text{ h} 48 \text{ min}$$

EXAMPLE 18. Suppose the Sun could collapse into a neutron star of radius 12.0 km. Your research team is in charge of sending a probe from Earth to study the transformed Sun, and the probe needs to end up in a circular orbit 4500 km from the neutronSun's centre. (a) Calculate the orbital speed of the probe. (b) Later on, plans call for construction of a permanent spaceport in that same orbit. To transport equipment and supplies, scientists on Earth need you to determine the escape speed for rockets launched from the spaceport (relative to the spaceport) in the direction of the spaceport's orbital velocity at takeoff time. What is that speed, and how does it compare to the escape speed at the surface of Earth?

APPROACH We can apply Newton's law to the probe orbiting the Sun to determine its orbital speed. Using the escape-speed equation will allow us to find the escape speed for rockets launched from the spaceport.

SOLUTION (a) Apply Newton's second law to the probe of mass m in orbit about the Sun:

$$\sum F_{\text{radial}} = \frac{GmM_{\text{Sun}}}{r^2} = m\frac{v_{\text{orbital}}^2}{r}$$

where r is the orbital radius.

$$v_{\text{orbital}} = \sqrt{\frac{GM_{\text{Sun}}}{r}}$$

Substitute numerical values and evaluate v_{orbital} :

$$v_{\text{orbital}} = \sqrt{\frac{(6.673 \times 10^{-11} \text{ N} \cdot \text{m}^2/\text{kg}^2)(1.99 \times 10^{30} \text{ kg})}{4.50 \times 10^6 \text{ m}}}$$
$$= 5.432 \times 10^6 \text{ m/s} = 5.43 \times 10^6 \text{ m/s}$$

(b) The escape speed (relative to the spaceport) for rockets launched from the spaceport is given by:

$$v_{\substack{\text{rel to} \\ \text{spaceport}}} = v_{\text{e}} - v_{\text{orbital}}$$

The escape speed at a distance r from the center of the neutron-Sun is given by:

$$v_{\text{e}} = \sqrt{\frac{2GM_{\text{neutron-Sun}}}{r}} = \sqrt{2}\sqrt{\frac{GM_{\text{neutron-Sun}}}{r}}$$
$$= \sqrt{2}\,v_{\text{orbital}}$$

Substituting for v_{e} in equation (1) yields:

$$v_{\substack{\text{rel to} \\ \text{spaceport}}} = \sqrt{2}\,v_{\text{orbital}} - v_{\text{orbital}}$$
$$= (\sqrt{2} - 1)v_{\text{orbital}}$$

Substitute numerical values and evaluate $v_{\substack{\text{rel to} \\ \text{spaceport}}}$:

$$v_{\substack{\text{rel to} \\ \text{spaceport}}} = (\sqrt{2} - 1)(5.431 \times 10^6 \text{ m/s}) = 2.25 \times 10^6 \text{ m/s}$$

Express the ratio of $v_{\substack{\text{rel to} \\ \text{spaceport}}}$ to $v_{\text{e, Earth}}$:

$$\frac{v_{\substack{\text{rel to} \\ \text{spaceport}}}}{v_{\text{e, Earth}}} = \frac{2.25 \times 10^6 \text{ m/s}}{11.2 \text{ km/s}} \approx 201$$

1. A planet of mass m is revolving round the sun (of mass M_s) in an elliptical orbit. If $\vec{v}$ is the velocity of the planet when its position vector from the sun is $\vec{r}$, then areal velocity of the planet is.
   (A) $\vec{v} \times \vec{r}$ 　　(B) $\vec{r} \times \vec{v}$
   (C) $\frac{1}{2}(\vec{v} \times \vec{r})$ 　　(D) $\frac{1}{2}(\vec{r} \times \vec{v})$

   SOLUTION (D) Areal velocity $= \dfrac{\Delta \vec{A}}{\Delta t} = \dfrac{\vec{L}}{2m}$
   $$= \frac{1}{2m}[m(\vec{r} \times \vec{v})] = \frac{1}{2}(\vec{r} \times \vec{v})$$

2. A system consists of n identical particles each of mass m. The total number of interactions possible is.
   (A) $n(n+1)$ 　　(B) $\frac{1}{2}n(n+1)$
   (C) $n(n-1)$ 　　(D) $\frac{1}{2}n(n-1)$

   SOLUTION (D) There is gravitational interaction between each pair of particles, so, total number of interactions
   $$= {}^n C_2 = \frac{n(n-1)}{2}$$

3. The ratio of the time period of a simple pendulum of length l_0 with a pendulum of infinite length is.
   (A) zero 　　(B) $\sqrt{\frac{l_0}{R}}$
   (C) $\sqrt{\frac{l_0+R}{R}}$ 　　(D) $\sqrt{\frac{R}{l_0+R}}$
   here, R is the radius of earth

   SOLUTION (B) General formula for time period of simple pendulum is-
   $$T = 2\pi\sqrt{\frac{1}{g\left(\frac{1}{R} + \frac{1}{l}\right)}} \qquad \text{(Note)}$$
   For simple pendulum of length l_0 , we have $R >> l_0$
   therefore, $T_1 = 2\pi\sqrt{\frac{l_0}{g}}$
   If $l \to \infty$, then $T_2 = 2\pi\sqrt{\frac{R}{g}}$
   Therefore, $\qquad \dfrac{T_1}{T_2} = \sqrt{\dfrac{l_0}{R}}$

4. The rotation of the earth having R radius about its axis speeds up to a value such that a man at latitude angle $60°$ feels weightlessness. The duration of the day in such a case is.
   (A) $2\pi\sqrt{\frac{R}{g}}$ 　　(B) $4\pi\sqrt{\frac{R}{g}}$
   (C) $2\pi\sqrt{\frac{g}{R}}$ 　　(D) $4\pi\sqrt{\frac{g}{R}}$

   SOLUTION (B) From Eq.1.20, the effective gravitational acceleration at latitude θ, is $g' = g - R\omega^2\cos^\theta$, therefore
   $$0 = g - R\omega^2\cos^2 60°$$
   $$\Rightarrow \qquad \frac{R^2\omega^2}{4} = g$$
   $$\Rightarrow \qquad \omega = 2\sqrt{\frac{g}{R}}$$
   $$\Rightarrow \qquad T = \pi\sqrt{\frac{R}{g}}$$

5. At what height the gravitational field reduces by 75% the gravitational field at the surface of earth?

  (A) R   (B) 2R   (C) 3R   (D) 4R

SOLUTION (A) $\left(g - \frac{3g}{4}\right) = \frac{gR^2}{(R+h)^2}$

$$\Rightarrow \quad h = R$$

6. In a certain region of space gravitational field intensity is given by $E = -\left(\frac{k}{r}\right)$. Taking the reference point to be at $r = r_o$ with $V = V_o$, the potential (V) is given by

  (A) $V = K \log \frac{r}{r_o} + v_o$ (B) $V = K \log \frac{r}{r_o} - v_o$

  (C) $V = K \log \frac{r_o}{r} + v_o$ (D) $V = K \log \frac{r_o}{r} - v_o$

SOLUTION (C) We know that intensity of gravitational field, $g = -\frac{dV}{dr}$

$$\Rightarrow \quad -\frac{k}{r} = -\frac{dV}{dr} \Rightarrow \int_{V_o}^{V} dV = K \int_{r_o}^{r} \frac{dr}{r}$$

$$\Rightarrow \quad V = K \log \frac{r}{r_o} + V_o$$

7. Two different planets have same density but different radii. The acceleration due to gravity (g) on the surface of the planets is dependent on its radius (R) as.

  (A) $g \propto \frac{1}{R^2}$ (B) $g \propto R^2$

  (C) $g \propto \frac{1}{R}$ (D) $g \propto R$

Solution: (D) $g = \frac{GM}{R^2} = \frac{G\left(\frac{4\pi R^3}{3}\right)\rho}{R^2} = \frac{4\pi RG\rho}{3}$

8. A particle of mass m lies at a distance r from the center of earth. The force of attraction between the particle and earth is $F(r)$.

  (A) $F(r) \propto \frac{1}{r^2}$ for r < R   (B) $F(r) \propto \frac{1}{r^2}$ for r ≥ R

  (C) $F(r) \propto r$ for r < R   (D) $F(r) \propto \frac{1}{r}$ for r < R

SOLUTION (B, C) $F = \begin{cases} \frac{GMm}{r^2} & r \geq R \\ \frac{4\pi G\rho rm}{3} & r < R \end{cases}$

where ρ is density of earth.

9. A shell of mass M and radius R has another point mass m placed at a distance r from its centre $(r > R)$. The force of attraction between the shell and point mass is-

  (A) $F = \frac{GMm}{r}$ (B) $F = \frac{GMm}{r^2}$

  (C) F = zero   (D) None of above

SOLUTION (B) $F = \begin{cases} \frac{GMm}{r^2} & r \geq R \text{ (outside and at surface)} \\ 0 & r < R \text{ (inside)} \end{cases}$

10. If g_h and g_d be the accelerations due to gravity at a height h and at depth d, above and below the surface of earth respectively. Assuming $h \ll R$ and $d \ll R$ and if $g_h = g_d$ then,

  (A) $d = h$

  (B) $d = 2h$

  (C) $h = 2d$

  (D) Data is insufficient to arrive at a conclusion.

Solution: (B) $g\left(1 - \frac{2h}{R}\right) \cong g\left(1 - \frac{d}{R}\right)$

$$\Rightarrow \quad d = 2h$$

11. A particle of mass m is located at a distance r from the centre of a shell of mass M and radius R. The force between the shell and mass is $F(r)$. The plot of $F(r)$ vs r is [Fig.1.181].

SOLUTION (A) $F = \begin{cases} \frac{GMm}{r^2}, & r \geq R \\ 0, & r < R \end{cases}$

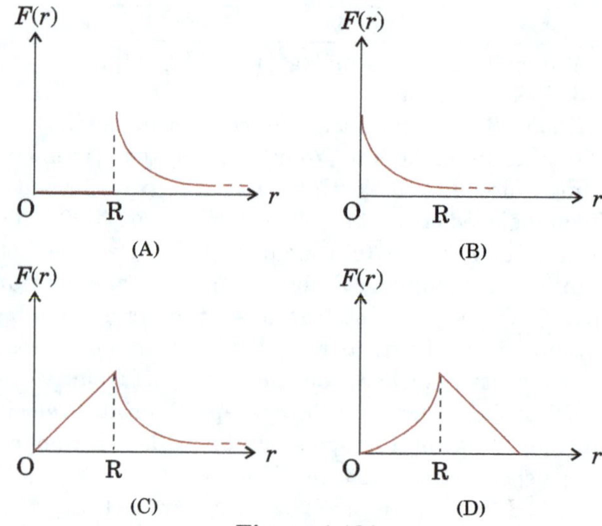

Figure 1.181

12. In objective problem 9, if the shell is replaced by a sphere of same mass and radius then the graph of $F(r)$ vs r will be (select your answer from answers of objective problem 9) [Fig.1.181].

SOLUTION (C) $F(r) = \begin{cases} \frac{GMm}{r^2} & r \geq R \\ \frac{4\pi G\rho rm}{3} & r < R \end{cases}$

where ρ is density of sphere.

13. A shell of mass M and radius R has a point mass m placed at a distance r from its center. The gravitational potential energy $U(r)$ vs r will be [Fig.1.182].

SOLUTION (C) $U(r) = \begin{cases} -\frac{GMm}{r}, & r \geq R \\ -\frac{4Gm}{R}, & r < R \end{cases}$

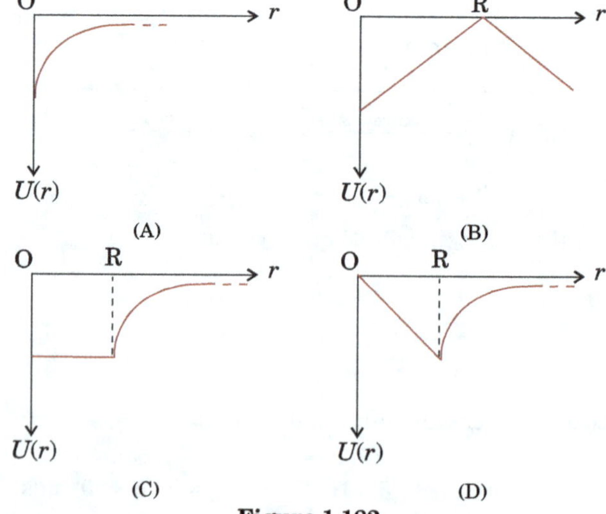

Figure 1.182

14. A satellite is revolving round the earth in an orbit of radius r with time period T. If the satellite is revolving round the earth in an orbit of radius $r + \Delta r (\Delta r \ll r)$ with time period $T + \Delta T (\Delta T \ll T)$ then.

  (A) $\frac{\Delta T}{T} = \frac{3}{2} \frac{\Delta r}{r}$ (B) $\frac{\Delta T}{T} = \frac{2}{3} \frac{\Delta r}{r}$

  (C) $\frac{\Delta T}{T} = \frac{\Delta r}{r}$ (D) $\frac{\Delta T}{T} = -\frac{\Delta r}{r}$

SOLUTION (A) Since, $T^2 = kr^3$

$$\Rightarrow \quad 2\frac{\Delta T}{T} = 3\frac{\Delta r}{r}$$

$$\Rightarrow \quad \frac{\Delta T}{T} = \frac{3}{2}\frac{\Delta r}{r}$$

15. A body of mass m is situated on the earth in the gravitational field of sun. For the body to escape from the gravitational pull of the solar system the body must be imparted an escape velocity of (assume earth to be stationary).

(A) 11.2 kms^{-1} (B) 22.4 kms^{-1}

(C) 33.6 kms^{-1} (D) 42 kms^{-1}

SOLUTION (D) By conservation of energy

$$-\frac{GM_s m}{R_s} - \frac{GM_e m}{R_e} + \frac{1}{2}mv^2 = 0$$

on simplifying for v, we get, $v = 42.174$ kms^{-1}

16. In objective problem 41, the gravitational potential at the point P is -

(A) Zero   (B) $-\sigma G$

(C) $2\pi\sigma Gr$ (D) $-4\pi\sigma G$

SOLUTION (C) $E = -\frac{dV}{dr}$

17. Consider a circular disc of mass M, radius R with surface mass density σ. The gravitational potential due to the disc at a point P lying on its axis at distance r from the centre is

(A) $-2\pi\sigma G$ (B) $-4\pi\sigma G$

(C) $-2\pi\sigma Gr\left(1 - \frac{r}{\sqrt{r^2+R^2}}\right)$ (D) $-4\pi\sigma Gr\left(1 - \frac{r}{\sqrt{r^2+R^2}}\right)$

SOLUTION (C) $E = -2\pi\sigma G\displaystyle\int_0^\alpha \sin\theta\, d\theta$

$$\Rightarrow \quad E = -2\pi\sigma G(1 - \cos\alpha)$$

$$\Rightarrow \quad E = -2\pi\sigma G\left(1 - \frac{r}{\sqrt{r^2 + R^2}}\right)$$

$$\Rightarrow \quad V = Er = -2\pi\sigma rG\left(1 - \frac{r}{\sqrt{r^2 + R^2}}\right)$$

18. A satellite is orbiting round the earth. While in orbit a small part separates from the satellite. The separated part

(A) falls directly to the earth.

(B) moves in a spiral path and reaches after few revolutions about the earth.

(C) Continue to move in the same orbit.

(D) Move gradually farther from the earth.

Ans. (C)

19. Two identical thin rings each of radius R are coaxially placed at a distance R. If the rings have a uniform mass distribution and each has mass m_1 and m_2 respectively, then the work done in moving a mass m from centre of one ring to that of the other is

(A) Zero   (B) $\frac{Gm(m_1-m_2)(\sqrt{2}-1)}{\sqrt{2}R}$

(C) $\frac{Gm\sqrt{2}(m_1+m_2)}{R}$ (D) $\frac{Gmm_1(\sqrt{2}+1)}{m_2 R}$

SOLUTION (B) $V_A = \left(\begin{array}{c}\text{Potential at}\\ \text{A due to A}\end{array}\right) + \left(\begin{array}{c}\text{Potntial at}\\ \text{A due to B}\end{array}\right)$

$$\Rightarrow \quad V_A = -\frac{Gm_1}{R} - \frac{Gm_2}{\sqrt{2}R}$$

Similarly, $V_B = \left(\begin{array}{c}\text{Potential at}\\ \text{B due to A}\end{array}\right) + \left(\begin{array}{c}\text{Potntial at}\\ \text{B due to B}\end{array}\right)$

Since, $W_{A\to B} = m(V_B - V_A)$

$$\Rightarrow \quad W_{A\to B} = \frac{Gm(m_1 - m_2)(\sqrt{2} - 1)}{\sqrt{2}R}$$

20. A point $P(R\sqrt{3}, 0, 0)$ lies on the axis of a ring of mass M and radius R. The ring is located in y-z plane with its centre at origin O. A small particle of mass m starts from P and reaches O under gravitational attraction only. Its speed at O will be.

(A) $\sqrt{\frac{GM}{R}}$ (B) $\sqrt{\frac{Gm}{R}}$ (C) $\sqrt{\frac{GM}{\sqrt{2}R}}$ (D) $\sqrt{\frac{Gm}{\sqrt{2}R}}$

Solution: (A) $\left(\begin{array}{c}\text{Total}\\ \text{Mechanical}\\ \text{Energy}\end{array}\right)_P = \left(\begin{array}{c}\text{Total}\\ \text{Mechanical}\\ \text{Energy}\end{array}\right)_O \Rightarrow$

$$\frac{1}{2}m(0)^2 - \frac{GMm}{\sqrt{(\sqrt{3}R)^2 + R^2}} = \frac{1}{2}mv^2 - \frac{GMm}{R}$$

$$\Rightarrow \quad -\frac{GMm}{2R} = \frac{1}{2}mv^2 - \frac{GMm}{R}$$

$$\Rightarrow \quad v = \sqrt{\frac{GM}{R}}$$

21. An artificial satellite moving in circular orbit around the earth has a total (kinetic + potential) energy E_0. Its potential energy and kinetic energy respectively are

(A) $2E_0$ and $-2E_0$ (B) $-2E_0$ and $3E_0$

(C) $2E_0$ and $-E_0$ (D) $-2E_0$ and $-E_0$

Solution: (C) $KE = \frac{GMm}{2r} = -E_0$, and PE$= -\frac{GMm}{r} = 2E_0$

$$\Rightarrow \quad TE = KE + PE = -\frac{GMm}{2r} = E_0$$

22. The ratio of Earth's orbital angular momentum (about the Sun) to its mass is 4.4×10^{15} m^{2}s^{-1}. The area enclosed by the earth's orbit is approximately.

(A) 1×10^{22} m^2 (B) 3×10^{22} m^2

(C) 5×10^{22} m^2 (D) 7×10^{22} m^2

Solution: (D) Since areal velocity

$$= \frac{\text{Area Swept}}{\begin{array}{c}\text{Time for one Revolution of}\\ \text{Earth about the sun}\end{array}}$$

Further, areal velocity $= \dfrac{L}{2M}$

$$\Rightarrow \quad \text{Area swept} = \left(\frac{L}{2M}\right)\left(\begin{array}{c}\text{Time for one}\\ \text{Revolution of Earth}\\ \text{about the sun}\end{array}\right)$$

$$= \frac{1}{2}(4.4 \times 10^{15})(365 \times 24 \times 60 \times 60)$$

$$\Rightarrow \quad \text{Area swept} = 7 \times 10^{22}\, m^2$$

23. A particle is projected vertically upwards from the surface of earth (radius R_e) with a kinetic energy equal to half of the minimum value needed for it to escape. The height to which it rises above the surface of earth is

(A) R_e (B) $2R_e$ (C) $3R_e$ (D) $4R_e$

Solution: (A) $K_{\text{escape}} = \frac{1}{2}m\left(\sqrt{\frac{2GM}{R_e}}\right)^2 = \frac{GMm}{R_e}$

Since, the particle is projected with kinetic energy equal to half of minimum kinetic energy needed to escape the gravity, therefore, initial kinetic energy of the particle-

$$K_i = \frac{1}{2}K_{escape} = \frac{1}{2}\frac{GMm}{R_e}$$

By law of conservation of mechanical energy, we can write

$$\left(\begin{array}{c}\text{Total Initial}\\\text{Mechanical}\\\text{Energy}\end{array}\right) = \left(\begin{array}{c}\text{Total Final}\\\text{Mechanical}\\\text{Energy}\end{array}\right)$$

$$\Rightarrow \quad (K + U)_{\text{surface}} = (K + U)_{\text{at height } h}$$

$$\Rightarrow \quad \frac{1}{2}\frac{GMm}{R_e} - \frac{GMm}{R_e} = 0 - \frac{GMm}{R_e + h}$$

$$[\because \text{velocity at maximum height is zero}]$$

$$\Rightarrow \quad h = R_e$$

Paragraph for objective problems 24 to 29

A satellite of mass 5000 kg is projected in space with an initial speed of 4000 m/s making an angle of 30° with the radial direction from a distance 3.6×10^7 m away from the centre of the earth

24. The angular momentum of satellite

    (A) 3.6×10^7 joule×sec   (B) 4.9×10^7 joule ×sec

    (C) 9.2×10^7 joule×sec   (D) 3.6×10^{14} joule ×sec

Solution: (D) Given, $r = 3.6\times10^7$ m, $m = 5000$ kg, $v = 4000$ ms^{-1}, $\phi = 30°$

Angular momentum of satellite,

$$L = mvr\sin\phi$$
$$= 5000\times4000\times3.6\times10^7\times\sin30°$$
$$= 3.6\times10^{14} \text{ joule} \times \text{second.}$$

25. The semi-major axis of the orbit of satellite

    (A) 6.6×10^7 m   (B) 14.9×10^7 m

    (C) 19.2×10^{17} m   (D) 1.6×10^4 m

Solution: (A) If the energy of satellite is E, then Energy of the satellite E $= \frac{1}{2}mv^2 - \frac{GMm}{r}$

$$= \frac{1}{2} \times 500 \times (4000)^2 - \frac{(6.67 \times 10^{-11}) \times (5.97 \times 10^{24}) \times 5000}{3.6 \times 10^7}$$

$$= -1.5\times10^{10} \text{ joule}$$

Now, semi−major axis,

$$a = \frac{-GMm}{2E} = -\frac{6.67\times10^{-11}\times5.97\times10^{24}\times5000}{2\times(-1.5\times10^{10})}$$
$$= 6.6 \times 10^7 \text{ m.}$$

26. Semi-minor axis of the orbit of satellite

    (A) 16.6×10^7 m   (B) 3.92×10^7 m

    (C) 10.2×10^{17} m   (D) 2.6×10^4 m

Solution: (B) Eccentricity of the orbit

$$e = \left(1 + \frac{2EL^2}{G^2M^2m^3}\right)$$

$$= 1 + \frac{2 \times (-1.5 \times 10^{10}) \times (3.6 \times 10^{14})^2}{(6.67 \times 10^{-11})^2 \times (5.97 \times 10^{24})^2 \times (5000)^3} = 0.804$$

Hence, semi−minor axis,

$$b = a\sqrt{1 - e^2} = 6.6 \times 10^7\sqrt{1 - (0.804)^2}$$
$$= 3.92 \times 10^7 m$$

27. The minimum distance of satellite from earth

    (A) 66.6×10^7 m   (B) 14.9×10^7 m

    (C) 1.29×10^7 m   (D) 1.6×10^4 m

Solution: (C) Minimum distance, i.e., perigee distance,

$$r_{\min} = a(1 - e) = 6.6\times10^7\times(1 - 0.804) = 1.29\times10^7 \text{ m.}$$

28. The maximum distance of satellite from earth.

    (A) 6.6×10^7 m   (B) 24.9×10^7 m

    (C) 11.9×10^7 m   (D) 1.6×10^4 m

Solution: (C) Maximum distance, i.e., apogee distance,

$$r_{max} = a(1 + e) = 6.6 \times 10^7 \times (1 + 0.804) = 11.9 \times 10^7 \text{ m}$$

29. The energy of satellite

    (A) 1.6×10^7 joule   (B) 4.9×10^7 joule

    (C) 0.2×10^7 joule   (D) -1.5×10^{10} joule

Solution: (D)

30. Imagine a light planet revolving around a very massive star in a circular orbit of radius R with a period of revolution T. If the gravitational force of attraction between the planet and the star is proportional to $R^{-5/2}$, then,

    (A) T^2 is proportional to R^2

    (B) T^2 is proportional to $R^{7/2}$

    (C) T^2 is proportional to $R^{3/2}$

    (D) T^2 is proportional to $R^{3.75}$

Solution: (B) $mR\omega^2 \propto R^{-5/2}$

$$\Rightarrow \quad T^2 \propto R^{7/2}$$

31. If the earth suddenly stopped in its orbit (assume orbit to be circular) the time that would elapse before it falls into the sun is

    (A) $\frac{1}{\sqrt{2}}T$ (B) $\frac{1}{2\sqrt{2}}T$

    (C) $\frac{1}{4\sqrt{2}}T$ (D) $\frac{1}{8\sqrt{2}}T$

Solution: (C) The time t_0 taken by earth to fall into the sun can be calculated by considering a very elongated ellipse having major axis equal to the radius of orbit of earth about sun (r).

$$\Rightarrow \quad 2a = r$$

According to Kepler's third law, we have

$$(2t_0)^2 \propto a^3$$

$$\Rightarrow \quad (2t_0)^2 \propto \left(\frac{r}{2}\right)^3$$

Since, $T^2 \propto r^3$, therefore, $\dfrac{2t_0}{T} = \dfrac{1}{\sqrt{8}}$

$$\Rightarrow \quad t_0 = \frac{T}{4\sqrt{2}} \frac{365}{4\sqrt{2}} = 64.53 \text{ days}$$

32. A projectile is fired upwards from the surface of the earth with a velocity kv_e where v_e is the escape velocity and $k < 1$. If r is the maximum distance from the centre of the earth to which it rises and R is the radius of earth, then r is

    (A) $\frac{R}{k^2}$ (B) $\frac{2R}{1-k^2}$ (C) $\frac{2R}{k^2}$ (D) $\frac{R}{1-k^2}$

Solution: (D)
$$\left(\begin{array}{c}\text{Total}\\\text{Mechanical}\\\text{Energy}\end{array}\right)_{\text{surface}} = \left(\begin{array}{c}\text{Total}\\\text{Mechanical}\\\text{Energy}\end{array}\right)_r$$

$$\Rightarrow \quad -\frac{GMm}{R} + \frac{1}{2}m(kv_e)^2 = -\frac{GMm}{r} + 0$$

where $v_e = \sqrt{\dfrac{2GM}{R}}$

$$\Rightarrow \quad r = \frac{R}{1 - k^2}$$

33. Two satellites S_1 and S_2 revolve round a planet in coplanar circular orbits in the same sense. Their periods of revolution are 1 hour and 8 hour respectively. The radius of the orbit of S_1 is 10^4 km. The speed of S_2 relative to S_1 when they are closest (in kmh^{-1}) is

(A) $10^4\pi$ (B) $2 \times 10^4\pi$

(C) $\frac{1}{2} \times 10^4\pi$ (D) $4 \times 10^4\pi$

Solution: (A) Since $T^2 \propto r^3$

$$\Rightarrow \quad \frac{r_1^3}{r_2^3} = \frac{T_1^2}{T_2^2} = \frac{1}{64}$$

$$\Rightarrow \quad \frac{r_1}{r_2} = \frac{1}{4}$$

$$\Rightarrow \quad r_2 = 4 \times 10^4 \text{ km}$$

$$v_1 = \frac{2\pi r_1}{T_1} \text{ and } v_2 = \frac{2\pi r_2}{T_2}$$

$$\Rightarrow \quad v_1 = \frac{2\pi \times 10^4}{1} \text{ and } v_2 = \frac{2\pi \times 4 \times 10^4}{8} \text{ (in km/h)}$$

$$\Rightarrow \quad v_1 = 2\pi \times 10^4 \, kmh^{-1} \text{ and } v_2 = \pi \times 10^4 \, kmh^{-1}$$

So, speed of S_2 with respect to S_1 is $\pi \times 10^4 \, kmh^{-1}$.

34. Two satellites S_1 and S_2 revolve around a planet in coplanar circular orbits in the same sense. Their periods of revolution are 1 hour and 8 hour respectively. The radius of the orbit of S_1 is 10^4 km., the angular speed of S_2 as observed by an astronaut in S_1 is

(A) $\frac{\pi}{2}$ (B) $\frac{\pi}{3}$ (C) $\frac{\pi}{4}$ (D) $\frac{\pi}{6}$

Solution: (B) $\omega = \left|\frac{v_2 - v_1}{r_2 - r_1}\right| \quad \Rightarrow \quad \omega = \left|\frac{10^4\pi}{3 \times 10^4}\right|$

$$\Rightarrow \quad \omega = \frac{\pi}{3} \text{rads}^{-1}$$

35. Two particles having masses m_1 and m_2 start moving towards each other from the state of rest from infinite separation. Their relative velocity of approach when they are interacting gravitationally at a separation r will be

(A) $\sqrt{\frac{G(m_1+m_2)}{r}}$ (B) $\sqrt{\frac{2G(m_1+m_2)}{r}}$

(C) $\sqrt{\frac{3G(m_1+m_2)}{r}}$ (D) $\sqrt{\frac{4G(m_1+m_2)}{r}}$

Solution: (B) Let m_1, be at rest and think that m_2 has been replaced by μ (reduced mass) and is moving with velocity v. Then by Law of Conservation of Energy, we have

$$\frac{1}{2}m_1(0)^2 + \frac{1}{2}\mu v^2 = -\frac{Gm_1m_2}{r} + 0$$

here, reduced mass of the system, $\mu = \dfrac{m_1m_2}{m_1+m_2}$

$$\Rightarrow \quad v = \sqrt{\frac{2G(m_1+m_2)}{r}}$$

36. The gravitational potential on the surface of a planet of radius R mass M is

(A) g (B) $\frac{GM}{R}$ (C) $-GM$ (D) $-gR$

Solution: (D) $V = -\dfrac{GM}{R}$

$$\Rightarrow \quad V = -\frac{GM}{R^2}R \quad \Rightarrow \quad V = -gR$$

37. A body is imparted a velocity v from the surface of the

earth. If v_0 is orbital velocity and v_e be the escape velocity then for

(A) $v = v_0$, the body follows a circular track around the earth.

(B) $v > v_0$ but $<v_e$, the body follows elliptical path around the earth.

(C) $v<v_0$, the body follows elliptical path and returns to surface of earth.

(D) $v>v_e$, the body follows hyperbolic path and escapes the gravitational pull of the earth.

(A) A, B   (B) B, C

(C) A, B, C   (D) A, B, C, D

Solution: (D) See table 1.5.

38. A particle is launched from the surface of earth with speed v. For the particle to move as a satellite, which statement is correct?

(A) $\frac{v_e}{2}<v<v_e$ (B) $\frac{v_e}{\sqrt{2}}<v<v_e$

(C) $v_e<v<\sqrt{2v_e}$ (D) $\frac{v_e}{\sqrt{2}}<v<\frac{v_e}{2}$

Solution: (B) For a particle to move as a satellite, we have

$$v_o < v < v_e$$

here, v is the speed of satellite, v_o is the orbital speed and v_e is escape speed.

$$\Rightarrow \quad \frac{v_e}{\sqrt{2}} < v < v_e$$

$$\left\{\because v_o = \sqrt{\frac{GM}{r}} \text{ and } v_e = \sqrt{\frac{2GM}{r}}\right\}$$

39. Two bodies of masses m and M are placed a distance d apart. The gravitational potential at the position where the gravitational field due to them is zero is-

(A) $V = -\frac{G}{d}(m+M)$ (B) $V = -\frac{Gm}{d}$

(C) $V = -\frac{GM}{d}$ (D) $V = -\frac{G}{d}\left(\sqrt{m}+\sqrt{M}\right)^2$

Solution: (D) Let gravitational field be zero at a point lying at distance x from M. Then,

$$\frac{GM}{x^2} = \frac{Gm}{(d-x)^2}$$

$$\Rightarrow \quad \frac{d-x}{x} = \sqrt{\frac{m}{M}}$$

$$\Rightarrow \quad \frac{d}{x} - 1 = \sqrt{\frac{m}{M}}$$

$$\Rightarrow \quad x = \left(\frac{\sqrt{M}}{\sqrt{M}+\sqrt{m}}\right)d \qquad (1.155)$$

$$\Rightarrow \quad (d-x) = \left(\frac{\sqrt{M}}{\sqrt{M}+\sqrt{m}}\right)d \qquad (1.156)$$

Now, $V_p = -\dfrac{Gm}{d-x} - \dfrac{Gm}{x}$

Substituting (1.155) and (1.156) in above expression, we get

$$V_p = -\frac{G}{d}\left(\sqrt{m}+\sqrt{M}\right)^2$$

40. The orbital period of revolution of a planet around the sun is T_0. Suppose, we make a model of Solar system scaled down in the ratio η but of materials of the same mean density as the actual material of planet and the sun. The new orbital period is-

(A) ηT_0 (B) $\eta^2 T_0$

(C) $\eta^3 T_0$ (D) T_0

Solution: (D) $T^2 = 4\pi^2\left(\dfrac{r^3}{GM}\right)$

41. P is a point at a distance r from the center of a solid sphere of radius R_0. The gravitational potential at P is V. If V is plotted as a function of r, then the curve representing the plot correctly is [Fig.1.183]

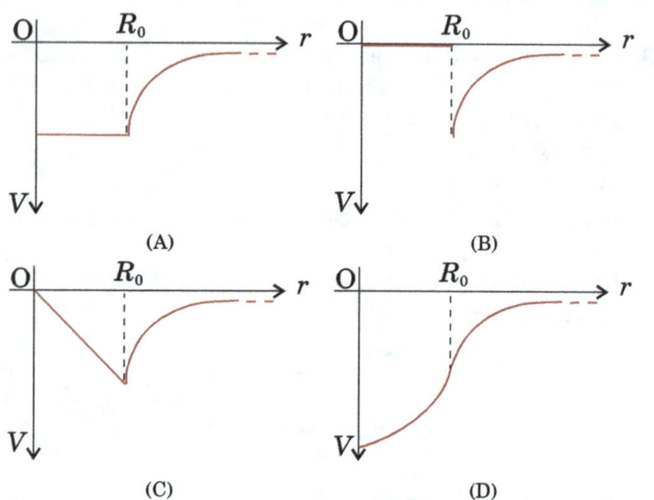

Figure 1.183

Solution: (D) $V = \begin{cases} -\dfrac{GM}{r}, & r \geq R_0 \\ -\dfrac{GM}{2R_0^3}(3R_0^2 - r^2,) & r < R \end{cases}$

42. A point P is lying at a distance r (< a) from the centre of shell of radius a. If E and V be the gravitational field and potential at the point P then E = ?

(A) $E = 0$ (B) $E = -\dfrac{GM}{r^2}$

(C) $v = 0$ (D) $V = -\dfrac{GM}{a}$

Solution: (A) The field inside the shell is zero.

43. A point P is lying at a distance r (< a) from the centre of shell of radius a. If E and V be the gravitational field and potential at the point P then, V = ?

(A) $E = 0$ (B) $E = -\dfrac{GM}{r^2}$

(C) $v = 0$ (D) $V = -\dfrac{GM}{a}$

Solution: (D) The field inside the shell is zero and so potential inside the shell is constant equal to the value that exists at the surface i.e. $-\dfrac{GM}{a}$.

44. Consider a thin uniform spherical layer of mass M and radius R. The potential energy of gravitational interaction of matter forming this shell is

(A) $-\dfrac{GM^2}{R}$ (B) $-\dfrac{1}{2}\dfrac{GM^2}{R}$

(C) $-\dfrac{3}{5}\dfrac{GM^2}{R}$ (D) $-\dfrac{2}{3}\dfrac{GM^2}{R}$

Solution: (B) Let us consider the shell when a mass m is already piled on it by the agency. If V is the potential on the shell, then

$$V = -\dfrac{Gm}{R}$$

To add a mass dm further we have, $dW = V dm$

$$\Rightarrow \qquad dW = -\dfrac{Gm}{R}dm$$

$$\Rightarrow \qquad W = -\dfrac{G}{R}\int_0^M m\,dm$$

$$\Rightarrow \qquad W = -\dfrac{1}{2}\dfrac{GM^2}{R}$$

$$= \text{Potential Energy of Interaction.}$$

45. Consider a thin uniform spherical layer of mass M and radius R, if we consider a solid sphere of mass M and radius R, then the potential energy of gravitational interaction of matter forming this solid sphere is

(A) $-\dfrac{GM^2}{R}$ (B) $-\dfrac{1}{2}\dfrac{GM^2}{R}$

(C) $-\dfrac{3}{5}\dfrac{GM^2}{R}$ (D) $-\dfrac{3}{2}\dfrac{GM^2}{R}$

Solution: (C)

$$\Rightarrow \qquad dU = -\dfrac{Gmdm}{r}$$

$$\Rightarrow \qquad dU = -\dfrac{G\left(\frac{4}{3}\pi r^3 \rho\right)\left(4\pi r^2 dr\rho\right)}{r}$$

$$\Rightarrow \qquad dU = -\dfrac{16\pi^2 G\rho^2}{3}r^4 dr$$

$$\Rightarrow \qquad U = -\dfrac{16}{3}\pi^2 G\left(\dfrac{M}{\frac{4}{3}\pi R^3}\right)^2\int_0^R r^4 dr$$

$$\Rightarrow \qquad U = -\left(\dfrac{16}{3}\pi^2 G\right)\left(\dfrac{M^2}{\frac{16}{9}\pi^2 R^6}\right)\left(\dfrac{R^5}{5}\right)$$

$$\Rightarrow \qquad U = -\dfrac{3}{5}\dfrac{GM^2}{R}$$

46. What should be the period of rotation of earth so as to make any object on the equator weigh half of its present value?

(A) 2 hrs   (B) 24 hrs

(C) 8 hrs   (D) 12 hrs

Solution: (A) $g_e = g - R\omega^2 \quad\Rightarrow\quad g/2 = g - R\omega^2$

$$\Rightarrow \qquad \omega^2 R = \dfrac{g}{2} \quad\Rightarrow\quad \omega = \sqrt{\dfrac{g}{2R}}$$

$$\Rightarrow \qquad T = 2\pi\sqrt{\dfrac{2R}{g}},$$

putting $R = 6.4 \times 10^6$ m and $g = 9.8$ m/s^2, we obtain,

$$T = 1.99 \text{ h} \approx 2\text{h}$$

47. An artificial satellite is describing an equatorial orbit at 3600 km above the earth's surface. Calculate its period of revolution ?

(A) 8.71 hrs   (B) 9.71 hrs

(C) 10.71 hrs   (D) 11.71 hrs

Solution: (A) By Kepler's third law, the time period of satellite is given by

$$T^2 = \dfrac{4\pi^2}{GM}(R+h)^3 = \dfrac{4\pi^2}{g}\dfrac{(R+h)^3}{R^2}$$

$$\Rightarrow \qquad T = \dfrac{2\pi}{R}\sqrt{\dfrac{(R+h)^3}{g}} = \dfrac{2\pi}{6400\times10^3}\sqrt{\dfrac{10^{21}}{9.8}}$$

$$= 31360.78 \text{ sec} = 8.71 \text{ hr}$$

$$[\text{as } R+h = (6400 + 3600)\times10^3\text{m} = 10^7\text{m}]$$

48. An artificial satellite is describing an equatorial orbit at

3600 km above the earth's surface. Calculate its orbital speed.

(A) 6.335 km/sec (B) 7.335 km/sec
(C) 8.335 km/sec (D) 9.335 km/sec

Solution: (A) The time period of satellite is given by

$$T^2 = \frac{4\pi^2}{GM}(R+h)^3 = \frac{4\pi^2}{g}\frac{(R+h)^3}{R^2}$$

$$T = \frac{2\pi}{R}\sqrt{\frac{(R+h)^3}{g}} = \frac{2\pi}{6400\times10^3}\sqrt{\frac{10^{21}}{9.8}}$$

$$[\text{as } R+h = (6400+3600)\times10^3\,\text{m} = 10^7\,\text{m}]$$
$$= 31360.78\,\text{sec} = 8.71\,\text{hrs}$$

Orbital speed is,

$$v_o = \sqrt{\frac{GM}{(R+h)}} = \sqrt{\frac{gR^2}{(R+h)}} = \sqrt{\frac{9.8\times(6400\times10^3)^2}{10^7}}\,\text{m/s}$$
$$= 6335\,\text{m/s} = 6.335\,\text{km/s}$$

49. Three particles each having a mass of 100 gm are placed on the vertices of an equilateral triangle of side 20 cm. The work done in increasing the side of the triangle to 40 cm is $\left[G = 6.67\times10^{-11}\frac{N-m^2}{kg^2}\right]$

(A) 5.0×10^{-12}J (B) 2.25×10^{-10}J
(C) 4.0×10^{-11}J (D) 6.0×10^{-15}J

Solution: (A) W = $U_{\text{final}} - U_{\text{initial}}$
$$= 3\left[-\left\{\frac{G(m)(m)}{r_f}\right\} - \left\{\frac{G(m)(m)}{r_i}\right\}\right]$$
Putting $r_f = 0.4$ m, $r_i = 0.2$ m and
$G = 6.67\times10^{-11}\frac{N-m^2}{kg^2}$ and $m = 0.1$ kg
We get $W = 5.0\times10^{-12}$J

50. If the time of revolution of a satellite is T, then Kinetic energy is proportional to

(A) 1/T (B) $1/T^2$ (C) $1/T^3$ (D) $T^{-2/3}$

Solution: Orbital velocity $v = \frac{2\pi r}{T}$
Hence, KE $\propto \frac{1}{T^2}$

51. The magnitude of gravitational potential energy of the earth-satellite system is U with zero potential energy at infinite separation. The kinetic energy of satellite is K. Mass of satellite << mass of the earth. Then

(A) $K = 2U$ (B) $K = \frac{U}{2}$ (C) $K = U$ (D) $K = 4U$

Solution: (A) Kinetic energy of the satellite $K = \frac{GMm}{2r}$ while magnitude of the potential energy $U = \frac{GMm}{r}$
Hence $K = \frac{U}{2}$

52. A Saturn year is 29.5 times the earth year. How far is the Saturn from the sun if the earth is 1.5×10^8 km away from the sun?

(A) 1.43×10^9 km (B) 2.43×10^9 km
(C) 3.43×10^9 km (D) 4.43×10^9 km

Solution: (A) It is given that, $T_s = 29.5T_e$; $R_e = 1.5\times10^{11}$ m
Now, according to Kepler's third law
$$\frac{T_s^2}{T_e^2} = \frac{R_s^3}{R_e^3} \Rightarrow R_s = R_e\left(\frac{T_s}{T_e}\right)^{\frac{2}{3}} = 1.5\times10^{11}\left(\frac{29.5T_e}{T_e}\right)^{\frac{2}{3}}$$
$$= 1.43\times10^{12}\,\text{m} = 1.43\times10^9\,\text{km}$$

53. A spherical planet for out in space has a mass M_o and diameter D_o. A particle of mass m falling freely near the surface of this planet will experience an acceleration due to gravity which is equal to

(A) $\frac{GM_o}{D_o^2}$ (B) $\frac{4GmM_o}{D_o^2}$
(C) $\frac{4GM_o}{D_o^2}$ (D) $\frac{GmM}{D_o^2}$

Solution: (C) Let g_p be the acceleration on the surface of the planet, then
$$mg_p = \frac{GM_o m}{\left(\frac{D_o}{2}\right)^2} \quad\Rightarrow\quad g_p = \frac{4GM_o}{D_o^2}$$

54. A body falls freely towards the earth from a height $2R$, above the surface of the earth, where initially it was at rest. If R is the radius of the earth then its velocity on reaching the surface of the earth is

(A) $\sqrt{\frac{4}{3}gR}$ (B) $\sqrt{\frac{2}{3}gR}$
(C) $\frac{4}{3}gR$ (D) $2gR$

Solution: (A) Initial energy of the body $= -\frac{GMm}{(R+2R)}$

Final energy of the body $= -\frac{GMm}{R} + \frac{1}{2}mv^2$
where m is the mass of the body and M is the mass of the earth.
On applying the law of conservation of mechanical energy, we get
$$-\frac{GMm}{3R} = -\frac{GMm}{R} + \frac{1}{2}mv^2$$
$$\Rightarrow \quad \frac{1}{2}mv^2 = \frac{2GMm}{3R}$$
$$\Rightarrow \quad v^2 = \frac{4GM}{3R} \Rightarrow v = \sqrt{\frac{4GM}{3R}} = \sqrt{\frac{4GMR}{3R^2}} = \sqrt{\frac{4}{3}gR}$$

55. Two satellite (I) and (II) are moving round a planet in circular orbit having radii R and $3R$ respectively, if the speed of satellite (I) is v the speed of satellite II will be

(A) $v/3$ (B) $v/\sqrt{3}$
(C) $3v$ (D) data insufficient

Solution: (C) $T = 2\pi r/v$
$$\Rightarrow \quad T_1 = 2\pi R/v_1 \text{ and } T_2 = 2\pi 3R/v_2$$
$$\Rightarrow \quad \frac{T_1}{T_2} = \frac{1}{3}\left(\frac{v_2}{v_1}\right)$$
$$\Rightarrow \quad \left(\frac{T_1}{T_2}\right)^2 = \frac{1}{9}\left(\frac{v_2}{v_1}\right)^2$$
But by Kepler's third law, $\frac{T_1^2}{T_2^2} = \left(\frac{R_1}{R_2}\right)^3 = \left(\frac{1}{3}\right)^3$
$$\therefore \quad \left(\frac{1}{3}\right)^3 = \frac{1}{9}\left(\frac{v_2}{v_1}\right)^2 \Rightarrow v_2 = v_1/\sqrt{3}$$

56. The radius of a planet is R. A satellite revolves around it in a circle of radius r with angular speed ω. The acceleration due to gravity on planet's surface will be:

(A) $\frac{r^3\omega}{R}$ (B) $\frac{r^2\omega^3}{R}$ (C) $\frac{r^3\omega^2}{R^2}$ (D) $\frac{r^2\omega^2}{R}$

Solution: (C) Let M be the mass of the planet and m the mass of satellite. Then
$$mr\omega^2 = \frac{GMm}{r^2} \Rightarrow GM = r^3\omega^2$$
So,
$$g = \frac{GM}{R^2} = \frac{r^3\omega^2}{R^2}$$

57. The gravitational field in a region is given by $\vec{g} = (4\hat{i}+\hat{j})$ N/kg. Work done by this field is zero when the particle is moved along the line

(A) $y + 4x = 2$ (B) $4y + x = 6$
(C) $x + y = 5$ (D) all of the above

Solution: (A) Work done will be zero when displacement is perpendicular to the field.

The field makes an angle, $\theta_1 = \tan^{-1}\left(\frac{1}{4}\right)$ with positive x-axis

While the line $y + 4x = 2$ makes an angle $\theta_2 = \tan^{-1}(-4)$ with positive x–axis

$$\theta_1 + \theta_2 = \tan^{-1}\frac{1}{4} + \tan^{-1}4$$

$$= \tan^{-1}\frac{\frac{1}{4}+4}{1-\frac{1}{4}4} = \tan^{-1}\frac{17/4}{0} = \tan^{-1}\infty = \frac{\pi}{2}$$

i.e., the line $y + 4x = 2$ is perpendicular to $\vec{g}$

58. A particle of mass m is placed inside a spherical shell, away from its centre. The mass of the shell is M.
   (A) The particle will move towards the centre.
   (B) The particle will move away from the centre, towards the nearest wall.
   (C) The particle will move towards the centre if $m < M$ and away from the centre if $m > M$.
   (D) The particle will remain stationary.

Solution: (D) As the gravitational field inside the shell is zero, the particle will remain stationary

59. Three solid spheres each of mass m and radius R are released from the position shown in Fig.1.184. The speed of any one sphere at the time of collision would be
   (A) $\sqrt{Gm\left(\frac{1}{d}-\frac{3}{R}\right)}$ (B) $\sqrt{Gm\left(\frac{3}{d}-\frac{1}{R}\right)}$
   (C) $\sqrt{Gm\left(\frac{2}{R}-\frac{1}{d}\right)}$ (D) $\sqrt{Gm\left(\frac{1}{R}-\frac{2}{d}\right)}$

Solution (D) From conservation of mechanical energy

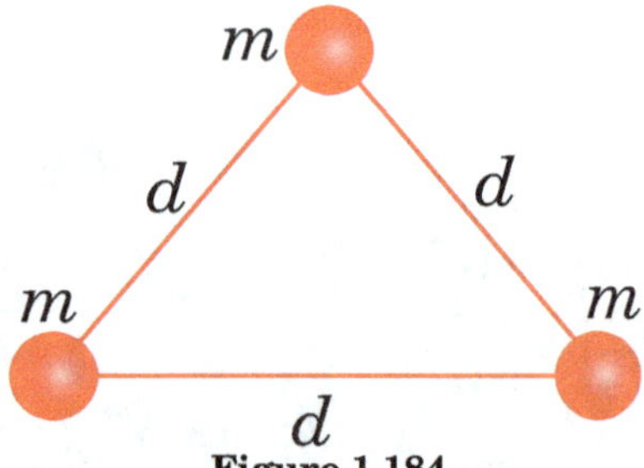

Figure 1.184

$$3\left[\frac{1}{2}mv^2\right] = 3\left[\frac{Gm^2}{2R} - \frac{Gm^2}{d}\right]$$

$$\Rightarrow \quad v^2 = Gm\left[\frac{1}{R} - \frac{2}{d}\right]$$

$$\therefore \quad v = \sqrt{Gm\left[\frac{1}{R} - \frac{2}{d}\right]}$$

60. Two particles of equal mass m go round a circle of radius R under the action of their mutual gravitational attraction. The speed of each particle is
   (A) $\frac{1}{2R}\sqrt{\frac{1}{Gm}}$ (B) $\sqrt{\frac{Gm}{2R}}$
   (C) $\frac{1}{2}\sqrt{\frac{Gm}{R}}$ (D) $\sqrt{\frac{4Gm}{R}}$

Sol For circular motion, the required centripetal force provided by the gravitational force of attraction between two particles [Fig.1.185], i.e.,

$$\frac{mv^2}{R} = \frac{Gm \times m}{(2R)^2} \Rightarrow v = \frac{1}{2}\sqrt{\frac{Gm}{R}}$$

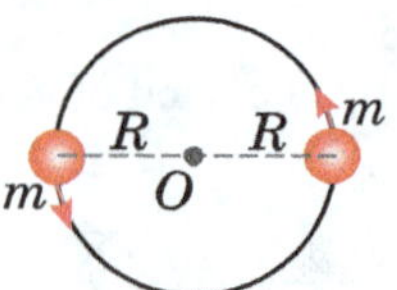

Figure 1.185

61. The escape speed for a planet is v_e. A particle starts from rest at a large distance from the planet, reaches the planet only under gravitational attraction, and passes through a smooth tunnel through its centre. Its speed at the centre of the planet will be-
   (A) $\sqrt{1.5}v_e$ (B) $\frac{v_e}{\sqrt{2}}$ (C) v_e (D) zero

Solution Ans.   (A) From conservation of mechanical energy, we have:

$$0 + 0 = \frac{1}{2}mv^2 - \frac{3GMm}{2R} \Rightarrow v = \sqrt{\frac{3GM}{R}} = \sqrt{1.5}v_e$$

62. A particle is projected vertically upwards the surface of the earth (radius R) with a speed equal to one fourth of escape speed. What is the maximum height attained by it from the surface of the earth ?
   (A) $\frac{16}{15}R$ (B) $\frac{R}{15}$
   (C) $\frac{4}{15}R$ (D) None of these

Solution (B) From conservation of mechanical energy

$$\frac{1}{2}mv^2 = \frac{GMm}{R} - \frac{GMm}{r}$$

where, r = maximum distance from centre of the earth. Also,

$$v = \frac{1}{4}v_e = \frac{1}{4}\sqrt{\frac{2GM}{R}}$$

$$\Rightarrow \quad \frac{1}{2}\,m \times \frac{1}{16} \times \frac{2GM}{R} = \frac{GMm}{R} - \frac{GMm}{r}$$

$$\Rightarrow \quad r = \frac{16}{15}R \quad \Rightarrow \quad h = r - R = \frac{R}{15}$$

63. A mass 6×10^{24} kg (= mass of earth) is to be compressed in a sphere in such a way that the escape speed from its surface is 3×10^8 m/s (equal to that of light). What should be the radius of the sphere?
   (A) 9 mm   (B) 8 mm   (C) 7 mm   (D) 6 mm

Sol As, $v_e = \sqrt{\left(\frac{2GM}{R}\right)}, R = \left(\frac{2GM}{v_e^2}\right)$,

$$\therefore \quad R = \frac{2\times6.67\times10^{-11}\times6\times10^{24}}{(3\times10^8)^2} = 9 \times 10^{-3} \text{ m} = 9 \text{ mm}$$

64. Calculate the mass of the sun if the mean radius of the earth's orbit is 1.5×10^8 km and $G = 6.67 \times 10^{-11}$ N m^2/kg^2.
   (A) M $\simeq 2 \times 10^{30}$ kg   (B) M $\simeq 3 \times 10^{30}$ kg
   (C) M $\simeq 2 \times 10^{15}$ kg   (D) M $\simeq 3 \times 10^{15}$ kg

Solution $M = \dfrac{4\times\pi^2\times(1.5\times10^{11})^3}{6.67\times10^{-11}\times(3.15\times10^7)^2}$

$$[\text{as } T = 1 \text{ year} = 3.15 \times 10^7 \text{ s}]$$

i.e., $M \simeq 2 \times 10^{30}$ kg

65. Gravitational potential difference between a point on surface of planet and another point 10 m above is 4 J/kg. Considering gravitational field to be uniform, how much work is done in moving a mass of 2 kg from the surface to a point 5 m above the surface?

(A) 4 J   (B) 5 J   (C) 6 J   (D) 7 J

Solution

Ans. (A)

Gravitational field $g = -\frac{\Delta V}{\Delta x} = -\left(\frac{-4}{10}\right) = \frac{4}{10}$ J/kgm

Work done in moving a mass of 2 kg from the surface to a point 5 m above the surface,

$W = mgh = (2\text{ kg})\left(\frac{4}{10}\frac{J}{\text{kgm}}\right)(5\text{ m}) = 4$ J

66. A particle is projected from point A, that is at a distance 4R from the center of the Earth, with speed v_1 in a direction making angle 30° with the line joining the centre of the Earth and point A, as shown in Fig.1.186. Find the speed v_1 of particle (in m/s) if particle passes grazing the surface of the earth. Consider gravitational interaction only between these two. (use $\frac{GM}{R} = 6.4 \times 10^7$ m²/s²)

(A) $\frac{8000}{\sqrt{2}}$ (B) 800

(C) $800\sqrt{2}$ (D) None of these

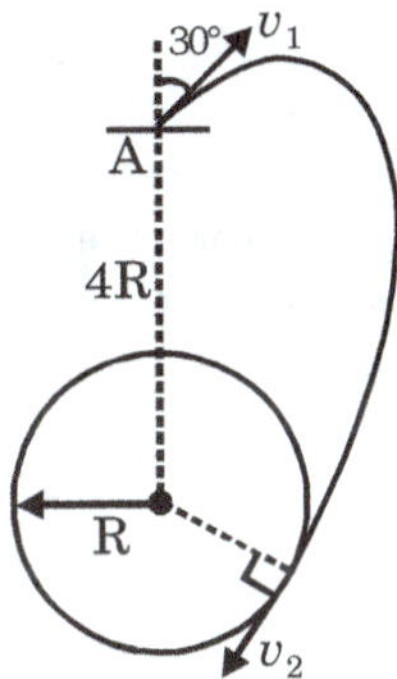

Figure 1.186

Sol Conserving angular momentum:

$m\,(v_1\cos 60)\,4R = mv_2R \Rightarrow \frac{v_2}{v_1} = 2.$

Conserving energy of the system:

$-\frac{GMm}{4R} + \frac{1}{2}mv_1^2 = -\frac{GMm}{R} + \frac{1}{2}mv_2^2$

$\Rightarrow \quad \frac{1}{2}v_2^2 - \frac{1}{2}v_1^2 = \frac{3}{4}\frac{GM}{R}$

$\Rightarrow \quad v_1^2 = \frac{1}{2}\frac{GM}{R} \Rightarrow v_1 = \frac{1}{\sqrt{2}}\sqrt{64\times10^6} = \frac{8000}{\sqrt{2}}$ m/s

67. If the law of gravitation be such that the force of attraction between two particles vary inversely as the 5/2th power of their separation, then the graph of orbital velocity v_0 plotted against the distance r of a satellite from the earth's centre on a log-log scale is shown alongside. The slope of line will be-

(A) $-\frac{5}{4}$ (B) $-\frac{5}{2}$ (C) $-\frac{3}{4}$ (D) -1

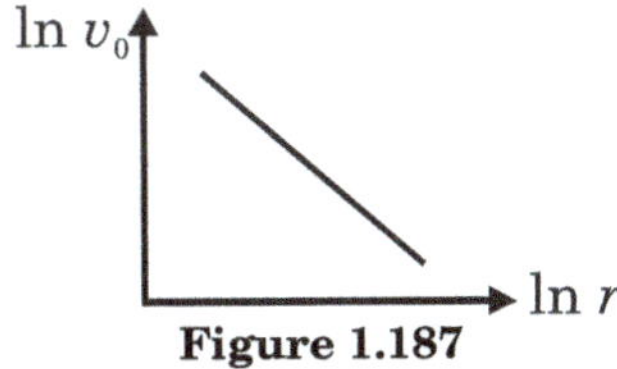

Figure 1.187

Solution $\frac{mv_0^2}{r} = \frac{GMm}{r^{5/2}} \Rightarrow v_0 = \frac{\sqrt{GM}}{r^{3/4}}$

$\Rightarrow \quad \ln v_0 = \ln\sqrt{GM} - \frac{3}{4}\ln r$

68. Two point objects of masses m and $4\,m$ are at rest at an infinite separation. They move towards each other under mutual gravitational attraction. If G is the universal gravitational constant, then at a separation r

(A) the total mechanical energy of the two objects is zero

(B) their relative velocity is $\sqrt{\frac{10Gm}{r}}$

(C) the total kinetic energy of the objects is $\frac{4Gm^2}{r}$

(D) their relative velocity is zero

Solution Ans. (A,B,C)

Initially, the point objects were at rest at infinite separation. So, their total mechanical energy was zero. Since, there is no force other than gravitational force, therefore by conservation of mechanical energy, the mechanical energy will always be zero. So, option (A) is correct.

By applying law of conservation of momentum [Fig.1.188],

$$mv_1 - 4mv_2 = 0 \quad \Rightarrow \quad v_1 = 4v_2 \qquad \text{... (1)}$$

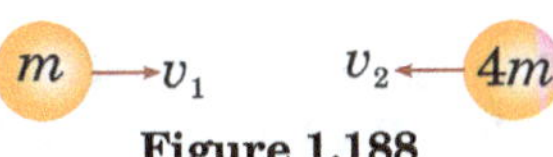

Figure 1.188

By applying conservation of mechanical energy, we have
Net mechanical energy at separation r = net mechanical energy at separation ∞

$$K_r + U_r = K_\infty + U_\infty$$

$\Rightarrow \quad \left[\frac{1}{2}mv_1^2 + \frac{1}{2}(4m)v_2^2\right] + \left[-\frac{Gm(4m)}{r}\right] = 0$

$\Rightarrow \quad \frac{1}{2}mv_1^2 + \frac{1}{2}(4m)v_2^2 = \frac{Gm(4m)}{r} = \frac{4Gm^2}{r} \qquad \text{... (2)}$

$\therefore$ Total kinetic energy $= \frac{4Gm^2}{r}$;

Therefore, option (B) is also correct.

Putting the value of v_1 from Eq.(1), in Eq.(2), we get

$\Rightarrow \quad 10mv_2^2 = \frac{G4m^2}{r} \Rightarrow v_2 = 2\sqrt{\frac{Gm}{10r}}$

Relative velocity for the particle,

$$v_{\text{rel}} = |\vec{v}_1 - \vec{v}_2| = 5v_2 = \sqrt{\frac{10Gm}{r}} \qquad (\because \ v_1 = 4v_2)$$

Therefore, mechanical energy of system = 0 = constant.
By using reduced mass concept
You can also find the kinetic energy of the system by using the concept of reduced mass-
In terms of reduced mass, the net kinetic energy is given by-

$$K_{\text{system}} = \frac{1}{2}\mu v^2_{\text{rel}}$$

here, $\mu = \frac{(m)(4\,m)}{m+4\,m} = \frac{4}{5}\,m$ and $v_{\text{rel}} = \sqrt{\frac{10Gm}{r}}$

Therefore, kinetic energy of the system

$$K_{\text{system}} = \frac{1}{2}\left(\frac{4}{5}m\right)\cdot\left(\frac{10Gm}{r}\right) = \frac{4Gm^2}{r}$$

69. Which of the following statements are true about acceleration due to gravity?

(A) 'g' decreases in moving away from the centre if $r > R$

(B) 'g' decreases in moving away from the centre if $r < R$

(C) 'g' is zero at the centre of earth

(D) 'g' decreases if earth stops rotating on its axis

Solution (A, C) Variation of g with distance, is shown in Fig.1.189:

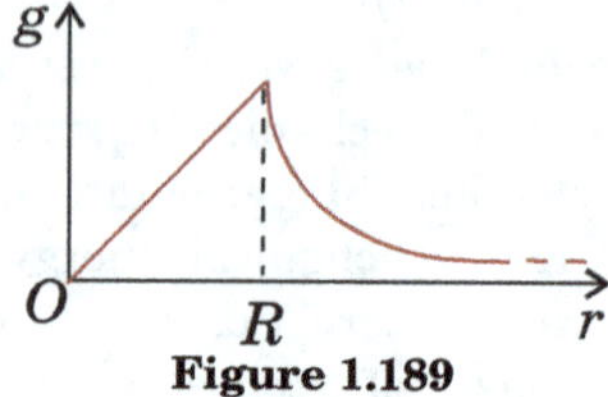

Figure 1.189

From Eq.1.19, the variation of g with ω:

$$g' = g - \omega^2 R \cos^2\theta \qquad \dots (1)$$

At poles, $\theta = \pm\pi/2$, therefore, from Eq.(1), $g' = g$. If earth stops rotating, i.e., $\omega = 0$, then again from Eq.(1), the effective gravitational acceleration increases to g everywhere except poles, where it was already g.

70. An astronaut, inside an earth satellite experiences weightlessness because:

(A) he is falling freely

(B) no external force is acting on him

(C) no reaction is exerted by floor of the satellite

(D) he is far away from the earth surface

Solution Ans. (AC)

As astronaut's acceleration $= g$, so he is falling freely. Since, he is in a state of free fall, therefore, no reaction is exerted by the floor of the satellite.

71. If a satellite orbits as close to the earth's surface as possible

(A) its speed is maximum

(B) time period of its revolution is minimum

(C) the total energy of the 'earth plus satellite' system is minimum

(D) the total energy of the 'earth plus satellite' system is maximum

Solution Ans. (ABC) For (A) : orbital speed, $v_0 = \sqrt{\dfrac{GM}{r}}$

For (B) : Time period of revolution, $T^2 \propto r^3$

For (C/D): Total energy $= -\dfrac{GMm}{2r}$

72. A planet is revolving around the sun is an elliptical orbit as shown in Fig.1.190.

Select correct alternative(s)

(A) Its total energy is negative at D.

(B) Its angular momentum is constant

(C) Net torque on planet about sun is zero

(D) Linear momentum of the planet is conserved

Sol

For (A) : For bounded system, the total energy is always

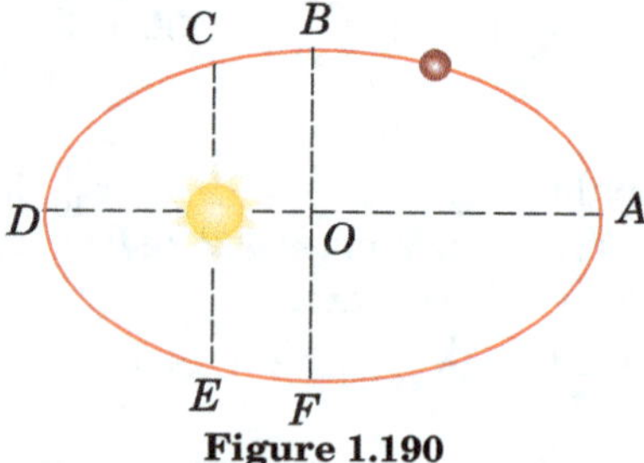

Figure 1.190

negative.

For (B) : For central force field, angular momentum is always conserved.

For (C) :For central force field, torque = 0.

For (D) : In presence of external force, linear momentum is not conserved.

Example 73 **to** 75 A triple star system consists of two stars, each of mass m, in the same circular orbit about central star with mass M = 2×10^{30} kg [Fig.1.191]. The two outer stars always lie at opposite ends of a diameter of their common circular orbit. The radius of the circular orbit is $r = 10^{11}$ m and the orbital period of each star is $1.6 \, 10^7$ s. [Take $\pi^2 = 10$ and G = $\frac{20}{3} \times 10^{-11}$ Nm^2 kg^{-2}]

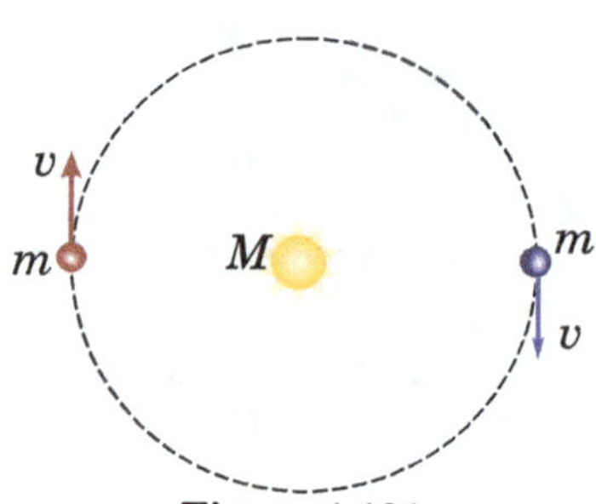

Figure 1.191

73. The mass m of the outer stars is

(A) $\frac{16}{15} \times 10^{30}$ kg   (B) $\frac{11}{8} \times 10^{30}$ kg

(C) $\frac{15}{16} \times 10^{30}$ kg   (D) $\frac{8}{11} \times 10^{30}$ kg

Sol (B) F_{mm} = Gravitational force between two outer stars = $\frac{Gm^2}{4r}$

F_{mM} = Gravitational force between central star and outer star = $\frac{GmM}{r^2}$

For circular motion of outer star, $\frac{mv^2}{r} = F_{mm} + F_{mM}$

$\therefore \qquad v^2 = \frac{G(m+4M)}{4r}$

T = period of orbital motion = $\frac{2\pi r}{v}$

$\therefore \quad m = \frac{16\pi^2 r^3}{GT^2} - 4M = \left(\frac{150}{16} - 8\right)10^{30} = \frac{11}{8} \times 10^{30} \, kg$

74. The orbital velocity of each star is

(A) $\frac{5}{4}\sqrt{10} \times 10^3$ m/s   (B) $\frac{5}{4}\sqrt{10} \times 10^5$ m/s

(C) $\frac{5}{4}\sqrt{10} \times 10^2$ m/s   (D) $\frac{5}{4}\sqrt{10} \times 10^4$ m/s

Sol Ans. (D) $T = \frac{2\pi r}{v}$

$\Rightarrow v = \frac{2\pi r}{T} = \frac{(2)(\sqrt{10})(10^{11})}{1.6 \times 10^7} = \frac{5}{4}\sqrt{10} \times 10^4 \, m/s$

75. The total mechanical energy of the system is

(A) $-\frac{1375}{64} \times 10^{35}$ J   (B) $-\frac{1375}{64} \times 10^{38}$ J

(C) $-\frac{1375}{64} \times 10^{34}$ J   (D) $-\frac{1375}{64} \times 10^{37}$ J

Sol. (B) Total mechanical energy = KE + PE

$$= 2\left(\tfrac{1}{2}mv^2\right) - \frac{2GMm}{r} - \frac{Gm^2}{2r}$$

$$= m\left[\frac{G(4M+m)}{4r} - \frac{2GM}{r} - \frac{Gm}{2r}\right]$$

$$= -\frac{Gm}{r}\left[M + \frac{m}{4}\right]$$

$$= -\left(\tfrac{20}{3}\times 10^{-11}\right)\left(\tfrac{11}{8}\times 10^{30}\right)\times \tfrac{1}{10^{11}}\left(2\times 10^{30} + \tfrac{11}{32}\times 10^{30}\right)$$

$$= -\frac{1375}{64}\times 10^{38}\ J$$

Question 76 to 78 A solid sphere of mass M and radius R is surrounded by a spherical shell of same same mass M and radius $2R$ as shown in Fig.1.192. A small particle of mass m is released from rest from a height $h(<< R)$ above the shell. There is a hole in the shell.

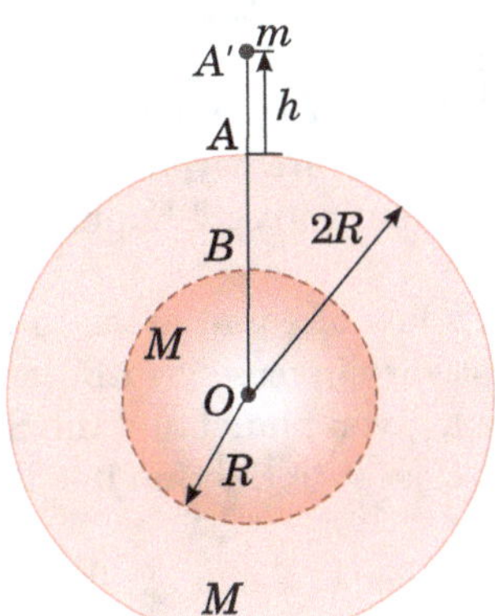

Figure 1.192

76. In what time will it enter the hole at $A :-$

(A) $2\sqrt{\dfrac{hR^2}{GM}}$ (B) $\sqrt{\dfrac{2hR^2}{GM}}$

(C) $\sqrt{\dfrac{hR^2}{GM}}$ (D) none of these

Sol (A) $a = \dfrac{F}{m} = \dfrac{1}{m}\left(\dfrac{GMm}{(2R+h)^2} + \dfrac{GMm}{(2R+h)^2}\right)$

$$\Rightarrow \quad a = \frac{GM}{(2R)^2} + \frac{GM}{(2R)^2} = \frac{GM}{2R^2}$$

[since, $h << R$, therefore, $2R + h = 2R$]

By Newton's equation of motion, we have

$$y - y_0 = v_0 t + \frac{1}{2}a_y t^2$$

$$\Rightarrow \quad h = 0 + \frac{1}{2}at^2$$

$$\Rightarrow \quad t = \sqrt{\frac{2h}{a}} = \sqrt{\frac{2h\times 2R^2}{GM}} = 2\sqrt{\frac{hR^2}{GM}}$$

77. What time will it take to move from A to B ?

(A) $= \dfrac{4R^2}{\sqrt{GMR}}$ (B) $> \dfrac{4R^2}{\sqrt{GMR}}$

(C) $< \dfrac{4R^2}{\sqrt{GMR}}$ (D) none of these

Sol Ans. (C) Given that $(h << R)$, so the velocity of particle at point A will be equal to the velocity at A', which is zero. Therefore, velocity of the particle at A will also be zero.

We can see here that the acceleration always increases from $2R$ to R and its value must be greater than $a = \dfrac{GM}{4R^2}$(at A)

$$\therefore \quad t < \frac{v}{a} \Rightarrow t < \sqrt{\frac{GM}{R}}\times\frac{4R^2}{GM} \Rightarrow t < \frac{4R^2}{\sqrt{GMR}}$$

78. With what approximate speed will it collide at B ?

(A) $\sqrt{\dfrac{2GM}{R}}$ (B) $\sqrt{\dfrac{GM}{2R}}$

(C) $\sqrt{\dfrac{3GM}{2R}}$ (D) $\sqrt{\dfrac{GM}{R}}$

Sol Ans. (D)

Given that $(h << R)$, so velocity of particle at point A is equal to velocity of the particle at point A', i.e., zero.

By conservation of mechanical energy,

$$U_A + K_A = U_B + K_B$$

$$\Rightarrow \quad -\frac{GMm}{2R} - \frac{GMm}{2R} + 0 = -\frac{GMm}{2R} - \frac{GMm}{R} + \frac{1}{2}mv^2$$

$$\Rightarrow \quad \frac{GMm}{2R} = \frac{1}{2}mv^2 \quad \Rightarrow \quad v = \sqrt{\frac{GM}{R}}$$

79. Imagine a light planet revolving around a very massive star in a circular orbit of radius R. If the gravitational force of attraction between the planet and the star is proportional to $R^{-5/2}$, then match the following

(A) Time period of revolution is proportional to	(P)	R^0
(B) Kinetic energy of planet is proportional to	(Q)	$R^{7/4}$
(C) Orbital velocity of planet is proportional to	(R)	$R^{-1/2}$
(D) Total mechanical energy of planet is proportional to	(S)	$R^{-3/2}$
	(T)	$R^{-3/4}$

Ans. (A) Q (B) S(C) T (D) S

Sol According to question $F = \dfrac{C}{R^{5/2}}$

where C is a constant,

so, $\dfrac{mv^2}{R} = \dfrac{C}{R^{5/2}} \Rightarrow mv^2 \propto \dfrac{1}{R^{3/2}}$

$\Rightarrow$ KE $\propto R^{-3/2}$

Also $v \propto R^{-3/4}$ and $PE \propto R^{-3/2}$.

Total mechanical energy $= KE + PE \propto R^{-3/2}$

Time period, $T = \dfrac{2\pi R}{v} \propto R^{7/4}$

80. An artificial satellite (mass m) of a planet (mass M) revolves in a circular orbit whose radius is n times the radius R of the planet. In the process of motion, the satellite experiences a slight resistance due to cosmic dust. Assuming the force of resistance on satellite to depend on velocity as $F = av^2$ where 'a' is a constant, calculate how long the satellite will stay in the space before it falls onto the planet's surface.

Solution Air resistance $F = -av^2$,

where orbital velocity $v = \sqrt{\dfrac{GM}{r}}$

$r =$ the distance of the satellite from planet's centre

$\Rightarrow \quad F = -\dfrac{GMa}{r}$

The work done by the resistance force

$$dW = Fdx = Fvdt = \frac{GMa}{r}\sqrt{\frac{GM}{r}}dt = \frac{(GM)^{3/2}a}{r^{3/2}}dt \qquad \ldots (i)$$

The loss of energy of the satellite $= dE$

$$\therefore \quad \frac{dE}{dr} = \frac{d}{dr}\left[-\frac{GMm}{2r}\right] = \frac{GMm}{2r^2} \Rightarrow dE = \frac{GMm}{2r^2}dr \qquad \ldots (ii)$$

Since $dE = -dW$ (work energy theorem), therefore

$$-\frac{GMm}{2r^2}dr = \frac{(GM)^{3/2}}{r^{3/2}}dt$$

$$\Rightarrow \quad t = -\frac{m}{2a\sqrt{GM}}\int_{nR}^{R}\frac{dr}{\sqrt{r}} = \frac{m\sqrt{R}(\sqrt{n}-1)}{a\sqrt{GM}} = (\sqrt{n}-1)\frac{m}{a\sqrt{gR}}$$

81. At what height from the surface of earth the weight of the body is 1/3rd of its weight at the surface?

   (A) 5000 km   (B) 5562.5 km

   (C) 4684.8 km   (D) 3600 km

Sol (C) We know that the gravitational acceleration at height h,

$$g' = \frac{g}{\left(1 + \frac{h}{R}\right)^2} \qquad \ldots (1)$$

For weight to be 1/3rd of the weight on the earth surface g' should be 1/3rd of g. So from Eq.(1), we have-

$$\frac{g}{3} = \frac{g}{\left(1 + \frac{h}{R}\right)^2} \qquad \text{or} \qquad \left(1 + \frac{h}{R}\right)^2 = 3$$

or $\qquad \left(1 + \frac{h}{R}\right) = \sqrt{3} \quad \Rightarrow \quad \frac{h}{R} = 1.732 - 1$

or $\qquad h = 0.732 \times R \ (R = 6400 \text{ km})$

$\Rightarrow \qquad h = 0.732 \times 6400 \text{ km} \qquad \text{or} \qquad h = 4684.8 \text{ km}$

82. Two satellites revolve around a planet, with radius 3200 km and 800 km what is their ratio of orbital speeds?

   (A) 2 : 3   (B) 2 : 1   (C) 3 : 2   (D) 1 : 2

SOLUTION (D) From Eq.1.120, the orbital speed of a satellite is given by,

$$v = \sqrt{\frac{GM}{r}}$$

Therefore, $\frac{v_1}{v_2} = \sqrt{\frac{r_2}{r_1}} = \sqrt{\frac{800}{3200}} = \sqrt{\frac{1}{4}} = \frac{1}{2}$

83. Particles A, B, C of mass 100 kg are in a straight line where distance between A, B and B, C is 13 m. A fourth particle P is placed on the perpendicular bisector of AC when $BP = 13$ m. If net force on P is F. Find F in terms of G (Gravitational constant)

   (A) $\frac{G \times 10^4}{13^2}\left(\frac{1}{\sqrt{4}} + 2\right)$ (B) $\frac{G \times 10^4}{13^2}\left(\frac{1}{\sqrt{2}} + 1\right)$

   (C) $\frac{G \times 10^4}{13^2}\left(\frac{1}{\sqrt{5}} + 3\right)$ (D) $\frac{G \times 10^4}{13^2}\left(\frac{1}{\sqrt{6}} + 1\right)$

SOLUTION (B) From Fig.1.193, we have

$$|F_{net}| = \left|\vec{F}_{PA} + \vec{F}_{PB} + \vec{F}_{PC}\right|$$

$$= 2\left(\frac{G(100)(100)}{(13\sqrt{2})^2}\cos 45°\right) + \left(\frac{G(100)(100)}{13^2}\right)$$

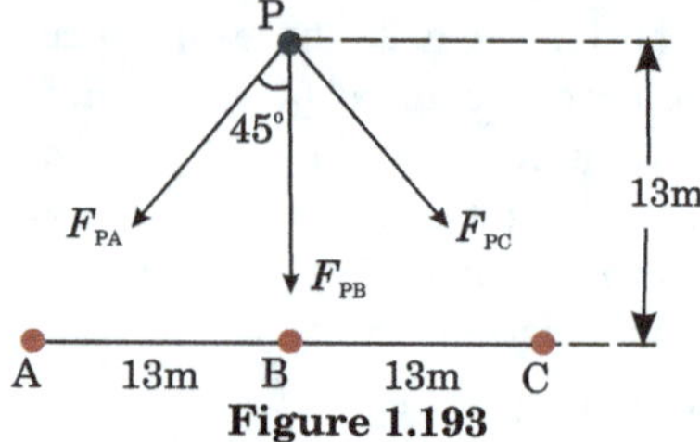

Figure 1.193

$$= \frac{G \times 10^4}{13^2}\left(\frac{1}{\sqrt{2}} + 1\right)$$

84. A body is taken from surface of earth to a distance of $5R/4$ from the centre of earth, where R is radius of earth. Percentage decrease in weight of body at this height is

   (A) 33.33%   (B) 64%   (C) 25%   (D) 36%

SOLUTION (D) From Eq.1.9, the acceleration due to gravity at distance $r \ (>R)$ from the centre of earth is

$$g' = \frac{GM}{r^2} \qquad \ldots (1)$$

Given that, $r = 5R/4$, therefore, we have

$$g' = \frac{GM}{r^2} = \frac{GM}{\left(5\frac{R}{4}\right)^2} = \frac{16}{25}g$$

$\%$ decrease $= \frac{g - g'}{g} \times 100\%$

$$= \left[\left(1 - \frac{16}{25}\right)\right] \times 100\% = 36\%$$

1.25   Questions and Problems

1.25.1   Conceptual Questions

1. A system consists of five particles. How many terms appear in the expression for the total gravitational potential energy of the system?

2. It is often said that astronauts in orbit experience weightlessness because they are beyond the pull of Earth's gravity. Is this statement correct? Explain.

3. True or false:

    (a) For Kepler's law of equal areas to be valid, the force of gravity must vary inversely with the square of the distance between a given planet and the Sun.

    (b) The planet closest to the Sun has the shortest orbital period.

    (c) The orbital period of a planet allows accurate determination of that planet's mass.

4. Does an apple exert a gravitational force on the Earth? If so, how large a force? Consider an apple (a) attached to a tree and (b) falling.

5. During what season in the northern hemisphere does Earth attain its maximum orbital speed about the Sun? What season is related to its minimum orbital speed? Hint: Earth is at perihelion in early January.

6. Imagine bringing the tips of your index fingers together. Each finger contains a certain finite mass, and the distance between them goes to zero as they come into contact. From the force law $F = Gm_1m_2/r^2$ one might conclude that the attractive force between the fingers is infinite, and, therefore, that your fingers must remain forever stuck together. What is wrong with this argument?

7. The Sun's gravitational pull on the Earth is much larger than the Moon's. Yet the Moon's is mainly responsible for the tides. Explain. [Hint: Consider the difference in gravitational pull from one side of the Earth to the other.]

8. Venus has no natural satellites. However artificial satellites have been placed in orbit around it. To use one of their orbits to determine the mass of Venus, what orbital parameters would you have to measure? How would you then use them to do the mass calculation?

9. Will an object weigh more at the equator or at the poles? What two effects are at work? Do they oppose each other?

10. Why is more fuel required for a spacecraft to travel from the Earth to the Moon than it does to return from the Moon to the Earth?

11. A majority of the asteroids are in approximately circular orbits in a belt between Mars and Jupiter. Do they all have the same orbital period about the Sun? Explain.

12. The gravitational force on the Moon due to the Earth is

only about half the force on the Moon due to the Sun. Why isn't the Moon pulled away from the Earth?

13. How did the scientists of Newton's era determine the distance from the Earth to the Moon, despite not knowing about spaceflight or the speed of light? [Hint: Think about why two eyes are useful for depth perception.]

14. If it were possible to drill a hole all the way through the Earth along a diameter, then it would be possible to drop a ball through the hole. When the ball was right at the centre of the Earth, what would be the total gravitational force exerted on it by the Earth?

15. If you had been working for NASA in the 1960's and planning the trip to the moon, you would have determined that there exists a unique location somewhere between Earth and the moon, where a spaceship is, for an instant, truly weightless. [Consider only the moon, Earth and the Apollo spaceship, and neglect other gravitational forces.] Explain this phenomenon and explain whether this location is closer to the moon, midway on the trip, or closer to Earth.

16. Why is it not possible to put a satellite in geosynchronous orbit above the North Pole?

17. Which pulls harder gravitationally, the Earth on the Moon, or the Moon on the Earth? Which accelerates more?

18. Would it require less speed to launch a satellite (*a*) toward the east or (*b*) toward the west? Consider the Earth's rotation direction.

19. An antenna loosens and becomes detached from a satellite in a circular orbit around the Earth. Describe the antenna's motion subsequently. If it will land on the Earth, describe where; if not, describe how it could be made to land on the Earth.

20. Describe how careful measurements of the variation in g in the vicinity of an ore deposit might be used to estimate the amount of ore present.

21. In India, rockets are launched into space from "Shriharikota" in an easterly direction. Is there an advantage to launching to the east versus launching to the west? Explain.

22. The Sun is below us at midnight, nearly in line with the Earth's centre. Are we then heavier at midnight, due to the Sun's gravitational force on us, than we are at noon? Explain.

23. When will your apparent weight be the greatest, as measured by a scale in a moving elevator: when the elevator (*a*) accelerates downward, (*b*) accelerates upward, (*c*) is in free fall, or (*d*) moves upward at constant speed? In which case would your apparent weight be the least? When would it be the same as when you are on the ground?

24. If the Earth's mass were double what it actually is, in what ways would the Moon's orbit be different?

25. If you light a candle on the International Space Station—which would not be a good idea—would it burn the same as on the Earth? Explain.

26. The source of the Mississippi River is closer to the center of the Earth than is its outlet in Louisiana (since the Earth is fatter at the equator than at the poles). Explain how the Mississippi can flow "uphill."

27. People sometimes ask, "What keeps a satellite up in its orbit around the Earth?" How would you respond?

28. Explain how a runner experiences "free fall" or "apparent weightlessness" between steps.

29. If you were in a satellite orbiting the Earth, how might you cope with walking, drinking, or putting a pair of scissors on a table?

30. Is the centripetal acceleration of Mars in its orbit around the Sun larger or smaller than the centripetal acceleration of the Earth?

31. The mass of the planet Pluto was not known until it was discovered to have a moon. Explain how this enabled an estimate of Pluto's mass.

32. The Earth moves faster in its orbit around the Sun in January than in July. Is the Earth closer to the Sun in January, or in July? Explain. [Note: This is not much of a factor in producing the seasons - the main factor is the tilt of the Earth's axis relative to the plane of its orbit.]

33. Kepler's laws tell us that a planet moves faster when it is closer to the Sun than when it is farther from the Sun. What causes this change in speed of the planet?

34. Does your body directly sense a gravitational field? (Compare to what you would feel in free fall.)

35. Discuss the conceptual differences between $\vec{g}$ as acceleration due to gravity and $\vec{g}$ as gravitational field.

36. Suppose that, using a telescope in your backyard, you discovered a distant object approaching the Sun, and were able to determine both its distance from the Sun and its speed. How would you be able to predict whether the object will remain "bound" to the Solar System, or if it is an interstellar interloper and would come in, turn around and escape, never to return?

37. Is the earth's gravitational force on the sun larger than, smaller than, or equal to the sun's gravitational force on the earth? Explain.

38. The gravitational force of a star on orbiting planet 1 is F_1. Planet 2 , which is twice as massive as planet 1 and orbits at twice the distance from the star, experiences gravitational force F_2. What is the ratio F_1/F_2?

39. A 1000 kg satellite and a 2000 kg satellite follow exactly the same orbit around the earth.
(a) What is the ratio F_1/F_2 of the force on the first satellite to that on the second satellite?
(b) What is the ratio a_1/a_2 of the acceleration of the first satellite to that of the second satellite?

40. How far away from the earth must an orbiting spacecraft be for the astronauts inside to be weightless? Explain.

41. A space shuttle astronaut is working outside the shuttle as it orbits the earth. If he drops a hammer, will it fall to earth? Explain why or why not.

42. The free-fall acceleration at the surface of planet 1 is 20 m/s^2. The radius and the mass of planet 2 are twice those of planet 1 . What is g on planet 2?

43. Why is the gravitational potential energy of two masses negative? Note that saying "because that's what the equation gives" is not an explanation.

44. Near the end of their useful lives, several large Earth-orbiting satellites have been maneuvered so they burn up as they enter Earth's atmosphere. These maneuvers have to be done carefully so large fragments do not impact populated land areas. You are in charge of such a project. Assuming the satellite of interest has on-board propulsion, in what direction would you fire the rockets for a short burn time to start this downward spiral? What would happen to the kinetic energy, gravitational potential energy and total mechanical energy following the burn as the satellite came closer and closer to Earth?

45. The mass of Jupiter is 300 times the mass of the earth. Jupiter orbits the sun with $T_{\text{Juniter}} = 11.9\text{yr}$ in an orbit with $r_{\text{Jupiter}} = 5.2r_{\text{earth}}$. . Suppose the earth could be moved to the distance of Jupiter and placed in a circular orbit around the sun. Which of the following describes the earth's new period? Explain.

a. 1y

b. Between 1yr and 11.9yr

c. 11.9yr

d. More than 11.9yr

e. It would depend on the earth's speed.

f. It's impossible for a planet of earth's mass to orbit at the distance of Jupiter.

46. Satellites in near-earth orbit experience a very slight drag due to the extremely thin upper atmosphere. These satellites slowly but surely spiral inward, where they finally burn up as they reach the thicker lower levels of the atmosphere. The radius decreases so slowly that you can consider the satellite to have a circular orbit slowly that you can consider the satellite to have a circular orbit at all times. As a satellite spirals inward, does it speed up, slow down, or maintain the same speed? Explain.

47. Can two particles be in equilibrium under the action of their mutual gravitational force? Can three particles be ? Can one of the three particles be?

48. Is there any meaning of "Weight of the earth"?

49. If heavier bodies are attracted more strongly by the earth, why don't they fall faster than the lighter bodies?

50. Can you think of two particles which do not exert gravitational force on each other?

51. The earth revolves round the sun because the sun attracts the earth. The sun also attracts the moon and this force is about twice as large as the attraction of the earth on the moon. Why does the moon not revolve round the sun? Or does it?

52. Suppose the gravitational potential due to a small system is k/r^2 at a distance r from it. What will be the gravitational field? Can you think of any such system ? What happens if there were negative masses?

53. The gravitational potential energy of a two-particle system is derived in this chapter as $U = -\frac{Gm_1 m_2}{r}$. Does it follow from this equation that the potential energy for $r = \infty$ must be zero? Can we choose the potential energy for $r = \infty$ to be 20 J and still use this formula? If no, what formula should be used to calculate the gravitational potential energy at separation r?

54. The weight of an object is more at the poles than at the equator. Is it beneficial to purchase goods at equator and sell them at the pole? Does it matter whether a spring balance is used or an equal-beam balance is used?

55. If the radius of the earth decreases by 1% without changing its mass, will the acceleration due to gravity at the surface of the earth increase or decrease? If so, by what per cent?

56. Is it necessary for the plane of the orbit of a satellite to pass through the centre of the earth?

57. Consider earth's satellites in circular orbits. A geostationary satellite must be at a height of about 36000 km from the earth's surface. Will any satellite moving at this height be a geostationary satellite? Will any satellite moving at this height have a time period of 24 hours?

58. During a trip back from the moon, the Apollo spacecraft fires its rockets to leave its lunar orbit. Then it coasts back to Earth where it enters the atmosphere at high speed, survives a blazing re-entry and parachutes safely into the ocean. In what direction do you fire the rockets to initiate this return trip? Explain the changes in kinetic energy, gravitational potential and total mechanical energy that occur to the spacecraft from the beginning to the end of this journey.

59. No part of India is situated on the equator. Is it possible to have a geostationary satellite which always remains over New Delhi?

60. As the earth rotates about its axis, a person living in his house at the equator goes in a circular orbit of radius equal to the radius of the earth. Why does he/she not feel weightless as a satellite passenger does?

61. Two satellites going in equatorial plane have almost same radii. As seen from the earth one moves from east to west and the other from west to east. Will they have the same time period as seen from the earth? If not, which one will have less time period?

62. The force exerted by the Sun on the Moon is more than twice the force exerted by the Earth on the Moon. Should the Moon be thought of as orbiting the Earth or the Sun? Explain.

63. The Earth and Moon exert gravitational forces on one another as they orbit the Sun. As a result, the path they follow is not the simple circular orbit you would expect if either one orbited the Sun alone. Occasionally you will see a suggestion that the Moon follows a path like a sine wave centred on a circular path, as in the upper part of Figure 1.194. This is incorrect. The Moon's path is qualitatively like that shown in the lower part of Figure 1.194. Explain. (Refer to Conceptual Question 62.)

1.25.2   Problems

Newton's law of Universal Gravitation

1. ••A typical Cavendish balance, for measuring the gravitational constant G, uses lead spheres with masses of 1.50 kg and 15.0 g whose centres are separated by about 4.50 cm. Calculate the gravitational force between these spheres, treating each as a particle located at the sphere's centre.

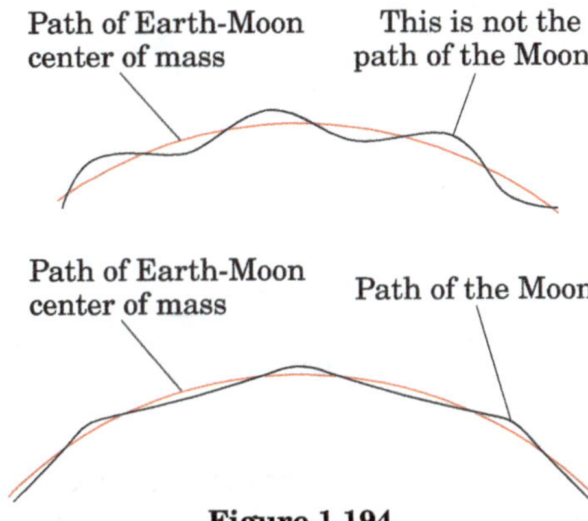

Figure 1.194

2. •Determine the order of magnitude of the gravitational force between two persons, each of mass 70 kg, standing 2 m away from each other.

3. ••A 200 kg object and a 500 kg object are separated by 4.00 m . (a) Find the net gravitational force exerted by these objects on a 50.0 kg object placed midway between them. (b) At what position (other than an infinitely remote one) can the 50.0 kg object be placed so as to experience a net force of zero from the other two objects?

4. ••Two objects, each of mass m, hang from strings of different lengths on a balance at the surface of the Earth, as shown in Fig. 1.195. If the strings have negligible mass and differ in length by h, (a) show that the error in weighing, associated with the fact that W' is closer to the Earth than W, is $W' - W = 8\pi G\rho mh/3$ in which ρ is the mean density of the Earth (5.5 g/cm^3). (b) Find the difference in length that will give an error of one part in a million.

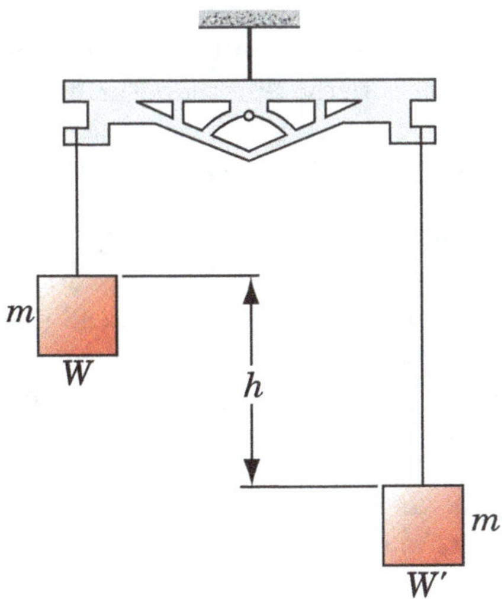

Figure 1.195

5. ••An object is suspended on a spring balance in a ship sailing along the equator with a speed v. Show that the scale reading will be very close to $W_0(1 \pm 2\omega v/g)$, where ω is the angular speed of the Earth and W_0 is the scale reading when the ship is at rest. Explain the plus or minus.

6. ••During a solar eclipse, the Moon, the Earth, and the Sun all lie on the same line, with the Moon between the Earth and the Sun. (a) What force is exerted by the Sun on the Moon? (b) What force is exerted by the Earth on the Moon? (c) What force is exerted by the Earth on the Moon? (c) What force is exerted by the Sun on the Earth? (d) Compare the answers to parts away from the Earth?

7. ••Two ocean liners, each with a mass of 40000 metric tons, are moving on parallel courses 100 m apart. What is the magnitude of the acceleration of one of the liners toward the other due to their mutual gravitational attraction? Model the ships as particles.

8. ••Three uniform spheres of masses $m_1 = 2.00$ kg, $m_2 = 4.00$ kg, and $m_3 = 6.00$ kg are placed at the corners of a right triangle as shown in Figure 1.196. Calculate the resultant gravitational force on the object of mass m_2, assuming the spheres are isolated from the rest of the Universe.

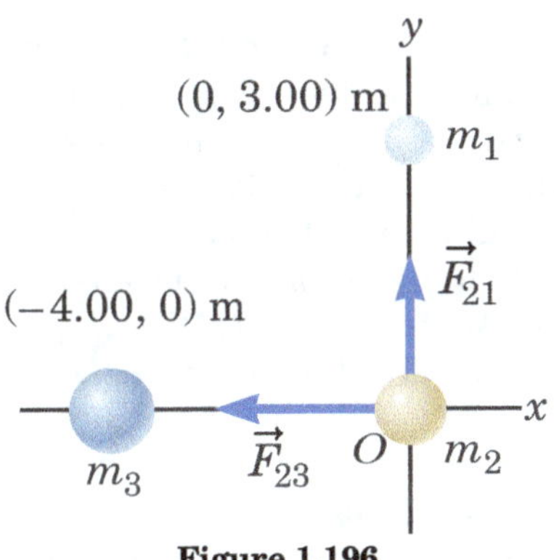

Figure 1.196

9. ••Two identical isolated particles, each of mass 2.00 kg, are separated by a distance of 30.0 cm. What is the magnitude of the gravitational force exerted by one particle on the other?

10. ••Why is the following situation impossible? The centres of two homogeneous spheres are 1.00 m apart. The spheres are each made of the same element from the periodic table. The gravitational force between the spheres is 1.00 N.

11. ••Two objects attract each other with a gravitational force of magnitude 1.00×10^{-8} N when separated by 20.0 cm. If the total mass of the two objects is 5.00 kg, what is the mass of each?

12. • A superconducting gravity meter can measure changes in gravity on the order $\Delta g/g = 1.00 \times 10^{-11}$. (a) You are hiding behind a tree holding the meter, and your 80 kg friend approaches the tree from the other side. How close to you can your friend come before the meter detects a change in g due to his presence? (b) You are in a hot air balloon and using the gravity meter to determine the rate of ascent (assume the balloon has constant acceleration). What is the smallest change in altitude that results in a detectable change in the gravitational field of Earth?

13. •••A student proposes to study the gravitational force by suspending two 100.0 kg spherical objects at the lower ends of cables from the ceiling of a tall tower and measuring the deflection of the cables from the vertical. The 45.00 m-long cables are attached to the ceiling 1.000 m

apart. The first object is suspended, and its position is carefully measured. The second object is suspended, and the two objects attract each other gravitationally. By what distance has the first object moved horizontally from its initial position due to the gravitational attraction to the other object? Suggestion: Keep in mind that this distance will be very small and make appropriate approximations.

14. ••Neutron stars are extremely dense objects formed from the remnants of supernova explosions. Many rotate very rapidly. Suppose the mass of a certain spherical neutron star is twice the mass of the Sun and its radius is 10.0 km. Determine the greatest possible angular speed it can have so that the matter at the surface of the star on its equator is just held in orbit by the gravitational force.

Free-Fall Acceleration and the Gravitational Force

15. ••When a falling meteoroid is at a distance above the Earth's surface of 3.00 times the Earth's radius, what is its acceleration due to the Earth's gravitation?

16. ••The free-fall acceleration on the surface of the Moon is about one-sixth that on the surface of the Earth. The radius of the Moon is about $0.250R_E$ (R_E = Earth's radius $= 6.37 \times 10^6$ m). Find the ratio of their average densities, $\rho_{\text{Moon}}/\rho_{\text{Earth}}$.

Particle in a Gravitational Field

17. ••(a) Compute the vector gravitational field at a point P on the perpendicular bisector of the line joining two objects of equal mass separated by a distance $2a$ as shown in Figure 1.197. (b) Explain physically why the field should approach zero as $r \to 0$. (c) Prove mathematically that the answer to part (a) behaves in this way. (d) Explain physically why the magnitude of the field should approach $2GM/r^2$ as $r \to \infty$. (e) Prove mathematically that the answer to part (a) behaves correctly in this limit.

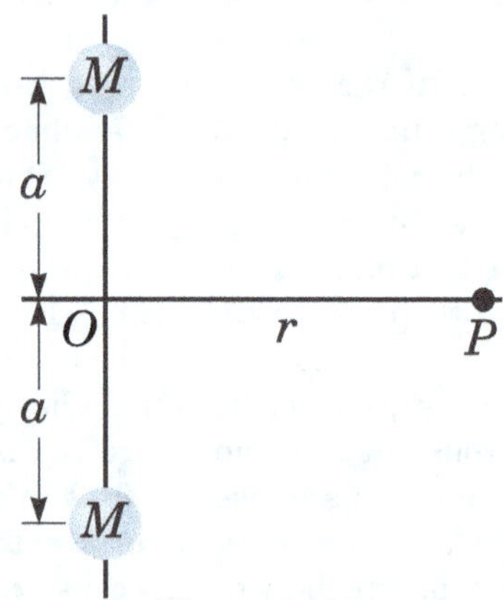

Figure 1.197

18. ••Three objects of equal mass are located at three corners of a square of edge length l as shown in Figure 1.198. Find the magnitude and direction of the gravitational field at the fourth corner due to these objects.

19. •••A spacecraft in the shape of a long cylinder has a

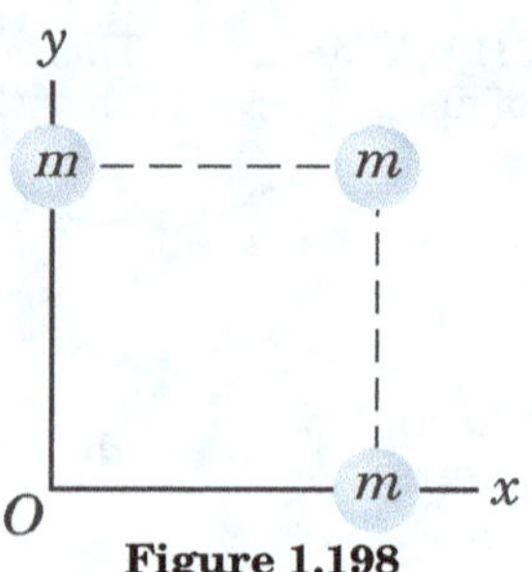

Figure 1.198

length of 100 m, and its mass with occupants is 1000 kg. It has strayed too close to a black hole having a mass 100 times that of the Sun (Fig. 1.199). The nose of the spacecraft points toward the black hole, and the distance between the nose and the center of the black hole is 10.0 km. (a) Determine the total force on the spacecraft. (b) What is the difference in the gravitational fields acting on the occupants in the nose of the ship and on those in the rear of the ship, farthest from the black hole? (This difference in accelerations grows rapidly as the ship approaches the black hole. It puts the body of the ship under extreme tension and eventually tears it apart.)

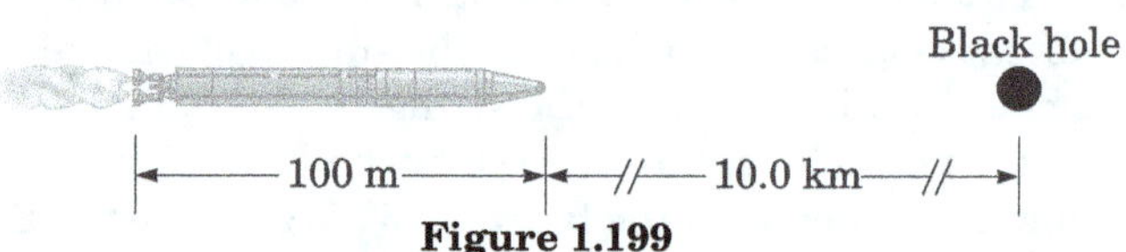

Figure 1.199

Kepler's Laws and the Motion of Planets

20. ••The speed of an asteroid is 20 km/s at perihelion and 14 km/s at aphelion. (a) Determine the ratio of the aphelion to perihelion distances. (b) Is this asteroid farther from the Sun or closer to the Sun than Earth, on average? Explain.

21. ••An artificial satellite circles the Earth in a circular orbit at a location where the acceleration due to gravity is 9.00 m/s^2. Determine the orbital period of the satellite.

22. ••A minimum-energy transfer orbit to an outer planet consists of putting a spacecraft on an elliptical trajectory with the departure planet corresponding to the perihelion of the ellipse, or the closest point to the Sun, and the arrival planet at the aphelion, or the farthest point from the Sun. (a) Use Kepler's third law to calculate how long it would take to go from Earth to Mars on such an orbit as shown in Figure 1.200. (b) Can such an orbit be undertaken at any time? Explain.
Given: Perihelion distance of Earth from the centre of Sun $= 1.496 \times 10^{11}$ m and aphelion distance of Mars from the centre of Sun $= 2.28 \times 10^{11}$ m, also 1 AU $= 1.496 \times 10^{11}$ m.

23. ••A particle of mass m moves along a straight line with constant velocity $\vec{v}_0$ in the x direction, a distance b from the x axis (Fig. 1.201). (a) Does the particle possess any angular momentum about the origin? (b) Explain why the amount of its angular momentum should change or should stay constant. (c) Show that Kepler's second law

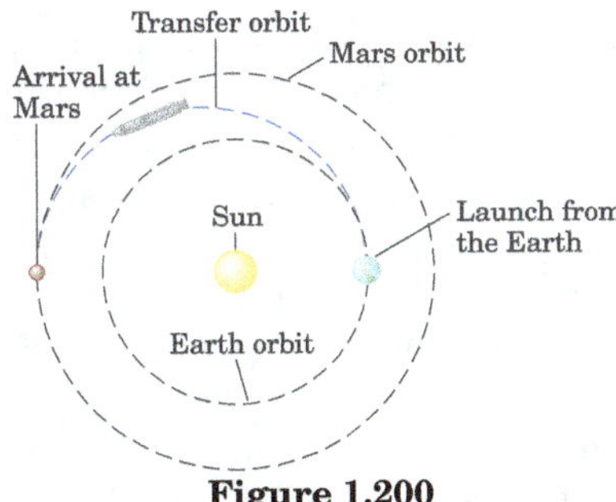

Figure 1.200

is satisfied by showing that the two shaded triangles in the figure have the same area when

$$t_{\text{D}} - t_{\text{C}} = t_{\text{B}} - t_{\text{A}}$$

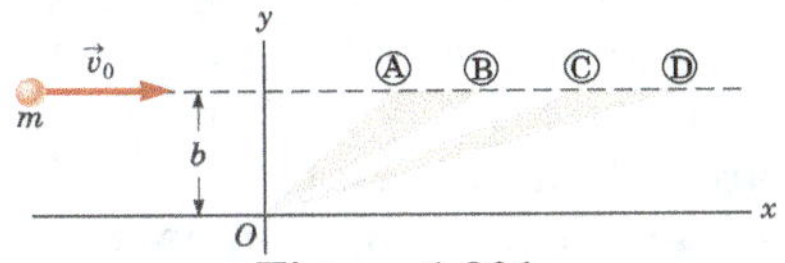

Figure 1.201

24. ••Plaskett's binary system consists of two stars that revolve in a circular orbit about a centre of mass midway between them. This statement implies that the masses of the two stars are equal (Fig. 1.202). Assume the orbital speed of each star is $|\vec{v}| = 220$ km/s and the orbital period of each is 14.4 days. Find the mass M of each star. (For comparison, the mass of our Sun is 1.99×10^{30} kg.)

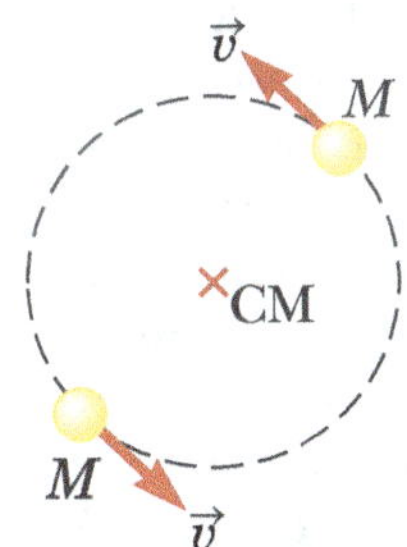

Figure 1.202

25. •••Two planets X and Y travel counterclockwise in circular orbits about a star as shown in Figure 1.203. The radii of their orbits are in the ratio 3 : 1. At one moment, they are aligned as shown in Figure 1.203a, making a straight line with the star. During the next five years, the angular displacement of planet X is 90.0° as shown in Figure 1.203b. What is the angular displacement of planet Y at this moment?

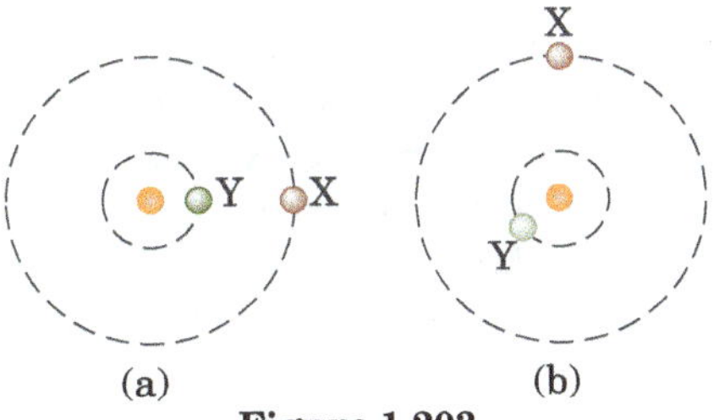

Figure 1.203

26. ••Comet Halley (Fig.1.204) approaches the Sun to within 0.570AU, and its orbital period is 75.6yr. (AU is the symbol for astronomical unit, where $1\text{AU} = 1.50 \times 10^{11}$ m is the mean Earth-Sun distance.) How far from the Sun will Halley's comet travel before it starts its return journey?

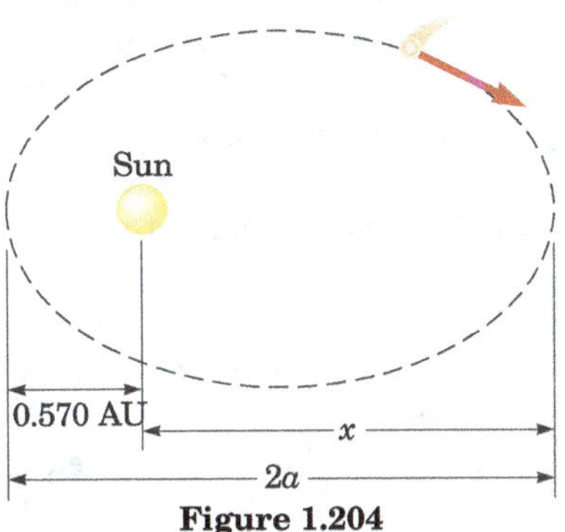

Figure 1.204

27. ••The Explorer VIII satellite, placed into orbit November 3, 1960, to investigate the ionosphere, had the following orbit parameters: perigee, 459 km; apogee, 2289 km (both distances above the Earth's surface); period, 112.7 min. Find the ratio v_p/v_a of the speed at perigee to that at apogee.

28. ••Use Kepler's third law to determine how many days it takes a spacecraft to travel in an elliptical orbit from a point 6670 km from the Earth's centre to the Moon, 385000 km from the Earth's centre.

29. ••A synchronous satellite, which always remains above the same point on a planet's equator, is put in orbit around Jupiter to study that planet's famous red spot. Jupiter rotates once every 9.84 h. Use the data of Table 13.2 to find the altitude of the satellite above the surface of the planet.

30. ••(a) Given that the period of Moon's orbit about the Earth is 27.32 days and the nearly constant distance between the centre of the Earth and the centre of the Moon is 3.84×10^8 m, use Eq.1.150 to calculate the mass of the Earth.

31. ••Suppose the Sun's (mass $= 1.99 \times 10^{30}$ kg) gravity were switched off. The planets would leave their orbits and fly away in straight lines as described by Newton's first law. (a) Would Mercury ever be farther from the Sun than Pluto? (b) If so, find how long it would take Mercury to achieve this passage. If not give a convincing argument that Pluto is always farther from the Sun than is Mercury(Orbits radii: for Mercury, $r_M = 5.79 \times 10^{11}$ m, for Pluto, $r_p = 5.91 \times 10^{12}$ m).

32. ••A satellite is in a circular orbit around the Earth (mass $= M$, radius $= R$) at an altitude of 2.80×10^6 m. Find (a) the period of the orbit, (b) the speed of the satellite, and (c) the acceleration of the satellite.

33. The mass of Saturn is 5.69×10^{26} kg. (a) Find the period of its moon 'Mimas', whose mean orbital radius is 1.86×10^8 m. (b) Find the mean orbital radius of its moon 'Titan', whose period is 1.38×10^6 s.

34. ••Calculate the mass of Earth by using following data:

period of the moon, $T = 27.3$ days; mean orbital radius of moon, $r_\mathrm{m} = 3.84 \times 10^8$ m; and the known value of $G = 6.6726 \times 10^{-11} \mathrm{N \cdot m^2/kg^2}$.

Gravitational Potential Energy

Note: In Problems 35 through 41, assume $U = 0$ at $r = \infty$.

35. ••A satellite in Earth orbit has a mass of 100 kg and is W at an altitude of 2.00×10^6 m. (a) What is the potential energy of the satellite-Earth system? (b) What is the magnitude of the gravitational force exerted by the Earth on the satellite? (c) What force, if any, does the satellite exert on the Earth?

36. ••An object is projected vertically from the surface of Earth at less than the escape speed. Show that the maximum height reached by the object is $H = R_E H'/(R_E - H')$, where H' is the height that it would reach if the gravitational field were constant. Neglect any effects due to air resistance.

37. •• How much work is done by the Moon's gravitational field on a 1000 kg meteor as it comes in from outer space and impacts on the Moon's surface?

38. ••How much energy is required to move a $1000-\mathrm{kg}$ object from the Earth's surface to an altitude twice the Earth's radius?

39. ••After the Sun exhausts its nuclear fuel, its ultimate fate will be to collapse to a white dwarf state. In this state, it would have approximately the same mass as it has now, but its radius would be equal to the radius of the Earth. Calculate (a) the average density of the white dwarf, (b) the surface free-fall acceleration, and (c) the gravitational potential energy associated with a $1.00-\mathrm{kg}$ object at the surface of the white dwarf.

40. ••An object is released from rest at an altitude h above the surface of the Earth (mass $= M$, radius $= R$). (a) Show that its speed at a distance r from the Earth's centre, where $R \le r \le R + h$, is

$$v = \sqrt{2GM\left(\frac{1}{r} - \frac{1}{R+h}\right)}$$

(b) Assume the release altitude is 500 km. Perform the integral

$$\Delta t = \int_i^f dt = -\int_i^f \frac{dr}{v}$$

to find the time of fall as the object moves from the release point to the Earth's surface. The negative sign appears because the object is moving opposite to the radial direction, so its speed is $v = -dr/dt$. Perform the integral numerically.

41. ••A system consists of three particles, each of mass 5.00 g, located at the corners of an equilateral triangle with sides of 30.0 cm. (a) Calculate the potential energy of the system. (b) Assume the particles are released simultaneously. Describe the subsequent motion of each. Will any collisions take place? Explain.

Energy Considerations in Planetary and Satellite Motion

42. ••A space probe is fired as a projectile from the Earth's surface with an initial speed of 2.00×10^4 m/s. What will its speed be when it is very far from the Earth? Ignore atmospheric friction and the rotation of the Earth.

43. ••A 500 kg satellite is in a circular orbit at an altitude of 500 km above the Earth's surface. Because of air friction, the satellite eventually falls to the Earth's surface, where it hits the ground with a speed of 2.00 km/s. How much energy was transformed into internal energy by means of air friction?

44. ••A "treetop satellite" moves in a circular orbit just above the surface of a planet, assumed to offer no air resistance. Show that its orbital speed v and the escape speed from the planet are related by the expression $v_\mathrm{esc} = \sqrt{2}v$.

45. ••A 1000 kg satellite orbits the Earth at a constant altitude of 100 km. (a) How much energy must be added to the system to move the satellite into a circular orbit with altitude 200 km ? What are the changes in the system's (b) kinetic energy and (c) potential energy?

46. ••A comet of mass 1.20×10^{10} kg moves in an elliptical orbit around the Sun. Its distance from the Sun ranges between 0.500AU and 50.0AU. (a) What is the eccentricity of its orbit? (b) What is its period? (c) At aphelion, what is the potential energy of the comet-Sun system? Note: 1AU = one astronomical unit = the average distance from the Sun to the Earth $= 1.496 \times 10^{11}$ m.

47. ••Derive an expression for the work required to move an Earth satellite of mass m from a circular orbit of radius $2R$ to one of radius $3R$.

48. ••(a) Determine the amount of work that must be done on a 100 kg payload to elevate it to a height of 1000 km above the Earth's surface. (b) Determine the amount of additional work that is required to put the payload into circular orbit at this elevation.

49. ••(a) What is the minimum speed, relative to the Sun, necessary for a spacecraft to escape the solar system if it starts at the Earth's orbit? (b) Voyager 1 achieved a maximum speed of 125000 km/h on its way to photograph Jupiter. Beyond what distance from the Sun is this speed sufficient to escape the solar system?

50. ••A satellite of mass 200 kg is placed into Earth orbit at a height of 200 km above the surface. (a) Assuming a circular orbit, how long does the satellite take to complete one orbit? (b) What is the satellite's speed? (c) Starting from the satellite on the Earth's surface, what is the minimum energy input necessary to place this satellite in orbit? Ignore air resistance but include the effect of the planet's daily rotation.

51. ••A satellite of mass m, originally on the surface of the Earth, is placed into Earth orbit at an altitude h. (a) Assuming a circular orbit, how long does the satellite take to complete one orbit? (b) What is the satellite's speed? (c) What is the minimum energy input necessary to place this satellite in orbit? Ignore air resistance but include the effect of the planet's daily rotation. Represent the mass and radius of the Earth as M and R, respectively.

52. ••Ganymede is the largest of Jupiter's moons. Consider a rocket on the surface of Ganymede, at the point farthest from the planet (Fig. 1.205). Model the rocket as a parti-

cle. (a) Does the presence of Ganymede make Jupiter exert a larger, smaller, or same size force on the rocket compared with the force it would exert if Ganymede were not interposed? (b) Determine the escape speed for the rocket from the planet-satellite system. The radius of Ganymede is 2.64×10^6 m, and its mass is 1.495×10^{23} kg. The distance between Jupiter and Ganymede is 1.071×10^9 m, and the mass of Jupiter is $1.90 \times 10^{27} kg$. Ignore the motion of Jupiter and Ganymede as they revolve about their center of mass.

Figure 1.205

53. ••A satellite moves around the Earth in a circular orbit of radius r. (a) What is the speed v_i of the satellite? (b) Suddenly, an explosion breaks the satellite into two pieces, with masses m and $4m$. Immediately after the explosion, the smaller piece of mass m is stationary with respect to the Earth and falls directly toward the Earth. What is the speed v of the larger piece immediately after the explosion? (c) Because of the increase in its speed, this larger piece now moves in a new elliptical orbit. Find its distance away from the centre of the Earth when it reaches the other end of the ellipse.

54. ••At the Earth's surface, a projectile is launched straight up at a speed of 10.0 km/s. To what height will it rise? Ignore air resistance.

Additional Problems

55. ••A neutron star is a highly condensed remnant of a massive star in the last phase of its evolution. It is composed of neutrons (hence the name), because the star's gravitational force causes electrons and protons to "coalesce" into the neutrons. Suppose, at the end of its current phase, the Sun collapsed into a neutron star (it cannot actually do this because it does not have enough mass) of radius 12.0 km, without losing any mass in the process. (a) Calculate the ratio of the gravitational acceleration at the surface of the Sun following the collapse compared to the value at the surface of the Sun today. (b) Calculate the ratio of the escape speed from the surface of the neutron-Sun to the Sun's value today.

56. •••A rocket is fired straight up through the atmosphere from the South Pole, burning out at an altitude of 250 km when travelling at 6.00 km/s. (a) What maximum distance from the Earth's surface does it travel before falling back to the Earth? (b) Would it's maximum distance from the surface be larger if the same rocket were fired with the same fuel load from a launch site on the equator? Why or why not?

57. ••A cylindrical habitat in space 6.00 km in diameter and 30.0 km long has been proposed (by G.K. O'Neill, 1974). Such a habitat would have cities, land, and lakes on the inside surface and air and clouds in the centre. They would all be held in place by rotation of the cylinder about its long axis. How fast would the cylinder have to rotate to imitate the Earth's gravitational field at the walls of the cylinder?

58. •••Voyager 1 and Voyager 2 surveyed the surface of Jupiter's moon Io and photographed active volcanoes spewing liquid sulfur to heights of 70 km above the surface of this moon. Find the speed with which the liquid sulfur left the volcano. Io's mass is 8.9×10^{22} kg, and its radius is 1820 km.

59. ••Why is the following situation impossible? A spacecraft is launched into a circular orbit around the Earth and circles the Earth once an hour.

60. ••Let Δg_M represent the difference in the gravitational fields produced by the Moon at the points on the Earth's surface nearest to and farthest from the Moon. Find the fraction $\Delta g_M/g$, where g is the Earth's gravitational field. (This difference is responsible for the occurrence of the lunar tides on the Earth.)

61. •••A sleeping area for a long space voyage consists of two cabins each connected by a cable to a central hub as shown in Figure 1.206. The cabins are set spinning around the hub axis, which is connected to the rest of the spacecraft to generate artificial gravity in the cabins. A space traveller lies in a bed parallel to the outer wall as shown in Figure 1.206. (a) With $r = 10.0$ m, what would the angular speed of the $60.0-$ kg traveller need to be if he is to experience half his normal Earth weight? (b) If the astronaut stands up perpendicular to the bed, without holding on to anything with his hands, will his head be moving at a faster, a slower, or the same tangential speed as his feet? Why? (c) Why is the action in part (b) dangerous?

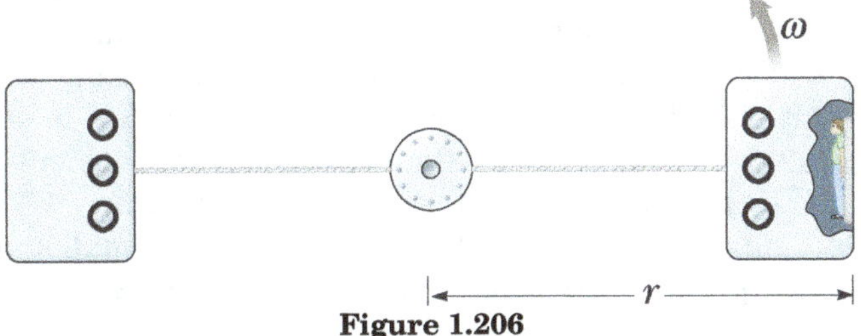

Figure 1.206

62. •••(a) A space vehicle is launched vertically upward from the Earth's surface with an initial speed of 8.76 km/s, which is less than the escape speed of 11.2 km/s. What maximum height does it attain? (b) A meteoroid falls toward the Earth. It is essentially at rest with respect to the Earth when it is at a height of 2.51×10^7 m above the Earth's surface. With what speed does the meteorite (a meteoroid that survives to impact the Earth's surface) strike the Earth?

63. •••(a) A space vehicle is launched vertically upward from the Earth's surface with an initial speed of v_i that is comparable to but less than the escape speed v_e. What maximum height does it attain? (b) A meteoroid falls toward the Earth. It is essentially at rest with respect to the

Earth when it is at a height h above the Earth's surface. With what speed does the meteorite (a meteoroid that survives to impact the Earth's surface) strike the Earth? (c) What If? Assume a baseball is tossed up with an initial speed that is very small compared to the escape speed. Show that the result from part (a) is consistent with Equation: $h = \dfrac{v_i^2 \sin^2 \theta_i}{2g}$.

64. ••Assume you are agile enough to run across a horizontal surface at 8.50 m/s, independently of the value of the gravitational field. What would be (a) the radius and (b) the mass of an airless spherical asteroid of uniform density 1.10×10^3 kg/m^3 on which you could launch yourself into orbit by running? (c) What would be your period? (d) Would your running significantly affect the rotation of the asteroid? Explain.

65. •••Two spheres having masses M and $2M$ and radii R and $3R$, respectively, are simultaneously released from rest when the distance between their centres is $12R$. Assume the two spheres interact only with each other and we wish to find the speeds with which they collide.
(a) What two isolated system models are appropriate for this system?
(b) Write an equation from one of the models and solve it for $\vec{v}_1$, the velocity of the sphere of mass M at any time after release in terms of $\vec{v}_2$, the velocity of $2M$.
(c) Write an equation from the other model and solve it for speed v_1 in terms of speed v_2 when the spheres collide. (d) Combine the two equations to find the two speeds v_1 and v_2 when the spheres collide.

66. ••Two hypothetical planets of masses m_1 and m_2 and radii r_1 and r_2, respectively, are nearly at rest when they are an infinite distance apart. Because of their gravitational attraction, they head toward each other on a collision course. (a) When their centre-to-centre separation is d, find expressions for the speed of each planet and for their relative speed. (b) Find the kinetic energy of each planet just before they collide, taking $m_1 = 2.00 \times 10^{24}$ kg, $m_0 = 8.00 \times 10^{24}$ kg, $r_1 = 3.00 \times 10^6$ m, and $r_2 = 5.00 \times 10^6$ m. Note: Both the energy and momentum of the isolated two-planet system are constant.

67. •••(a) Show that the rate of change of the free-fall acceleration with vertical position near the Earth's surface is
$$\frac{dg}{dr} = -\frac{2GM}{R^3}$$
This rate of change with position is called a gradient. (b) Assuming h is small in comparison to the radius of the Earth, show that the difference in free-fall acceleration between two points separated by vertical distance h is
$$|\Delta g| = \frac{2GMh}{R^3}$$
(c) Evaluate this difference for $h = 6.00$ m, a typical height for a two-story building.

68. •••A ring of matter is a familiar structure in planetary and stellar astronomy. Examples include Saturn's rings and a ring nebula. Consider a uniform ring of mass 2.36×10^{20} kg and radius 1.00×10^8 m. An object of mass 1000 kg is placed at a point A on the axis of the ring, 2.00×10^8 m from the centre of the ring (Fig. 1.207). When the object is released, the attraction of the ring makes the object move along the axis toward the centre of the ring (point B). (a) Calculate the gravitational potential energy of the object-ring system when the object is at A. (b) Calculate the gravitational potential energy of the system when the object is at B. (c) Calculate the speed of the object as it passes through B.

Figure 1.207

69. •••A spacecraft of mass 1.00×10^4 kg is in a circular orbit at an altitude of 500 km above the Earth's surface. Mission Control wants to fire the engines in a direction tangent to the orbit so as to put the spacecraft in an elliptical orbit around the Earth in an apogee distance of 2.0×10^4 km, measured from the Earth's centre. How much energy must be used from the fuel to achieve this orbit? (Assume that all the fuel energy goes into increasing the orbital energy. This model will give a lower limit to the required energy because some of the energy from the fuel will appear as internal energy in the hot exhaust gases and engine parts.)

70. •••As an astronaut, you observe a small planet to be spherical. After landing on the planet, you set off, walking always straight ahead, and find yourself returning to your spacecraft from the opposite side after completing a lap of 25.0 km. You hold a hammer and a falcon feather at a height of 1.40 m, release them, and observe that they fall together to the surface in 29.2 s. Determine the mass of the planet.

71. •••A certain quaternary star system consists of three stars, each of mass m, moving in the same circular orbit of radius r about a central star of mass M [Fig.1.261a]. The stars orbit in the same sense and are positioned one-third of a revolution apart from one another. Show that the period of each of the three stars is given by
$$T = 2\pi\sqrt{\frac{r^3}{G(M + m/\sqrt{3})}}$$

72. •••Studies of the relationship of the Sun to our galaxy the Milky Way-have revealed that the Sun is located near the outer edge of the galactic disc, about 30000 ly $\left(1\text{ly} = 9.46 \times 10^{15} \text{ m}\right)$ from the centre. The Sun has an orbital speed of approximately 250 km/s around the galactic centre. (a) What is the period of the Sun's galactic motion? (b) What is the order of magnitude of the mass of the Milky Way galaxy? (c) Suppose the galaxy is made mostly of stars of which the Sun is typical. What is the order of magnitude of the number of stars in the Milky Way?

73. •••Two identical hard spheres, each of mass m and radius r, are released from rest in otherwise empty space with their centres separated by the distance R. They are al-

lowed to collide under the influence of their gravitational attraction. (a) Show that the magnitude of the impulse received by each sphere before they make contact is given by $\left[Gm^3(1/2r - 1/R)\right]^{1/2}$. (b) What If? Find the magnitude of the impulse each receives during their contact if they collide elastically.

74. $\bullet\bullet\bullet$The maximum distance from the Earth to the Sun (at aphelion) is 1.521×10^{11} m, and the distance of closest approach (at perihelion) is 1.471×10^{11} m. The Earth's orbital speed at perihelion is 3.027×10^4 m/s. Determine (a) the Earth's orbital speed at aphelion (b) The kinetic and potential energies at perihelion and (c) at aphelion. (d) Is the total energy of the system constant? Explain. Ignore the effect of the Moon and other planets.

75. $\bullet\bullet\bullet$Many people assume air resistance acting on a moving object will always make the object slow down. It can, however, actually be responsible for making the object speed up. Consider a 100-kg Earth satellite in a circular orbit at an altitude of 200 km. A small force of air resistance makes the satellite drop into a circular orbit with an altitude of 100 km. (a) Calculate the satellite's initial speed. (b) Calculate its final speed in this process. (c) Calculate the initial energy of the satellite-Earth system. (d) Calculate the final energy of the system. (e) Show that the system has lost mechanical energy and find the amount of the loss due to friction. (f) What force makes the satellite's speed increase? Hint: You will find a free-body diagram useful in explaining your answer.

76. $\bullet\bullet\bullet$X-ray pulses from Cygnus X-1, the first black hole to be identified and a celestial x-ray source, have been recorded during high-altitude rocket flights. The signals can be interpreted as originating when a blob of ionized matter orbits a black hole with a period of 5.0 ms. If the blob is in a circular orbit about a black hole whose mass is $20M_{\text{Sun}}$, what is the orbit radius?

77. $\bullet\bullet$Show that the minimum period for a satellite in orbit around a spherical planet of uniform density ρ is independent of the planet's radius.

$$T_{\min} = \sqrt{\frac{3\pi}{G\rho}}$$

78. Astronomers detect a distant meteoroid moving along a straight line that, if extended, would pass at a distance $3R$ from the centre of the Earth, where R is the Earth's radius. What minimum speed must the meteoroid have if it is not to collide with the Earth?

79. $\bullet\bullet$Two stars of masses M and m, separated by a distance d, revolve in circular orbits about their centre of mass (Fig. 1.208). Show that each star has a period given by

$$T^2 = \frac{4\pi^2 d^3}{G(M+m)}$$

80. $\bullet\bullet$Two identical particles, each of mass 1000 kg, are coasting in free space along the same path, one in front of the other by 20.0 m. At the instant their separation distance has this value, each particle has precisely the same velocity of $800\ \hat{i}$ m/s. What are their precise velocities when they are 2.00 m apart?

81. $\bullet\bullet\bullet$Consider an object of mass m, not necessarily small

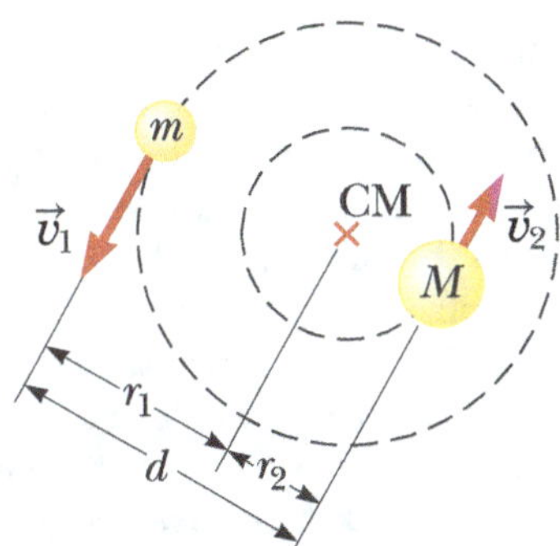

Figure 1.208

compared with the mass of the Earth, released at a distance of 1.20×10^7 m from the centre of the Earth. Assume the Earth and the object behave as a pair of particles, isolated from the rest of the Universe. (a) Find the magnitude of the acceleration a_{rel} with which each starts to move relative to the other as a function of m. Evaluate the acceleration (b) for $m = 5.00$ kg, (c) for $m = 2000$ kg, and (d) for $m = 2.00 \times 10^{24}$ kg. (e) Describe the pattern of variation of a_{rel} with m.

82. $\bullet\bullet\bullet$As thermonuclear fusion proceeds in its core, the Sun loses mass at a rate of 3.64×10^9 kg/s. During the 5000 -yr period of recorded history, by how much has the length of the year changed due to the loss of mass from the Sun? Suggestions: Assume the Earth's orbit is circular. No external torque acts on the Earth-Sun system, so the angular momentum of the Earth is constant.

Challenging Problems

83. $\bullet\bullet\bullet$The Solar and Heliospheric Observatory (SOHO) spacecraft has a special orbit, located between the Earth and the Sun along the line joining them, and it is always close enough to the Earth to transmit data easily. Both objects exert gravitational forces on the observatory. It moves around the Sun in a near-circular orbit that is smaller than the Earth's circular orbit. Its period, however, is not less than 1yr but just equal to 1 yr. Show that its distance from the Earth must be 1.48×10^9 m. In 1772 , Joseph Louis Lagrange determined theoretically the special location allowing this orbit. Suggestions: Use data that are precise to four digits. The mass of the Earth is 5.974×10^{24} kg. You will not be able to easily solve the equation you generate; instead, use a calculator to verify that 1.48×10^9 m is the correct value.

84. $\bullet\bullet\bullet$The oldest artificial satellite still in orbit is Vanguard I, launched March 3,1958 . Its mass is 1.60 kg. Neglecting atmospheric drag, the satellite would still be in its initial orbit, with a minimum distance from the centre of the Earth of 7.02 Mm and a speed at this perigee point of 8.23 km/s. For this orbit, find (a) the total energy of the satellite-Earth system and (b) the magnitude of the angular momentum of the satellite. (c) At apogee, find the satellite's speed and its distance from the centre of the Earth. (d) Find the semi-major axis of its orbit. (e) Determine its period.

85. $\bullet\bullet\bullet$A spacecraft is approaching Mars after a long trip from the Earth. Its velocity is such that it is travelling along a parabolic trajectory under the influence of the gravitational force from Mars. The distance of closest approach will be 300 km above the Martian surface. At this point of closest approach the engines will be fired to slow down the spacecraft and place it in a circular orbit 300 km above the surface. (a) By what percentage must the speed of the spacecraft be reduced to achieve the desired orbit? (b) How would the answer to part (a) change if the distance of closest approach and the desired circular orbit altitude were 600 km instead of 300 km ? (Note: The energy of the spacecraft-Mars system for a parabolic orbit is $E = 0$.)

1.25.3 Multiple Choice Assignments

1.25.3.1 Level 1

Newton's Law of Universal Gravitation

1. If R is the radius of the earth and g the acceleration due to gravity on the earth's surface, the mean density of the earth is
   (A) $4\pi G/3gR$
   (B) $3\pi R/4gG$
   (C) $3g/4\pi RG$
   (D) $\pi Rg/12G$

2. The weight of an object in the coal mine, sea level, at the top of the mountain are W_1, W_2 and W_3 respectively, then
   (A) $W_1 < W_2 > W_3$
   (B) $W_1 = W_2 = W_3$
   (C) $W_1 < W_2 < W_3$
   (D) $W_1 > W_2 > W_3$

3. The height above surface of earth where the value of gravitational acceleration is one fourth of that at surface, will be
   (A) $R_e/4$ (B) $R_e/2$ (C) $3R_e/4$ (D) R_e

4. At the surface of a certain planet acceleration due to gravity is one quarter of that on earth. If a brass ball is transported to this planet, then which one of the following statements is not correct?
   (A) the mass of the brass ball on this planet is a quarter of its mass as measured on the earth
   (B) the weight of the brass ball on this planet is a quarter of the weight as measured on the earth
   (C) the brass ball has same mass on the another planet as on the earth
   (D) the brass ball has the same volume on the other planet as on earth

5. The weight of a person in a lift accelerating upwards-
   (A) is zero
   (B) decrease
   (C) increases
   (D) remains same

6. The decrease in the value of g on going to a height $\frac{R}{2}$ above the earth's surface will be -
   (A) $g/2$ (B) $\frac{5g}{9}$ (C) $\frac{4g}{9}$ (D) $\frac{g}{3}$

7. Value of g at any place will -
   (A) remain same
   (B) decrease
   (C) increase
   (D) none of these

8. If the rotational motion of earth increases, then the weight of the body -
   (A) will remain same
   (B) will increase
   (C) will decrease
   (D) none of these

9. If 'R' is the radius of earth and 'g' the acceleration due to gravity then mass of earth will be :
   (A) $\frac{gR^2}{G}$
   (B) $\frac{g^2R}{G}$
   (C) $\frac{Rg}{G}$
   (D) $\frac{GR^2}{g}$

10. The dimensions of G are -
    (A) ML^3T^{-2}
    (B) $M^{-1}LT^{-2}$
    (C) $M^{-1}L^3T^{-2}$
    (D) $M^{-1}L^3T^{-2}$

11. Up to which distance the gravitational law is applicable
    (A) 10^8 m
    (B) 1 m
    (C) 10^{-10} m
    (D) at all distances.

12. On doubling the distance between two masses the gravitational force between them will -
    (A) remain unchanged
    (B) become one-fourth
    (C) become half
    (D) become double

13. Newton's law of gravitation is true for:
    (A) only uncharged particles
    (B) only for planets
    (C) only for heavenly bodies
    (D) all the bodies

14. The first successful determination of G in the laboratory was carried out by
    (A) Sir Airy
    (B) Maskylene
    (C) Cavendish
    (D) Faraday

15. The gravitational force between two bodies is
    (A) repulsive at short distances
    (B) attractive at large distances
    (C) repulsive at large distances
    (D) attractive at all distances

16. The weight of a body at the centre of the earth will -
    (A) be greater than that at earth's surface
    (B) be equal to zero
    (C) be less than that at earth's surface
    (D) become infinite

17. Which of the following formula is correct- (d is the density of earth)
    (A) $g = \frac{4}{3}\pi Gd$
    (B) $g = \frac{4}{3}\pi RGd$
    (C) $g = \frac{4}{3}\frac{\pi dG}{R}$
    (D) $g = \frac{4}{3}\frac{\pi dG}{R^2}$

18. A bomb explodes on the moon. On the earth-
    (A) we will hear the sound after 10 minutes
    (B) we will hear the sound after 2 hours 18 minutes
    (C) we will hear the sound after 37 minutes
    (D) we can not hear the sound of explosion

19. Newton's law of gravitation is called universal law because -
    (A) force is always attractive
    (B) it is applicable to lighter and heavier bodies
    (C) it is applicable at all times
    (D) it is applicable at all places of universe for all distances between all particles.

20. If the acceleration due to gravity inside the earth is to be kept constant, then the relation between the density d and the distance r from the centre of earth will be -
    (A) $d \propto r$
    (B) $d \propto r^{1/2}$
    (C) $d \propto 1/r$
    (D) $d \propto \frac{1}{r^2}$

21. On increasing the angular velocity of the earth, the value of g-
    (A) will decrease
    (B) will increase
    (C) will remain unchanged
    (D) can increase or decrease.

22. The value of g on the surface of earth is 9.8 m/s^2 and the

radius of earth is 6400 km. The average density of earth in kg/m^3 will be -
(A) 5.29×10^3 (B) 2.64×10^3
(C) 7.60×10^3 (D) 1.46×10^3

23. The acceleration due to gravity of that planet whose mass and radius are half those of earth, will be (g is acceleration due to gravity at earth's surface)
(A) $2\,g$ (B) g (C) $g/2$ (D) $g/4$

24. There is a body lying on the surface of the earth and suppose the earth suddenly loses its power of attraction then
(A) the mass of the body will be reduced to zero
(B) the weight of the body will be reduced to zero
(C) both will be reduced to zero
(D) it becomes infinite

Gravitational potential & Gravitational based on potential Energy

25. A body is in a state of rest at infinite distance. If it is made a satellite of earth, which of the following physical quantities will be reduced -
(A) gravitational force
(B) kinetic energy
(C) potential energy
(D) mass

26. Two different masses are dropped from same heights, then just before these strike the ground, the following is same:
(A) kinetic energy
(B) potential energy
(C) linear momentum
(D) Acceleration

27. A body of mass m rises to height $h = R/5$ from the earth's surface, where R is earth's radius. If g is acceleration due to gravity at earth's surface, the increase in potential energy is
(A) mg/h (B) $\frac{5}{6}mgh$
(C) $\frac{3}{5}mgh$ (D) $\frac{6}{7}mgh$

28. A body of mass m is taken to a height h from the surface of the earth (radius R) The gain in its gravitational potential energy is
(A) mgR (B) mgh
(C) $mgh\left(\frac{R}{R+h}\right)$ (D) $mgh\left(\frac{R+h}{R}\right)$

29. Binding energy of earth-moon system can be expressed as :
(A) $GM_eM_m/2r$ (B) $-GM_eM_m/r$
(C) GM_eM_m/r (D) $-GM_eM_m/2r$

30. Weight of a person is 800 newton. If he runs 4 m in vertical ladder in 2 seconds then he needs a power of
(A) 3200 kW   (B) 3.2 kW   (C) 1.6 kW   (D) zero

31. A planet has mass 1/10 of that of earth, while radius is 1/3 that of earth. If a person can throw a stone on earth surface to a height of 90 m, then he will be able to throw the stone on that planet to a height
(A) 90 m   (B) 40 m   (C) 100 m   (D) 45 m

32. Work done in taking a body of mass m to a height nR above surface of earth will be : (R = radius of earth)
(A) $mgnR$ (B) $mgR(n/n+1)$
(C) $mgR\frac{(n+1)}{n}$ (D) $\frac{mgR}{n(n+1)}$

33. Mass of a planet is 5×10^{24} kg and radius is 6.1×10^6 m. The energy needed to send a 2 kg body from its surface in space is
(A) 9 J   (B) 10 J
(C) 2.2×10^8 J   (D) 1.1×10^8 J

34. The potential energy due to gravitational field of earth will be maximum at -
(A) infinite distance   (B) the poles of earth
(C) the centre of earth   (D) the equator of earth

35. A particle falls on earth : (i) from infinity (ii) from a height 10 times the radius of earth. The ratio of the velocities gained on reaching at the earth's surface is
(A) $\sqrt{11} : \sqrt{10}$ (B) $\sqrt{10} : \sqrt{11}$ (C) $10 : 11$ (D) $11 : 10$

Escape Speed

36. The escape speed on earth's surface, is -
(A) $2gR$ (B) gR (C) $\sqrt{gR}$ (D) $\sqrt{2gR}$

37. A missile is launched with a velocity less than the escape speed. The sum of kinetic energy and potential energy will be -
(A) positive
(B) negative
(C) negative or positive, uncertain
(D) zero

38. There is no atmosphere on moon because -
(A) it is near the earth
(B) it is orbiting around the earth
(C) there was no gas at all
(D) the escape speed of gas molecules is less than their root-mean square velocity

39. The relation between the escape speed from the earth and the speed of a satellite orbiting near the earth's surface is
(A) $v_e = v$ (B) $v_e = v\sqrt{2}$
(C) $v_e = 2v$ (D) $v_e = v/\sqrt{2}$

40. The escape velocity from the earth does not depend upon
(A) mass of earth
(B) mass of the body
(C) radius of earth
(D) acceleration due to gravity

41. If the kinetic energy of a satellite orbiting around the earth is doubled then -
(A) the satellite will escape into the space.
(B) the satellite will fall down on the earth
(C) radius of its orbit will be doubled
(D) radius of its orbit will become half

42. The escape speed from a planet is v_e. The escape speed from a planet having twice the radius but same density will be-
(A) $0.5v_e$ (B) v_e (C) $2v_e$ (D) $4v_e$

43. The potential energy of a body of mass 3 kg on the surface of a planet is 54 joule. The escape speed will be -
(A) 18 m/s   (B) 162 m/s   (C) 36 m/s   (D) 6 m/s

44. A body of mass m is situated at a distance $4R_e$ above the earth's surface, where R_e is the radius of earth. How much minimum energy be given to the body so that it may escape -
(A) mgR (B) $2mgR$ (C) $\frac{mgR}{5}$ (D) $\frac{mgR}{16}$

45. If the radius of earth is to decrease by 4% and its density

remains same, then its escape speed will
  (A) remain same   (B) increase by 4%
  (C) decrease by 4%   (D) increase by 2%

Energy Relations

46. The velocity of a satellite at a height h above the earth's surface is
  (A) $\sqrt{\frac{GM}{R}}$ (B) $\sqrt{\frac{GM}{R^2}}$ (C) $\sqrt{\frac{GM}{h}}$ (D) $\sqrt{\frac{GM}{R+h}}$

47. Kinetic energy of a satellite will be
  (A) $\frac{GmM}{2r^2}$ (B) $\frac{GmM}{2r}$ (C) $\frac{GmM}{r^2}$ (D) $\frac{GmM}{r}$

48. Binding energy of moon and earth is -
  (A) $\frac{GM_eM_m}{r_{em}}$ (B) $\frac{GM_eM_m}{2r_{em}}$
  (C) $-\frac{GM_eM_m}{r_{em}}$ (D) $-\frac{GM_eM_m}{2r_{em}}$

49. A satellite is moving in a circular orbit around earth with a speed v. If its mass is m, then its total energy will be-
  (A) $\frac{3}{4}mv^2$ (B) mv^2 (C) $\frac{1}{2}mv^2$ (D) $-\frac{1}{2}mv^2$

50. A satellite of earth is moving in its orbit with a constant speed v. If the gravity of earth suddenly vanishes, then this satellite will -
  (A) continue to move in the orbit with speed v.
  (B) start moving with speed v in a direction tangential to the orbit
  (C) fall down with increased speed
  (D) be lost in outer space.

51. The ratio of kinetic energy of a body orbiting near the earth's surface and the kinetic energy of the same body escaping the earth's gravitational field is
  (A) 1   (B) 2   (C) $\sqrt{2}$ (D) 0.5

52. The ratio of kinetic and potential energies of a satellite is -
  (A) 1 : 4   (B) 4 : 1   (C) 1 : 2   (D) 2 : 1

53. If mass of earth is 5.98×10^{24} kg and earth moon distance is 3.8×10^5 km, the orbital period of moon, in days is -
  (A) 27 days   (B) 2.7 days
  (C) 81 days   (D) 8.1 days

54. The ratio of distances of satellites A and B above the earth's surface is 1.4 : 1, then the ratio of energies of satellites B and A will be -
  (A) 1.4 : 1   (B) 2 : 1
  (C) 1 : 3   (D) 4 : 1

55. A satellite of earth can move only in those orbits whose plane coincides with -
  (A) the plane of great circle of earth
  (B) the plane passing through the poles of earth
  (C) the plane of a circle at any latitude of earth
  (D) none of these

56. A body is dropped by a satellite in its geostationary orbit
  (A) it will burn on entering into the atmosphere
  (B) it will remain in the same place with respect to the earth
  (C) it will reach the earth in 24 hours
  (D) it will perform uncertain motion

57. An earth satellite is moved from one stable circular orbit to another higher stable circular orbit. Which one of the following quantities increases for the satellite as a result of this change
  (A) gravitational force
  (B) gravitational potential energy
  (C) centripetal acceleration
  (D) Linear orbital speed

58. A geostationary satellite
  (A) revolves about the polar axis
  (B) has a time period less than that of the near earth satellite
  (C) moves faster than a near earth satellite
  (D) is stationary in the space

59. One satellite is revolving around the earth in an elliptical orbit. It's speed will :
  (A) be same at all the points of orbit
  (B) be maximum at the point farthest from the earth
  (C) be maximum at the point nearest from the earth
  (D) depend on mass of satellite.

60. The orbital velocity of an artificial satellite in a circular orbit just above the earth's surface is v. For a satellite orbiting at an altitude of half of the earth's radius, the orbital speed is
  (A) $\frac{3}{2}v$ (B) $\sqrt{\frac{3}{2}}v$ (C) $\sqrt{\frac{2}{3}}v$ (D) $\frac{2}{3}v$

61. Two satellites of masses m_1 and $m_2(m_1 > m_2)$ are revolving round the earth in circular orbits of radius r_1 and $r_2(r_1 > r_2)$ respectively. Which of the following statements is true regarding their speed v_1 and v_2 ?
  (A) $v_1 = v_2$ (B) $v_1 < v_2$ (C) $v_1 > v_2$ (D) $\frac{v_1}{r_1} = \frac{v_2}{r_2}$

62. Orbital speed of earth's satellite near the surface is 7Km/sec. When the radius of the orbit is 4 times than that of earth's radius then orbital speed in that orbit is
  (A) 3.5Km/sec   (B) 7Km/sec
  (C) 14Km/sec   (D) 72Km/sec

63. If the earth-sun distance is held constant and the mass of the sun is doubled, then the period of revolution of the earth around the sun will change to-
  (A) 2 years   (B) $\frac{1}{2}$ years
  (C) $\frac{1}{\sqrt{2}}$ years   (D) $\sqrt{2}$ years

64. The velocity of a satellite orbiting near the earth's surface is -
  (A) $\sqrt{GR}$ (B) $\sqrt{gR}$ (C) $\sqrt{\frac{GM}{R^2}}$ (D) $\sqrt{2gR}$

65. A satellite of mass m is revolving around the earth of mass M in a path of radius r, the angular velocity of the satellite will be
  (A) $M\sqrt{Gr}$ (B) $\sqrt{\frac{GM}{r}}$ (C) $M\sqrt{\frac{G}{r}}$ (D) $\sqrt{\frac{GM}{r^3}}$

66. A satellite moves in a path of radius r with a speed v, the mass of earth will be -
  (A) $\frac{vr}{G}$ (B) $\frac{v^2r}{G}$ (C) $\frac{Gr}{v^2}$ (D) $\frac{G}{vr^2}$

67. Following quantity is conserved in the motion of a satellite
  (A) angular velocity   (B) linear velocity
  (C) angular momentum   (D) linear momentum

68. In an artificial satellite a person will have -
  (A) zero mass   (B) zero weight
  (C) some weight   (D) infinite weight

69. If a satellite is orbiting very near to the earth's surface then its orbital velocity depends upon-
  (A) mass of the satellite.
  (B) mass of earth only
  (C) radius of earth only
  (D) mass and radius of earth

70. The period of revolution of a communication satellite of earth is

(A) zero   (B) 4.8 hours
(C) 36 hours   (D) 24 hours

71. The formula of period of revolution of a satellite is

(A) $T = 2\pi\sqrt{\dfrac{R}{GM}}$ (B) $T = 2\pi\sqrt{\dfrac{R^2}{GM}}$

(C) $T = 2\pi\sqrt{\dfrac{GM}{R^3}}$ (D) $T = 2\pi\sqrt{\dfrac{R^3}{GM}}$

72. An astronaut feels weightlessness because -
   (A) gravity is zero there
   (B) atmosphere is not there
   (C) energy is zero in the chamber of a rocket.
   (D) the fictitious force in rotating frame of reference cancels the effect of weight.

73. Orbital velocity of INSAT 1-B is nearly -
   (A) 11.2 km/s   (B) 7.9 km/s
   (C) 2.6 km/s   (D) 1.8 km/s

74. Two artificial satellites A and B are at a distances r_A and r_B above the earth's surface. If the radius of earth is R, then the ratio of their speeds will be -

(A) $\left(\dfrac{r_B+R}{r_A+R}\right)^{1/2}$ (B) $\left(\dfrac{r_B+R}{r_A+R}\right)^{2}$

(C) $\left(\dfrac{r_B}{r_A}\right)^{2}$ (D) $\left(\dfrac{r_B}{r_A}\right)^{1/2}$

Kepler's laws

75. If the distance between sun and earth is made 3 times of the present value then gravitational force between them will become :
   (A) 9 times   (B) $\frac{1}{9}$ times
   (C) $\frac{1}{3}$ times   (D) 3 times

76. According to Kepler the period of revolution of a planet (T) and its mean distance from the sun (r) are related by the equation
   (A) $T^2 r = $ constant   (B) $T^2 r^3 = $ constant
   (C) $T^2 r^{-3} = $ constant   (D) $T^3 = $ constant

77. If the earth is to be at half of the present distance from sun, then number of days in one year would be
   (A) 92 days   (B) 129 days
   (C) 183 days   (D) 365 days

78. If a body is carried from surface of earth to moon, then
   (A) the weight of a body will continuously increase,
   (B) the mass of a body will continuously increase,
   (C) the weight of a body will decrease first, become zero and then increase,
   (D) the mass of a body will decrease first, become zero and then increase.

79. The radii of paths of two planets moving around the sun are R_1 and R_2 and their periods are T_1 and T_2 respectively, T_1/T_2 will be -

(A) $\left(\dfrac{R_2}{R_1}\right)^{3/2}$ (B) $\left(\dfrac{R_1}{R_2}\right)^{3/2}$ (C) $\left(\dfrac{R_2}{R_1}\right)^{1/2}$ (D) $\left(\dfrac{R_1}{R_2}\right)^{1/2}$

80. The paths of planets moving around the sun are -
   (A) circular   (B) elliptical
   (C) parabolic   (D) hyperbolic

81. Orbit traced out by a planet around the sun is in general
   (A) circular   (B) elliptical
   (C) parabolic   (D) none of the above

82. The period of revolutions of two satellites are 3 hours and 24 hours. The ratio of their speeds will be -
   (A) $1:8$ (B) $1:2$ (C) $2:1$ (D) $4:1$

1.25.3.2   Level 2

1. In the Fig.1.209, motion of a planet around the sun in elliptical path is shown with sun at one of its foci. Two shaded areas shown in figure are equal. If the times taken by the planet to move from A to B and from C to D are t_1 and t_2 respectively, then
   (A) $t_1 < t_2$ (B) $t_1 > t_2$
   (C) $t_1 = t_2$ (D) there is no relation between t_1 and t_2.

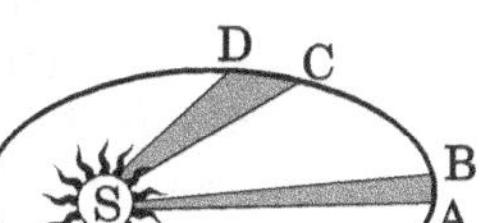

Figure 1.209

2. The escape speed of a body at earth is 11.2 km/s, the escape speed of the body thrown at an angle $45°$ with the horizontal will be -
   (A) 11.2 km/s   (B) 22.4 km/s
   (C) $\frac{11.2}{\sqrt{2}}$ km/s   (D) $11.2\sqrt{2}$ km/s

3. The loss in weight of a body taken from earth's surface to a height h is 1%. The change in weight taken into a mine of depth h will be
   (A) 1% loss   (B) 1% gain
   (C) 0.5% gain   (D) 0.5 loss

4. Clock A based on spring oscillations and a clock B based on oscillations of simple pendulum are synchronised on earth. Both are taken to mars whose mass is 0.1 times the mass of earth and radius is half that of earth. Which of the following statement is correct -
   (A) Both will show same time
   (B) Time measured in clock A will be greater than that in clock B.
   (C) Time measured in clock B will be greater than that in clock A.
   (D) Clock A will stop and clock B will show time as it shows on earth.

5. The weight of a man is equivalent to 50 kg wt. If keeping the density of earth constant the radius of the earth is doubled, then the weight of the man will become -
   (A) 100 kg wt   (B) same   (C) half   (D) zero

6. Two satellites A and B (each of mass m) of earth are put in circular orbits around the centre of earth. The height of satellite A above the earth's surface is equal to the radius R of earth and that of satellite B is $3R$. The ratio of potential energies of satellites A and B is -
   (A) $1:2$ (B) $2:1$ (C) $3:1$ (D) $1:3$

7. A particle is carried from A to B along different paths in a gravitational field as shown in Fig.1.210, then
   (A) Work done along path IV will be maximum
   (B) Work done along path I will be maximum
   (C) Work done along all paths will be same
   (D) Work done along path III will be minimum

8. If the angular speed of earth is increased so much that the objects start flying from the equator, then the length of the day will be nearly -

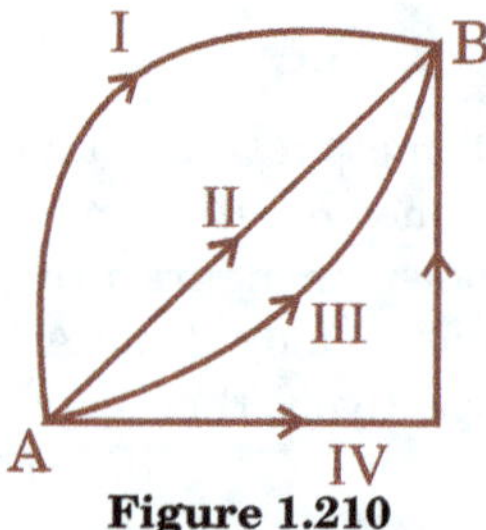

Figure 1.210

(A) $1\frac{1}{2}$ hours   (B) 8 hours
(C) 18 hours   (D) 24 hours

9. Imagine the acceleration due to gravity on earth is 10 m/s^2 and on mars is 4 m/s^2. A traveller of mass 60 kg goes from earth to mars by a rocket moving with constant velocity. If effect of other planets is assumed to be negligible, which one of the following graphs shown [Fig.1.211] the variation of weight of traveller with time -
   (A) A   (B) B   (C) C   (D) D

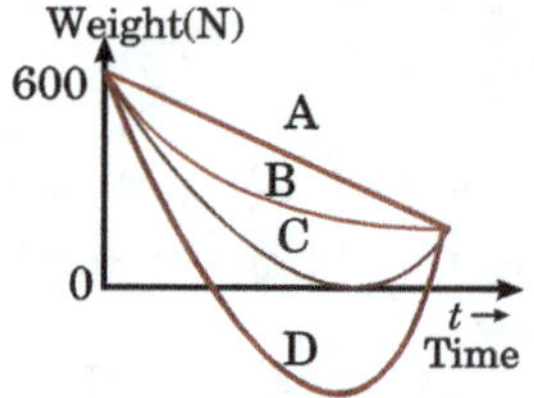

Figure 1.211

10. Two satellites of same mass m are revolving around the earth (mass $= M$) in the same orbit of radius r. Rotational directions of the two are opposite therefore, they can collide and stick with each other [Fig.1.212]. Total mechanical energy of the system (both satellites and earths) is ($m << M$)
   (A) $-\dfrac{GMm}{r}$ (B) $-\dfrac{2GMm}{r}$
   (C) $-\dfrac{GMm}{2r}$ (D) zero

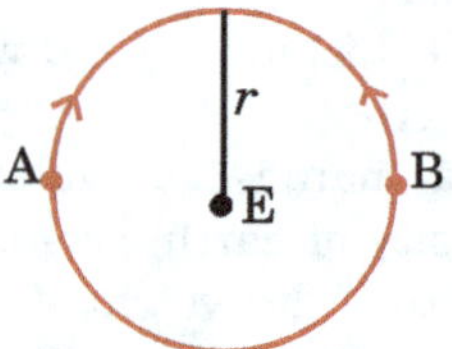

Figure 1.212

11. A missile which missed its target, went into orbit around the earth. The radius of the orbit is four times the radius of the parking orbit of a satellite. The period of the missile as a satellite is -
   (A) 2 days   (B) 4 days
   (C) 8 days   (D) 16 days

12. The acceleration due to gravity on the moon is one-sixth that on earth. If the average density of moon is three-fifth that of earth, the moon's radius in terms of earth's radius R is -
   (A) $0.16R$ (B) $0.27R$ (C) $0.32R$ (D) $0.36R$

13. The ratio of radii of two satellites is p and the ratio of their accelerations due to gravity is q. The ratio of their escape speeds will be -
   (A) $\left(\dfrac{q}{p}\right)^{1/2}$ (B) $\left(\dfrac{p}{q}\right)^{1/2}$ (C) pq   (D) $\sqrt{pq}$

14. The acceleration due to gravity of that planet whose mass and radius are half those of earth, will be (g is acceleration due to gravity at earth's surface)
   (A) $2g$ (B) g (C) $g/2$ (D) $g/4$

15. A satellite is revolving around earth in a circular orbit. The radius of orbit is half of the radius of the orbit of moon. Satellite will complete one revolution in -
   (A) $2^{-3/2}$ lunar month   (B) $2^{-2/3}$ lunar month
   (C) $2^{3/2}$ lunar month   (D) $2^{2/3}$ lunar month

16. Assuming earth as a sphere of radius R, the weight of mass 1 kg at a distance 2R from the centre of earth is 2.5 N. The weight of same mass at a distance 3R will be -
   (A) 4.75 N   (B) 3.75 N
   (C) 2.5 N   (D) 1.1 N

17. If a body is taken up to a height of 1600 km from the earth's surface, then the percentage loss of gravitational force acting on that body will be (Radius of earth $R = 6400$ km)
   (A) 50%   (B) 25%   (C) 36%   (D) 10%

18. The escape speed for a projectile in the case of earth is 11.2 km/sec. A body is projected from the surface of earth with a speed which is equal to twice the escape speed. The speed of the body when at infinite distance from the centre of the earth is:
   (A) 11.2 km/sec   (B) 22.4 km/sec
   (C) $11.2\sqrt{3}$ km/sec   (D) $11.2\sqrt{2}$ km/sec

19. Two identical satellites are moving in the same circular orbit around the earth but in opposite senses of rotation, Assuming an inelastic collision to take place, so that the wreckage remains as one piece of tangled material, the wreckage
   (A) will be moving in a circular orbit of half the radius
   (B) will be moving in a circular orbit on one fourth the radius
   (C) will spin with no translatory motion
   (D) falls directly down

20. Consider the earth to be a homogeneous sphere. Scientist A goes deep down in a mine and scientist B goes high up in a balloon. The gravitational field measured by
   (A) A goes on decreasing and that by B goes on increasing
   (B) B goes on decreasing and that by A goes on increasing
   (C) each remains unchanged
   (D) each goes on decreasing

21. An iron ball and a wooden ball of the same radius are released from a height h in vacuum. The times taken by both of these to reach the ground are -
   (A) unequal   (B) exactly equal
   (C) roughly equal   (D) zero

22. A body of mass m is taken from the surface of earth to a height $R/4$ above the surface. The change in its potential energy is -
   (A) $\dfrac{5mgR}{4}$ (B) $\dfrac{3mgR}{4}$ (C) $\dfrac{mgR}{4}$ (D) $\dfrac{mgR}{5}$

23. An artificial satellite moving in a circular orbit around the earth has a total (kinetic + potential) energy E_0. Its potential energy is -

(A) $-E_0$ (B) $5E_0$ (C) $2E_0$ (D) E_0

24. The diameters of two planets are in ratio $4:1$. Their mean densities have ratio $1:2$. The ratio of 'g' on the plants will be:

(A) $1:2$ (B) $1:4$ (C) $2:1$ (D) $4:1$

25. The weight of a man on earth is 600 N. The acceleration due to gravity of moon is $\frac{1}{6}$th the acceleration due to gravity on earth. If the acceleration due to gravity on earth is 10 m/s^2. then the weight of the man on moon will be-

(A) 100 N   (B) 300 N   (C) 600 N   (D) 900 N

26. If the density of a planet is constant, then the graph between the g on its surface and its radius r [Fig.1.213] will be -

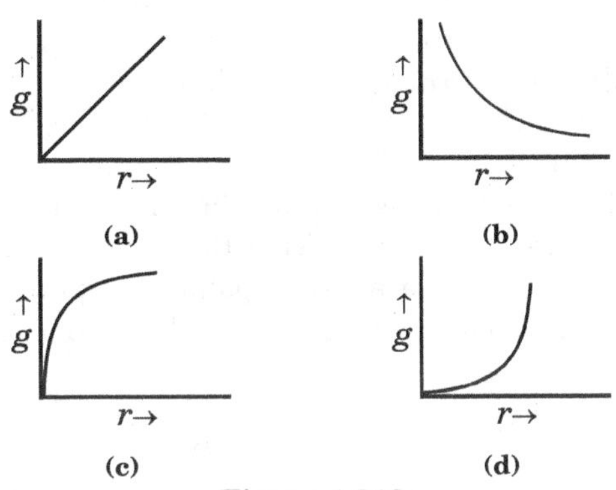

Figure 1.213

27. The mass of moon is $\frac{1}{81}$ of the mass of earth and the acceleration due to gravity is $\frac{1}{6}$ th of acceleration due to gravity on earth. The ratio of radii of moon and earth is -

(A) 6/81   (B) 4/729   (C) $\sqrt{2/27}$ (D) $\sqrt{2/48}$

28. If a person can jump on the earth's surface upto a height of 2 m, his jump on a satellite where acceleration due to gravity is 1.96 m/s^2, will be -

(A) 5 m   (B) 10 m   (C) 20 m   (D) 2 m

29. If acceleration due to gravity at any point is $g/2$, then the intensity of gravitational field at that point will be-

(A) $g/2$ (B) $2g$ (C) g (D) zero

30. Two boys of masses 50 kg and 60 kg are 1 m apart. The gravitational force of attraction between them $\left(G = 6.67 \times 10^{-11} \text{ N} \times \text{m}^2/\text{kg}^2\right)$ is

(A) 2×10^{-10} N   (B) 2×10^{-7} N

(C) 2×10^{-4} N   (D) 2×10^{-1} N

31. There are two bodies of masses 100 kg and 10000 kg separated by a distance 1 m. At what distance from the smaller body, the intensity of gravitational field will be zero ?

(A) $\frac{1}{9}$ m   (B) $\frac{1}{10}$ m   (C) $\frac{1}{11}$ m   (D) $\frac{10}{11}$ m

32. Escape speed of a 1 kg body on a planet is 100 m/s. Potential energy of body at that planet is

(A) -5000 J   (B) -1000 J

(C) -2400 J   (D) -10000 J

33. The escape speed for a rocket from earth is 11.2 km/sec. Its value on a planet where acceleration due to gravity is double that on the earth and diameter of the planet is twice that of earth will be in km/sec

(A) 11.2   (B) 5.6   (C) 22.4   (D) 53.6

34. The escape speed from a planet is v. If its mass and radius becomes four and two times respectively, then the escape speed will become

(A) v (B) $2v$ (C) $0.5v$ (D) $\sqrt{2}v$

35. A satellite is projected with a speed $\sqrt{1.5}$ times the orbital speed just above the surface of earth. Initial speed of the satellite is parallel to the surface of earth. The maximum distance of the satellite from earth will be -

(A) $2R$ (B) $8R$ (C) $4R$ (D) $3R$

36. The average radii of orbits of mercury and earth around the sun are 6×10^7 km and 1.5×10^8 km respectively, The ratio of their orbital speeds will be-

(A) $\sqrt{5}:\sqrt{2}$ (B) $\sqrt{2}:\sqrt{5}$

(C) $2.5:1$ (D) $1:25$

37. With what speed a satellite is to be projected to establish it at a height $6R$ above the earth's surface - $\left(g = 9.8 \text{ m/s}^2 \right.$ and $R = 6400$ km)

(A) 11.2 km/s   (B) 10.4 km/s

(C) 8.6 km/s   (D) 7.9 km/s

38. Imagine a light planet revolving around a very massive star in a circular orbit of radius R with a period of revolution T. If the gravitational force of attraction between planet and star is proportional to $R^{-5/2}$, then T^2 is proportional to

(A) R^3 (B) $R^{7/2}$ (C) $R^{5/2}$ (D) $R^{3/2}$

39. Consider a particle moving in a circular orbit under the action of an attractive central force $F \propto \frac{1}{r}$. Its orbital period T will depend on r as -

(A) $T \propto r^{3/2}$ (B) $T \propto r$

(C) $T \propto r^{2/3}$ (D) $T \propto \frac{1}{r}$

1.25.3.3 Level 3

1. A geostationary satellite is revolving at a height $6R$ above the earth's surface, where R is the radius of earth. The period of revolution of a satellite orbiting at a height $2.5R$ above the earth's surface will be-

(A) 24 h   (B) 12 h   (C) 6 h   (D) $6\sqrt{2}$ h

2. The value of g at any place is 9.8 m/s^2. If the size of the earth suddenly shrinks to half but the density remains unchanged, then the value of g at that place will be-

(A) 4.9 m/s^2 (B) 9.8 m/s^2 (C) 3.1 m/s^2 (D) 19.6 m/s^2

3. Mass of a planet is 5×10^{24} kg and radius is 6.1×10^6 m. The energy needed to send a 2 kg body into space from its surface, would be -

(A) 9 joule   (B) 18 joule

(C) 2.2×10^8 joule   (D) 1.1×10^8 joule

4. How much energy will be needed for a body of mass 100 kg to escape from the earth$\left(g = 10 \text{ m/s}^2 \right.$ and radius of earth $= 6.4 \times 10^6$ m)

(A) 6.4×10^9 joule   (B) 4×10^{16} joule

(C) 8×10^6 joule   (D) zero

5. The acceleration due to gravity on the surface of earth is 10 m/s^2 and the radius of earth is 6400 km. With what minimum speed must a body be thrown from the surface of the earth so that is reaches a height of 6400 km ?

(A) 8 km/s   (B) 64 km/s
(C) 1 km/s   (D) 32 km/s

6. The mass of planet mars is $\frac{1}{10}$ of the mass of earth and radius is $\frac{1}{2}$ of the radius of earth. If the escape speed at the earth is 11.2 km/s, then the escape speed at Mars will be -
(A) 10 km/s   (B) 5 km/s
(C) 20 km/s   (D) 40 km/s

7. If the radius of earth is 6400 km and the acceleration due to gravity g is 10 m/s^2, the angular speed of earth for which the weight of a body would become zero at equator, will be -
(A) $\frac{1}{400}$ rad/s   (B) $\frac{1}{800}$ rad/s
(C) $\frac{1}{1600}$ rad/s   (D) impossible

8. The speed of a satellite of mass 500 kg at a height 10^3 km above the earth's surface is 7.36×10^3 m/s. If the orbit of satellite is circular then the gravitational force due to earth on satellite is (in N)
(A) 4×10^2 (B) 3.66×10^3
(C) 3.75×10^4 (D) 4.5×10^2

9. Angular momentum of a planet of mass m orbiting around sun is J, areal velocity of its radius vector will be
(A) $\frac{1}{2}$ mJ   (B) $\frac{J}{2m}$ (C) $\frac{m}{2J}$ (D) $\frac{1}{2mJ}$

10. A satellite is set in a circular orbit of radius R. Another satellite is set in a circular orbit of radius 1.01 R. The percentage difference of the period of the second satellite with respect to first satellite will be-
(A) 1% increased   (B) 1% decreased
(C) 1.5% increased   (D) 1.5 decreased

11. The potential energy of a body of mass m is $U = ax + by$. The magnitude of acceleration of the body will be-
(A) $\frac{ab}{m}$ (B) $\frac{a+b}{m}$
(C) $\frac{\sqrt{a^2+b^2}}{m}$ (D) $\frac{a^2+b^2}{m}$

12. If the length of the day is T, the height of that TV satellite above the earth's surface which always appears stationary from earth, will be -
(A) $h = \left(\frac{4\pi^2 GM}{T^2}\right)^{1/3}$ (B) $h = \left(\frac{4\pi^2 GM}{T^2}\right)^{1/3} - R$
(C) $h = \left(\frac{GMT^2}{4\pi^2}\right)^{1/3} - R$ (D) $h = \left(\frac{GMT^2}{4\pi^2}\right)^{1/3} + R$

13. A number of particles each of mass 0.75 kg are placed at distances 1 m, 2 m, 4 m, 8 m etc. from origin along positive X axis. the intensity of gravitational field at the origin will be-
(A) G (B) $\frac{3}{4}G$ (C) $\frac{1}{2}G$ (D) zero

14. What should be the angular velocity of earth about its own axis so that the weight of the body at the equator would become $\frac{3}{5}$th of its present value. (Radius of earth at equator = R)
(A) $\sqrt{\frac{5g}{2R}}$ (B) $\sqrt{\frac{2g}{5R}}$ (C) $\sqrt{\frac{3g}{5R}}$ (D) $\sqrt{\frac{5g}{3R}}$

15. Two particles of masses m and M are initially at infinity and at rest. Both particles move towards each other due to mutual interaction. When the distance between them is d, the relative velocity of approach will be -
(A) $\left[\frac{2Gd}{M+m}\right]^{1/2}$ (B) $\left[\frac{2G(M+m)}{d}\right]^{1/2}$
(C) $2G\left(\frac{M+m}{d}\right)$ (D) $\frac{2Gd}{(M+m)}$

16. A satellite whose mass is m, is orbiting around the earth at a height R above the earth's surface. If the intensity of gravitational field at the earth's surface is g and radius of earth is R, then the kinetic energy of the satellite will be-
(A) $mgR/4$ (B) $mgR/2$ (C) mgR (D) $2mgR$

17. The masses of moon and earth are 7.36×10^{22} kg and 5.98×10^{24} kg respectively and their mean separation is 3.82×10^5 km. The energy required to break the earth-moon system is -
(A) 12.4×10^{32} J   (B) 7.68×10^{28} J
(C) 5.36×10^{24} J   (D) 2.96×10^{20} J

18. Three identical point masses, each of mass 1 kg lie in the $x - y$ plane at points (0, 0), (0, 0.2m) and (0.2m, 0). The gravitational force on the mass at the origin is-
(A) $1.67 \times 10^{-19}(\hat{i}+\hat{j})N$ (B) $3.34 \times 10^{-10}(\hat{i}+\hat{j})N$
(C) $1.67 \times 10^{-9}(\hat{i}-\hat{j})N$ (D) $3.34 \times 10^{-10}(\hat{i}-\hat{j})N$

19. The value of g at any point is 9.8 m/s^2. If earth is reduced to half of its present size by shrinking without loss of mass but the point remains at the same position, then the value of g at that point will be-
(A) 4.9 m/s^2 (B) 3.1 m/s^2
(C) 9.8 m/s^2 (D) 19.6 m/s^2

20. One planet is orbiting around the sun, At a point P it is at minimum distance d_1 from the sun and at that time speed is v_1. If at a another point Q it is at maximum distance d_2 from the sun then at this point the speed of planet will be -
(A) $d_1^2 v_1/d_2^2$ (B) $d_2 v_1/d_1$
(C) $d_1 v_1/d_2$ (D) $d_2^2 v_1/d_1^2$

21. The radius of a planet is 74×10^3 km. A satellite of this planet completes one revolution in 16.7 days. If the radius of the orbit of satellite is 27 times the radius of planet, then the mass of planet is
(A) 2.3×10^{27} kg   (B) 2.3×10^{30} kg
(C) 2.3×10^{32} kg   (D) 2.3×10^{42} kg

22. If M_e is the mass of earth and M_m is the mass of moon ($M_e = 81 M_m$). The potential energy of an object of mass m situated at a distance R from the centre of earth and r from the centre of moon, will be -
(A) $-G m M_m\left(\frac{R}{81}+r\right)\frac{1}{R^2}$ (B) $-G m M_e\left(\frac{81}{R}+\frac{1}{R}\right)$
(C) $-G m M_m\left(\frac{81}{R}+\frac{1}{r}\right)$ (D) $G m M_m\left(\frac{81}{R}-\frac{1}{r}\right)$

23. A body of mass M is divided into two parts of masses m and $(M-m)$. If they are kept at a constant distance, then the relation between m and M for maximum force of attraction between them, will be -
(A) $m = \frac{M}{4}$ (B) $m = \frac{2M}{3}$ (C) $m = \frac{M}{3}$ (D) $m = \frac{M}{2}$

24. A body placed at a distance r from the centre of earth, starts moving from rest. The velocity of the body on reaching at the earth's surface will be: (Given that- R = radius of earth and M = mass of earth)
(A) $GM\left(\frac{1}{R}-\frac{1}{r}\right)$ (B) $2GM\left(\frac{1}{R}-\frac{1}{r}\right)$
(C) $GMe\sqrt{\frac{1}{R}-\frac{1}{r}}$ (D) $\sqrt{2GM\left(\frac{1}{R}-\frac{1}{r}\right)}$

25. The masses and the radii of earth and moon are respectively M_e, R_e and M_m, R_m. The distance between the centres is r. At what minimum velocity a particle of mass m be projected from the mid point of the distance between their centres so that it may escape into space ?
(A) $\sqrt{\frac{4G}{r}(M_e+M_m)}$ (B) $\frac{4G}{r}\sqrt{(M_e+M_m)}$
(C) $\sqrt{\frac{2G}{r}(M_e+M_m)}$ (D) $\frac{2G}{r}\sqrt{(M_e+M_m)}$

26. Three particles each of mass 100 gm are brought from a very large distance to the vertices of an equilateral triangle whose side is 20 cm in length. The work done will be -
   (A) 0.33×10^{-11} joule
   (B) -0.33×10^{-11} joule
   (C) 1.0×10^{-11} joule
   (D) -1.00×10^{-11} joule

27. A tunnel is dug along the diameter of the earth. If a particle of mass m is situated in the tunnel at a distance x from the centre of earth then gravitational force acting on it, will be -
   (A) $\frac{GM_e m}{R_e^3} x$
   (B) $\frac{GM_e m}{R_e^2}$
   (C) $\frac{GM_e m}{x^2}$
   (D) $\frac{GM_e m}{(R_e + x)^2}$

28. Four particles of masses m, $2m$, $3m$ and $4m$ are kept in sequence at the corners of a square of side a. The magnitude of gravitational force acting on a particle of mass m placed at the centre of the square will be-
   (A) $\frac{24m^2 G}{a^2}$ (B) $\frac{6m^2 G}{a^2}$ (C) $\frac{4\sqrt{2}Gm^2}{a^2}$ (D) zero

29. If the change in the value of g at a height h above the surface of the earth is the same as at a depth x below it. When both x and h are much smaller than the radius of the earth. Then
   (A) $x = h$ (B) $x = 2h$ (C) $x = h/2$ (D) $x = h^2$

30. The gravitational force on a body inside the surface of the earth varies as r^a, where r is the distance from the center of the earth and a is some constant. If the density of the earth is assumed uniform, then -
   (A) $a = 1$ (B) $a = -1$ (C) $a = 2$ (D) $a = -2$

31. Distance between the centres of two stars is $10\,a$. The masses of these stars are M and $16\,M$ and their radii a and $2a$ respectively [Fig.1.214]. A body of mass m is fired straight from the surface of the larger star towards the smaller star. What should be its minimum initial speed to reach the surface of the smaller star? Obtain the expression in terms of G, M and a.
   (A) $v_{\min} = \frac{3}{2}\sqrt{\left(\frac{5GM}{a}\right)}$
   (B) $v_{\min} = \frac{5}{2}\sqrt{\left(\frac{5GM}{a}\right)}$
   (C) $v_{\min} = \frac{5}{2}\sqrt{\left(\frac{5M}{Ga}\right)}$
   (D) None

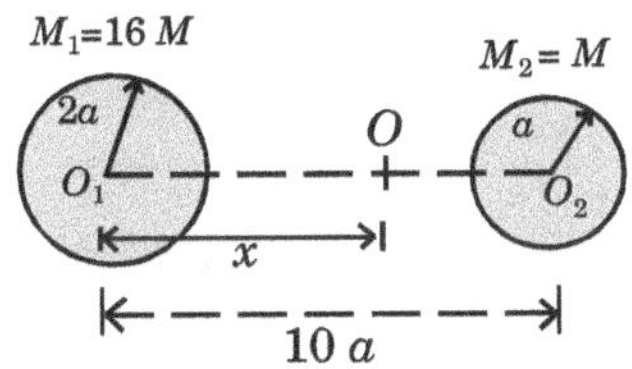

Figure 1.214

Assertion & Reason type Questuions

Each of the questions given below consist of Statement-I and Statement-II. Use the following Key to choose the appropriate answer.
(A) If both Statement-I and Statement-II are true, and Statement-II is the correct explanation of Statement-I.
(B) If both Statement-I and Statement-II are true but Statement-II is not the correct explanation of Statement-I.
(C) If Statement-I is true but Statement-II is false.
(D) If Statement-I is false but Statement-II is true.

32. Statement I: Escape speed of a tennis ball from the surface of earth is the same as the escape velocity of a cricket ball from the surface of earth.
Statement II: Escape speed of a body is independent of the mass of the body

33. Statement I: The weight of a body at the centre of earth is zero
Statement II: The mass of a body decreases with increase in depth below the surface of earth.

34. Statement I: Earth continuously attracts the moon towards its centre and yet the moon does not move towards the centre of earth.
Statement II: The gravitational force of attraction between the moon and earth provides the necessary centripetal force.

35. Assertion: At height h from ground and at depth h below ground, where h is approximately equal to $0.62\,R$, the value of g acceleration due to gravity is same.
Reason: Value of g decreases both sides, in going up and down.

36. Assertion: If radius of earth (assuming to be perfect sphere) is decreased keeping its mass constant, effective value of g will increase at poles while may increase or decrease at equator.
Reason: Value of g on the surface of earth is given by $g = \frac{GM}{R^2}$

37. Assertion: Radius of circular orbit of a satellite is made two times, then its areal velocity will also become two times.
Reason: Areal velocity is given as $\frac{dA}{dt} = \frac{L}{2m} = \frac{mvr}{2m}$.

38. Assertion: If ice on the pole melts, length of the day will shorten.
Reason: Ice will flow towards equator and will result in the decrease in moment of inertia of earth resulting in decrease in frequency of rotation of earth.

39. Assertion: An astronaut in a satellite feels weightlessness:
Reason: As observed by another astronaut's in the same satellite, force of gravity and centrifugal force balance each other.

Passage for questions (40 to 42)
Gravitational Potential energy depends on the product of the masses of two interacting bodies in the same way as the gravitational force. However the mathematical construct of gravitational potential energy differs from law of gravitation in its dependence on distance. The existence of gravitational potential energy is a good proof of the fact that gravitational forces are conservative forces. Again, since gravitational forces are attractive therefore gravitational potential energy is negative.

40. Consider a particle of mass m on the axis of a ring, of mass M and radius r, at a distance r from the centre of the ring. This particle, moving under the gravitation attraction of the ring, reaches the centre of the ring. The velocity of this particle at the centre of the ring will be:

(A) $\sqrt{\dfrac{GM}{r}}$ (B) $\sqrt{\dfrac{2GM}{r}}$

(C) $\sqrt{\dfrac{2GM}{r(\sqrt{2}+1)}}$ (D) $\sqrt{\dfrac{2GM}{r}\left[1-\dfrac{1}{\sqrt{2}}\right]}$

41. Three particles of equal mass m are situated at the vertices of an equilateral of side l. The work done in increasing the side of the triangle to $2l$ will be:

(A) $\dfrac{3G^2m}{2l}$ (B) $\dfrac{Gm^2}{2l}$ (C) $\dfrac{3Gm^2}{2l}$ (D) $\dfrac{3Gm^2}{l}$

42. A stationary object is released from a point P a distance $3R$ from the centre O the Moon which has radius R and mass M. Which one of the following expressions gives the speed of the Moon?

(A) $\left(\dfrac{3GM}{3R}\right)^{1/2}$ (B) $\left(\dfrac{4GM}{3R}\right)^{1/2}$

(C) $\left(\dfrac{4GM}{4}\right)^{1/2}$ (D) $\left(\dfrac{GM}{R}\right)^{1/2}$

Column Matching Type Questions

43. On the surface of earth acceleration due to gravity is g and gravitational potential is V. Match the following:

Column 1	Column 1
(A) At height $h = R$, value of g	(P) decreases by a factor 1/4
(B) At depth $h = R/2$, value of g	(Q) decreases by a factor 1/2
(C) At height $h = R$, value of V	(R) increases by a factor 11/8
(D) at depth $h = R/2$, value of V	(S) increases by a factor 2
	(T) None

44. A particle is projected from the surface of earth with speed v. Suppose it travels a distance x when its speed becomes v to $\dfrac{v}{2}$ and y when speed changes from $\dfrac{v}{2}$ to 0. Similarly, the corresponding times are suppose t_1 and t_2. Then:

Column 1	Column 1
(A) $\dfrac{x}{y}$	(P) 1
(B) $\dfrac{t_1}{t_{f2}}$	(Q) > 1
	(R) < 1

45. Density of a planet is two times the density of earth. Radius of this planet is half. Match, the followings. (As compared to earth)

Column 1	Column 1
(A) Acceleration due to gravity on this planet's surface	(P) Half
(B) Gravitational potential on the surface	(Q) Same
(C) Gravitational potential at centre	(R) Two times
(D) Gravitational field strength at centre	(S) Four times

46. In elliptical orbit of a planet, as the planet moves from apogee position to perigee position, match the following table:

Column 1	Column 1
(A) speed of planet	(P) remains same
(B) distance of planet from center of Sun	(Q) decreases
(C) potential energy	(R) increases
(D) angular momentum about center of sun	(S) can not say

47. Match the following:

Column 1	Column 1
(A) Kepler's first law	(P) $T^2 \propto r^3$
(B) Kepler's second law	(Q) areal velocity constant
(C) Kepler's third law law	(R) orbit of planet is elliptical

48. Match the following:

Column 1	Column 1
(A) Time period of an earth satellite in circular orbit	(P) Independent of of satellite
(B) Orbital speed of satellite	(Q) Independent of radius of orbit
(C) Mechanical energy	(R) Independent of mass of earth
	(S) None

49. If earth decreases its rotational speed. Match the following:

Column 1	Column 1
(A) Value of g at pole	(P) will remain same
(B) Value of g at equator	(Q) will increase
(C) Distance of geostationary satellite	(R) will decrease
(D) Energy of geostationary satellite	(S) can not say

50. Match the following (for a satellite in circular orbit)

Column 1	Column 1
(A) Kinetic energy	(P) $-\dfrac{GMm}{2r}$
(B) Potential energy	(Q) $\sqrt{\dfrac{GM}{r}}$
(C) Total energy	(R) $-\dfrac{GMm}{r}$
(D) Orbital velocity	(S) $\dfrac{GMm}{2r}$

1.25.3.4   Level 4

(Questions asked in JEE Mains and Advanced)

Section - A : JEE Mains

1. A mass m is raised from a distance $2R$ from surface of earth to $3R$. Work done, to do so against gravity will be-
                             [AIEEE-2002]

(A) $\dfrac{MgR}{10}$ (B) $\dfrac{MgR}{11}$ (C) $\dfrac{MgR}{12}$ (D) $\dfrac{MgR}{14}$

2. The escape speed of a body of mass m from earth depends on-
                             [AIEEE-2002]

(A) m^2 (B) m^1

(C) m^0 (D) None of above

3. If suddenly gravitational force on a satellite becomes zero it will -
                             [AIEEE-2002]

(A) go in tangential direction of orbit

(B) fall on earth

(C) follow hellical path towards earth

(D) follow hellical path away from earth

4. The kinetic energy needed to project a body of mass m from the earth surface (radius R) to infinity is-
   [AIEEE-2002]
   (A) $\frac{mgR}{2}$ (B) $2mgR$ (C) mgR (D) $\frac{mgR}{4}$

5. The escape velocity for a body projected vertically upwards from the surface of earth is 11 km/s. If the body is projected at an angle of $45°$ with the vertical, the escape velocity will be-
   [AIEEE-2003]
   (A) 22 km/s   (B) 11 km/s
   (C) $11/\sqrt{2}$ km/s   (D) $11\sqrt{2}$

6. The time period of a satellite of earth is 5 hours. If the separation between the earth and the satellite is increased to 4 times the previous value, the new time period will become-
   [AIEEE-2003]
   (A) 80 h   (B) 40 h   (C) 20 h   (D) 10 h

7. Two spherical bodies of mass M and $5M$ and radii R and $2R$ respectively are released in free space with initial separation between their centres equal to $12R$. If they attract each other due to gravitational force only, then the distance covered by the smaller body just before collision is -
   [AIEEE-2003]
   (A) $4.5R$ (B) $7.5R$ (C) $1.5R$ (D) $2.5R$

8. A satellite of mass m revolves around the earth of radius R at a height x from its surface. If g is the acceleration due to gravity on the surface of the earth, the orbital speed of the satellite is
   [AIEEE-2004]
   (A) gx (B) $\frac{gR}{R-x}$ (C) $\frac{gR^2}{g+x}$ (D) $\left(\frac{gR^2}{R+x}\right)^{1/2}$

9. The time period of an earth satellite in circular orbit is independent of -
   [AIEEE-2004]
   (A) The mass of the satellite
   (B) Radius of its orbit
   (C) Both the mass and radius of the orbit
   (D) Neither the mass of the satellite nor the radius of its orbit

10. If 'g' is the acceleration due to gravity on the earth's surface, the gain in the potential energy of an object of mass 'm' raised from the surface of the earth to a height equal to the radius 'R' of the earth is-   [AIEEE-2004]
    (A) $2mgR$ (B) $1/2mgR$ (C) $1/4mgR$ (D) mgR

11. Suppose the gravitational force varies inversely as the nth power of distance. Then the time period of a planet in circular orbit of radius 'R' around the sun will be proportional to-
    [AIEEE-2004]
    (A) $R^{\left(\frac{n+1}{2}\right)}$ (B) $R^{\left(\frac{n-1}{2}\right)}$
    (C) R^n (D) $R^{\left(\frac{n-2}{2}\right)}$

12. Average density of the earth   [AIEEE-2005]
    (A) does not depend on g
    (B) is a complex function of g
    (C) is directly proportional to g
    (D) is inversely proportional to g

13. The change in the value of 'g' at a height 'h' above the surface of the earth is the same as at a depth 'd' below the surface of earth. When both 'd' and 'h' are much smaller than the radius of earth, then which one of the following is correct?   [AIEEE-2005]
    (A) $d = h/2$ (B) $d = 2h$ (C) $d = 3h/2$ (D) $d = h$

14. A particle of mass 10 g is kept on the surface of a uniform sphere of mass 100 kg and radius 10 cm. Find the work to be done against the gravitational force between them to take the particle far away from the sphere (you may take $G = 6.67 \times 10^{-11}\text{Nm}^2/\text{kg}^2$)   [AIEEE-2005]
    (A) 13.34×10^{-10} J   (B) 6.67×10^{-9} J
    (C) 3.33×10^{-10} J   (D) 6.67×10^{-10} J

15. A planet in a distant solar system is 10 times more massive than the earth and its radius is 10 times smaller. Given that the escape velocity from the earth is 11 km s^{-1}, the escape velocity from the surface of the planet would be   [AIEEE-2008]
    (A) 11 km s^{-1} (B) 110 km s^{-1}
    (C) 0.11 km s^{-1} (D) 1.1 km s^{-1}

16. The height at which the acceleration due to gravity becomes $\frac{g}{9}$ (where g = the acceleration due to gravity on the surface of the earth) in terms of R, the radius of the earth, is -   [AIEEE-2009]
    (A) $\frac{R}{\sqrt{2}}$ (B) $R/2$ (C) $\sqrt{2}R$ (D) $2R$

17. Two bodies of masses m and $4m$ are placed at distance r. The gravitational potential at a point on the line joining them where the gravitational field is zero is [2011]
    (A) zero   (B) $-\frac{4Gm}{r}$ (C) $-\frac{6Gm}{r}$ (D) $-\frac{9Gm}{r}$

18. The mass of a spaceship is 1000 kg. It is to be launched from the earth's surface out into free space. The value of g and R (radius of earth) are 10 and 6400 km respectively. The required energy for this work will be   [2012]
    (A) 6.4×10^8 joules   (B) 6.4×10^9 joules
    (C) 6.4×10^{10} joules   (D) 6.4×10^{11} joules

19. What is the minimum energy required to launch a satellite of mass m from the surface of a planet of mass M and radius R in a circular orbit at an altitude of $2R$?
    [2013]
    (A) $\frac{GmM}{3R}$ (B) $\frac{5GmM}{6R}$ (C) $\frac{2GmM}{3R}$ (D) $\frac{GmM}{2R}$

20. Four particles, each of mass M and equidistant from each other, move along a circle of radius R under the action of their mutual gravitational attraction. The speed of each particle is   [2014]
    (A) $\frac{1}{2}\sqrt{\frac{GM}{R}(1+2\sqrt{2})}$ (B) $\sqrt{\frac{GM}{R}}$
    (C) $\sqrt{2\sqrt{2}\frac{GM}{R}}$ (D) $\sqrt{\frac{GM}{R}(1+2\sqrt{2})}$

21. From a solid sphere of mass M and radius R, a spherical portion of radius $\frac{R}{2}$ is removed, as shown in the Fig.1.215. Taking gravitational potential $V = 0$ at $r = \infty$, the potential at the center of the cavity thus formed is: (G = gravitational constant)   [2015]
    (A) $-\frac{GM}{2R}$ (B) $\frac{GM}{R}$ (C) $-\frac{2GM}{3R}$ (D) $-\frac{2GM}{R}$

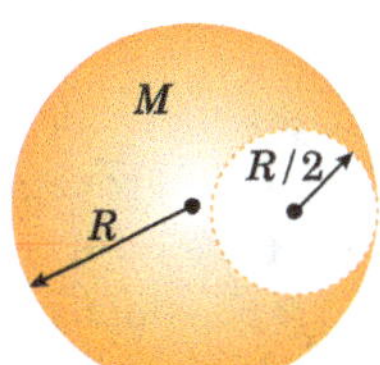

Figure 1.215

22. The variation of acceleration due to gravity g with distance d from centre of the earth is best represented by [Fig.1.216] (R = Earth's radius):   [2017]

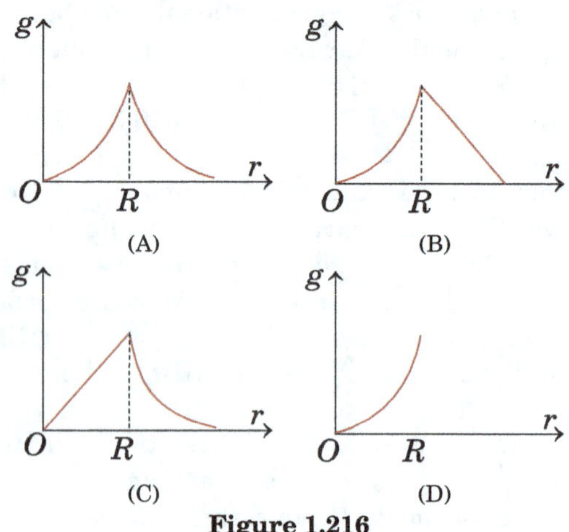

Figure 1.216

23. A particle is moving with a uniform speed in a circular orbit of radius R in a central force inversely proportional to the nth power of R. If the period of rotation of the particle is T, then [2018]
(A) $T \propto R^{n/2}$
(B) $T \propto R^{3/2}$ for any value of n
(C) $T \propto R^{\frac{n}{2}+1}$
(D) $T \propto R^{\frac{n+1}{2}}$

24. The ratio of the weights of a body on the earth's surface, to that on the surface of a planet is $9 : 4$. The mass of the planet is $\frac{1}{9}$ th of that of the earth. If R is the radius of the earth, what is the radius of the planet? (Take, the planets to have the same mass density) [2019]
(A) $\frac{R}{3}$
(B) $\frac{R}{4}$
(C) $\frac{R}{9}$
(D) $\frac{R}{2}$

25. The value of acceleration due to gravity at earth's surface is 9.8 ms^{-2}. The altitude above its surface at which the acceleration due to gravity decreases to 4.9 ms^{-2}, is close to (Take, radius of earth $= 6.4 \times 10^6$ m) [2019]
(A) 9.0×10^6 m
(B) 2.6×10^6 m
(C) 6.4×10^6 m
(D) 1.6×10^6 m

26. A spaceship orbits around a planet at a height of 20 km from its surface. Assuming that only gravitational field of the planet acts on the spaceship, what will be the number of complete revolutions made by the spaceship in 24 hours around the planet? [Take, mass of planet $= 8 \times 10^{22}$ kg, radius of planet $= 2 \times 10^6$ m, gravitational constant $G = 6.67 \times 10^{-11}$ N$-$m^2/kg^2] [2019]
(A) 11
(B) 17
(C) 13
(D) 9

27. A solid sphere of mass M and radius a is surrounded by a uniform concentric spherical shell of thickness $2a$ and $2M$. The gravitational field at distance $3a$ from the centre will be [2019]
(A) $\frac{GM}{9a^2}$
(B) $\frac{2GM}{9a^2}$
(C) $\frac{GM}{3a^2}$
(D) $\frac{2GM}{3a^2}$

28. A rocket has to be launched from earth in such a way that it never returns. If E is the minimum energy delivered by the rocket launcher, what should be the minimum energy that the launcher should have, if the same rocket is to be launched from the surface of the moon? Assume that the density of the earth and the moon are equal and that the earth's volume is 64 times the volume of the moon. [2019]
(A) $\frac{E}{64}$
(B) $\frac{E}{16}$
(C) $\frac{E}{32}$
(D) $\frac{E}{4}$

29. A test particle is moving in a circular orbit in the gravitational field produced by mass density $\rho(r) = \frac{K}{r^2}$. Identify the correct relation between the radius R of the particle's orbit and its period T [2019]
(A) $\frac{T^2}{R^3}$ is a constant
(B) $\frac{T}{R^2}$ is a constant
(C) TR is a constant
(D) $\frac{T}{R}$ is a constant

30. Two satellites A and B have masses m and $2m$ respectively. A is in a circular orbit of radius R and B is in a circular orbit of radius $2R$ around the earth. The ratio of their kinetic energies, T_A/T_B is [2019]
(A) $\frac{1}{2}$
(B) 2
(C) $\sqrt{\frac{1}{2}}$
(D) 1

31. A satellite is revolving in a circular orbit at a height h from the earth surface such that $h \ll R$, where R is the radius of the earth. Assuming that the effect of earth's atmosphere can be neglected, the minimum increase in the speed required so that the satellite could escape from the gravitational field of earth is [2019]
(A) $\sqrt{\frac{gR}{2}}$
(B) $\sqrt{gR}$
(C) $\sqrt{2gR}$
(D) $\sqrt{gR}(\sqrt{2}-1)$

32. A satellite is moving with a constant speed v in a circular orbit about the earth. An object of mass m is ejected from the satellite such that it just escapes from the gravitational pull of the earth. At the time of its ejection, the kinetic energy of the object is [2019]
(A) $\frac{1}{2}mv^2$
(B) mv^2
(C) $\frac{3}{2}mv^2$
(D) $2mv^2$

33. The energy required to take a satellite to a height 'h' above earth surface (where, radius of earth $= 6.4 \times 10^3$ km) is E_1 and kinetic energy required for the satellite to be in a circular orbit at this height is E_2. The value of h for which E_1 and E_2 are equal, is [2019]
(A) 3.2×10^3 km
(B) 1.28×10^4 km
(C) 6.4×10^3 km
(D) 1.6×10^3 km

34. If the angular momentum of a planet of mass m, moving around sun in a circular orbit is L about the centre of the sun, its areal velocity is [2019]
(A) $\frac{4L}{m}$
(B) $\frac{2L}{m}$
(C) $\frac{L}{2m}$
(D) $\frac{L}{m}$

35. Four identical particles of mass M are located at the corners of a square of side a [Fig.1.217]. What should be their speed, if each of them revolves under the influence of other's gravitational field in a circular orbit circumscribing the square? [2019]
(A) $1.35\sqrt{\frac{GM}{a}}$
(B) $1.16\sqrt{\frac{GM}{a}}$
(C) $1.21\sqrt{\frac{GM}{a}}$
(D) $1.41\sqrt{\frac{GM}{a}}$

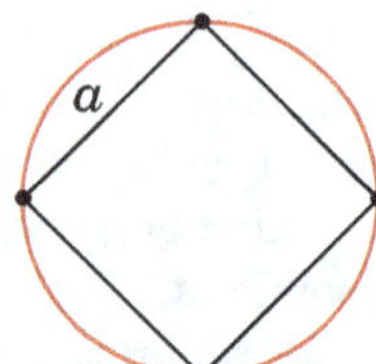

Figure 1.217

36. The mass and the diameter of a planet are three times the respective values for the earth. The period of oscillation of a simple pendulum on the earth is 2 s. The period of oscillation of the same pendulum on the planet would be [2019]
(A) $\frac{2}{\sqrt{3}}$ s
(B) $\frac{3}{2}$ s
(C) $2\sqrt{3}$ s
(D) $\frac{\sqrt{3}}{2}$ s

37. A satellite of mass M is in a circular orbit of radius R about the centre of the earth. A meteorite of the same mass falling towards the earth collides with the satellite completely inelastically. The speeds of the satellite and the meteorite are the same just before the collision. The subsequent motion of the combined body will be [2019]
  (A) in the same circular orbit of radius R
  (B) in an elliptical orbit
  (C) such that it escapes to infinity
  (D) in a circular orbit of a different radius

38. A straight rod of length L extends from $x = a$ to $x = L + a$. The gravitational force it exerts on a point mass m at $x = 0$, if the mass per unit length of the rod is $A + Bx^2$, is given by [2019]
  (A) $Gm \left[A \left(\frac{1}{a+L} - \frac{1}{a} \right) - BL \right]$ (B) $Gm \left[A \left(\frac{1}{a+L} - \frac{1}{a} \right) + BL \right]$
  (C) $Gm \left[A \left(\frac{1}{a} - \frac{1}{a+L} \right) + BL \right]$ (D) $Gm \left[A \left(\frac{1}{a} - \frac{1}{a+L} \right) - BL \right]$

39. Four particles, each of mass M and equidistant from each other, move along a circle of radius R under the action of their mutual gravitational attraction, the speed of each particles [2014]
  (A) $\sqrt{\frac{GM}{R}}$ (B) $\sqrt{2\sqrt{2}\frac{GM}{R}}$
  (C) $\sqrt{\frac{GM}{R}(1+2\sqrt{2})}$ (D) $\frac{1}{2}\sqrt{\frac{GM}{R}(1+2\sqrt{2})}$

40. A satellite is revolving in a circular orbit at a height h from the Earth's surface (radius of earth is R and $h << R$). The minimum increase in its orbital velocity required, so that the satellite could escape from the Earth's gravitational field, is close to (Neglect the effect of atmosphere) [2016]
  (A) $\sqrt{2gR}$ (B) $\sqrt{gR}$
  (C) $\sqrt{gR/2}$ (D) $\sqrt{gR}(\sqrt{2}-1)$

41. A rocket is launched normal to the surface of the Earth, away from the Sun, along the line joining the Sun and the Earth. The Sun is 3×10^5 times heavier than the Earth and is at a distance 2.5×10^4 times larger than the radius of Earth. The escape velocity from Earth's gravitational field is $v_e = 11.2$ km s^{-1}. The minimum initial velocity (v_s) required for the rocket to be able to leave the Sun-Earth system is closest to (Ignore the rotation and revolution of the Earth and the presence of any other planet) [2017]
  (A) $v_s = 72$ km s^{-1} (B) $v_s = 22$ km s^{-1}
  (C) $v_s = 42$ km s^{-1} (D) $v_s = 62$ km s^{-1}

42. Two stars of masses 3×10^{31} kg each and at distance 2×10^{11} m, rotate in a plane about their common centre of mass O. A meteorite passes through O moving perpendicular to the star's rotation plane. In order to escape from the gravitational field of this double star, the minimum speed that meteorite should have at O, is (Take, gravitational constant, $G = 6.67 \times 10^{-11}$ N m^{2}kg^{-2}) [2019]
  (A) 2.8×10^5 m/s   (B) 3.8×10^4 m/s
  (C) 2.4×10^4 m/s   (D) 1.4×10^5 m/s

43. A satellite is moving in a low nearly circular orbit around the earth. It's radius is roughly equal to that of the earth's radius, 'R_e'. By firing rockets attached to it, its speed is instantaneously increased in the direction of its motion, so that it becomes $\sqrt{\frac{3}{2}}$ times larger. Due to this, the farthest distance from the centre of the earth that the satellite reaches is R. Value of R is [2020]
  (A) $4R$ (B) $2.5R$ (C) $3R$ (D) $2R$

44. At what height from the surface of earth the weight of the body is 1/3rd of its weight at the surface? [24-06-2022 Shift I]
  (A) 5000 km   (B) 5562.5 km
  (C) 4684.8 km   (D) 3600 km

45. Two satellites revolve around a planet, with radius 3200 km and 800 km what is their ratio of orbital speeds? [25-06-2022 Shift II]
  (A) $2:3$ (B) $2:1$ (C) $3:2$ (D) $1:2$

46. Particles A, B, C of mass 100 kg are in a straight line where distance between A, B and B, C is 13 m. A fourth particle P is placed on the perpendicular bisector of AC when $BP = 13$ m. If net force on P is F. Find F in terms of G (Gravitational constant) [25-07-2022 Shift I]
  (A) $\frac{G \times 10^4}{13^2} \left(\frac{1}{\sqrt{4}} + 2 \right)$ (B) $\frac{G \times 10^4}{13^2} \left(\frac{1}{\sqrt{2}} + 1 \right)$
  (C) $\frac{G \times 10^4}{13^2} \left(\frac{1}{\sqrt{5}} + 3 \right)$ (D) $\frac{G \times 10^4}{13^2} \left(\frac{1}{\sqrt{6}} + 1 \right)$

47. A body is taken from surface of earth to a distance of $5R/4$ from the centre of earth, where R is radius of earth. Percentage decrease in weight of body at this height is [25-07-2022 Shift II]
  (A) 33.33%   (B) 64%   (C) 25%   (D) 36%

48. A body is projected from surface of earth with velocity $\frac{1}{3}$rd of escape velocity. Find maximum height achieved. [26-07-2022 Shift II]
  (A) $\frac{R}{2}$ (B) $\frac{R}{6}$ (C) $\frac{R}{8}$ (D) $\frac{R}{10}$

49. Find percentage change in 'g'. If radius of earth shrinks by 2%, mass remains constant [29-07-2022 Shift I]
  (A) 2   (B) 3   (C) 4   (D) 5

50. Gain in potential energy in bringing a body of mass 1 g to a height of 3 R (R = radius of earth) from the surface of earth is: [29-07-2022 Shift II]
  (A) 48 mJ   (B) 24 mJ   (C) 30 mJ   (D) 26 mJ

Section - B : JEE Advanced

1. A solid sphere of uniform density and radius 4 units is located with its centre at the origin O of coordinates. Two spheres of equal radii 1 unit, with their centers at $A(-2,0,0)$ and $B(2,0,0)$ respectively, are taken out of the solid leaving behind spherical cavities as shown in the Fig.1.218. Then [IIT- 1993]
  (A) the gravitational force due to this object at the orgin is zero
  (B) the gravitational potential is the same at all points of the circle $y^2 + z^2 = 36$
  (C) the gravitational potential is the same at all points on the circle $y^2 + z^2 = 4$
  (D) All of these

2. The magnitudes of the gravitational field at distances r_1 and r_2 from the centre of a uniform sphere of radius R and mass M are F_1 and F_2 respectively. Then [IIT - 1994]
  (A) $\frac{F_1}{F_2} = \frac{r_1}{r_2}$ if $r_1 < R$ and $r_2 < R$
  (B) $\frac{F_1}{F_2} = \frac{r_2^2}{r_1^2}$ if $r_1 > R$ and $r_2 > R$
  (C) $\frac{F_1}{F_2} = \frac{r_1}{r_2}$ if $r_1 > R$ and $r_2 > R$
  (D) $\frac{F_1}{F_2} = \frac{r_1^2}{r_2^2}$ if $r_1 < R$ and $r_2 < R$

3. If the distance between the earth and the sun were half

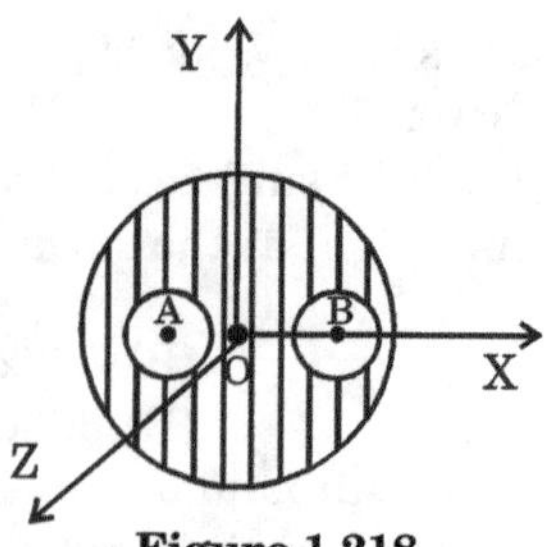

Figure 1.218

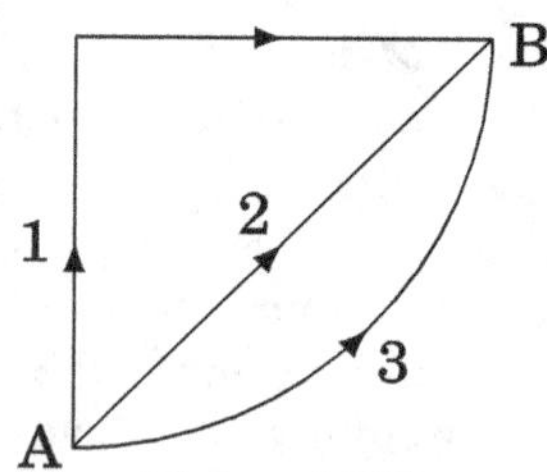

Figure 1.219

its present value, the number of days in a year would have been   [1996]

(A) 64.5   (B) 129   (C) 182.5   (D) 730

4. A particle is projected vertically upward from the surface of earth (radius R) with a kinetic energy equal to half of the minimum vlaue needed for it to escape. The height to which it rises above the surface of earth is   [IIT- 1997]

(A) $h = R$ (B) $h = R/2$ (C) $h = 2R$ (D) $h = 3R$

5. An artificial satellite moving in a circular orbit around the Earth has a total (kinetic + potential) energy E_0. Its potential energy is   [IIT- 1997]

(A) $-E_0$ (B) $2E_0$ (C) $1.5E_0$ (D) E_0

6. The ratio of Earth's orbital angular momentum (about the Sun) to its mass is 4.4×10^{15} m²/s. The area enclosed by Earth's orbit is approximately.   [IIT- 1997]

(A) 6.94×10^{22} m²   (B) 6.94×10^{20} m²

(C) 6.94×10^{18} m²   (D) 6.94×10^{24} m²

7. A satellite S is moving in an elliptical orbit around the earth. The mass of the satellite is very small compared to the mass of the earth.   [IIT -1998]

(A) The acceleration of S is always directed towards the centre of the earth

(B) The angular momentum of about the centre of the earth changes in direction, but its magnitude remains constant

(C) The total mechanical energy of S varies periodically with time

(D) The linear momentum of S remains constant in magnitude

8. A simple pendulum has a time period T_1 when on the earth's surface, and T_2 when taken to a height R above the earth's surface, where R is the radius of the earth. The value of T_2/T_1 [IIT - 2001]

(A) 1   (B) $\sqrt{2}$ (C) 4   (D) 2

9. A geostationary satellite orbits around the earth in a circular orbit of radius 36000 km. Then, the time period of a spy satellite orbiting a few hundred km above the earth's surface ($R_{earth} = 6400$ km) will approximately be   [IIT - 2002]

(A) 1/2 h   (B) 1 h   (C) 2 h   (D) 4 h

10. In a region of only gravitational field a particle of mass 'M' is shifted from A to B via three different paths as shown in the Fig.1.219. The work done in different paths are W_1, W_2, W_3 respectively, then [IIT - 2003]

(A) $W_1 > W_2 > W_3$ (B) $W_1 = W_2 = W_3$

(C) $W_3 > W_2 > W_1$ (D) $W_2 > W_3 > W_1$

11. A binary star system consists of two stars of masses m_1 and m_2, rotating about common centre of mass in orbits of radii r_1 and r_2, with periods T_1 and T_2 respectively, then   [IIT - 2006]

(A) $\left(\frac{T_1}{T_2}\right) = \left(\frac{r_1}{r_2}\right)^{3/2}$

(B) $T_1 > T_2$ if $r_1 > r_2$

(C) $T_1 > T_2$ if $m_1 > m_2$

(D) $T_1 = T_2$

12. A spherically symmetric gravitational system of particles has a mass density $\rho = \begin{cases} \rho_0 & \text{for } r \le R \\ 0 & \text{for } r > R \end{cases}$ where ρ_0 is a constant. A test mass can undergo circular motion under the influence of the gravitational field of particles. Its speed V as a function of distance r ($0 < r < \infty$) from the center of the system is represented by [Fig.1.220]

         [IIT-2008]

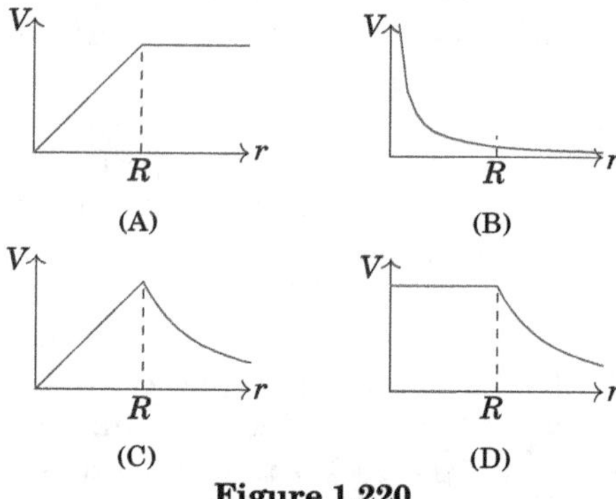

Figure 1.220

13. **Statement-1** An astronaut in an orbiting space station above the Earth experiences weightlessness. [IIT - 2008]

and

Statement-2 An object moving around the Earth under the influence of Earth's gravitational force is in a state of 'free-fall'.

(A) Statement-1 is True, Statement-2 is True; Statement -2 is a correct explanation for Statement-1

(B) Statement-1 is True, Statement-2 is True; Statement -2 is NOT a correct explanation for Statement-1.

(C) Statement-1 is True, Statement-2 is False

(D) Statement-1 is False, Statement-2 is True

14. A thin uniform annular disc (see Fig.1.221) of mass M has outer radius $4R$ and inner radius $3R$. The work required to take a unit mass from point P on its axis to infinity is   [2010]

(A) $\frac{2GM}{7R}(4\sqrt{2} - 5)$ (B) $-\frac{2GM}{7R}(4\sqrt{2} - 5)$

(C) $\frac{GM}{4R}$ (D) $\frac{2GM}{5R}(\sqrt{2} - 1)$

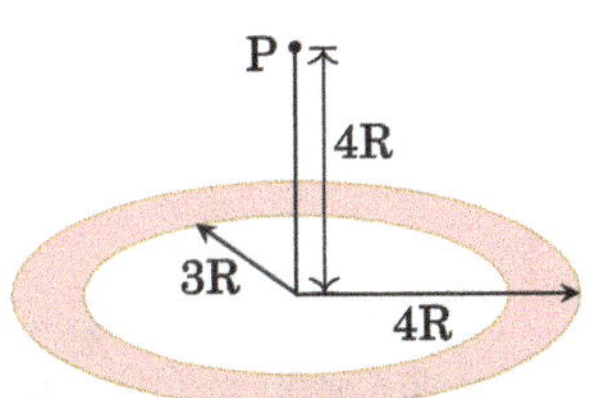

Figure 1.221

15. A satellite is moving with a constant speed v in a circular orbit about earth. An object of mass m is ejected from the satellite such that it just escapes from the gravitational pull of the earth. At the time of its injection, the kinetic energy of the object is [2011]
(A) $\frac{1}{2}mv^2$ (B) mv^2 (C) $\frac{3}{2}mv^2$ (D) $2mv^2$

16. A planet of radius $R = \frac{1}{10} \times$ (radius of Earth) has the same mass density as Earth. Scientists dig well of depth $\frac{R}{5}$ on it and lower a wire of the same length and of linear mass density 10^{-3} kg/m into it. If the wire is not touching anywhere, the force applied at the top of the wire by a person holding it in place is [take the radius of Earth $= 6 \times 10^6$ m and the acceleration due to gravity on Earth $= 10$ m/s^2] [2014]
(A) 96 N   (B) 108 N   (C) 120 N   (D) 150 N

17. A rocket is launched normal to the surface of the earth, away from the sun, along the line joining the sun and the earth. The sun is 3×10^5 times heavier than the earth and is at a distance 2.5×10^4 times larger than the radius of the earth. The escape velocity from the earth's gravitational field is $v_e = 11.2$ km/s. The minimum initial velocity (v_s) required for the rocket to be able to leave the sun-earth system is closest to [Ignore the rotation and revolution of the earth and the presence of any other planet.] [2017]
(A) $v_s = 72$ km/s   (B) $v_s = 22$ km/s
(C) $v_s = 42$ km/s   (D) $v_s = 62$ km/s

18. A planet of mass M, has two natural satellites with masses m_1 and m_2. The radii of their circular orbits are R_1 and R_2, respectively. Ignore the gravitational force between the satellites. Define v_1, L_1, K_1 and T_1 to be respectively, the orbital speed, angular momentum, kinetic energy and time period of revolution of satellite 1; and v_2, L_2, K_2 and T_2 to be the corresponding quantities of satellite 2. Given, $m_1/m_2 = 2$ and $R_1/R_2 = 1/4$, match the ratios in column-I to the numbers in column-II. [2018]

	Column-I		Column-II
A.	v_1/v_2	P.	1/8
B.	L_1/L_2	Q.	1
C.	K_1/K_2	R.	2
D.	T_1/T_2	S.	8

(A) A $\to$ R; B $\to$ Q; C $\to$ P; D $\to$ R
(B) A $\to$ R; B $\to$ Q; C $\to$ S; D $\to$ P
(C) A $\to$ Q; B $\to$ R; C $\to$ P; D $\to$ S
(D) A $\to$ Q; B $\to$ R; C $\to$ S; D $\to$ P

19. Two spherical stars A and B have densities ρ_A and ρ_B, respectively. A and B have the same radius, and their masses M_A and M_B are related by $M_B = 2M_A$. Due to an interaction process, star A loses some of its mass, so that its radius is halved, while its spherical shape is retained, and its density remains ρ_A. The entire mass lost by A is deposited as a thick spherical shell on B with the density of the shell being ρ_A. If v_A and v_B are the escape velocities from A and B after the interaction process, the ratio $\frac{v_B}{v_A} = \sqrt{\frac{10n}{15^{1/3}}}$. The value of n is [2022]

1.26 Answer keys and Solutions

1.26.1 Checkpoint 1

1. **APPROACH** First of all, find the gravitational force acting between the particles for general separation r. Now, by applying Newton's second law of motion, $F = ma$, find the accelerations of both the particle and then find their relative acceleration. The integration of relative acceleration gives the relative velocity of approach.
SOLUTION The gravitational force of attraction on m_1 due to m_2 at a separation r is
$$F_1 = \frac{Gm_1m_2}{r^2}$$
Therefore, the acceleration of m_1 is
$$a_1 = \frac{F_1}{m_1} = \frac{Gm_2}{r^2}$$
Similarly, the acceleration of m_2 due to m_1 is
$$a_2 = -\frac{Gm_1}{r^2}$$
Here, negative sign shows that, a_2 is directed opposite to a_1.
The relative acceleration of approach is
$$a = a_1 - a_2 = \frac{G(m_1+m_2)}{r^2}$$
If v is the relative velocity, then
$$a = \frac{dv}{dt} = \frac{dv}{dr}\frac{dr}{dt}$$
But $-dr/dt = v$ (negative sign shows that r decreases with increasing t).
$$\therefore \qquad a = -\frac{dv}{dr}v$$
$$v\,dv = -\frac{G(m_1+m_2)}{r^2}dr$$
On integrating, we get,
$$\frac{v^2}{2} = \frac{G(m_1+m_2)}{r} + C \qquad \qquad \dots (1)$$
Since, it is given that, at $r = \infty$, $v = 0$, therefore, from Eq. (1), we get
$$\frac{0^2}{2} = \frac{G(m_1+m_2)}{\infty} + C$$
or $\qquad C = 0.$
Substituting, this value of C in Eq. (1), we get
$$v^2 = \frac{2G(m_1+m_2)}{r}$$
Let $v = v_R$ when $r = R$. Then
$$v_R = \sqrt{\frac{2(Gm_1+m_2)}{R}}$$

2. **APPROACH** It is given that, on placing third ball of mass m on the line joining the two balls, the net force on first ball becomes $6Gm^2/d^2$. No information about the direction of resultant is given. So, the resultant may be directed either towards right or towards left.
From Eq.1.1, the gravitational force applied by the

second ball of mass $2m$ on the first ball of mass m is:
$$F_{12} = \frac{Gm(2m)}{d^2} = \frac{2Gm^2}{d^2} \qquad \ldots (1)$$
If third ball of mass m, is placed at distance x from the first ball towards right of it (Fig.1.222a), then both third ball of mass m and second ball of mass $2m$ apply the gravitational pulling force on the first ball towards right. If, F_{12} and F_{13} are magnitudes of gravitational forces on first ball due to second and third ball respectively, then
$$F_{12} + F_{13} = \frac{6Gm^2}{d^2} \qquad \ldots (2)$$
Here,
$$F_{13} = \frac{Gm^2}{x^2} \qquad \ldots (3)$$
Now, substitution of F_{12} and F_{13} from Eq.(1) and (3) in Eq.(2), will give you the required values of x. Since, we have already mentioned the position of third ball towards the right of first ball, therefore, only the positive value of x will be the acceptable position.

If third ball is placed towards left of the first ball at distance x from it (Fig.1.222b), then the gravitational pull of third ball on the first ball will be leftwards which is opposite to the gravitational pull of second ball of mass $2m$. In case of opposite forces, the larger force is always greater than the resultant. But, from Eq. (1), the force applied by the second ball of mass $2m$ on the first ball of mass m, is $F_{12} = \dfrac{2Gm^2}{d^2}$, which is less than the given resultant $\dfrac{6Gm^2}{d^2}$, so, in such case, the net gravitational force on the first ball cannot be rightward and F_{13} must be greater than F_{23}.

In this case, we have
$$F_{13} - F_{23} = G\frac{6m^2}{d^2} \qquad \ldots (4)$$
Substitution of the values of F_{12} and F_{13} in Eq.(4) will give you the required position of third ball.

SOLUTION **Case 1. If the net force on the first**

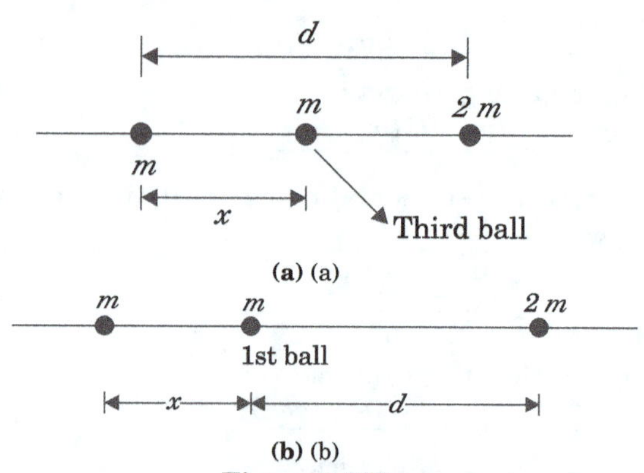

Figure 1.222

ball is toward right

In this case (Fig.1.222a), the third ball must be towards right of first ball.

On substituting the values of F_{12} and F_{13} from Eq.(1) and (3) in Eq.(2), we get,
$$\frac{2Gm^2}{d^2} + \frac{Gm^2}{x^2} = \frac{6Gm^2}{d^2}$$

or
$$\frac{2}{d^2} + \frac{1}{x^2} = \frac{6}{d^2}$$
or
$$\frac{1}{x^2} = \frac{4}{d^2}$$
or
$$x^2 = \frac{d^2}{4}$$
or
$$x = \pm \frac{d}{2}$$
Since, x is directed towards right of first ball of mass m, so we consider only positive values of x. Therefore, the required value of x is $d/2$ towards right of first ball.

Case 2. If the net force on the first ball is toward left

In this case (Fig.1.222b), the third ball must be left of first ball. So, substitution of F_{12} and F_{13} in Eq.(4) gives
$$G\frac{m^2}{x^2} - G\frac{2m^2}{d^2} = G\frac{6m^2}{d^2}$$
or
$$G\frac{m^2}{x^2} = G\frac{2m^2}{d^2} + G\frac{6m^2}{d^2}$$
or
$$\frac{1}{x^2} = \frac{8}{d^2} \quad \Rightarrow \quad x = \pm\frac{d}{2\sqrt{2}}$$
Since, x is positive in left side of first ball, therefore we can ignore the negative root value of x. So, we can write-
$$x = \frac{d}{2\sqrt{2}}$$

3. **APPROACH** Let us consider an equilateral triangle $\triangle ABC$ of side a [Fig.1.223]. A point mass m is placed at each vertex of this triangle.

Suppose, we want to calculate the net gravitational force on the point mass at B.

To determine, net gravitational force on the mass at B, we first calculate the gravitational forces applied by point masses at A and C and then apply the principle of superposition of forces[7].

SOLUTION Since, masses at A and C are same, there-

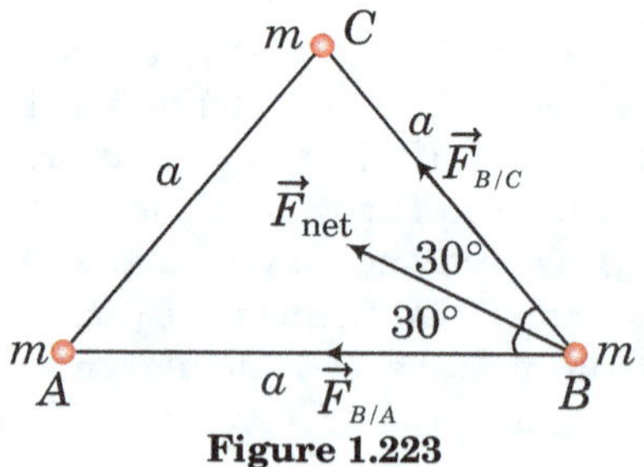

Figure 1.223

fore, the magnitude of forces applied by both point masses on the mass placed at B, will be same. If $\vec{F}_{B/A}$ and $\vec{F}_{B/C}$ are the forces applied by masses at A and C respectively on mass at B, then, we have
$$F_{B/A} = F_{B/C} = F = \frac{G(m)(m)}{a^2} = \frac{Gm^2}{a^2}$$
Therefore, net force acting on the mass at B, is given by:
$$F_{net} = \sqrt{F_{B/A}{}^2 + F_{B/C}{}^2 + 2F_{B/A}F_{B/C}\cos 60°}$$
or
$$F_{net} = \sqrt{F^2 + F^2 + 2FF\cos 60°}$$

---

[7]According to principle of superposition of forces, if there are more than number of forces acting on a particle, then net force acting on the particle will be equal to vector sum of all these forces.

or $\qquad F_{\text{net}} = F\sqrt{2 + 2(1/2)} = F\sqrt{3}$

or $\qquad F_{\text{net}} = \dfrac{\sqrt{3}Gm^2}{a^2}$

Direction of net force: If θ is the angle made by $\vec{F}_{\text{net}}$ with $\vec{F}_{A/C}$, then by vector method, we have

$$\tan\theta = \frac{F_{B/C}\sin 60°}{F_{B/A} + F_{B/C}\cos 60°} = \frac{F(\sqrt{3}/2)}{F + (F/2)} = \frac{1}{\sqrt{3}}$$

or $\qquad \theta = 30°$

So, the net force vector makes angle of $30°$ from force vector $\vec{F}_{B/A}$

Remarks Since, the two forces are equal in magnitudes, therefore their resultant divides the angle between them in two equal parts and hence passes through the centre as shown in Fig.1.223.

4. **APPROACH** (a) Resultant of forces due to point mass placed at A and point mass placed at C on the point mass placed at D, is towards left. So, to make zero resultant on D, we have to put an extra identical point mass at such a point that the resultant of gravitational force applied by this particle together with the particle placed at B on point mass placed at D becomes equal and opposite (i.e. rightwards) to above resultant. In this case, the net gravitational force on particle at D will be zero. In Fig.1.224, such point is represented by E.

SOLUTION (a) The net force on mass at D will be zero if an identical mass is placed at E [Fig.1.224]. In this case the resultant force due to masses at A and C will be leftward and that due to masses at B and E will be rightward.

    APPROACH (b) Since, the particle at E, makes the

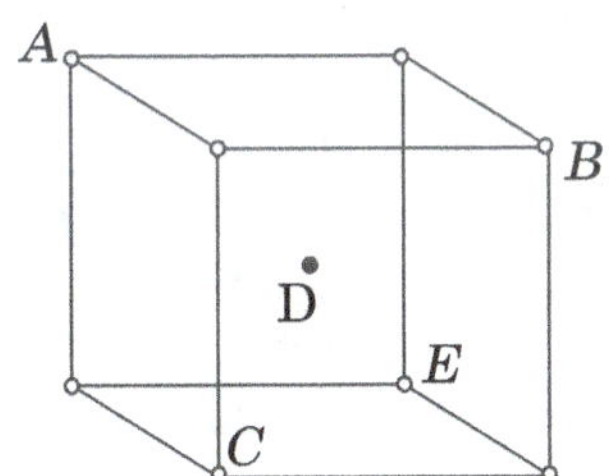

Figure 1.224

resultant zero on D, therefore,

Required force = Force equal and opposite to that applied by the mass placed at E

SOLUTION $\qquad F_{\text{net}} = \dfrac{Gm^2}{(\sqrt{3}a/2)^2} = \dfrac{4Gm^2}{3a^2}$

5. **APPROACH** (a) Fig.1.225 shows the given situation. Now, write the expression of gravitational force acting between the point masses m and M by using Eq.1.1:

$$F = \frac{GmM}{x^2} \qquad \ldots (1)$$

Since, this gravitational force acts on both the masses m and M, therefore, you can find their absolute accelerations by using Newton's second law of motion, i.e., $F = ma$. If a_1 and a_2 are the accelerations of masses m and M respectively, then

$$a_1 = F/m \text{ and } a_1 = F/M$$

Now, the relative acceleration of mass m with respect to M is given by

$$a_{1/2} = a_1 - a_2 \qquad (2)$$

(b) Integration of above relative acceleration with respect to time, gives the relative velocity of m with respect to M.

(c) Acceleration of center of mass of the system of masses is equal to the net external force on the system of masses divided by total mass of the system, i.e.,

$$a_{CM} = \frac{F_{\text{ext}}}{m + M}$$

SOLUTION (a) Force between the two masses $F = \frac{GmM}{x^2}$

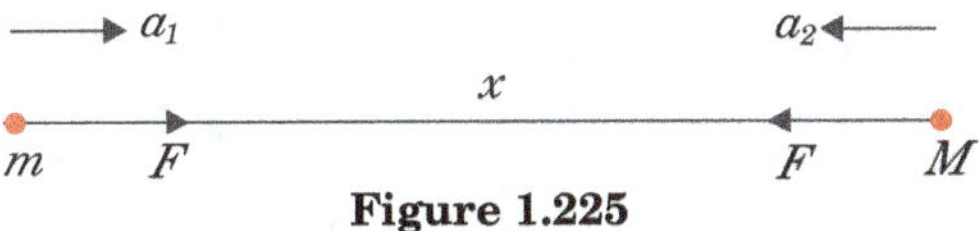

Figure 1.225

If we consider the positive direction of x axis, towards right, then, acceleration of m,

$$a_1 = \frac{F}{m} = \frac{GM}{x^2}$$

Acceleration of M, $a_2 = -\dfrac{F}{M} = -\dfrac{Gm}{x^2}$

Therefore, acceleration of m with respect to M,

$$a_{1/2} = a_1 - a_2 = \frac{G(M+m)}{x^2}$$

Therefore, the magnitude of relative acceleration of the particles:

$$a = \frac{G(M+m)}{x^2} \qquad (3)$$

(b) Since, $a = \dfrac{dv}{dt} = \dfrac{dv}{dx}\cdot\dfrac{dx}{dt} = v\dfrac{dv}{dx}$ [v = relative velocity]

Therefore, from Eq.(3), $-v\dfrac{dv}{dx} = \dfrac{G(M+m)}{x^2}$

here $-$ve sign has been placed because v is increasing with decreasing x]

or $\qquad v\,dv = -G(M+m)\dfrac{dx}{x^2}$

$\therefore \qquad \displaystyle\int_0^v v\,dv = -G(M+m)\int_\infty^x \frac{dx}{x^2}$

or $\qquad \dfrac{v^2}{2} = -G(M+m)\left[-\dfrac{1}{x}\right]_\infty^x = \dfrac{G(M+m)}{x}$

$\therefore \qquad v = \sqrt{\dfrac{2G(M+m)}{x}}$

(c) The centre of mass (COM) has no acceleration as there is no external force. Hence, velocity of COM is zero throughout.

6. **APPROACH** Angular momentum of the asteroid about the center of mass of the planet,

$$\vec{L} = \vec{r} \times m\vec{v}$$

Figure 1.226

here, $\vec{r}$ is the position vector of the asteroid with respect to centre of the planet and $\vec{v}$ is the velocity of the planet at position $\vec{r}$.

Since, the gravitational force of the planet on the asteroid passes through the centre of planet, therefore, the torque of gravitational force on the asteroid about the center of the planet is zero, so the angular momentum of the asteroid about the center of the planet will always be conserved.

As, initially, $\vec{v}$ and $\vec{r}$ are not along the same line, therefore, initial angular momentum is non zero. So, by conservation of angular momentum, the angular momentum of the asteroid about the center of the planet must always be non zero.

If the asteroid would have been collided with the surface of planet normally, then the velocity vector would be directed toward the centre of the planet which means it's angular momentum about the centre of the planet would be zero. Which is the violation of principle of conservation of angular momentum, as initially it was non zero.

SOLUTION Asteroid cannot hit the surface of the planet normally.

7. **(a)APPROACH** To move all three particles on a circular path, the net force on each particle must be directed towards the center of the circle and it's magnitude must also provide the required centripetal force for circular motion. So, to find the speed of each particle, we first find the net gravitational force on any one particle due to other two and then equate it to the required centripetal force for that particle.

SOLUTION Let us consider the circular motion of particle 'B'. It is revolving in circle of radius $r = a/\sqrt{3}$. Resultant force on it towards the centre of the circle is

$$2F\cos 30° = 2\frac{Gm^2}{a^2}\frac{\sqrt{3}}{2} = \sqrt{3}\frac{Gm^2}{a^2}$$

For circular motion of the particle B, this force provides the required centripetal force, i.e.,

$$\therefore \qquad \frac{mv^2}{r} = \frac{\sqrt{3}Gm^2}{a^2}$$

$$\Rightarrow \qquad v^2 = \frac{Gm}{a} \qquad \left[\because r = \frac{a}{\sqrt{3}}\right]$$

$$\Rightarrow \qquad v = \sqrt{\frac{Gm}{a}}$$

(b) Acceleration of center of mass of a system of particles can be written as

$$a_{COM} = \frac{F_{ext}}{m_{system}}$$

here, m_{system} is the mass of system of particles under consideration, F_{ext} is the external force on the system of particles, i.e., the force applied by particles other than system particles and a_{COM} is the acceleration of system of particles.

SOLUTION Consider our system to be made up of B and C. External force on this system is due to A. Net external force

$$F_{ext} = 2F\sin 60° = \sqrt{3}F = \frac{\sqrt{3}Gm^2}{a^2}$$

$$\therefore \qquad a_{COM} = \left(\frac{\sqrt{3}Gm^2}{a^2}\right) \times \frac{1}{2m} = \frac{\sqrt{3}}{2}\frac{Gm}{a^2}$$

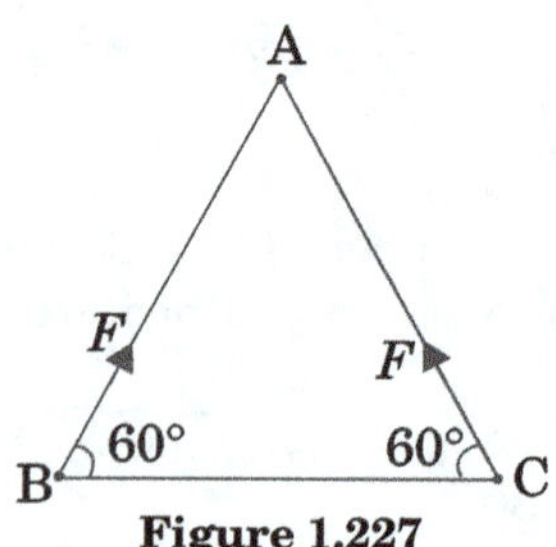

Figure 1.227

It will be directed towards the centre of the triangle

(c)APPROACH Suppose, particle A, suddenly looses the ability of attracting B and C, then the net external force on the system formed by particles B and C will become zero[8]. As a result of this, the net linear momentum of the system will get conserved. So, by applying the conservation of linear momentum, you can find the velocity of center of mass of the system of particles containing B and C.

SOLUTION From Fig.1.228, the linear momentum of the system of particles B and C, during circular motion just before A looses the attraction power is,

$$p_{0,\,system} = mv\cos 60° + mv\cos 60° = mv \text{ [along } CB\,]$$

If v_{COM} is the speed of center of mass of system, just after

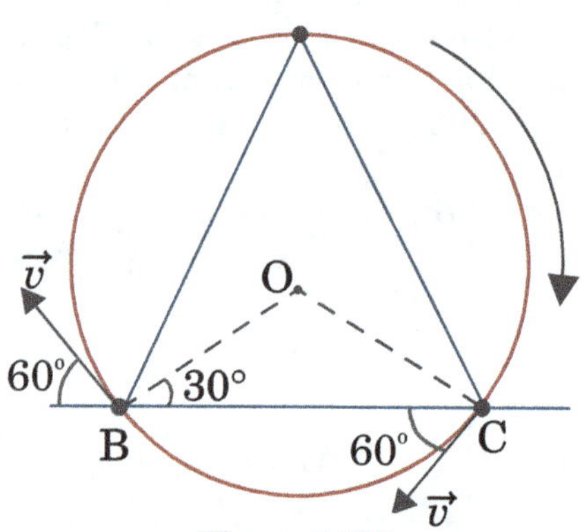

Figure 1.228

A looses the ability of attraction, then linear momentum of the system

$$p_{system} = 2mv_{COM}$$

By conservation of linear momentum, we have

$$p_{system} = p_{0,\,system}$$

$$\text{or} \qquad 2mv_{COM} = mv$$

$$\text{or} \qquad v_{COM} = \frac{mv}{2m} = \frac{v}{2} = \frac{1}{2}\sqrt{\frac{Gm}{a}}$$

The velocity of COM does not change after A stops attracting.

8. **APPROACH** Fig.1.229, shows earth-spaceship-astronaut system. Spaceship is revolving around the earth in an orbit of radius R, such that $R = 2R_e$. The astronaut is connected with a cable of length l to the spaceship such that his distance from the center of earth

---

[8]If $\vec{p}$ is the linear momentum of the system and $\vec{F}_{ext}$ is the applied external force, then $\vec{F}_{ext} = d\vec{p}/dt$. In the absence of external force, $d\vec{p}/dt = 0$ or $\vec{p} =$ constant.

is $R' = 2R_e + l$.

Circular Motion of the spaceship

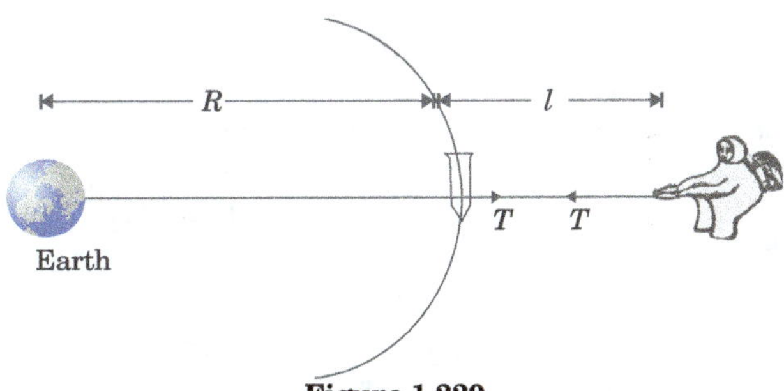

Figure 1.229

The forces acting on the space ship are,

1. Gravitational force of earth, $F_1 = \dfrac{GMm_0}{(R)^2}$: towards the center of earth.

2. Tension, T, applied by the cable: away from the earth. The net force on the spaceship, provides required centripetal force for it's circular motion, i.e,

$$F_1 - T = m_0 R^2 \omega$$

or $\quad \dfrac{GMm_0}{R^2} - T = m_0 R^2 \omega \qquad \text{... (1)}$

here, M = mass of earth, m_0 = mass of spaceship, R = the radius of earth and ω is the angular speed of spaceship-astronaut system around the earth.

Circular Motion of the astronaut

In Fig.1.30, the forces acting on the astronaut are:

1. the gravitational pulling force applied by earth:

$$F_g = \dfrac{GMm}{(R')^2},$$

here, m is the mass of the astronaut and $R' = R + l$ is the distance of astronaut from the centre of earth.

2. Tension (T) applied by the cable

Both of these forces are acting towards the center of the earth and hence their sum is providing the net centripetal force to the astronaut for his circular motion of radius R' around the earth, i.e.,

$$\dfrac{GMm}{(R')^2} + T = m\omega^2(R') \qquad \text{... (2)}$$

Now, simplify equations (1) and (2) for T.

SOLUTION Equating the values of ω^2 obtained from Eq. (1) and (2), we get

$$\dfrac{GM}{R^3} - \dfrac{T}{m_0 R} = \dfrac{GM}{(R')^3} + \dfrac{T}{mR}$$

$$\Rightarrow \quad T\left[\dfrac{1}{mR'} + \dfrac{1}{m_0 R}\right] = GM\left(\dfrac{1}{R^3} - \dfrac{1}{R'^3}\right)$$

$$\Rightarrow \quad T = \left(\dfrac{GMmm_0}{m_0 R + mR'}\right)\left(\dfrac{1}{R^3} - \dfrac{1}{R'^3}\right)$$

This gives $R'^3 - R^3 = (R' - R)(R'^2 + RR' + R^2)$

$$\simeq l\left(3R^2\right)$$

$$\therefore \quad T \simeq \dfrac{GMmm_0 3lR^2}{(m_0 + m)R^5} = \dfrac{GMm}{R^3} \cdot \dfrac{3m_0 l}{m_o + m}$$

$$\simeq \dfrac{GM}{(2R_e)^3} \cdot 3ml \quad [\because m_o + m \simeq m_o]$$

$$= \dfrac{GM}{R_e^2} \cdot \dfrac{3ml}{8R_e} = g \cdot \dfrac{3}{8}\dfrac{ml}{R_e}$$

$$= 9.8 \times \dfrac{3}{8} \times \dfrac{100 \times 200}{6400 \times 1000} = 0.01 \text{ N}$$

9. **APPROACH** Follow the approach given in section 1.3.1. In this case (Fig.1.230), only the values of masses are interchanged. Separation r is replaced by a and length of the rod is L.

Ans. $\dfrac{GMm}{a(a+L)}$

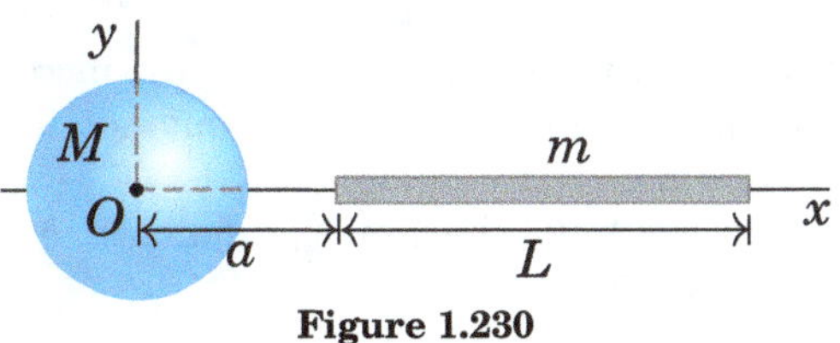

Figure 1.230

Multiple Choice Questions

10. (A): Since two astronauts are floating in gravitational free space. The only force acting on the two astronauts is the gravitational pull of their masses, $F = \dfrac{Gm_1 m_2}{r^2}$, which is attractive in nature.
Hence they move towards each other.

11. (B): Since, the gravitational force does not depend upon the medium in which objects are placed, therefore it will not change.

12. (B)

13. (B): $F_{\text{surface}} = G\dfrac{Mm}{R_e^2}$

$$F_{R_e/2} = G\dfrac{Mm}{(R_e + R_e/2)^2} = \dfrac{4}{9} \times F_{\text{surface}} = \dfrac{4}{9} \times 72 = 32\text{N}$$

14. (A) : Force between the two particles of identical masses (each of mass m),

$$F = -G\dfrac{mm}{4R^2}.$$

This force will provide the necessary centripetal force for the masses to go around a circle, so

$$\dfrac{Gmm}{4R^2} = \dfrac{mv^2}{R} \Rightarrow v^2 = \dfrac{Gm}{4R} \Rightarrow v = \dfrac{1}{2}\sqrt{\dfrac{Gm}{R}}$$

15. (A) : Given that: mass $(m) = 6 \times 10^{24}$kg; angular velocity $(\omega) = 2 \times 10^{-7}$rad/s and radius $(r) = 1.5 \times 10^8$km $= 1.5 \times 10^{11}$m.
Force exerted on the earth $= mR\omega^2$
$$= \left(6 \times 10^{24}\right) \times \left(1.5 \times 10^{11}\right) \times \left(2 \times 10^{-7}\right)^2 = 36 \times 10^{21}\text{N}$$

16. (B) : Given that, gravitational force is proportional to $1/R$(and not as $1/R^2$), i.e.
$$F = \dfrac{GMm}{R}$$
Since, there is no other force than gravitaional one, so this the only force which provides the required centripetal force for circular motion of the particle. So,
$$\dfrac{mv^2}{R} = \dfrac{GMm}{R} \qquad \Rightarrow \qquad v = \sqrt{GM}.$$
Thus, velocity v is independent of R.

1.26.2   Check Point 2

1. **APPROACH** From Eq. 1.12, the gravity above the surface of earth at height 'h' is given by:

$$g' = g\left(1 + \frac{h}{R}\right)^{-2} \qquad \dots (1)$$

Here, $g = GM/R^2$, is the gravity for $h = 0$

$$g' = \frac{GM}{R^2}\left(1 + \frac{h}{R}\right)^{-2} \qquad \dots (2)$$

From above equation, we see that with increase in the value of 'h' value of g' decreases but not linearly. At $h \to \infty$, $g' \to 0$.

Now, from Eq.1.16, the gravity at depth h below the surface of earth:

$$g' = g\left(1 - \frac{h}{R}\right) \qquad \dots (3)$$

Since, $g = GM/R^2$, therefore above equation can also be written as

$$g' = \frac{GM}{R^2}\left(1 - \frac{h}{R}\right) \qquad \dots (4)$$

i.e. g' decreases linearly with depth. It is directed towards the center of the earth. At the center of earth, depth, $h = R$, therefore from Eq.(4), $g' = 0$

SOLUTION Using equations (2) and (4), the graph

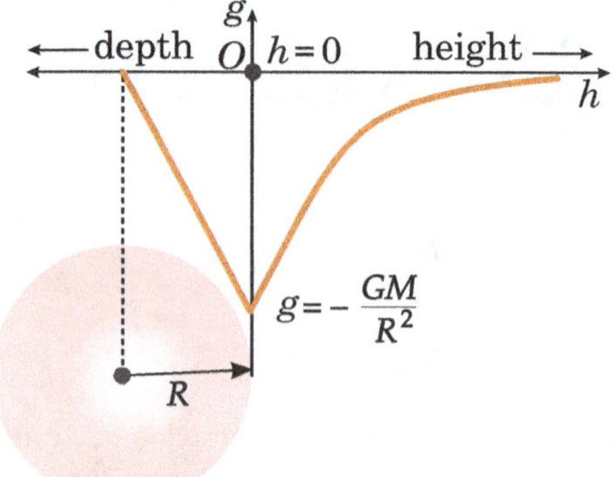

Figure 1.231

drawn is shown in Fig.1.231
In graph − ve sign is used for g. It shows that g is directed towards the centre of earth.

2. **APPROACH** In this problem earth and moon will be treated as point masses as they are far apart and as in case of a point mass $g = GM/r^2$, the field due to both the masses will be zero at infinite. In addition to it, the field will also be zero at a point in between earth and moon where the pull of earth balances the pull of moon. This is the required point of the given problem. To find this point assume it's distance from earth, equal to x. Now, at his point the gravitational field produced by earth will be equal and opposite to gravitational field produced by moon. Find the magnitude of gravitational field at this position due to earth and moon both and then equate and simplify for x.

SOLUTION Let M_E is the mass of earth and M_M is the mass of moon. If the required point is at a distance x from the earth, then at this point

$$\frac{GM_E}{x^2} = \frac{GM_M}{(3.8 \times 10^8 - x)^2}$$

or $\quad\dfrac{81}{x^2} = \dfrac{1}{(3.8 \times 10^8 - x)^2} \qquad$ [as $M_E = 81\,M_M$]

or $\quad 9\left(3.8 \times 10^8 - x\right) = x$

or $\quad x = 3.42 \times 10^8 m = 3.42 \times 10^5$ km

3. **APPROACH** You can use either Eq.1.12 or Eq.1.14
Sol (a) From Eq.1.12, we have

$$g' = \frac{g}{\left(1 + \frac{h}{R}\right)^2}$$

Given that, $g' = g - 0.36g = 0.64g$

$$\therefore \quad 0.64 = \frac{1}{\left(1 + \frac{h}{R}\right)^2}$$

$$\Rightarrow \quad 1 + \frac{h}{R} = \frac{1}{0.8} \Rightarrow \quad \frac{h}{R} = \frac{5}{4} - 1$$

$$\Rightarrow \quad h = \frac{R}{4} = 1600 \text{ km}$$

(b) $g' = \dfrac{g}{\left(1 + \frac{h}{R}\right)^2} = g\left(1 + \frac{h}{R}\right)^{-2}$

$$\simeq g\left[1 - \frac{2h}{R}\right]$$

[expanding binomially and neglecting higher order terms]

Therefore, $\quad \dfrac{g'}{g} = 1 - \dfrac{2h}{R} \quad \Rightarrow \quad \dfrac{2h}{R} = \dfrac{g - g'}{g}$

So, $\quad \dfrac{2h}{R} \times 100 = \dfrac{\Delta g}{g} \times 100$

or $\quad \dfrac{2h}{R} \times 100 = 0.36$

$$\Rightarrow \quad h = \frac{0.18}{100} \times 6400 \text{ km} = 11.52 \text{ km}$$

4. **APPROACH** Given conditions:

$$\frac{1}{3}\left(g_{\text{pole}}\right)_{\text{earth}} = \left(g_{\text{pole}}\right)_{\text{planet}} \qquad \dots (1)$$

From Eq.1.20, we have

$$g' = g\left(1 - \frac{\omega^2 R}{g}\cos^2\theta\right)$$

At equator, $\theta = 0$, therefore, $g'_P = g_P\left(1 - \dfrac{\omega^2 R_P}{g_P}\right)$ Astronaut found himself weightless at equator of the planet, i.e., $g'_P = 0$, therefore above equation gives

$$\omega^2 R_P = g_P,$$

or $\quad \omega^2 R_P = \dfrac{GM_P}{R_P^2} \qquad \dots (2)$

here, $\omega = 2\pi/T$, and $T \to$ time period of the planet.
Now, calculate ω and hence T by using equations (1) and (2)

SOLUTION From Eq.(1), we have

$$\frac{1}{3}\frac{GM_e}{R_e^2} = \frac{GM_P}{R_P^2}$$

$$\Rightarrow \quad \frac{1}{3} \cdot \frac{\frac{4}{3}\pi R_e^3 \cdot \rho_0}{R_e^2} = \frac{\frac{4}{3}\pi R_P^3 \cdot \rho_P}{R_P^2}$$

$$\therefore \quad \rho_0 R_e = 3\rho_P R_P$$

And from Eq.(2), $\omega^2 R_P = \dfrac{GM_P}{R_P^2}$

$$\Rightarrow \qquad \omega^2 = G\frac{4}{3}\pi\rho_P \;\; = \frac{4}{3}G\pi\frac{1}{3}\frac{\rho_0 R_e}{R_P}$$

$$= \frac{4}{9}G\pi\rho_0(2) \qquad\qquad \left[\because \frac{R_e}{R_P}=2\right]$$

$$\therefore \qquad \frac{4\pi^2}{T^2} = \frac{8}{9}G\pi\rho_0$$

$$\text{or}\qquad T^2 = \frac{9\pi}{2G\rho_0} \quad\Rightarrow\quad T = \sqrt{\frac{9\pi}{2G\rho_0}}$$

5. APPROACH Apply Eq.1.14, and simplify for altitude h
SOLUTION From Eq.1.14, we have

$$\frac{\Delta g}{g}\times 100 = \frac{2h}{R}\times 100\%$$

$$\text{or}\qquad \frac{\Delta g}{g} = \frac{2h}{R}$$

Here, it is given that, $\frac{\Delta g}{g}=10^{-9}\%=10^{-11}$, therefore

$$10^{-11} = \frac{2h}{R}$$

$$\text{or}\qquad \frac{2h}{R} = 10^{-11}$$

$$\text{or}\qquad 2h = 6.4\times 10^{6}\text{ m}\times 10^{-11}$$

$$\text{or}\qquad h = 3.2\times 10^{-5}\text{ m}\qquad\text{or}\qquad h = 32\ \mu m$$

6. APPROACH Your weight is the local gravitational force exerted on you. We can use the definition of density to relate the mass of the planet to the mass of Earth and the law of gravity to relate your weight on the planet to your weight on Earth.
SOLUTION Using the definition of density, relate the mass of Earth to its radius:

$$M_E = \rho V_E = \tfrac{4}{3}\rho\pi R_E^3$$

Relate the mass of the planet to its radius:

$$M_P = \rho V_P = \tfrac{4}{3}\rho\pi R_P^3$$

$$= \tfrac{4}{3}\rho\pi(10R_E)^3$$

Divide the second of these equations by the first to express M_P in terms of M_E:

$$\frac{M_P}{M_E} = \frac{\tfrac{4}{3}\rho\pi(10R_E)^3}{\tfrac{4}{3}\rho\pi R_E^3} \Rightarrow M_P = 10^3 M_E$$

If w' is your weight on the planet, then

$$w' = mg_P = m\left(\frac{GM_P}{R_P^2}\right)$$

$$= \frac{Gm\left(10^3 M_E\right)}{(10R_E)^2}$$

$$= 10\frac{GmM_E}{R_E^2} = 10\ mg = 10\ w$$

Thus, your weight would be ten times your weight on Earth.

7. APPROACH We can relate the acceleration due to gravity of a test object at the surface of the new planet to the acceleration due to gravity at the surface of Earth through use of the law of gravity and Newton's second law of motion.
SOLUTION If 'a' represent the acceleration due to gravity at the surface of this new planet and m the mass of a test object, then Newton's second law and the law of gravity give

$$\sum F_{\text{radial}} = \frac{GmM_E}{\left(\frac{1}{2}R_E\right)^2} = ma \quad\Rightarrow a = \frac{GM_E}{\left(\frac{1}{2}R_E\right)^2}$$

Simplify this expression to obtain:

$$a = 4\left(\frac{GM_E}{R_E^2}\right) = 4g = 4\left(9.8\text{ m/s}^2\right) = 39.2\text{ m/s}^2$$

8. APPROACH To find gravitational field g at any point, we draw a hypothetical Gaussian surface symmetrical to the source mass. If mass enclosed with in the Gaussian surface is M_{encl}, then

$$g = \frac{GM_{\text{encl}}}{r^2}$$

here, r is the radius of Gaussian surface.
Sol (a) Let, g_1 and g_2 are the gravitational accelerations at the surface of earth and planet respectively. As for earth and the planet, the mass enclosed within the Gaussian surface that passes through the given point on the surface, is M, therefore

$$g_1 = \frac{GM}{R^2}; \quad g_2 = \frac{GM}{R^2}$$

$$\therefore \qquad \frac{g_1}{g_2} = 1$$

(b) $\int_0^R \rho 4\pi r^2 dr = M$

$$\Rightarrow\quad 4\pi\rho_0\int_0^R r^3 dr = M \quad\Rightarrow \pi\rho_0 R^4 = M$$

$$\text{or}\qquad \rho_0 = \frac{M}{\pi R^4}$$

9. APPROACH If we consider the rotation of earth, then the apparent weight of any object can be obtained by Eq.1.21:

$$W = mg - m\omega^2 R\cos^2\theta \qquad\qquad \text{... (1)}$$

Here, R is the radius of earth, θ is the latitude and ω is the absolute angular velocity of the object about the axis of rotation of earth.
At equator, $\theta=0$, therefore, Eq.(1) gives-

$$W = mg - m\omega^2 R \qquad\qquad \text{... (2)}$$

At pole, $\theta=90°$, therefore, if W_0 is the weight of the object at pole, then Eq. (1) gives-

$$W_0 = mg \qquad\qquad \text{... (3)}$$

Substituting the value of mg in Eq.(2), gives

$$W = W_0 - m\omega^2 R \qquad\qquad \text{... (4)}$$

If the object is stationary with respect to earth and the angular velocity of earth is ω_e, then

$$\omega = \omega_e$$

If the object is moving with respect to earth with velocity, v, from west to east, i.e., in the same sense as that of earth's rotation as seen from a point over the northern hemisphere, then

$$\omega = \omega_{\text{earth}} + \omega_{\text{object/earth}}$$

$$\text{or}\qquad \omega = \omega_e + \frac{v}{R}$$

If the object is moving from east to west, i.e., in opposite sense as that of earth's rotation as seen from a point over the northern hemisphere, then

$$\omega = \omega_e - \frac{v}{R}$$

Now, put these values of ω and calculate the weight of object in each case. Finally calculate the difference in the weights.
SOLUTION **Case 1.** When the train (i.e. object) is

moving from west to east with respect to earth, with velocity v, then

$$\omega = \omega_e + \frac{v}{R}$$

Therefore, from Eq.(2), the apparent weight of the object

$$W_1 = W_0 - m\left(\omega_e + \frac{v}{R}\right)^2 R$$

Case 2. In return journey, the train is moving from east to west with respect to earth, which is opposite to the linear velocity of earth. In this case, the absolute angular velocity of train (i.e., object) with respect to the axis of rotation of earth

$$\omega = \omega_e - \frac{v}{R}$$

Therefore, from Eq.(2), new apparent weight of the object

$$W_2 = W_0 - m\left(\omega_e - \frac{v}{R}\right)^2 R$$

Therefore, $W_2 - W_1 = m\left(\omega_e + \frac{v}{R}\right)^2 R - m\left(\omega_e - \frac{v}{R}\right)^2 R$

or $\quad W_2 - W_1 = 4m\omega_e v \qquad\qquad$ ... (5)

Since, $W_0 = mg$, therefore, $m = W_0/g$

Substituting this value of m in above expression, we get

$$W_2 - W_1 = 4W_0\omega_e v/g$$

10. **APPROACH** If we consider the rotation of planet, then the apparent weight of any object on it can be obtained by Eq.1.21

$$W = mg - m\omega^2 R\cos^2\theta \qquad\qquad \text{... (1)}$$

Here, all variables are taken for planet not earth.

If object is at rest on the surface of planet, then ω will be equal to angular speed of planet.

From above equation, we can say that, if angular speed of planet keeps on increasing continuously, then apparent weight of the object also keeps on decreasing. For a certain value of angular speed (ω), it becomes zero and as soon as ω exceeds this value, the objects on its surface start flying off from it.

Let for $\omega = \omega_0$, the object feels the condition of weightlessness, i.e., $W = 0$, for $\omega = \omega_0$, then, above equation gives

$$0 = mg - m\omega_0^2 R\cos^2\theta$$

or $\quad g = \omega_0^2 R\cos^2\theta$

or $\quad \omega_0 = \dfrac{1}{\cos\theta}\sqrt{\dfrac{g}{R}}$

Since, ω_0 is inversely proportional to $\cos\theta$, therefore the needed value of ω_0 is minimum corresponding to maximum value of $\cos\theta$, i.e.,

$$\omega_{0,\ min} = \frac{1}{(\cos\theta)_{max}}\sqrt{\frac{g}{R}}$$

Since, $(\cos\theta)_{max} = 1$, which is corresponding to position of equator, i.e., for $\theta = 0$, therefore,

$$\omega_{0,\ min} = \sqrt{\frac{g}{R}} \qquad\qquad \text{... (2)}$$

For this value of angular speed, only the objects at equator feels weightlessness. Objects at other positions will still have some apparent weight.

If angular speed of planet exceeds this value of ω_0, then objects on the equator of planet's surface starts flying off. Now, use $T = 2\pi/\omega$, and substitute g in terms of density of the planet.

SOLUTION Objects will not flying off from the surface of planet if angular velocity of planet is less than the value given in Eq.(2).

Thus, the condition for objects to not flying off the surface of planet is:

$$\omega \le \omega_{0,\ min}$$

If T is the time period of planet, then $\omega = 2\pi/T$, on substituting it in above equation, we get

$$\frac{2\pi}{T} \le \sqrt{\frac{g}{R}}$$

or $\qquad\qquad T \ge 2\pi\sqrt{\dfrac{R}{g}}$

So, the condition for rocks to not fly away from the equator of the planet becomes-

$$T \ge 2\pi\sqrt{\frac{R}{g}} = 2\pi\sqrt{\frac{R}{GM/R^2}}$$

$$= 2\pi\sqrt{\frac{R^3}{G\cdot\frac{4}{3}\pi R^3\cdot\rho}} = \sqrt{\frac{3\pi}{G\rho}}$$

11. **(a)** **APPROACH** From Fig.1.77, the centripetal acceleration of the man

$$a = \omega^2 r = \omega^2 R\cos\theta \qquad\qquad \text{... (1)}$$

[since, the radius of rotation at latitude $\theta = 40°$ is $r = R\cos\theta$]

Now, put given values in Eq.(1), and simplify for a

SOLUTION On substituting, $\omega = 7.27\times10^{-5}$ rad s^{-1}, $R = 6.37\times10^6$m and $\cos\theta = \cos40° = 0.77$ in Eq.(1), we get

$$a = \left(7.27\times10^{-5}\right)^2\left(6.37\times10^6\right)(0.77) = 0.026\ \text{ms}^{-2}$$

(b) **APPROACH** From Eq.1.20, the effective value of g is given by:

$$g' = g - \omega^2 R\cos^2\theta \qquad\qquad \text{... (2)}$$

SOLUTION If earth suddenly stops rotating, then $\omega = 0$. In this case, Eq.(2) gives

$$g' = g = 9.80\text{m/s}^{-2}$$

12. **APPROACH** First of all find the gravitational field at $R/2$ for radius of star $r \ge R/2$ and then for $r < R/2$. By using these expressions, you can draw g vs r graph.

SOLUTION Let at any instant t, the radius of the star is r. For, $r > R/2$, the point will be inside the sphere. In this case, the gravitational field is given by

$$g = \frac{GM}{r^3}x = \frac{GM}{r^3}\left(\frac{R}{2}\right)$$

$\therefore \qquad\qquad g \propto \dfrac{1}{r^3}$

In the beginning, i.e., at $t = 0$, $r = R$, therefore

$$g = \frac{GM}{2R^2}$$

When $r = \dfrac{R}{2}$; $g = \dfrac{4GM}{R^2}$

After the star shrinks to a radius $r < R/2$, the point is now outside the sphere.

In this case, the gravitational field at the point becomes constant. It is given by

$$g = \frac{GM}{(R/2)^2} = \frac{4GM}{R^2}$$

g vs r graph is shown in Fig.1.232

13. **APPROACH** Write the values of gravitational field for

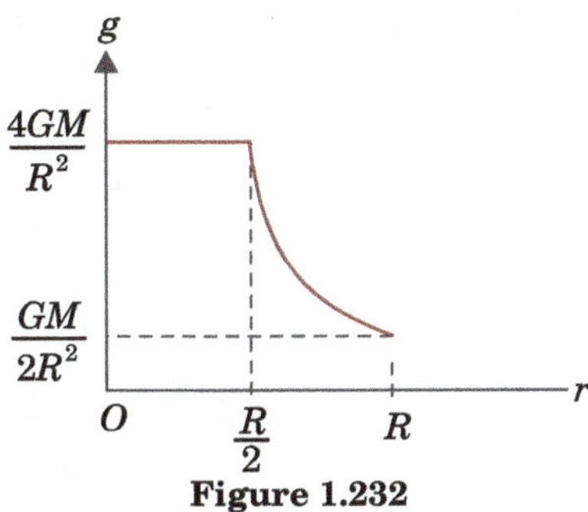

Figure 1.232

inside and out side points to earth. Now, apply the given condition.

SOLUTION Acceleration inside the earth at a distance r from the centre

$$g_1 = \frac{GM}{R^3} \cdot r = \frac{GM}{R^2}\left(\frac{R-h_1}{R}\right) = g\left(1 - \frac{h_1}{R}\right)$$

$$\Delta g_1 = \left(\frac{dg_1}{dh_1}\right)\Delta h_1 = -\frac{g}{R}\Delta h_1$$

Given that, $\Delta h_1 = +1km$

$$\therefore \qquad \Delta g_1 = -\frac{g}{R}(1km) \qquad \qquad \dots (1)$$

This is independent of depth h_1

At a distance r from the centre outside the earth

$$g_2 = \frac{GM}{r^2} = \frac{GM}{R^2}\frac{R^2}{r^2} = g\frac{R^2}{(R+h_2)^2}$$

$$\therefore \qquad \Delta g_2 = \left(\frac{dg_2}{dh_2}\right)(\Delta h_2) = gR^2\left[-2(R+h_2)^{-2-1}\right]\Delta h_2$$

$$= \frac{-2gR^2}{(R+h_2)^3}\Delta h_2$$

Given that $\Delta h_2 = 1\text{km}$

So, $$\Delta g_2 = -\frac{2gR^2}{(R+h_2)^3}(1km) \qquad \qquad \dots (2)$$

Since, $\Delta g_1 = \Delta g_2$

$$\therefore \qquad -\frac{g}{R}(1km) = -\frac{2gR^2}{(R+h_2)^3}(1km)$$

or $\qquad (R+h_2)^3 = 2R^3$

$$\Rightarrow \qquad 1 + \frac{h_2}{R} = (2)^{1/3}$$

or $\qquad h_2 = R\left[2^{1/3} - 1\right]$

14. **APPROACH** We can find the mass of the rod by integrating dm over its length. The gravitational field at $x_0 > L$ can be found by integrating $d\vec{g}$ at x_0 over the length of the rod.

SOLUTION (a) The total mass of the stick is given by:

$$M = \int_0^L \lambda dx$$

Substitute for λ and evaluate the integral to obtain

$$M = C\int_0^L xdx = \frac{1}{2}CL^2$$

(b) Express the gravitational field due to an element of the stick of mass dm :

$$d\vec{g} = -\frac{Gdm}{(x_0-x)^2}\hat{i} = -\frac{G\lambda dx}{(x_0-x)^2}\hat{i}$$

$$= -\frac{GCxdx}{(x_0-x)^2}\hat{i}$$

Integrate this expression over the length of the stick to obtain:

$$\vec{g} = -GC\int_0^L \frac{xdx}{(x_0-x)^2}\hat{i}$$

$$= \frac{2GM}{L^2}\left[\ln\left(\frac{x_0}{x_0-L}\right) - \left(\frac{L}{x_0-L}\right)\right]\hat{i}$$

15. **APPROACH** The elements of the rod of mass dm and length dx produce a gravitational field at any point P located a distance $x_0 > L$ from the origin. We can calculate the total field by integrating the magnitude of the field due to dm from $x = 0$ to $x = L$.

SOLUTION (a) Express the gravitational field at P due to the element dm :

$$d\vec{g}_x = -\frac{Gdm}{r^2}\hat{i}$$

Relate dm to dx :

$$dm = \frac{M}{L}dx$$

Express the distance r between dm and point P in terms of x and x_0 :

$$r = x_0 - x$$

Substitute these results to express $d\vec{g}_x$ in terms of x and x_0 :

$$d\vec{g}_x = \left\{-\frac{GM}{L(x_0-x)^2}dx\right\}\hat{i}$$

(b) Integrate from $x = 0$ to $x = L$ to find the total gravitational field at point P :

$$\vec{g}_x = -\frac{GM}{L}\int_0^L \frac{dx}{(x_0-x)^2}\hat{i}$$

$$= \left\{-\frac{GM}{L}\left[\frac{1}{x_0-x}\right]_0^L\right\}\hat{i}$$

$$= -\frac{GM}{x_0(x_0-L)}\hat{i}$$

(c) Use the definition of gravitational field and the result from Part (b) to express $\vec{F}_g$ at $x = x_0$:

$$\vec{F}_g = m_0\vec{g} = -\frac{GMm_0}{x_0(x_0-L)}\hat{i}$$

(d) Factor x_0 from the denominator of the expression for $\vec{g}_x$ to obtain:

$$\vec{g}_x = -\frac{GM}{x_0^2\left(1 - \frac{L}{x_0}\right)}\hat{i}$$

For $x_0 > L$ the second term in parentheses is very small and:

$$\vec{g}_x \approx -\frac{GM}{x_0^2}\hat{i}$$

which is the gravitational field of a point mass M located at the origin.

16. **APPROACH** Make FBD of a mass m placed on the surface of earth at latitude ϕ (see Fig.1.233). On rotating earth's frame, the centrifugal force $\vec{F}_c$ acting away from the axis of rotation of the earth, the true weight $m\vec{g}$

acts towards centre of earth (i.e., in vertically downward direction), and the effective force mg_{eff} at angle β from vertical. From FBD of Fig.1.233, we can say that corresponding to larger value of perpendicular component of $\vec{F}_c$, i.e., $F_c\sin\theta$, the angle β will be larger.

SOLUTION In Fig.1.233, the centrifugal force is given by

$$F_c = m\omega^2 r = m\omega^2 R\cos\theta$$

Therefore, component of $\vec{F}_c$, perpendicular to $m\vec{g}$ is

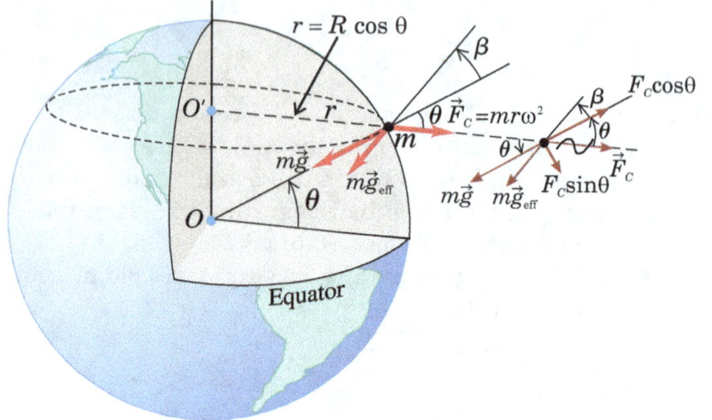

Figure 1.233

$$F_c\sin\theta = mr\omega^2\sin\theta$$
$$= mR\cos\theta\,\omega^2\sin\theta = \frac{1}{2}mR\omega^2\sin 2\theta$$

Therefore, $[F_c\sin\theta]_{max} = \left[\frac{1}{2}mR\omega^2\sin 2\theta\right]_{max}$

$$= \frac{1}{2}mR\omega^2[\sin 2\theta]_{max}$$

Now, $[\sin 2\theta]_{max} = 1 = \sin 90°$

$\Rightarrow \quad 2\theta = 90° \qquad \Rightarrow \qquad \theta = 45°$

So, corresponding to maximum β, we have, $\theta = 45°$.

17. APPROACH By Gauss's theorem, the gravitational field at distance r from the centre of the planet is given by Eq.1.50

$$g = \frac{GM_{\text{encl}}}{r^2}$$

here, M_{encl} is the enclosed mass within the Gaussian surface. Find net enclosed mass in terms of r and think how can g be independent on r

SOLUTION Let's first calculate the mass of the atmosphere between $R \le x < r$

$$m = \int_R^r \rho 4\pi x^2 dx$$

$$= 4\pi\sigma_\circ \int_R^r \frac{x^2}{x} dx \qquad \left[\because \rho = \frac{\sigma_\circ}{x}\right]$$

$$= 2\pi\sigma_0\left[r^2 - R^2\right]$$

Acceleration due to gravity at a distance r from the centre is

$$g = \frac{G(M+m)}{r^2}$$

Since, the whole mass distribution (planet + atmosphere is spherically symmetric) $\therefore \quad g =$
$\frac{GM}{r^2} + \frac{G2\pi\sigma_\circ}{r^2}\left[r^2 - R^2\right]$

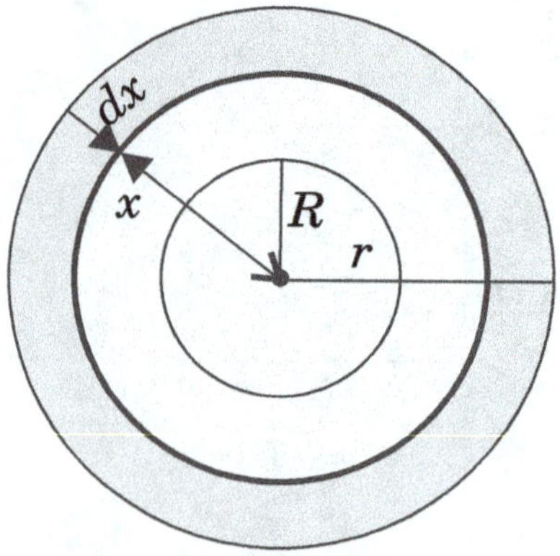

$$= \frac{GM}{r^2} - \frac{G2\pi\sigma_\circ R^2}{r^2} + G.2\pi\sigma_\circ$$

This expression is independent of r if

$$GM = G\cdot 2\pi\sigma_\circ R^2 \Rightarrow \sigma_\circ = \frac{M}{2\pi R^2}$$

Alter: Acceleration due to gravity on the surface

$$g = \frac{GM}{R^2} \Rightarrow \frac{\Delta g}{g} = \frac{\Delta M}{M} - \frac{2\Delta R}{R}$$

But $\Delta g = 0 \quad \therefore \quad \frac{\Delta M}{M} = \frac{2\Delta R}{R}$

or $\quad \dfrac{4\pi R^2\cdot\Delta R\frac{\sigma_\circ}{R}}{M} = \dfrac{2\Delta R}{R}$

$\Rightarrow \qquad \sigma_\circ = \dfrac{M}{2\pi R^2}$

18. APPROACH We can use conservation of angular momentum to relate the planet's speeds at aphelion and perihelion. Since, gravitational force applied by star on the planet is directed towards the centre of the star, therefore, the torque of gravitational force on the planet about the centre of the star will be zero and hence, the angular momentum of the planet about centre of the star will be conserved.

The angular momentum of planet about centre of the star, when the planet is at aphelion,

$$L_a = mv_a r_a$$

The angular momentum of planet about centre of the star, when the planet is at perihelion,

$$L_p = mv_p r_p$$

Using conservation of angular momentum, relate the angular momenta of the planet at aphelion and perihelion:

$$L_a = L_p \quad \text{or} \quad mv_a r_a = mv_p r_p \Rightarrow v_a = \frac{v_p r_p}{r_a}$$

On substituting numerical values, we get

$$v_a = \frac{(5.0\times 10^4 m/s)(1.0\times 10^{15}m)}{2.2\times 10^{15}m} = 2.3\times 10^4 m/s$$

19. APPROACH We can use Newton's law of gravity to express the gravitational force acting on an object at the surface of the neutron star in terms of the weight of the object. We can find gravitational acceleration at the surface of the neutron star by dividing the gravitational force by mass of the object.

SOLUTION Apply Newton's law of gravity to an object of mass m at the surface of the neutron star to obtain:

$$\frac{GM_N m}{R_N^2} = mg$$

where M_N, R_N are the mass and radius of neutron star and g represents the magnitude of the gravitational field at the surface of the neutron star.

Solve for g and substitute for the mass of the neutron star

$$g = \frac{GM_N}{R_N^2} = \frac{G(1.60M_S)}{R_N^2}$$

Here, M_S is mass of the Sun.

Substitute numerical values and evaluate g:

$$g = \frac{1.60\,(6.673 \times 10^{-11}N \cdot m^2/kg^2)(1.99 \times 10^{30}kg)}{(10.5 \text{ km})^2}$$

$$= 1.93 \times 10^{12} m/s^2$$

20. APPROACH Apply the approach used in **Example 12.**

SOLUTION Let λ is the linear mass density of the ring, then-

$$\lambda = \frac{m}{\pi R/2} = \frac{2m}{\pi R}.$$ Now consider a very small segment of the ring of mass dm subtending angle $d\theta$ at the centre where mass m_0 is placed (see Fig.1.235a). The gravitational acceleration due to this segment at the position of mass m_0 is given by

$$dg = \frac{Gdm}{R^2} = \frac{G\lambda(Rd\theta)}{R^2} = G\frac{2m}{\pi R}\frac{Rd\theta}{R^2} = \frac{2Gm}{\pi R^2}d\theta$$

The horizontal component of $d\vec{g}$ is given by

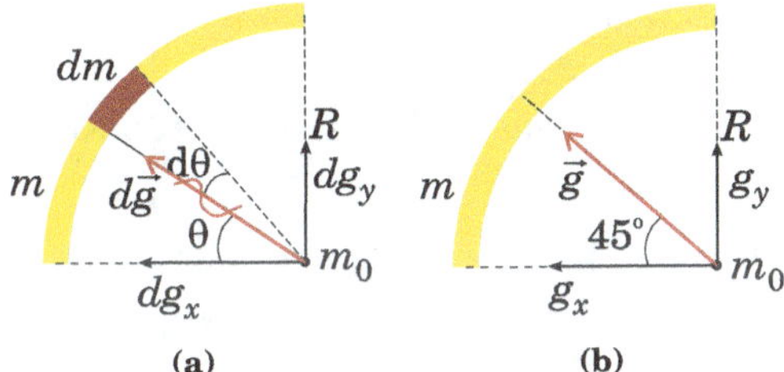

(a)   **(b)**

Figure 1.235

$$dg_x = dg\cos\theta = \frac{2Gm}{\pi R^2}\cos\theta\, d\theta$$

Therefore, net gravitational acceleration along x axis is given by

$$g_x = \int dg_x = \frac{2Gm}{\pi R^2}\int_0^{\pi/2}\cos\theta\, d\theta = \frac{2Gm}{\pi R^2}[\sin\theta]_0^{\pi/2}$$

$$= \frac{2Gm}{\pi R^2}\left[\sin\frac{\pi}{2} - \sin 0\right]$$

$$= \frac{2Gm}{\pi R^2}[1 - 0]$$

$$= \frac{2Gm}{\pi R^2}$$

Similarly, the vertical component of $d\vec{g}$ is given by

$$dg_y = dg\sin\theta = \frac{2Gm}{\pi R^2}\sin\theta\, d\theta$$

$$g_y = \int dg_y = \frac{2Gm}{\pi R^2}\int_0^{\pi/2}\sin\theta\, d\theta$$

$$= -\frac{2Gm}{\pi R^2}\left[\cos\frac{\pi}{2} - \cos 0\right]$$

$$= -\frac{2Gm}{\pi R^2}[0 - 1] = \frac{2Gm}{\pi R^2}$$

Therefore, the magnitude of net gravitational field at the center of quarter of ring,

$$g = \sqrt{g_x^2 + g_y^2} = \frac{2\sqrt{2}Gm}{\pi R^2}$$

Direction : If angle of $\vec{g}$ with g_x is ϕ, then

$$\tan\phi = g_y/g_x = 1$$

i.e., $\phi = 45°$

Now, net gravitational force acting on a particle of mass m_0 placed at the center of the quarter ring:

$$F = m_0 g = \frac{2\sqrt{2}Gmm_0}{\pi R^2}$$

21. APPROACH We can use the definitions of the gravitational fields at the surfaces of Earth and the moon to express the accelerations due to gravity at these locations in terms of the average densities of Earth and the moon. Expressing the ratio of these accelerations will lead us to the ratio of the densities.

SOLUTION The acceleration due to gravity at the surface of Earth in terms of Earth's average density:

$$g_e = \frac{GM_e}{R_e^2} = \frac{G\rho_e V_e}{R_e^2} = \frac{G\rho_e \frac{4}{3}\pi R_e^3}{R_e^2}$$

$$= \frac{4}{3}G\rho_e\pi R_e$$

The acceleration due to gravity at the surface of the moon in terms of the moon's average density is:

$$g_m = \frac{4}{3}G\rho_m\pi R_m$$

Dividing the second of these equations by the first to obtain:

$$\frac{g_m}{g_e} = \frac{\rho_m R_m}{\rho_e R_e} \quad\Rightarrow\quad \frac{\rho_m}{\rho_e} = \frac{g_m R_e}{g_e R_m}$$

Substitute numerical values and evaluate $\frac{\rho_m}{\rho_e}$:

$$\frac{\rho_m}{\rho_e} = \frac{(1.62 m/s^2)(6.37 \times 10^6 m)}{(9.81 m/s^2)(1.738 \times 10^6 m)} = 0.605$$

22. APPROACH Newton's second law of motion relates the weights of these two objects to their masses and the acceleration due to gravity.

SOLUTION (a) Apply Newton's second law to the standard object:

$$F_{net} = w_1 = m_1 g$$

Apply Newton's second law to the object of unknown mass:

$$F_{net} = w_2 = m_2 g$$

Eliminate g between these two equations and solve for m_2:

$$m_2 = \frac{w_2}{w_1}m_1$$

Substitute numerical values and evaluate m_2:

$$m_2 = \frac{56.6\text{ N}}{9.81\text{ N}}(1.00\text{ kg}) = 5.77\text{ kg}$$

(b) Because this result is determined by the effect on m_2 of Earth's gravitational field, it is the gravitational mass of the second object.

23. APPROACH The gravitational field inside a spherical shell is zero and the field at the surface of and outside the shell is given by $g = GM/r^2$.

(a) Because $0.50\text{ m} < R : g(0.50\text{ m}) = 0$

(b) Because $1.9\text{ m} < R : g(1.9\text{ m}) = 0$

(c) Because $2.5\text{ m} > R :$

$$g(2.5 \text{ m}) = \frac{GM}{r^2}$$

$$= \frac{\left(6.673 \times 10^{-11} \text{ N} \cdot \text{m}^2/\text{kg}^2\right)(300 \text{ kg})}{(2.5 \text{ m})^2}$$

$$= 3.2 \times 10^{-9} \text{ N/kg}$$

24. APPROACH We need to find the gravitational field in three regions: $r < R_1$, $R_1 < r < R_2$, and $r > R_2$.

For $r < R_1$: $g(r < R_1) = 0$

For $r > R_2$, $g(r)$ is the field due to the thick spherical shell of mass M centered at the origin:

$$g(r > R_2) = \frac{GM}{r^2}$$

For $R_1 < r < R_2$, $g(r)$ is determined by the mass within the shell of radius r:

$$g(R_1 < r < R_2) = \frac{Gm}{r^2} \qquad \dots (1)$$

where $m = \frac{4}{3}\pi\rho(r^3 - R_1^3)$ $\dots (2)$

Write the density of the spherical shell:

$$\rho = \frac{M}{V} = \frac{M}{\frac{4}{3}\pi(R_2^3 - R_1^3)}$$

Substitute for ρ in equation (2) and simplify to obtain:

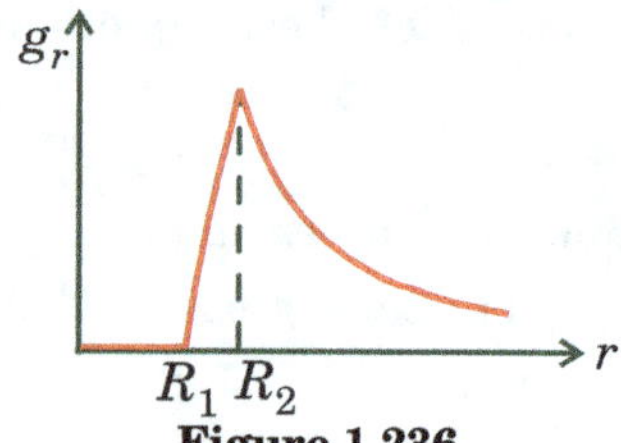

Figure 1.236

$$m = \frac{M(r^3 - R_1^3)}{R_2^3 - R_1^3}$$

Substituting this value of m in equation (1), we get

$$g_r = \frac{GM(r^3 - R_1^3)}{r^2(R_2^3 - R_1^3)}$$

A graph of magnitude of g_r with r, is shown in Fig.1.236.

25. APPROACH Take a differential element and find the gravitational field at an axial position due to this element. Integrate it for complete ring. Here, you will get g as a function of x alone. Now, Resolve g into components along x, y and z axes. Here, you will find that y and z components of g, due to symmetry, get cancelled. The effective components is the component along x axis. Let us denote it by g_x. Corresponding to the extreme value of g_x, put $dg_x/dt = 0$ and solve for x. Now, substitution of this value of x in the expression of g_x will give you the extreme value of g_x.

SOLUTION A ring of radius R is shown in Fig.1.237. Choose a coordinate system in which the origin is at the centre of the ring and x axis is as shown. An element of length dL and mass dm is responsible for the field dg at a distance x from the centre of the ring. We can express the x component of $d\vec{g}$ and then integrate over the circumference of the ring to find the total field as a function of x.

(a) Express the differential gravitational field at a distance x from the center of the ring in terms of the mass of an elemental segment of length dL:

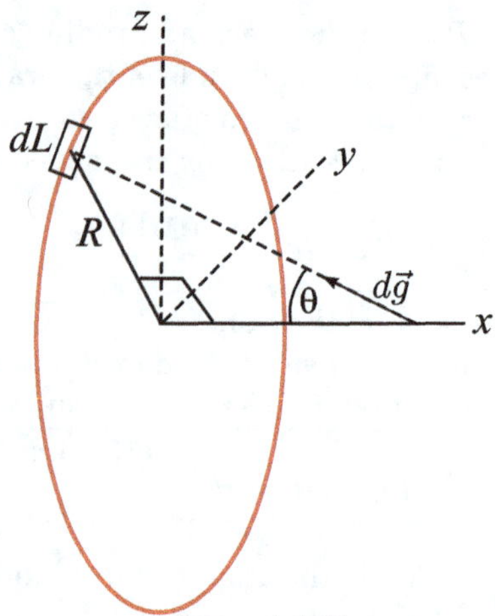

Figure 1.237

$$dg = \frac{Gdm}{R^2 + x^2} \qquad \dots (1)$$

Relate the mass of the element to its length:

$$dm = \lambda dL$$

where λ is the linear density of the ring.

Substituting the value of dm in Eq.(1), we get:

$$dg = \frac{G\lambda dL}{R^2 + x^2}$$

By symmetry, the y and z components of g vanish. The x component of dg is:

$$dg_x = dg\cos\theta = \frac{G\lambda dL}{R^2 + x^2}\cos\theta \qquad \dots (2)$$

From Fig.1.237, we get

$$\cos\theta = \frac{x}{\sqrt{R + x^2}}$$

Substituting this value of $\cos\theta$ in Eq. (2), we get

$$dg_x = \frac{G\lambda dL}{R^2 + x^2}\frac{x}{\sqrt{R^2 + x^2}} = \frac{G\lambda x dL}{(R^2 + x^2)^{3/2}}$$

Because $\lambda = \frac{M}{2\pi R}$: $dg_x = \frac{GMx dL}{2\pi R(R^2 + x^2)^{3/2}}$

Integrate to find g_x: $g_x = \frac{GMx}{2\pi R(R^2 + x^2)^{3/2}}\int_0^{2\pi R} dL$

or
$$\boxed{g_x = \frac{GM}{(R^2 + x^2)^{3/2}}x} \qquad (1.157)$$

A graph of g_x is shown in Fig.1.238.

(b) In Eq.1.157, G, M and R are constants so g_x is a

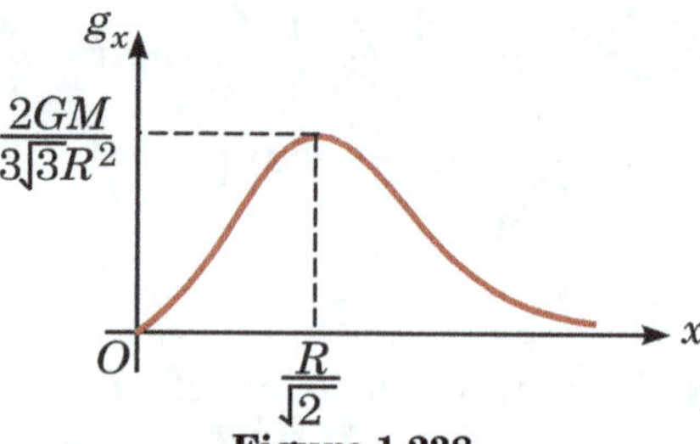

Figure 1.238

function of x. For extreme value of g_x $dg_x/dx = 0$

Therefore, from Eq.1.157,

$$\frac{dg_x}{dt} = 0$$

$$\Rightarrow \quad \frac{dg}{dx} = GM\left[\frac{(x^2 + R^2)^{3/2} - x\left(\frac{3}{2}\right)(x^2 + R^2)^{1/2}(2x)}{(R^2 + x^2)^3}\right] = 0$$

$$\Rightarrow \quad (x^2 + R^2)^{3/2} - 3x^2(x^2 + R^2)^{1/2} = 0$$

Solving for x yields: $x = \pm\frac{R}{\sqrt{2}}$

Because the curve is concave downward, we can conclude that this result corresponds to a maximum. Note that this result agrees with our graphical maximum.

Maximum Value of g_x On substituting $x = \frac{R}{\sqrt{2}}$, in Eq.1.157, we get

$$(g_x)_{\max} = \frac{GM}{\left(R^2 + \left(\frac{R}{\sqrt{2}}\right)^2\right)^{3/2}} \frac{R}{\sqrt{2}} = \frac{2GM}{3\sqrt{3}R^2}$$

Multiple Choice Questions

26. (A) : For a point inside the earth i.e., $r < R$

$$g = -\frac{GM}{R^3} r$$

where M and R be mass and radius of the earth respectively. At the centre, $r = 0$

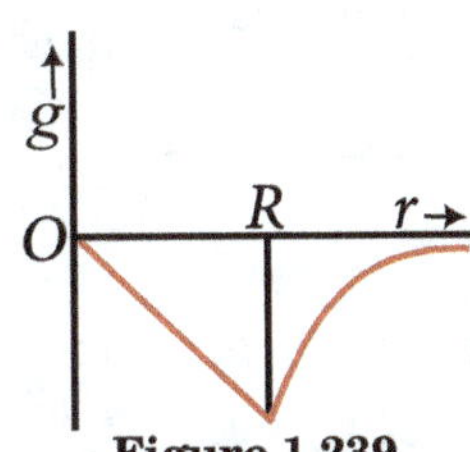

Figure 1.239

$\therefore \qquad g = 0$

For a point outside the earth i.e. $r > R$,

$$g = -\frac{GM}{r^2}$$

On the surface of the earth i.e. $r = R$,

$$g = -\frac{GM}{R^2}$$

The variation of g with distance r from the centre is as shown in the Figure 1.239.

27. (B) : Gravitational field due to the thin spherical shell Inside the shell, (for $r < R$) $g = 0$
On the surface of the shell, (for $r = R$)

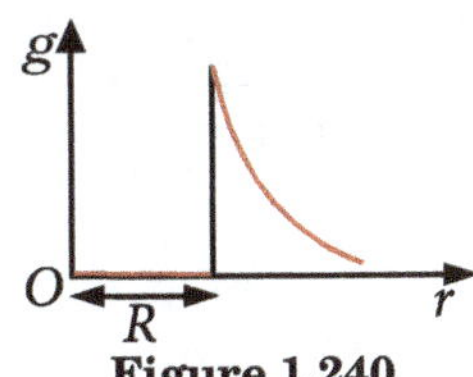

Figure 1.240

$$g = \frac{GM}{R^2}$$

Outside the shell, (for $r > R$)

$$g = \frac{GM}{r^2}$$

The variation of g with distance r from the centre is as shown in the Figure 1.240.

28. (D) If universal gravitational constant becomes ten times, then $G' = 10G$.
So, acceleration due to gravity increases. i.e, (D) is the wrong option.

29. (A) Gravitational force acting on particle of mass m is

$$F = \frac{GM_p m}{\left(D_p/2\right)^2}$$

Acceleration due to gravity experienced by the particle is

$$g = \frac{F}{m} = \frac{GM_p}{\left(D_p/2\right)^2} = \frac{4GM_p}{D_p^2}$$

30. (D) $g = \dfrac{GM}{r^2} = \dfrac{G}{r^2}\left(\dfrac{4}{3} \times \pi r^3 \rho\right) = \dfrac{4}{3} \times \pi \rho G r$

$$\frac{g'}{g} = \frac{3R}{R} \Rightarrow g' = 3g$$

31. (D) The gravitational acceleration at the surface of earth,

$$g_e = \frac{GM_e}{R_e^2} = \frac{G(4/3)\pi R_e^3}{R_e^2}\rho_e$$

$$g_e \propto R_e \rho_e$$

Similarly, the acceleration due to gravity of planet $g_p \propto R_p \rho_p$

$$R_e \rho_e = R_p \rho_p \Rightarrow R_e \rho_e = R_p 2\rho_e$$
$$\Rightarrow \qquad R_p = \tfrac{1}{2}R \qquad\qquad (\because R_e = R)$$

32. (B) Initial velocity of the mass on both the planets is same.

i.e., $\sqrt{2g'h'} = \sqrt{2gh}$
$\Rightarrow \sqrt{2 \times g' \times h'} = \sqrt{2 \times 9g' \times 2} \quad \Rightarrow \quad 2h' = 36$
$\Rightarrow \qquad h' = 18$ m

33. (B)

34. (A) Acceleration due to gravity

$$g = \frac{GM}{R^2} = \frac{G(4/3)\pi R^3 \rho}{R^2} = \frac{4}{3} G\pi R\rho$$

or $\qquad \rho = \dfrac{3g}{4\pi GR}$

35. (C) Given: radius of earth $(R_e) = 6400$ km; radius of mars $(R_m) = 3200$ km; mass of earth $(M_e) = 10M_m$ and weight of the object on earth $(W_e) = 200$ N.

$$\frac{W_m}{W_e} = \frac{mg_m}{mg_e} = \frac{M_m}{M_e} \times \left(\frac{R_e}{R_m}\right)^2 = \frac{1}{10} \times (2)^2 = \frac{2}{5}$$

or $\quad W_m = W_e \times \frac{2}{5} = 200 \times 0.4 = 80$ N

36. (B) Gravitational force at a height h,

$$mg' = \frac{mg_0}{\left(1 + \frac{h}{r}\right)^2} = \frac{72}{\left(1 + \frac{R/2}{R}\right)^2}$$

or $\qquad mg' = 32$ N
or $\qquad F_g = 32$ N

37. (A) Acceleration due to gravity at a depth d, $g_d = g\left(1 - \frac{d}{R}\right)$

For $d = R/2 \Rightarrow g_d = g\left(1 - \frac{R/2}{R}\right) = \frac{g}{2}$

Required weight $W' = mg_d = \frac{mg}{2} = \frac{W}{2} = \frac{200}{2} = 100$ N

38. (C) The acceleration due to gravity at a height h is given as

$$g_h = g\left(1 - \frac{2h}{R_e}\right)$$

where R_e is radius of earth.
The acceleration due to gravity at a depth d is given as

$$g_d = g\left(1 - \frac{d}{R_e}\right)$$

Given, $g_h = g_d$

$$\therefore \qquad g\left(1 - \frac{2h}{R_e}\right) = g\left(1 - \frac{d}{R_e}\right)$$

or $\qquad d = 2h = 2 \times 1 \text{ km} = 2 \text{ km} \quad (\because h = 1 \text{ km})$

39. (B) Acceleration due to gravity is given by

$$g = \begin{cases} \frac{4}{3}\pi\rho G r; & r \le R \\ \frac{4}{3}\frac{\pi\rho R^3 G}{r^2}; & r > R \end{cases}$$

The variation of g with r, is shown in Fig.1.241.

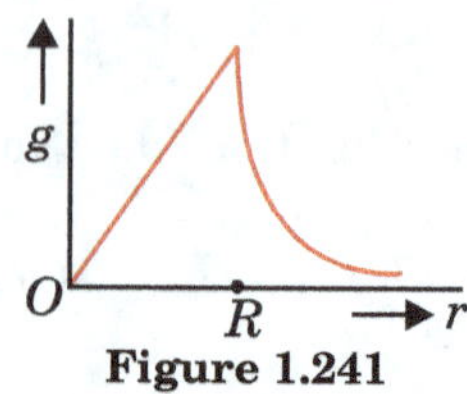

Figure 1.241

40. (C) Acceleration due to gravity at a height h, from the surface of earth is

$$g' = \frac{g}{\left(1 + \frac{h}{R}\right)^2} \qquad \qquad \text{... (1)}$$

where g is the acceleration due to gravity at the surface of earth and R is the radius of earth.

On multiplying both sides of Eq.(1), with 'm', (mass of the body), we get

$$mg' = \frac{mg}{\left(1 + \frac{h}{R}\right)^2}$$

$\therefore$ Weight of body at height $h, W' = mg'$

Weight of body at surface of earth, $W = mg$

$\because \quad W' = \dfrac{1}{16}W \quad \therefore \quad \dfrac{1}{16} = \dfrac{1}{\left(1 + \frac{h}{R}\right)^2}$

or $\qquad \left(1 + \dfrac{h}{R}\right)^2 = 16 \quad \Rightarrow \quad 1 + \dfrac{h}{R} = 4$

or $\qquad \dfrac{h}{R} = 3 \quad \Rightarrow \quad h = 3R$

41. (C) Use Gauss Theorem for Gravitation.

1.26.3 Check Point 3

1. APPROACH In this problem, both particles are moving under the action of their mutual gravitational pull. There is no external force on the system, so the linear momentum of the system will be conserved.

Since, the gravitational force is conservative, therefore total mechanical energy of the system will also be conserved.

So, apply both, linear momentum conservation law and energy conservation law and determine the individual velocities and then relative velocity of the particles.

SOLUTION Suppose, at separation r, the velocities of particles having masses m_1 and m_2 are v_1 and v_2 respectively (Fig.1.242).

Considering the direction of v_1 as positive, law of conservation of linear momentum gives-

$$m_1 v_1 - m_2 v_2 = 0$$

or $\qquad m_1 v_1 = m_2 v_2 \qquad \qquad \text{... (1)}$

By conservation of mechanical energy, we have

$$U_i + K_i = U_f + K_f \qquad \qquad \text{... (2)}$$

Initially, the particles were at rest and the separation

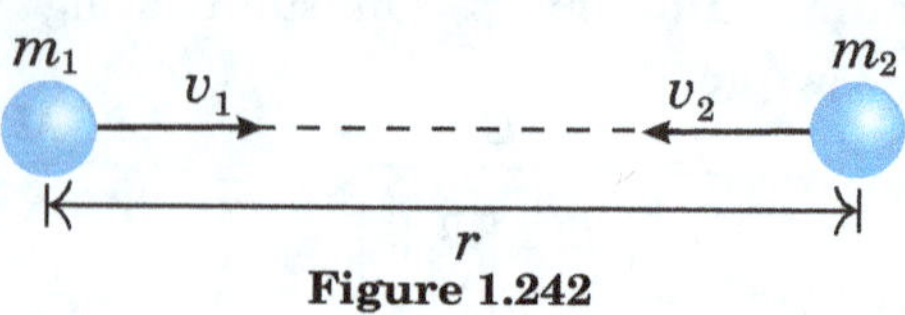

Figure 1.242

between them was infinite, therefore, $U_i = 0$, $K_i = 0$. At separation r, we have

So, $U_f = -\dfrac{Gm_1 m_2}{r}$ and $K_f = \dfrac{1}{2}m_1 v_1^2 + \dfrac{1}{2}m_2 v_2^2.$

On substituting these values, in Eq.(2), we get

or $\qquad 0 + 0 = -\dfrac{Gm_1 m_2}{r} + \dfrac{1}{2}m_1 v_1^2 + \dfrac{1}{2}m_2 v_2^2$

or $\qquad 0 = -\dfrac{Gm_1 m_2}{r} + \dfrac{1}{2}m_1 v_1^2 + \dfrac{1}{2}m_2 v_2^2$

or $\qquad \dfrac{2Gm_1 m_2}{r} = m_1 v_1^2 + m_2 v_2^2 \qquad \text{... (3)}$

On substituting the value of v_2 from Eq.(1), in (3), we get

$$\frac{2Gm_1 m_2}{r} = m_1 v_1^2 + m_2 \frac{m_1^2 v_1^2}{m_2^2} = m_1 v_1^2\left[1 + \frac{m_1}{m_2}\right]$$

or $\qquad \dfrac{2Gm_2}{r} = v_1^2\left[\dfrac{m_1 + m_2}{m_2}\right]$

or $\qquad v_1^2 = \dfrac{2Gm_2^2}{(m_1 + m_2)r}$

or $\qquad v_1 = m_2\sqrt{\dfrac{2G}{r(m_1 + m_2)}}$

Similarly, $v_2 = m_1\sqrt{\dfrac{2G}{r(m_1 + m_2)}}$

Therefore, relative velocity of approach

$$v_r = v_1 - (-v_2) = v_1 + v_2$$

$\qquad\qquad\qquad$ [since, v_2 is directed opposite to v_1]

$$v_r = m_2\sqrt{\frac{2G}{r(m_1 + m_2)}} + m_2\sqrt{\frac{2G}{r(m_1 + m_2)}}$$

or $\qquad v_r = \sqrt{\dfrac{2G(m_1 + m_2)}{r}}$

2. APPROACH Calculate initial and final potential energies and then their difference.

SOLUTION Let $m = 0.020$ kg and $d = 0.600$ m (the

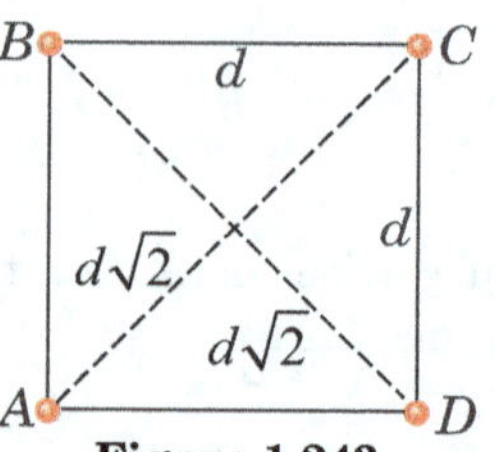

Figure 1.243

original edge-length, in terms of which the final edge-length is $d/3$). The total initial gravitational potential energy of the system (using Eq. 1.89 in Fig.1.243 and some elementary trigonometry) is

$$U_i = -\frac{4Gm^2}{d} - \frac{2Gm^2}{\sqrt{2}d}$$

After reducing the side to $d/3$, the potential energy becomes,

$$U_f = \frac{-4Gm^2}{d/3} - \frac{2Gm^2}{\sqrt{2}d/3}$$
$$= 3\left(-\frac{4Gm^2}{d} - \frac{2Gm^2}{\sqrt{2}d}\right) = 3U_i$$
$$\Rightarrow \quad \Delta U = U_f - U_i = 2U_i$$
$$= 2(4 + \sqrt{2})\left(-\frac{Gm^2}{d}\right) = -4.82 \times 10^{-13} J$$

3. **APPROACH** Let the zero of gravitational potential energy be at infinity, m represent the mass of the object, and h the maximum height reached by the object. We'll use conservation of energy to relate the initial potential and kinetic energies of the object-Earth system to the final potential energy.

SOLUTION Use conservation of energy to relate the initial potential energy of the system to its energy as the object is at its maximum height:

$$U_i + K_i = U_f + K_f \qquad \text{... (1)}$$

here, $U_i = -\dfrac{GMm}{R}$ (PE of object-earth system, when the object is at earth's surface), $K_i = \dfrac{1}{2}mv^2$ (KE of object at earth's surface), v is velocity of projection of the object at the surface of earth, $U_f = -\dfrac{GMm}{r} = -\dfrac{GMm}{(R+h)}$ (PE of earth-object system when the object is at height h over the surface of earth), $K_f = 0$ (KE of object at height h from the surface of earth where it's speed becomes zero). Substituting these values in Eq.(1), we get

$$\frac{1}{2}mv^2 - \frac{GMm}{R} + \frac{GMm}{R+h} = 0$$

Solving for h yields:

$$h = \frac{R}{\frac{2gR}{v^2} - 1}$$

Substitute numerical values and evaluate h :

$$h = \frac{6.37 \times 10^6 m}{\frac{2(9.81 m/s^2)(6.37 \times 10^6 m)}{(4.0 \times 10^3 m/s)^2} - 1}$$
$$= 9.4 \times 10^5 m$$

4. **APPROACH** Since, there is only gravitational force on the probe, so total mechanical energy of the system will be conserved and hence you can use the conservation of mechanical energy principle.

SOLUTION Energy conservation for this situation may be expressed as follows:

$$K_i + U_i = K_f + U_f \Rightarrow K_i - \frac{GmM}{r_i} = K_f - \frac{GmM}{r_f}$$

where $M = 5.0 \times 10^{23} kg, r_i = R = 3.0 \times 10^6 m$ and $m = 10\ kg$.

(a) If $K_i = 5.0 \times 10^7 J$ and $r_f = 4.0 \times 10^6 m$, then the above equation leads to

$$K_f = K_i + GmM\left(\frac{1}{r_f} - \frac{1}{r_i}\right) = 2.2 \times 10^7\ J$$

(b) In this case, we require $K_f = 0$ and $r_f = 8.0 \times 10^6 m$, and solve for K_i :

$$K_i = K_f + GmM\left(\frac{1}{r_i} - \frac{1}{r_f}\right) = 6.9 \times 10^7\ J.$$

5. **APPROACH** Since, work done by an external force against system's conservative force is equal to change in the potential energy of the system, therefore, work done by you will also be equal to change in potential energy of the system. Work done by gravity is equal to − ve of change in potential energy.

SOLUTION (a) The work done by you in moving the sphere of mass m_B equals the change in the potential energy of the three-sphere system. The initial potential of the system is

$$U_i = -\frac{Gm_A m_B}{d} - \frac{Gm_A m_C}{L} - \frac{Gm_B m_C}{L-d}$$

and the final potential energy is

$$U_f = -\frac{Gm_A m_B}{L-d} - \frac{Gm_A m_C}{L} - \frac{Gm_B m_C}{d}$$

The work done is

$$W = U_f - U_i$$
$$= Gm_B\left[m_A\left(\frac{1}{d} - \frac{1}{L-d}\right) + m_C\left(\frac{1}{L-d} - \frac{1}{d}\right)\right]$$
$$= Gm_B\left[m_A\frac{L-2d}{d(L-d)} + m_C\frac{2d-L}{d(L-d)}\right]$$
$$= Gm_B(m_A - m_C)\frac{L-2d}{d(L-d)}$$
$$= \left(6.67 \times 10^{-11}m^3/s^2 \cdot kg\right)(0.010 kg)$$
$$(0.080 kg - 0.020 kg)\frac{0.12m - 2(0.040m)}{(0.040m)(0.12 - 0.040m)}$$
$$= +5.0 \times 10^{-13} J$$

(b) The work done by the force of gravity is $-(U_f - U_i) = -5.0 \times 10^{-13}$ J.

6. **APPROACH** The work done by you against gravity to move the particle from a distance r_1 to r_2 is equal to the change in the particle's gravitational potential energy.

SOLUTION (a) Relate the work you must do to the change in the gravitational potential energy of Earth-particle system:

$$W = \Delta U = -\int_{r_1}^{r_2} \vec{F}_g \cdot d\vec{r}$$
$$= -\int_{r_1}^{r_2} F_g dr\,(\cos 180°) = \int_{r_1}^{r_2} F_g dr$$

Substitute for F_g and evaluate the integral to obtain:

$$W = GMm\int_{r_1}^{r_2}\frac{dr}{r^2} = -GMm\left(\frac{1}{r_2} - \frac{1}{r_1}\right)$$
$$= GMm\left(\frac{1}{r_1} - \frac{1}{r_2}\right)$$

(b) Substitute gR^2 for GM, R for r_1, and $R+h$ for r_2 to obtain:

$$W = mgR^2\left(\frac{1}{R} - \frac{1}{R+h}\right) \qquad \text{... (1)}$$

(c) Rewrite equation (1) with a common denominator and simplify to obtain:

$$W = mgR^2\left(\frac{R+h-R}{R(R+h)}\right) = mgh\left(\frac{R}{R+h}\right)$$
$$= mgh\left(\frac{1}{1+\frac{h}{R}}\right) \approx mgh, \quad \text{provided } h \ll R.$$

7. **APPROACH** Since, the radii of spheres are not mentioned, therefore we won't take internal potential energies of spheres into account and consider both spheres like point masses. The gravitational potential energy of two sphere system can be obtained by using Eq.1.79:

$$U(r) = -G\frac{m_1 m_2}{r} \quad \text{(zero at infinite r)} \qquad \text{... (1)}$$

Since, sphere B is moving under gravitational force only, which is conservative in nature, therefore we can apply the conservation of mechanical energy to determine the required kinetic energy of B.

SOLUTION (a) The initial gravitational potential energy of the two sphere system, is

$$U_i = -\frac{GM_A M_B}{r_i} = -\frac{\left(6.67 \times 10^{-11} m^3/s^2 \cdot kg\right)(20 kg)(10 kg)}{0.80 m}$$

$$= -1.67 \times 10^{-8} J \approx -1.7 \times 10^{-8} J$$

(b) We use conservation of energy (with $K_i = 0$):

$$U_i = K + U$$

$$\Rightarrow -1.7 \times 10^{-8} = K - \frac{\left(6.67 \times 10^{-11} m^3/s^2 \cdot kg\right)(20 kg)(10 kg)}{0.60 m}$$

which yields $K = 5.6 \times 10^{-9} J$.

Note that the value of r is the difference between $0.80 m$ and $0.20 m$.

8. **APPROACH** The two neutron stars are attracted toward each other due to their gravitational interaction. If we consider both stars in a single system, then net external force on it is zero. Therefore, the linear momentum of the two-star system is conserved, and since the stars have the same mass, their speeds and kinetic energies are the same. We use the principle of conservation of mechanical energy. The initial potential energy is $U_i = -GM^2/r_i$, where M is the mass of either star and r_i is their initial center-to-center separation. The initial kinetic energy is zero since the stars are at rest. The final potential energy is $U_f = -GM^2/r_f$, where the final separation is $r_f = r_i/2$. We write Mv^2 for the final kinetic energy of the system. This is the sum of two terms, each of which is $1/2Mv^2$. Conservation of energy yields

$$-\frac{GM^2}{r_i} = -\frac{2GM^2}{r_i} + Mv^2$$

SOLUTION (a) The solution for v is

$$v = \sqrt{\frac{GM}{r_i}} = \sqrt{\frac{\left(6.67 \times 10^{-11} m^3/s^2 \cdot kg\right)\left(10^{30} kg\right)}{10^{10} m}}$$

$$= 8.2 \times 10^4 m/s$$

(b) Now the final separation of the centers is $r_f = 2R = 2 \times 10^5 m$, where R is the radius of either of the stars. The final potential energy is given by $U_f = -GM^2/r_f$ and the energy equation becomes

$$-GM^2/r_i = -GM^2/r_f + Mv^2$$

The solution for v is

$$v = \sqrt{GM\left(\frac{1}{r_f} - \frac{1}{r_i}\right)}$$

$$= \sqrt{\left(6.67 \times 10^{-11} m^3/s^2 \cdot kg\right)\left(10^{30} kg\right)\left(\frac{1}{2 \times 10^5 m} - \frac{1}{10^{10} m}\right)}$$

$$= 1.8 \times 10^7 m/s$$

9. **APPROACH** Since, rocket is moving under gravitational force only, so you can apply the principle of conservation of mechanical energy.

SOLUTION Let v be the intial speed of the rocket which reaches a height h above the earth. The total energy of the rocket when it is on the Earth

$$KE + PE = \frac{1}{2}mv^2 + \left(-\frac{GMm}{R}\right)$$

At the highest point, the rocket will have zero speed and its potential energy will be $-\frac{GMm}{(R+h)}$. Applying the principle of conservation of energy, we get

$$\frac{1}{2}mv^2 - \frac{GMm}{R} = 0 - \frac{GMm}{(R-h)}$$

$$\Rightarrow \quad v^2 = \frac{2GMh}{R(R+h)}$$

But $GM = gR^2$, $\therefore \quad v^2 = \frac{2gR^2h}{R(R+h)} = \frac{2gh}{(R+h)}$

$$\Rightarrow \quad v^2 R + v^2 h = 2ghR$$

$$\Rightarrow \quad h = \frac{v^2}{2gr - v^2} = \frac{\left(5 \times 10^3\right)^2 \times 6.4 \times 10^6}{2 \times 9.8 \times 6.4 \times 10^6 - \left(5 \times 10^3\right)^2}$$

$$\Rightarrow \quad h = \frac{6.4 \times 25}{100.44} \times 10^6 = 1.6 \times 10^6 m$$

10. **APPROACH** Because we are to neglect the effects of the atmosphere on the asteroid, the mechanical energy of the asteroid-Earth system is conserved during the fall. Thus, the final mechanical energy (when the asteroid reaches Earth's surface) is equal to the initial mechanical energy. With kinetic energy K and gravitational potential energy U, we can write this as

$$K_f + U_f = K_i + U_i \qquad \qquad \text{... (1)}$$

Also, if we assume the system is isolated, the system's linear momentum must be conserved during the fall. Therefore, the momentum change of the asteroid and that of Earth must be equal in magnitude and opposite in sign. However, because Earth's mass is so much greater than the asteroid's mass, the change in Earth's speed is negligible relative to the change in the asteroid's speed. So, the change in Earth's kinetic energy is also negligible. Thus, we can assume that the kinetic energies in Eq. (1) are those of the asteroid alone.

SOLUTION Let m represent the asteroid's mass and M represent Earth's mass $(5.98 \times 10^{24}$ kg$)$. The asteroid is initially at distance $10R$ and finally at distance R, where R is Earth's radius $(6.37 \times 10^6$ m$)$. Substituting Eq. $U = -\frac{GMm}{r}$ for U and $\frac{1}{2}mv^2$ for K, we rewrite Eq. (1) as

$$\frac{1}{2}mv_f^2 - \frac{GMm}{R} = \frac{1}{2}mv_i^2 - \frac{GMm}{10R}$$

Rearranging and substituting known values, we find

$$v_f^2 = v_i^2 + \frac{2GM}{R}\left(1 - \frac{1}{10}\right)$$

$$= \left(12 \times 10^3 \text{m/s}\right)^2$$

$$+ \frac{2\left(6.67 \times 10^{-11} \text{m}^3/\text{kg} \cdot \text{s}^2\right)\left(5.98 \times 10^{24} \text{kg}\right)}{6.37 \times 10^6 \text{m}} 0.9$$

$$= 2.567 \times 10^8 \text{m}^2/\text{s}^2,$$

or $\quad v_f = 1.60 \times 10^4 \text{m/s} = 16$ km/s

At this speed, the asteroid would not have to be particularly large to do considerable damage at impact. If it were only $5\ m$ across, the impact could release about as much energy as the nuclear explosion at Hiroshima. Alarmingly, about 500 million asteroids of this size are near Earth's orbit, and in 1994 one of them apparently penetrated Earth's atmosphere and exploded 20 km above the South Pacific.

11. **APPROACH** Since, both stars are moving due to their mutual gravitational attraction only, therefore their total mechanical energy will be conserved.

SOLUTION Initial potential energy of the stars when they are 10^9 km apart

$$U_i = -\frac{GMm}{r} = -\frac{GMM}{10^{12}} = -\frac{GM^2}{10^{12}}$$

Final potential energy when the stars are just going to colloid

$$U_f = -\frac{GM^2}{2 \times 10^7}$$

[radius of each star is 10^4 km. Distance r between two stars when they collide $= 2R = 2 \times 10^4$ km $= 2 \times 10^7$ m]
Initial kinetic energy of the stars, $K_i = 0$
Final kinetic energy, $K_f = 2 \times \frac{1}{2}Mv^2 = Mv^2$
where v is the speed of each star just before collision.
According to the law of conservation of energy

$$(U_f + K_f)_{\text{system}} = (U_i + K_i)_{\text{system}}$$

$$\Rightarrow \quad -\frac{GM^2}{10^{12}} + 0 = -\frac{GM^2}{2 \times 10^7} + Mv^2$$

$$\Rightarrow \quad Mv^2 = \frac{GM^2}{2 \times 10^7} - \frac{GM^2}{10^{12}} \approx \frac{GM^2}{2 \times 10^7}$$

$$\Rightarrow \quad v = \sqrt{\frac{GM}{2 \times 10^7}}$$

$$= \sqrt{\frac{6.67 \times 10^{-11} \times 2 \times 10^{30}}{2 \times 10^7}} = 2.6 \times 10^6 m/s$$

12. **APPROACH** For gravitational field and potential, use superposition principle. If net gravitational field at P is zero, then the object placed at P will be in translational equilibrium. Now, displace the object from it's equilibrium position and observe whether the net force on this new location is acting towards point P or away from it. If it is towards the position of equilibrium point P, then the equilibrium is said to be stable and if it is away from P, then equilibrium is said to be unstable.

SOLUTION We know that gravitational field $g = \frac{GM}{r^2}$
If we have a point P mid-way between the two spheres A and B, the field at P due to A

$$= \frac{6.67 \times 10^{-11} \times 100}{(0.5)^2} = 2.67 \times 10^{-8} \text{ N, acting along } PA.$$

Similarly, field at P due to B

$$= \frac{6.67 \times 10^{-11} \times 100}{(0.5)^2} = 2.67 \times 10^{-8} \text{ N, acting along}$$

PB
Since the fields at P are equal in magnitude and opposite in direction, the strength of the gravitational field at P will be 0 (zero).
Now, potential at P due to A

$$= -\frac{GM}{r} = \frac{6.67 \times 10^{-11} \times 100}{0.5} = -1.334 \times 10^{-8} J/kg$$

Similarly, Potential at P due to B

$$= -\frac{GM}{r} = \frac{6.67 \times 10^{-11} \times 100}{0.5} = -1.334 \times 10^{-8} \text{ J/kg}$$

$\therefore$ Total gravitational potential at P due to spheres A and B

$$= -1.334 \times 10^{-8} + \left(-1.334 \times 10^{-8}\right) \text{ J/kg}$$
$$= -2.668 \times 10^{-8} \text{ J/kg} \approx 2.7 \times 10^{-9} \text{ J/kg}$$

Now, an object placed at P will be in equilibrium since there is no net force acting on it. But when it is slightly displaced from its position say towards A, the force acting on it due to A will be greater than that due to B. So it will be away from P and towards A. Hence the equilibrium at P will be an unstable equilibrium.

13. **APPROACH** Since, radius of asteroid is not mentioned in the problem, so we will consider it like a point mass and apply the law of conservation of mechanical energy.

SOLUTION Applying energy conservation, we have
Total energy of asteroid at 12 km = Total energy of asteroid at surface of earth

$$\Rightarrow \quad U_i + K_i = U_f + K_f$$

$$\Rightarrow \quad \frac{-GMm}{10R} + \frac{1}{2}mv_0^2 = \frac{-GMm}{R} + \frac{1}{2}mv^2$$

$$\Rightarrow \quad \frac{9}{10}\frac{GMm}{R} + \frac{1}{2}mv_0^2 = \frac{1}{2}mv^2$$

$$\Rightarrow \quad \frac{9}{10} \times \frac{2GM}{R} + v_0^2 = v^2$$

$$\Rightarrow \quad \frac{9}{10} \times \left(\sqrt{v_e}\right)^2 + v_0^2 = v^2$$

$$\Rightarrow \quad \frac{9}{10} \times (11.2)^2 + (12)^2 = v^2$$

$$\Rightarrow \quad v^2 = \sqrt{\frac{9}{10} \times (11.2)^2 + (12)^2} \Rightarrow v \approx 16 \text{ kms}^{-1}$$

14. (a) Gravitational potential at $A, V_A = -\frac{GM}{R}$ [Fig.1.244]
Potential at C

$$V_c = -\frac{GM}{R^3}\left[\frac{3}{2}R^2 - \frac{1}{2}\left(\frac{R}{2}\right)^2\right] = -\frac{11}{8}\frac{GM}{R}$$

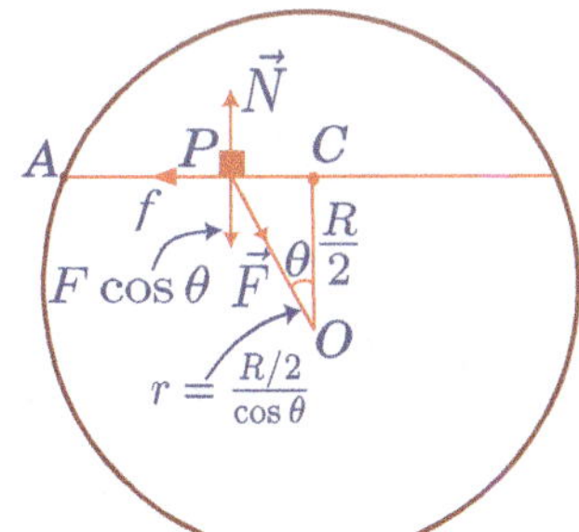

Figure 1.244

$\therefore$ Loss in gravitational PE of the block-earth system:

$$= -\frac{GMm}{R} - \left(-\frac{11GMm}{8R}\right) = \frac{3GMm}{8R}$$

$\therefore$ Work done by friction $= -\frac{3GMm}{8R}$
(b) At any intermediate position $P(\theta)$ shown in the Fig.1.244, $F\cos\theta = N, \vec{N} \to$ the normal reaction of tunnel. If g' is the gravitational acceleration at the intermediate point P (Fig.1.244), then-

$$mg' = \cos\theta = N \quad \Rightarrow \quad m\left(\frac{GM}{R^3}r\cos\theta = N\right)$$

$$\Rightarrow \quad \frac{GMm}{R^3}\left(\frac{R/2}{\cos\theta}\right)\cos\theta = N$$

$$\therefore \quad N = \frac{GMm}{2R^2} = a \text{ constant}$$

$$\therefore \quad \text{Work done by friction} = -\mu N(AC)$$

$$-\frac{3GMm}{8R} = -\mu\frac{GMm}{2R^2}\left(\frac{\sqrt{3}R}{2}\right)$$

$$\Rightarrow \qquad \mu = \frac{\sqrt{3}}{2}$$

Multiple Choice Questions

15. (D) Work done = Change in potential energy

$$= U_f - U_i = \frac{-GMm}{(R+h)} - \left(\frac{-GMm}{R}\right)$$

where M is the mass of earth and R is the radius of earth.

$$\therefore \qquad W = GMm\left[\frac{1}{R} - \frac{1}{(R+h)}\right]$$

Since, $h = R$

$$\therefore \qquad W = GMm\left[\frac{1}{R} - \frac{1}{2R}\right] = \frac{GMm}{2R}$$

$$\Rightarrow \qquad W = \frac{mgR}{2} \qquad \left[\because g = \frac{GM}{R^2}\right]$$

16. (C) : Gravitation potential at a height h from the surface of earth, $V_h = -5.4 \times 10^7$ J kg^{-1}.

At the same point acceleration due to gravity, $g_h = 6$ m s^{-2}, $R = 6400$ km $= 6.4 \times 10^6$ m
We know that , $V_h = -\frac{GM}{(R+h)}$,

$$\text{and} \quad g_h = \frac{GM}{(R+h)^2} = -\frac{V_h}{R+h} \Rightarrow R+h = -\frac{V_h}{g_h}$$

$$\therefore \quad h = -\frac{V_h}{g_h} - R = -\frac{(-5.4 \times 10^7)}{6} - 6.4 \times 10^6$$

$$= 9 \times 10^6 - 6.4 \times 10^6 = 2600 \text{ km}$$

17. (B) The resulting gravitational potential at the origin O due to each of mass 2 kg located at positions as shown in Fig.1.245, is

$$V = -\frac{G \times 2}{1} - \frac{G \times 2}{2} - \frac{G \times 2}{4} - \frac{G \times 2}{8} - \cdots$$

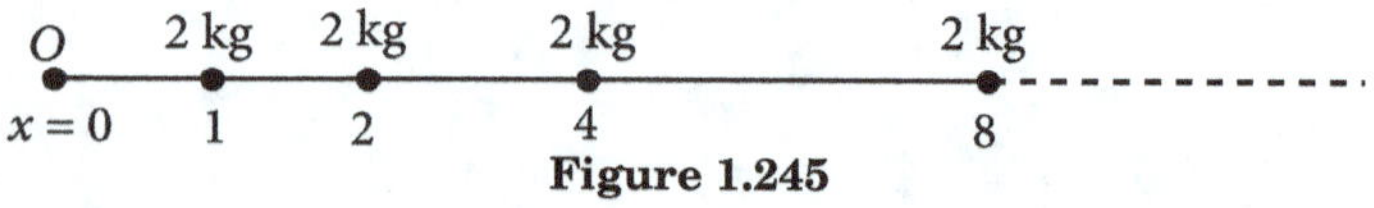

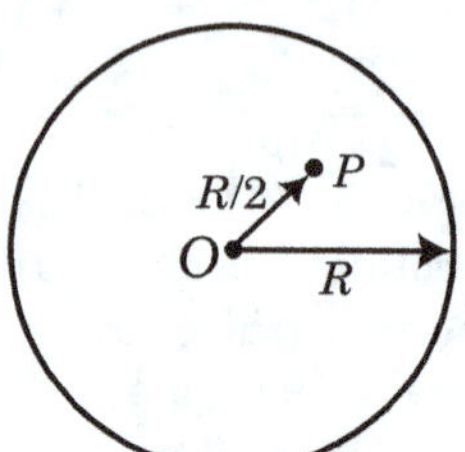

Figure 1.245

$$= -2G\left[1 + \frac{1}{2} + \frac{1}{4} + \frac{1}{8} + \ldots\ldots\right] = -2G\left[\frac{1}{1-\frac{1}{2}}\right]$$

$$= -2G\left[\frac{2}{1}\right] = -4G$$

18. (D) Gravitational potential energy at any point at a distance r from the centre of the earth is

$$U = -\frac{GMm}{r}$$

where M and m be masses of the earth and the body respectively. At the surface of the earth, $r = R; U_i = -\frac{GMm}{R}$
At a height h from the surface,

$$r = R + h = R + 2R = 3R \quad (h = 2R \text{ (Given)})$$

$$\therefore \qquad U_f = -\frac{GMm}{3R}$$

Change in potential energy, $\Delta U = U_f - U_i$

$$= -\frac{GMm}{3R} - \left(-\frac{GMm}{R}\right) = \frac{GMm}{R}\left(1 - \frac{1}{3}\right)$$

$$= \frac{2}{3}\frac{GMm}{R} = \frac{2}{3}mgR \qquad \left[\because g = \frac{GM}{R^2}\right]$$

19. (C) : Here, Mass of a particle $= M$,
Mass of a spherical shell $= M$,
Radius of a spherical shell $= R$
Let O be centre of a spherical shell[Fig.1.246].
    Gravitational potential at point P due to particle at O is

Figure 1.246

$$V_1 = -\frac{GM}{R/2}$$

Gravitational potential at point P due to spherical shell is

$$V_2 = -\frac{GM}{R}$$

Hence, total gravitational potential at point P is

$$V = V_1 + V_2$$
$$= \frac{-GM}{R/2} + \left(\frac{-GM}{R}\right) = \frac{-2GM}{R} - \frac{GM}{R} = \frac{-3GM}{R}$$
$$|V| = \frac{3GM}{R}$$

20. (C) Gravitational potential energy on earth's surface $= -\frac{GMm}{R}$, where M and R are the mass and radius of the earth respectively, m is the mass of the body and G is the universal gravitational constant.
Gravitational potential energy at a height $h = 3R$

$$= -\frac{GMm}{R+h} = -\frac{GMm}{R+3R} = -\frac{GMm}{4R}$$

$\therefore$ Change in potential energy,

$$= -\frac{GMm}{4R} - \left(-\frac{GMm}{R}\right) = -\frac{GMm}{4R} + \frac{GMm}{R}$$

$$= \frac{3}{4}\frac{GMm}{R} = \frac{3}{4}mgR$$

1.26.4   Check Point 4

1. **APPROACH** Choosing the zero of gravitational potential energy to be at infinite separation yields, as the potential energy of a two-body system in which the objects are separated by a distance r, $U(r) = -GMm/r$, where M and m are the masses of the two bodies. In order for an object to just escape a gravitational field from a particular location, it must have enough kinetic energy so that its total energy is zero.
SOLUTION (a) Letting $U(\infty) = 0$, express the gravitational potential energy of the Earth-object system:

$$U(r) = -\frac{GMm}{r} \qquad \ldots (1)$$

Substitute for GM and simplify to obtain:

$$U(R) = -\frac{GMm}{R} = -\frac{gR^2 m}{R} = -mgR$$

Substitute numerical values and evaluate $U(R_E)$:
$$U(R) = -(100 \text{ kg})(9.81 \text{ N/kg})(6.37 \times 10^6 \text{m})$$
$$= -6.25 \times 10^9 \text{J}$$

(b) Evaluate equation (1) with $r = 2R_E$
$$U(2R) = -\frac{GMm}{2R} = -\frac{gR^2 m}{2R}$$
$$= -\frac{1}{2}mgR$$

Substitute numerical values and evaluate $U(2R_E)$:
$$U(2R) = -\frac{1}{2}(100 kg)(9.81 N/kg)(6.37 \times 10^6 m)$$
$$= -3.124 \times 10^9 J = -3.12 \times 10^9 \ J$$

(c) Express the condition that an object must satisfy in order to escape from Earth's gravitational field from a height R_E above its surface:
$$K_e(2R) + U(2R) = 0$$
or
$$\frac{1}{2}mv_e^2 + U(2R) = 0$$

Solving for v_e yields:
$$v_e = \sqrt{\frac{-2U(2R)}{m}}$$

Substitute numerical values and evaluate v_e :
$$v_e = \sqrt{\frac{-2(-3.124 \times 10^9 J)}{100 kg}} = 7.90 \text{ km / s}$$

2. **APPROACH** We can use its definition to express the period of the spacecraft's motion and apply Newton's second law to the spacecraft to determine its orbital speed. We can then use this orbital speed to calculate the kinetic energy of the spacecraft. We can relate the spacecraft's angular momentum to its kinetic energy and moment of inertia.

SOLUTION (a) Express the period of the spacecraft's orbit about Earth:
$$T = \frac{2\pi r}{v} = \frac{2\pi(3R)}{v} = \frac{6\pi R}{v}$$
where v is the orbital speed of the spacecraft.

Use Newton's second law to relate the gravitational force acting on the spacecraft to its orbital speed:
$$F_{\text{radial}} = \frac{GMm}{(3R)^2} = \frac{mv^2}{3R} \Rightarrow v = \sqrt{\frac{gR}{3}}$$

Substitute for v in our expression for T to obtain:
$$T = 6\sqrt{3}\pi\sqrt{\frac{R}{g}}$$

Substitute numerical values and evaluate T
$$T = 6\sqrt{3}\pi\sqrt{\frac{6.37 \times 10^6 \text{ m}}{9.81 \text{ m/s}^2}}$$
$$= 2.631 \times 10^4 \text{s} \times \frac{1 \text{ h}}{3600 \text{ s}} = 7.31 \text{ h}$$

(b) Using its definition, express the spacecraft's kinetic energy:
$$K = \tfrac{1}{2}mv^2 = \tfrac{1}{2}m\left(\tfrac{1}{3}gR\right)$$

Substitute numerical values and evaluate K:
$$K = \tfrac{1}{6}(100 \text{ kg})(9.81 \text{m/s}^2)(6.37 \times 10^6 m)$$
$$= 1.041 \text{ GJ} = 1.04 \text{ GJ}$$

(c) Express the kinetic energy of the spacecraft in terms of its angular momentum:
$$K = \frac{L^2}{2l} \Rightarrow L = \sqrt{2IK}$$

Express the moment of inertia of the spacecraft with respect to an axis through the centre of Earth:
$$I = m(3R)^2 = 9mR^2$$

Substitute for I in the expression for L and simplify to obtain:
$$L = \sqrt{2(9mR^2)K} = 3R\sqrt{2mK}$$

Substitute numerical values and evaluate L:
$$L = 3(6.37 \times 10^6 \text{m})\sqrt{2(100 \text{ kg})(1041 \times 10^9 \text{ J})}$$
$$= 8.72 \times 10^{12} \text{ J.s}$$

3. **APPROACH** We can express the energy difference between these two orbits in terms of the total energy of a satellite at each elevation. The application of Newton's second law to the force acting on a satellite will allow us to express the total energy of each satellite as a function of its mass, the radius of Earth, and its orbital radius.

SOLUTION Express the energy difference:
$$\Delta E = E_{geo} - E_{1000} \qquad \qquad \dots (1)$$

Express the total energy of an orbiting satellite:
$$E_{\text{tot}} = K + U = \frac{1}{2}mv^2 - \frac{GMm}{r} \qquad \dots (2)$$
where r is the orbital radius.

Apply Newton's second law to a satellite to relate the gravitational force to the orbital speed:
$$F_{\text{radial}} = \frac{GMm}{r^2} = \frac{mv^2}{r}$$

Solving for v^2 yields:
$$v^2 = \frac{gR^2}{r}$$

Substitute in equation (2) to obtain:
$$E_{\text{total}} = \frac{mgR^2}{r} - \frac{gRm}{r} = -\frac{mgR^2}{2r}$$

Substituting in equation (1) and simplifying yields:
$$\Delta E = -\frac{mgR^2}{2r_{\text{geo}}} + \frac{mgR^2}{2r_{1000}} = \frac{mgR^2}{2}\left(\frac{1}{r_{1000}} - \frac{1}{r_{\text{geo}}}\right)$$
$$= \frac{mgR^2}{2}\left(\frac{1}{R + 1000 \text{ km}} - \frac{1}{R + 35790 \text{ km}}\right)$$

Substitute numerical values and evaluate ΔE:
$$\Delta E = \frac{(500 \text{kg})(9.81 \text{N/kg})(6.37 \times 10^6 \text{m})^2}{2}\left(\frac{1}{7.37 \times 10^6 \text{m}} - \frac{1}{4.22 \times 10^7 \text{m}}\right)$$
$$= 11.1 \text{ GJ}$$

4. **APPROACH** The escape speed from the moon is given by $v_{e,m} = \sqrt{2GM_m/R_m}$, where M_m and R_m represent the mass and radius of the moon, respectively.

SOLUTION The escape speed from the moon:
$$v_{\text{e.m}} = \sqrt{\frac{2GM_{\text{m}}}{R_{\text{m}}}} = \sqrt{2g_{\text{m}}R_{\text{m}}}$$

Because $g_{\text{m}} = 0.166g$ and $R_m = 0.273R$
$$v_{em} = \sqrt{2(0.166g)(0.273R)}$$

Substitute numerical values and evaluate $v_{e,m}$:
$$v_{\text{e.m}} = \sqrt{2(0.166)(9.81 \text{ m/s}^2)(0.273)(6.371 \times 10^6 \text{m})}$$
$$= 2.38 \text{ km/s}$$

5. **APPROACH** Let the zero of gravitational potential energy be at infinity, m represent the mass of the particle, M and R are the mass and radius of earth. When the particle is very far from Earth, the gravitational potential energy of the Earth-particle system is zero. We'll use conservation of energy to relate the initial potential and

kinetic energies of the particle-Earth system to the final kinetic energy of the particle.

SOLUTION Use conservation of mechanical energy to relate the initial energy of the system to its energy when the particle is very far away:

$$K_f + U_f = K_i + U_i$$

Since, $U_f = 0$, therefore,

$$K_f = K_i + U_i$$

$$\Rightarrow \quad \frac{1}{2}mv_\infty^2 = \frac{1}{2}mv_i^2 + \frac{GMm}{R}$$

Since, $GM = gR^2$,

$$\frac{1}{2}mv_\infty^2 - \frac{1}{2}mv_i^2 + mgR = 0$$

Solving for v_i yields:

$$v_i = \sqrt{v_\infty^2 + 2gR}$$

Substituting numerical values and evaluate v_i :

$$v_i = \sqrt{\left(11.2 \times 10^3 m/s\right)^2 + 2\left(9.81 m/s^2\right)\left(6.37 \times 10^6 m\right)}$$
$$= 15.8 \text{ km / s}$$

6. APPROACH Let the zero of gravitational potential energy be at infinity and let m represent the mass of the object. We'll use conservation of energy to relate the initial potential energy of the object-Earth system to the final potential and kinetic energies.

SOLUTION Use conservation of mechanical energy to relate the initial potential energy of the system to its energy as the object is about to strike Earth:

$$K_f + U_f = K_i + U_i = 0$$

Since, $K_i = 0$, therefore

$$K(R) + U(R) = U(R + h) \qquad \ldots (1)$$

where h is the initial height above Earth's surface and R is the radius of earth.

The potential energy of the object-Earth system when the object is at a distance r from the surface of Earth:

$$U(r) = -\frac{GMm}{r}$$

Substituting in equation (1), we get:

$$\frac{1}{2}mv^2 - \frac{GMm}{R} = \frac{GMm}{R+h} = 0$$

Solving for v yields:

$$v = \sqrt{2\left(\frac{GM}{R} - \frac{GM}{R+h}\right)}$$
$$= \sqrt{2gR\left(\frac{h}{R+h}\right)}$$

Substitute numerical values and evaluate v :

$$v = \sqrt{\frac{2\left(9.81 m/s^2\right)\left(6.37 \times 10^6 m\right)\left(4.0 \times 10^6 m\right)}{6.37 \times 10^6 m + 4.0 \times 10^6 m}}$$
$$= 6.9 \text{ km/s}$$

7. APPROACH Let the zero of gravitational potential energy be at infinity, m represent the mass of the particle, and M, R are respectively the mass and radius of earth. When the particle is very far from Earth, the gravitational potential energy of Earth-particle system will be zero. We'll use conservation of energy to relate the initial potential and kinetic energies of the particle-Earth system to the final kinetic energy of the particle.

SOLUTION Use conservation of energy to relate the initial energy of the system to its energy when the particle is very far from Earth:

$$K_f + U_f = K_i + U_i$$

since, $U_f = 0$, therefore

$$K(\infty) = K(R) + U(R) = 0 \qquad \ldots (1)$$

Substituting given values in equation (1), we get

$$\frac{1}{2}mv_\infty^2 = \frac{1}{2}m(2v_e)^2 - \frac{GMm}{R}$$

Since, $GM = gR^2$, therefore

$$\frac{1}{2}mv_\infty^2 = 2mv_e^2 - mgR = 0$$

Solving for v_∞ yields:

$$v_\infty = \sqrt{2\left(2v_e^2 - gR\right)}$$

On substituting numerical values in above equation, we get

$$v_\infty = \sqrt{2\left[2\left(11.2 \times 10^3 m/s\right)^2 - \left(9.81 m/s^2\right)\left(6.37 \times 10^6 m\right)\right]}$$
$$= 19.4 km/s$$

8. APPROACH The pictorial representation summarizes the initial positions of the Sun, Earth, and rocket.

SOLUTION From Problem 90, the escape speed from

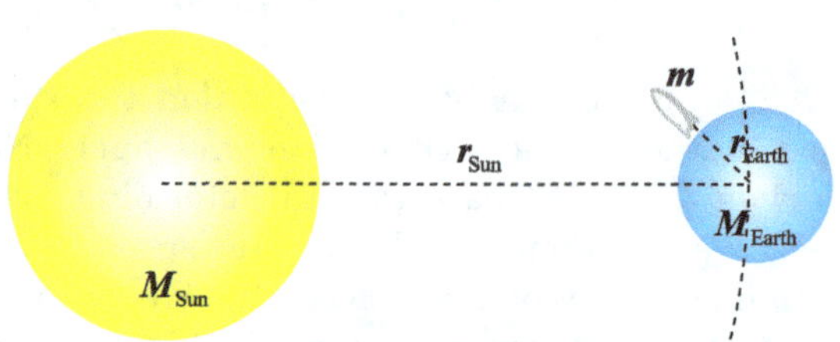

Figure 1.247

the Earth-Sun system is given by:

$$v_e^2 = \frac{2GM_{Earth}}{r_{Earth}} + \frac{2GM_{Sun}}{r_{Sun}}$$

Solving for v_e yields:

$$v_e = \sqrt{2G\left(\frac{M_{Earth}}{r_{Earth}} + \frac{M_{Sun}}{r_{Sun}}\right)}$$

Substitute numerical values and evaluate v_e :

$$v_e = \left[2\left(6.673 \times 10^{-11} N \cdot m^2/kg^2\right)\left(\frac{5.98 \times 10^{24} kg}{6.37 \times 10^6 m} + \frac{1.99 \times 10^{30} kg}{1.496 \times 10^{11} m}\right)\right]^{1/2}$$
$$= 43.6 \text{ km / s}$$

What we have just calculated is the escape speed from Earth's surface, at Earth's orbit. Because the launch will take place from a moving Earth, we need to consider Earth's motion and use it to our advantage. If we launch at sunrise from the Equator, zenith will be pointed directly along the direction of motion of Earth and the required launch speed will be the minimum launch speed.

The launch speed in terms of the escape speed calculated above and Earth's orbital speed:

$$v_{\min} = v_e - v_{orbital}$$

The orbital speed of Earth is given by:

$$v_{orbital} = \frac{2\pi r_{orbital}}{T_{orbital}}$$

Substituting for $v_{orbital}$ yields:

$$v_{\min} = v_e - \frac{2\pi r_{\text{orbital}}}{T_{\text{orbital}}}$$

Substitute numerical values and evaluate $v_{\min}$:

$$v_{\min} = 43.6\,km/s - \frac{2\pi\left(1.496\times 10^{11}m\right)}{1y\times\frac{365.24d}{y}\times\frac{8.64\times 10^4 s}{d}}$$

$$= 13.8 \text{ km}/\text{s}$$

9. From Fig.1.248, we have
$$r\cos 60^\circ = R \Rightarrow r = 2R$$

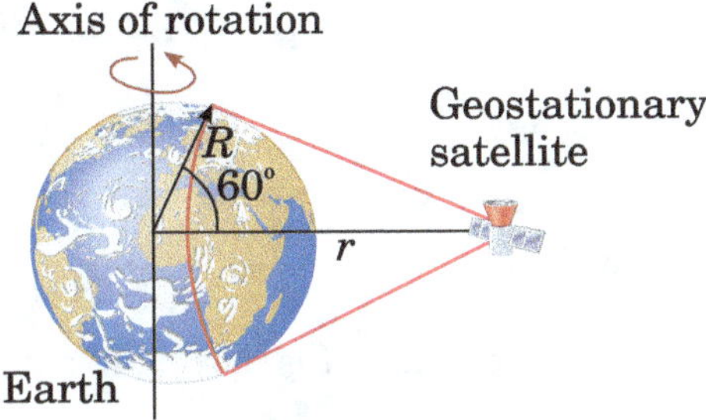

Figure 1.248

Orbital velocity, $v_0 = \sqrt{\dfrac{GM}{r}} = \sqrt{\dfrac{GM}{2R}} \Rightarrow \alpha = 2$

Multiple Choice Questions

10. (D) As escape velocity, $v = \sqrt{\dfrac{2GM}{R}}$

$$= \sqrt{\frac{2G}{R}\cdot\frac{4\pi R^3}{3}\rho} = R\sqrt{\frac{8\pi G}{3}\rho}$$

$$\therefore \frac{v_e}{v_p} = \frac{R_e}{R_p}\times\sqrt{\frac{\rho_e}{\rho_p}} = \frac{1}{2}\times\sqrt{\frac{1}{2}} = \frac{1}{2\sqrt{2}}$$

$$\left(\therefore R_p = 2R_e \text{ and } \rho_p = 2\rho_e\right)$$

11. (B) Here, $R_p = 2R_e, \rho_e = \rho_p$ Escape velocity of the earth,

$$v_e = \sqrt{\frac{2GM_e}{R_e}} = \sqrt{\frac{2G}{R_e}\left(\frac{4}{3}\pi R_e^3\rho_e\right)} = R_e\sqrt{\frac{8}{3}\pi G\rho_e} \quad \ldots(1)$$

Escape velocity of the planet,

$$v_p = \sqrt{\frac{2GM_p}{R_p}} = \sqrt{\frac{2G}{R_p}\left(\frac{4}{3}\pi R_p^3\rho_p\right)} = R_p\sqrt{\frac{8}{3}\pi G\rho_p} \quad \ldots(2)$$

Divide (1) by (2), we get $\qquad \dfrac{v_e}{v_p} = \dfrac{R_e}{R_p}\sqrt{\dfrac{\rho_e}{\rho_p}} =$

$\dfrac{R_e}{2R_e}\sqrt{\dfrac{\rho_e}{\rho_e}} = \dfrac{1}{2}$ or $v_p = 2v_e$

12. (A) The minimum speed with which the particle should be projected from the surface of the earth so that it does not return back is known as escape speed and it is given by

$$v_e = \sqrt{\frac{2GM}{(R+h)}}$$

Here, $h = 3R$

$$\therefore\qquad v_e = \sqrt{\frac{2GM}{(R+3R)}} = \sqrt{\frac{2GM}{4R}} = \sqrt{\frac{GM}{2R}}$$

13. (B) According to law of conservation of mechanical energy,

$$\frac{1}{2}mu^2 - \frac{GMm}{R} = 0 \text{ or } u^2 = \frac{2GM}{R}$$

$$u = \sqrt{\frac{2GM}{R}} = \sqrt{2gR} \quad \left(\because g = \frac{GM}{R^2}\right)$$

14. (C) Escape velocity of the body from the surface of earth is $\qquad v_e = \sqrt{2gR} \qquad \ldots(1)$

For escape velocity of the body from the platform, potential energy + kinetic energy = 0

$$-\frac{GMm}{2R} + \frac{1}{2}mv_e^2 = 0$$

$$\Rightarrow\qquad v_e = \sqrt{\frac{GM}{R^2}\cdot R} = \sqrt{gR}$$

$$\Rightarrow\qquad fv = \sqrt{gR} \qquad \ldots(2)$$

From equation (i) and (ii), we get $f = \dfrac{1}{\sqrt{2}}$

15. (A) By conservation of mechanical energy, we have

$$U_f + K_f = U_i + K_i$$

$$\Rightarrow\quad -\frac{GMm}{R+h} + 0 = -\frac{GMm}{R} + \frac{1}{2}mv^2$$

$$\Rightarrow\quad -\frac{GM}{2R} = -\frac{GM}{R} + \frac{1}{2}v^2 \qquad (\text{since, } h=R)$$

$$\Rightarrow\quad v^2 = \frac{GM}{R} \quad\Rightarrow\quad v = \sqrt{\frac{GM}{R}}$$

16. (B) $v_e = \sqrt{2gR_e} = \sqrt{\dfrac{2GM}{R_e}}$

$\because \quad R_P = \dfrac{1}{4}R_e$, therefore, $\quad v_p = 2v_e = 2\times 11.2 = 22.4$ km/s.

17. (A) Escape velocity of a body (v_e) = 11.2 km/s; New mass of the earth $M' = 2M$ and new radius of the earth $R' = 0.5R$.

Escape velocity $(v_e) = \sqrt{\dfrac{2GM}{R}} \propto \sqrt{\dfrac{M}{R}}$

Therefore $\dfrac{v_e}{v_e'} = \sqrt{\dfrac{M}{R}\times\dfrac{0.5R}{2M}} = \sqrt{\dfrac{1}{4}} = \dfrac{1}{2}$

or, $v_e' = 2v_e = 22.4$ km/s

18. (B) If earth is considered as an isolated system, then escape velocity does not depend on the angle of projection.

19. (A)

20. (A) From Eq.1.125, the time period of Geostationary satellite is,

$$T = 2\pi\sqrt{\frac{r^3}{GM}} \quad\Rightarrow\quad T^2 \propto a^3$$

$$\therefore\quad \frac{T_1^2}{T_2^2} = \frac{r_1^3}{r_2^3} \Rightarrow \frac{(24)^2}{T_2^2} = \frac{(7R)^3}{(3.5R)^3}$$

$$\Rightarrow\quad T_2^2 = \frac{(24)^2\times(3.5)^3}{(7)^3} \Rightarrow T_2 = \frac{\sqrt{(24)^2}}{\sqrt{8}} = 6\sqrt{2}\text{ h}.$$

21. (C) From Eq.1.122, the orbital speed of the satellite is,

$$v_o = R\sqrt{\frac{g}{(R+h)}}$$

where R is the earth's radius, g is the acceleration due to gravity on earth's surface and h is the height above the surface of earth.

Here, $R = 6.38\times 10^6$ m, $g = 9.8$ m s^{-2}, $h = 0.25\times 10^6$ m

$$\therefore v_o = \left(6.38 \times 10^6 \text{ m}\right) \sqrt{\frac{(9.8 \text{ m s}^{-2})}{(6.38 \times 10^6 \text{ m} + 0.25 \times 10^6 \text{ m})}}$$

$$= 7.76 \times 10^3 \text{ m s}^{-1} = 7.76 \text{ km s}^{-1}$$

22. (B) The gravitational force on the satellite S acts towards the centre of the earth, so the acceleration of the satellite S is always directed towards the centre of the earth.

23. (D) Escape velocity, $v_e = \sqrt{\frac{2GM}{R}}$ ...(1)

where M and R be the mass and radius of the earth respectively.

The orbital velocity of a satellite close to the earth's surface is

$$v_o = \sqrt{\frac{GM}{R}} \qquad \text{...(2)}$$

From (1) and (2), we get $v_e = \sqrt{2} v_o$

24. (B) Orbital speed of the satellite around the earth is

$$v = \sqrt{\frac{GM}{r}}$$

For satellite A, $r_A = 4R$, $v_A = 3v$ ...(1)

$$v_A = \sqrt{\frac{GM}{r_A}}$$

For satellite B, $r_B = R$, $v_B = ?$

$$v_B = \sqrt{\frac{GM}{r_B}} \qquad \text{...(2)}$$

Dividing equation (2) by equation (1), we get

$$\frac{v_B}{v_A} = \sqrt{\frac{r_A}{r_B}} \quad \text{or} \quad v_B = v_A \sqrt{\frac{r_A}{r_B}}$$

Substituting the given values, we get

$$v_B = 3v \sqrt{\frac{4R}{R}} \quad \text{or} \quad v_B = 6v$$

25. (C) Since no external torque is applied about the centre of earth, therefore, according to law of conservation of angular momentum, the ball will continue to move with the same angular velocity along the original orbit of the spacecraft.

26. (D) From Eq.1.125, the time period of a satellite at distance r from the centre of earth is,

$$T = 2\pi \sqrt{\frac{r^3}{Gm}}$$

From above it is clear that the time period of satellite does not depend on its mass. Therefore,

$$T^2 \propto r^3$$

$$\frac{T_A}{T_B} = \frac{r^{3/2}}{2^{3/2} r^{3/2}} = \frac{1}{2\sqrt{2}}$$

27. (B) : Total energy of satellite at height h from the earth's surface,

$$E = PE + KE = -\frac{GMm}{(R+h)} + \frac{1}{2}mv^2 \qquad \text{...(1)}$$

Also, $\dfrac{mv^2}{(R+h)} = \dfrac{GMm}{(R+h)^2}$

or $v^2 = \dfrac{GM}{R+h}$ ...(2)

From Eq.(1) and (2),

$$E = -\frac{GMm}{(R+h)} + \frac{1}{2}\frac{GMm}{(R+h)} = -\frac{1}{2}\frac{GMm}{(R+h)}$$

$$= -\frac{1}{2}\frac{GM}{R^2} \times \frac{mR^2}{(R+h)} = -\frac{mgR^2}{2(R+h)}$$

28. (D) Net energy of satellite-earth system in orbit of radius R_1, is

$$E_1 = U_1 + K_1 = -\frac{GMm}{R_1} + \frac{1}{2}\frac{GMm}{R_1} = -\frac{GMm}{2R_1}$$

Similarly, for orbit of radius R_2,

$$E_2 = -\frac{GMm}{2R_2}$$

So, needed energy for orbit transfer from R_1 to R_2,

$$\Delta KE = E_2 - E_1 = \frac{GMm}{2}\left(\frac{1}{R_1} - \frac{1}{R_2}\right)$$

29. (B) : The satellite of mass m is moving in a circular orbit of radius r.

$\therefore$ Kinetic energy of the satellite,

$$K = \frac{GMm}{2r} \qquad \text{...(1)}$$

Potential energy of the satellite,

$$U = \frac{-GMm}{r} \qquad \text{...(2)}$$

Orbital speed of satellite,

$$v = \sqrt{\frac{GM}{r}} \qquad \text{...(3)}$$

From Eq.1.125, the time-period of satellite,

$$T = \left[\left(\frac{4\pi^2}{GM}\right)r^3\right]^{1/2} \qquad \text{...(4)}$$

Given $m_{S_1} = 4m_{S_2}$

Since M, r is same for both the satellites S_1 and S_2.

$\therefore$ From Eq.(2), we get

$U \propto m$

$\therefore$ $\dfrac{U_{S_1}}{U_{S_2}} = \dfrac{m_{S_1}}{m_{S_2}} = 4$ or, $U_{S_1} = 4U_{S_2}$.

So, option (A) is wrong.

From Eq.(3), since v is independent of the mass of a satellite, the orbital speed is same for both satellites S_1 and S_2.

Hence option (B) is correct.

From (1), we get $K \propto m$

$\therefore$ $\dfrac{K_{S_1}}{K_{S_2}} = \dfrac{m_{S_1}}{m_{S_2}} = 4$ or $K_{S_1} = 4K_{S_2}$

Hence option (C) is wrong.

From Eq.(4), since T is independent of the mass of a satellite, therefore, time period is same for both the satellites S_1 and S_2. Hence option (D) is wrong.

30. (A) $K = \dfrac{GMm}{2R}$; $U = -\dfrac{GMm}{R}$

$\therefore$ $K = \dfrac{|U|}{2}$ or, $\dfrac{\text{K.E.}}{|U|} = \dfrac{1}{2}$

31. (D) Total energy $= -K = -\dfrac{1}{2}mv^2$

32. (A) $\dfrac{GMm}{r^2} = m\omega^2 r \Rightarrow r^3 = \dfrac{GM}{\omega^2} = \dfrac{gR^2}{\omega^2}$

$\therefore$ $r = \left(gR^2/\omega^2\right)^{1/3}$

1.26.5 Check Point 5

1. **APPROACH** The satellite moves in an elliptical orbit about Earth. An elliptical orbit can be characterized by its semi-major axis and eccentricity.

The greatest distance between the satellite and Earth's center (the apogee distance) and the least distance (perigee distance) are, respectively,

$$R_a = R_E + d_a = 6.37 \times 10^6 m + 360 \times 10^3 m = 6.73 \times 10^6 m$$

$$R_p = R_E + d_p = 6.37 \times 10^6 m + 180 \times 10^3 m = 6.55 \times 10^6 m$$

Here, d_a, d_p are respectively the aphelion and perihelion distances from earth's surface. $R_E = 6.37 \times 10^6 m$ is the radius of Earth.

SOLUTION From Eq.1.143, the semi-major axis is given by

$$a = \frac{R_a + R_p}{2} = \frac{6.73 \times 10^6 m + 6.55 \times 10^6 \text{ m}}{2} = 6.64 \times 10^6 \text{ m}$$

(b) From Eq.1.144, the eccentricity of orbit is given by-

$$e = \frac{R_a - R_p}{R_a + R_p} = \frac{6.73 \times 10^6 \text{m} - 6.55 \times 10^6 \text{m}}{6.73 \times 10^6 \text{m} + 6.55 \times 10^6 \text{m}} = 0.0136$$

Since e is very small, the orbit is nearly circular. On the other hand, if e is close to unity, then the orbit would be a long, thin ellipse.

2. **APPROACH** (a) The distance from the center of an ellipse to a focus is ae where a is the semimajor axis and e is the eccentricity. Thus, the separation of the foci (in the case of Earth's orbit) is

$$2ae = 2\left(1.50 \times 10^{11} \text{ m}\right)(0.0167) = 5.01 \times 10^9 \text{ m}$$

(b) To express this in terms of solar radii, we set up a ratio:

$$\frac{5.01 \times 10^9 \text{m}}{6.96 \times 10^8 \text{m}} = 7.20$$

3. **SOLUTION** From Kepler's law of periods (where $T = (2.4 \text{ h})(3600 \text{ s/h}) = 8640 \text{ s}$), we find the planet's mass M:

$$T^2 = \left(\frac{4\pi^2}{GM}\right) a^3$$

$$\Rightarrow \quad (8640 \text{ s})^2 = \left(\frac{4\pi^2}{GM}\right)\left(8.0 \times 10^6 \text{ m}\right)^3 \quad \Rightarrow \quad M = 4.06 \times 10^{24} \text{ kg}$$

However, we also know $g_p = GM/R^2 = 8.0 \text{ m/s}^2$ so that we are able to solve for the planet's radius:

$$R = \sqrt{\frac{GM}{g_p}}$$

$$= \sqrt{\frac{\left(6.67 \times 10^{-11} \text{m}^3/\text{kg} \cdot \text{s}^2\right)\left(4.06 \times 10^{24}\text{kg}\right)}{8.0 \text{ m/s}^2}}$$

$$= 5.8 \times 10^6 \text{ m}$$

4. **SOLUTION** When the distance is in astronomical units and time in years, then by Kepler's third law, we have

$$T_{\text{Pluto}}^2 = r_{\text{Pluto}}^3 \Rightarrow T_{\text{Pluto}} = \sqrt{r_{\text{Pluto}}^3}$$

Substitute numerical value of r, we get

$$T_{\text{Pluto}} = \sqrt{(39.5)^3} \approx 249 \text{ y}$$

5. The two stars are in circular orbits, not about each other, but about the two-star system's centre of mass (denoted as O), which lies along the line connecting the centres of the two stars. The gravitational force between the stars provides the centripetal force necessary to keep their orbits circular. Thus, for the visible star, Newton's second law gives

$$F = \frac{Gm_1 m_2}{r^2} = \frac{m_1 v^2}{r_1}$$

where r is the distance between the centres of the stars. To find the relation between r and r_1, we use the definition of centre of mass. By definition of centre of mass, we have

$$m_1 r_1 = m_2 r_2 \Rightarrow \frac{r_1}{m_2} = \frac{r_2}{m_1} = \frac{r_1 + r_2}{m_1 + m_2} = \frac{r}{m_1 + m_2}$$

$$\Rightarrow \quad r = \frac{m_1 + m_2}{m_2} r_1$$

On the other hand, since the orbital speed of m_1 is $v = 2\pi r_1/T$, therefore, $r_1 = vT/2\pi$ and the expression for r can be rewritten as

$$r = \frac{m_1 + m_2}{m_2} \frac{vT}{2\pi}$$

Substituting r and r_1 into the force equation, we get

$$F = \frac{4\pi^2 G m_1 m_2^3}{(m_1 + m_2)^2 v^2 T^2} = \frac{2\pi m_1 v}{T}$$

or

$$\frac{m_2^3}{(m_1 + m_2)^2} = \frac{v^3 T}{2\pi G}$$

$$= \frac{\left(2.7 \times 10^5 m/s\right)^3 (1.70 \text{ days })(86400s/ \text{ day })}{2\pi \left(6.67 \times 10^{-11} m^3/kg \cdot s^2\right)}$$

$$= 6.90 \times 10^{30} \text{ kg}$$

$$= 3.467 M_s,$$

where $M_s = 1.99 \times 10^{30} kg$ is the mass of the sun. With $m_1 = 6M_s$, we write $m_2 = \alpha M_s$ and solve the following cubic equation for α :

$$\frac{\alpha^3}{(6 + \alpha)^2} - 3.467 = 0$$

The equation has one real solution: $\alpha = 9.3$, which implies $m_2/M_s \approx 9$.

6. For any planet,

$$MR\omega^2 = \frac{GMM_S}{R^2} \quad \text{or} \quad \omega = \sqrt{\frac{GM_S}{R^3}}$$

So,

$$T = \frac{2\pi}{\omega} = \frac{2\pi R^{3/2}}{\sqrt{GM_S}}$$

(a) Thus, So,

$$\frac{T_J}{T_E} = \left(\frac{R_J}{R_E}\right)^{3/2}$$

$$\Rightarrow \quad \frac{R_J}{R_E} = \left(\frac{T_J}{T_E}\right)^{2/3} = (12)^{2/3} = 5.24$$

(b)

$$\frac{M_J v_J^2}{R_J} = \frac{GM_S M_J}{R_J^2}$$

$$\Rightarrow \quad v_J^2 = \frac{GM_S}{R_J} \quad \text{and} \quad R_J = \left(T\frac{\sqrt{GM_S}}{2\pi}\right)^{2/3}$$

So,

$$v_J^2 = \frac{(GM_S)^{2/3}(2\pi)^{2/3}}{T^{2/3}} \quad \text{or} \quad v_J = \left(\frac{2\pi GM_S}{T}\right)^{2/3}$$

where $T = 12$ years, M_S = mass of the Sun.

Putting the values, we get $v_J = 12.97$ km/s

$$\text{Acceleration} \quad = \quad \frac{v_J^2}{R_J} \quad = \quad \left(\frac{2\pi GM_S}{T}\right)^{2/3} \times$$

$$\left(\frac{2\pi}{T\sqrt{GM_S}}\right)^{2/3}$$

$$= \left(\frac{2\pi}{T}\right)^{4/3}(GM_s)^{1/3}$$

$$= 2.15 \times 10^{-4} \text{ km/s}^2$$

7. APPROACH This case is discussed in section 1.20.3. The time period of the planet around sun, is given by Eq.1.151b:

$$T = \left[\frac{\pi}{\sqrt{2GM}}\right](R_a + R_p)^{3/2} \qquad \ldots (1)$$

Now, replace R_a and R_p by given quantities and simplify for T.

SOLUTION It is given that, maximum distance (i.e., aphelion distance) of planet from Sun, $R_a = R$, and minimum distance (i.e., perihelion distance) $R_p = r$, therefore, equation (1) gives-

$$T = \left[\frac{\pi}{\sqrt{2GM}}\right](R + r)^{3/2}$$

8. Let x be the distance moved by small body in time 't', then by conservation of mechanical energy, we can write-

$$U_i + K_i = U_f + K_f$$

or $\quad -\dfrac{GM_S m}{r} + 0 = -\dfrac{GM_S m}{r-x} + \dfrac{1}{2}mv^2 \qquad \ldots (1)$

Here, r is the radius of earth's orbit around the Sun, which is also the initial distance of the body from Sun; v is the speed of the mass after moving distance x, M_S is the mass of Sun and m is the mass of the body.

From Eq.(1), we can write-

$$\frac{1}{2}v^2 = \frac{GM_S}{r-x} - \frac{GM_S}{r} = GM_S\left(\frac{1}{r-x} - \frac{1}{r}\right)$$

$$= GM_S\left(\frac{r-(r-x)}{r(r-x)}\right)$$

$\Rightarrow \quad v = \sqrt{\dfrac{2GM_S x}{r(r-x)}} \quad \Rightarrow \quad \dfrac{dx}{dt} = \sqrt{\dfrac{2GM_S x}{r(r-x)}}$

or $\quad dt = \sqrt{\dfrac{r(r-x)}{2GM_S x}}\,dx$

Therefore, time taken to cover the distance from $x = 0$ to $x = r$, is

$$t = \sqrt{\frac{r}{2GM_S}}\int_0^r \sqrt{\frac{r-x}{x}}\,dx \qquad \ldots (2)$$

[since, we have considered x from the initial position of the body, therefore, lower limit of integration is 0 and upper is r]

Now, integration $\int_0^r \sqrt{\dfrac{r-x}{x}}\,dx = \dfrac{\pi r}{2}$

[To prove above result use substitution method and put $x = r\sin\theta$]

therefore, Eq.(2) gives-

$$t = \frac{\sqrt{r}}{2GM_S}\frac{\pi}{2} \qquad \ldots (3)$$

But for earth-Sun system,

$$\frac{GM_S M_e}{r^2} = M_e\omega^2 r,$$

here, ω is the angular speed of earth

therefore, $\quad GM_S = \omega^2 r^3$

On substituting it in Eq.(3), we get

$$t = \sqrt{\frac{r}{2\omega^2 r^3}}\frac{\pi r}{r}$$

$$= \frac{1}{\omega r}\frac{\pi r}{(2\sqrt{2})}$$

$$= \frac{\pi}{\omega(2\sqrt{2})} = \frac{\pi}{\frac{2\pi}{T}(2\sqrt{2})} \qquad [\because \ \omega = 2\pi/T]$$

or $\qquad t = \dfrac{T}{4\sqrt{2}}$

Here, T is the time period of earth about the Sun.

Since, $T = 365$ days, therefore, $t = \frac{365}{4\sqrt{2}}$ days $= 64.5$ days.

Aliter Since, the body starts falling onto the Sun and finally it collapses in it, so Sun will be at far focus if we complete the orbit by extrapolating the path (Fig.1.249). Corresponding to zero initial speed, the perigee distance will be almost zero. Length of major axis, $2a = R \Rightarrow a = R/2$. If T is the time of free fall, then time period corresponding to it will be $2T$.

By Kepler's third law, $(2T)^2 \propto \left(\dfrac{R}{2}\right)^3$

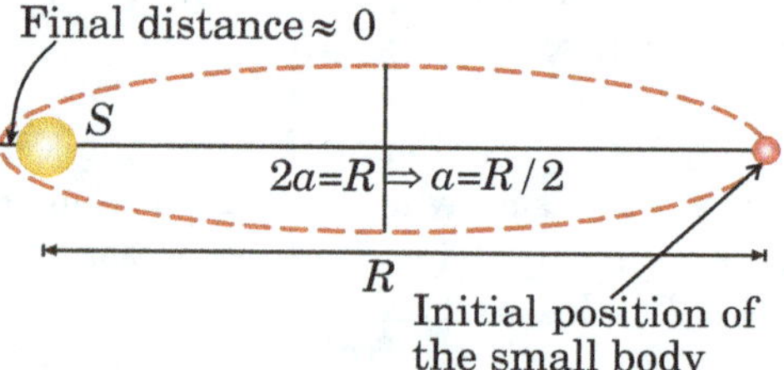

Figure 1.249

$\Rightarrow \qquad 4T^2 = k\dfrac{R^3}{8} \qquad \ldots (1)$

For earth orbiting around Sun, time period of earth T' can be given as-

$$T'^3 \propto R^3 \quad \Rightarrow \quad T'^3 = kR^3 \qquad \ldots (2)$$

Therefore, $\dfrac{4T^2}{T'^2} = \dfrac{kR^3/8}{kR^3} \Rightarrow T^2 = \dfrac{T'^2}{32} \quad$ or $\quad T = T'/4\sqrt{2}$

Since, $T' = 365$ days, therefore-

$$T = 365/4\sqrt{2} \text{ days} = 64.5 \text{ days}$$

9. APPROACH Apply Eq.1.150b,

$$T = \left[\frac{2\pi}{\sqrt{GM}}\right]a^{3/2} \qquad \ldots (1)$$

with $a = (R_a + R_p)/2$.

SOLUTION Given that, the distances are scaled down by a factor η, i.e., $R_a \to R_a/\eta$ and $R_p \to R_p/\eta$, therefore new length of semi-major axis will be

$a' = (R_a + R_p)/2\eta = a/\eta$ Now, if radius of Sun is R_S and it's density is ρ_S, then it's mass can be written as:

$$M = \frac{4}{3}\pi R_S^3 \rho_S$$

When $R_S \to R_S/\eta$ keeping ρ_S constant, then new mass of the Sun becomes:

$$M' = \frac{4}{3\eta^3}\pi R_S^3 \rho_S$$

$$= M/\eta^3$$

Therefore, from above Eq. (1), new time period becomes:

$$T' = \left[\frac{2\pi}{\sqrt{GM'}}\right]a'^{3/2}$$

$$= \left[\frac{2\pi}{\sqrt{GM/\eta^3}}\right](a/\eta)^{3/2}$$

$$= \left[\frac{2\pi}{\sqrt{GM}}\right]a^{3/2} = T$$

So, T does not change.

10. **SOLUTION** Kepler's second law is a consequence of the conservation of angular momentum.

Multiple Choice Problems

11. (B) Point A is perihelion and C is aphelion.

So, $v_A > v_B > v_C$

As kinetic energy $K = (1/2)mv^2$ or $K \propto v^2$

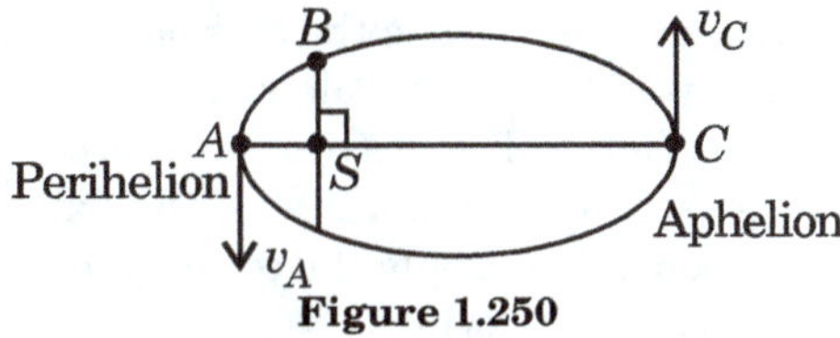

Figure 1.250

So, $K_A > K_B > K_C$.

12. (B) According to the law of conservation of angular momentum $L_1 = L_2$

$mv_1 r_1 = mv_2 r_2 \Rightarrow v_1 r_1 = v_2 r_2$ or $\frac{v_1}{v_2} = \frac{r_2}{r_1}$

13. (B) Equal areas are swept in equal time. As it is given that area $SCD = 2\times$ area of SAB, therefore, the time taken to go from C to D: $t_1 = 2t_2$, where t_2 is the time taken to go from A to B.

14. (A) Period of revolution of planet $A, (T_A) = 8T_B$. According to Kepler's III law of planetary motion $T^2 \propto a^3$. Therefore,

$\left(\frac{r_A}{r_B}\right)^3 = \left(\frac{T_A}{T_B}\right)^2 = \left(\frac{8T_B}{T_B}\right)^2 = 64$ or $\frac{r_A}{r_B} = 4$ or $r_A = 4r_B$

15. (B) Distance of two planets from sun, $r_1 = 10^{13}$ m and $r_2 = 10^{12}$ m

Relation between time period (T) and distance of the planet from the sun is $T^2 \propto r^3$ or $T \propto r^{3/2}$.

Therefore, $\frac{T_1}{T_2} = \left(\frac{r_1}{r_2}\right)^{3/2} = \left(\frac{10^{13}}{10^{12}}\right)^{3/2} = 10^{3/2} = 10\sqrt{10}$

16. (C) : In a circular or elliptical orbital motion a planet, angular momentum is conserved. In attractive field, potential energy and the total energy is negative. Kinetic energy increases with increase is velocity. If the motion is in a plane, the direction of L does not change.

17. (C) Applying the properties of ellipse, we have

$$\frac{2}{R} = \frac{1}{r_1} + \frac{1}{r_2} = \frac{r_1 + r_2}{r_1 r_2}; R = \frac{2r_1 r_2}{r_1 + r_2} \quad \text{(Note)}$$

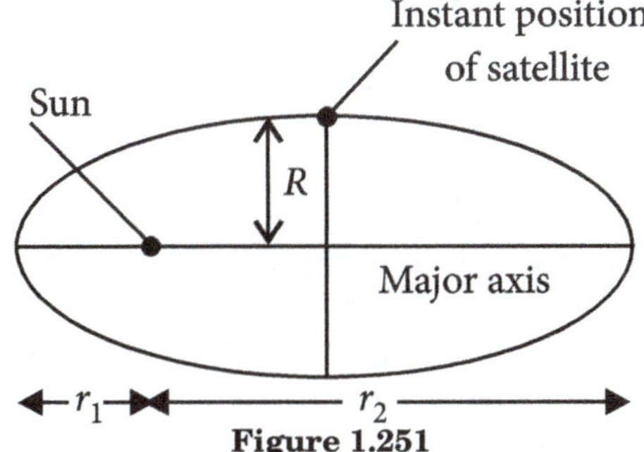

Figure 1.251

18. (B)

19. (C) According to Kepler's third law $T \propto r^{3/2}$

$$\therefore \quad \frac{T_2}{T_1} = \left(\frac{r_2}{r_1}\right)^{3/2} = \left(\frac{R + 2R}{R + 5R}\right)^{3/2} = \frac{1}{2^{3/2}}$$

Since, $T_1 = 24$ hours so,

$$\frac{T_2}{24} = \frac{1}{2^{3/2}} \text{ or } T_2 = \frac{24}{2^{3/2}} = \frac{24}{2\sqrt{2}} = 6\sqrt{2} \text{ hours}$$

20. (D) : Gravitational force of attraction between sun and planet provides centripetal force for the orbit of planet.

$$\therefore \quad \frac{GMm}{r^2} = \frac{mv^2}{r}; \quad v^2 = \frac{GM}{r} \quad \quad \dots (1)$$

Time period of the planet is given by

$$T = \frac{2\pi r}{v} \quad \Rightarrow \quad T^2 = \frac{4\pi^2 r^2}{v^2} = \frac{4\pi^2 r^2}{\left(\frac{GM}{r}\right)} \quad \text{(Using Eq.(1))}$$

$$\Rightarrow \quad T^2 = \frac{4\pi^2 r^3}{GM} \quad \quad \dots (2)$$

According to question, $T^2 = kr^3$ $\quad \dots (3)$

Comparing equations (2) and (3), we get

$$k = \frac{4\pi^2}{GM}, \quad \therefore \quad GMk = 4\pi^2$$

1.26.6 Conceptual Questions

1. Each pair of particles makes a term in gravitational potential energy calculation. If there are n particles in a system, then total terms used in gravitational energy calculations are $^nC_2$.

In 5 particles system, the number of terms:

$$^5C_2 = \frac{5!}{2!(5-2)!}$$
$$= \frac{5!}{2!(3)!} = 10$$

So, ten terms are needed in the potential energy:

$U = U_{12} + U_{13} + U_{14} + U_{15} + U_{23} + U_{24} + U_{25} + U_{34} + U_{35} + U_{45}$

2. No. The force of Earth's gravity is practically as strong in orbit as it is on the surface of the Earth. The astronauts experience weightlessness because they are in constant free fall.

3. (a) False. Kepler's law of equal areas is a consequence of the fact that the gravitational force acts along the line joining two bodies but is independent of the manner in which the force varies with distance.

(b) True. The periods of the planets vary with the three-halves power of their distances from the Sun. So, the shorter the distance from the Sun, the shorter the period of the planet's motion.

(c) False. The orbital period of a planet is independent of the planet's mass.

4. Whether the apple is attached to a tree or falling, it exerts a gravitational force on the Earth equal to the force the Earth exerts on it, which is the weight of the apple (Newton's third law).

5. Earth is closest to the Sun during winter in the northern hemisphere. This is the time of fastest orbital speed. Summer would be the time for minimum orbital speed.

Note: The earth's spin axis is tilted with respect to it's orbital plane. This is what causes the seasons. When the earth's axis points towards the Sun, it is summer for that hemisphere. When the earth's axis points away, there is winter for that hemisphere.

6. As the tips of the fingers approach one another, we can think of them as two small spheres (or we can replace the finger tips with two small marbles if we like). As we know, the net gravitational attraction outside a sphere of mass is the same as that of an equivalent point mass at its centre. Therefore, the two fingers simply experience the finite force of two point masses separated by a finite distance.

7. The tides are caused by the difference in gravitational pull on two opposite sides of the Earth. The gravitational pull from the Sun on the side of the Earth closest to it depends on the distance from the Sun to the close side of the Earth. The pull from the Sun on the far side of the Earth depends on this distance plus the diameter of the Earth. The diameter of the Earth is a very small fraction of the total Earth–Sun distance, so these two forces, although large, are nearly equal. The diameter of the Earth is a larger fraction of the Earth–Moon distance, and so the difference in gravitational force from the Moon to the two opposite sides of the Earth will be greater.

8. To obtain the mass M of Venus you need to measure the period T and semi-major axis a of the orbit of one of the satellites, substitute the measured values into $\frac{T^2}{a^3} = \frac{4\pi^2}{GM}$ (Kepler's third law), and solve for M.

9. The object will weigh more at the poles. The value of r^2 at the equator is greater, both from the Earth's center and from the bulging mass on the opposite side of the Earth. Also, the object has centripetal acceleration at the equator. The two effects do not oppose each other.

10. Since the Earth's mass is greater than the Moon's, the point at which the net gravitational pull on the spaceship is zero is closer to the Moon. A spaceship traveling from the Earth towards the Moon must therefore use fuel to overcome the net pull backwards for over half the distance of the trip. However, when the spaceship is returning to the Earth, it reaches the zero point at less than half the trip distance, and so spends more of the trip "helped" by the net gravitational pull in the direction of travel.

11. No. As described by Kepler's third law, the asteroids closer to the Sun have a shorter year and are orbiting faster.

12. The gravitational force from the Sun provides the centripetal force to keep the Moon and the Earth going around the Sun. Since the Moon and Earth are at the same average distance from the Sun, they travel together, and the Moon is not pulled away from the Earth.

13. As the Moon revolves around the Earth, its position relative to the distant background stars changes. This phenomenon is known as "parallax." As a demonstration, hold your finger at arm's length and look at it with one eye at a time. Notice that it "lines up" with different objects on the far wall depending on which eye is open. If you bring your finger closer to your face, the shift in its position against the background increases. Similarly, the Moon's position against the background stars will shift as we view it in different places in its orbit. The distance to the Moon can be calculated by the amount of shift.

14. At the centre of Earth, the gravitational field is zero, so, the net force on the object would be zero.

15. Between Earth and the moon, the gravitational pulls on the spaceship are oppositely directed. Because of the moon's relatively small mass compared to the mass of Earth, the location where the gravitational forces cancel (thus producing no net gravitational force, a weightless condition) is considerably closer to the moon.

16. A satellite in a geosynchronous orbit stays over the same spot on the Earth at all times. The satellite travels in an orbit about the Earth's axis of rotation. The needed centripetal force is supplied by the component of the gravitational force perpendicular to the axis of rotation. A satellite directly over the North Pole would lie on the axis of rotation of the Earth. The gravitational force on the satellite in this case would be parallel to the axis of rotation, with no component to supply the centripetal force needed to keep the satellite in orbit.

17. According to Newton's third law, the force the Earth exerts on the Moon has the same magnitude as the force the Moon exerts on the Earth. The Moon has a larger acceleration, since it has a smaller mass (Newton's second law, $F = ma$).

18. The satellite needs a certain speed with respect to the centre of the Earth to achieve orbit. The Earth rotates towards the east so it would require less speed (with respect to the Earth's surface) to launch a satellite towards the east (a). Before launch, the satellite is moving with the surface of the Earth so already has a "boost" in the right direction.

19. If the antenna becomes detached from a satellite in orbit, the antenna will continue in orbit around the Earth with the satellite. If the antenna were given a component of velocity toward the Earth (even a very small one), it would eventually spiral in and hit the Earth.

20. Ore normally has a greater density than the surrounding rock. A large ore deposit will have a larger mass than an equal amount of rock. The greater the mass of ore, the greater the acceleration due to gravity will be in its vicinity. Careful measurements of this slight increase in g can therefore be used to estimate the mass of ore present.

21. Yes. The rotational motion of the Earth is to the east, and therefore if you launch in that direction you are adding the speed of the Earth's rotation to the speed of your rocket.

22. Yes. At noon, the gravitational force on a person due to the Sun and the gravitational force due to the Earth are in the opposite directions. At midnight, the two forces point in the same direction. Therefore, your apparent weight at midnight is greater than your apparent weight at noon.

23. Your apparent weight will be greatest in case (b), when the elevator is accelerating upward. The scale reading (your apparent weight) indicates your force on the scale, which, by Newton's third law, is the same as the normal force of the scale on you. If the elevator is accelerating upward, then the net force must be upward, so the normal force (up) must be greater than your actual weight (down). When in an elevator accelerating upward, you "feel heavy."

Your apparent weight will be least in case (c), when the elevator is in free fall. In this situation your apparent weight is zero since you and the elevator are both accelerating downward at the same rate and the normal force is zero.

Your apparent weight will be the same as when you are on the ground in case (d), when the elevator is moving upward at a constant speed. If the velocity is constant, acceleration is zero and normal reaction, $N = mg$. (Note that it doesn't matter if the elevator is moving up or down or even at rest, as long as the velocity is constant.)

24. Since, orbital speed is given by, $v_o = \sqrt{\frac{GM}{r}}$. Therefore, if the Earth's mass were double what it is, the radius of the Moon's orbit would have to double (if the Moon's speed remained constant), or the Moon's speed in orbit would have to increase by a factor of the square root of 2 (if the radius remained constant). If both the radius and orbital speed were free to change, then the product rv^2 would have to double.

25. No. In the weightless environment of the International Space Station (ISS) there would be no convection, which is needed to bring fresh oxygen to the flame. Without convection a flame usually goes out very quickly. In carefully controlled experiments on the ISS, however, small flames have been maintained for considerable times. These "weightless" flames are spherical in shape, as opposed to the teardrop-shaped flames here on Earth.

26. If the Earth were a perfect, non-rotating sphere, then the gravitational force on each droplet of water in the Mississippi would be the same at the headwaters and at the outlet, and the river wouldn't flow. Since the Earth is rotating, the droplets of water experience a centripetal force provided by a part of the component of the gravitational force perpendicular to the Earth's axis of rotation. The centripetal force is smaller for the headwaters, which are closer to the North pole, than for the outlet, which is closer to the equator. Since the centripetal force is equal to $mg - N$ (apparent weight) for each droplet, N is smaller at the outlet, and the river will flow. This effect is large enough to overcome smaller effects on the flow of water due to the bulge of the Earth near the equator.

27. The satellite remains in orbit because it has a velocity. The instantaneous velocity of the satellite is tangent to the orbit. The gravitational force provides the centripetal force needed to keep the satellite in orbit, acting like the tension in a string when twirling a rock on a string. A force is not needed to keep the satellite "up"; a force is needed to bend the velocity vector around in a circle.

28. Between steps, the runner is not touching the ground. Therefore there is no normal force up on the runner and so he has no apparent weight. He is momentarily in free fall since the only force is the force of gravity pulling his back toward the ground.

29. If you were in a satellite orbiting the Earth, you would have no apparent weight (no normal force). Walking, which depends on the normal force, would not be possible. Drinking would be possible, but only from a tube or pouch, from which liquid could be sucked. Scissors would not sit on a table (no apparent weight = no normal force).

30. The centripetal acceleration of Mars in its orbit around the Sun is smaller than that of the Earth. For both planets, the centripetal force is provided by gravity, so the centripetal acceleration is inversely proportional to the square of the distance from the planet to the Sun:

$$\frac{m_p v^2}{r} = \frac{G m_s m_p}{r^2} \quad \text{so,} \quad \frac{v^2}{r} = \frac{G m_s}{r^2}$$

Since, Mars is at a greater distance from the Sun than Earth, it has a smaller centripetal acceleration. Note that the mass of the planet does not appear in the equation for the centripetal acceleration.

31. For Pluto's moon, we can equate the gravitational force from Pluto on the moon to the centripetal force needed to keep the moon in orbit:

$$\frac{m_m v^2}{r} = \frac{G m_p m_m}{r^2}$$

This allows us to solve for the mass of Pluto (m_p) if we know G, the radius of the moon's orbit, and the velocity of the moon, which can be determined from the period and orbital radius. Note that the mass of the moon cancels out.

32. The Earth is closer to the Sun in January. The gravitational force between the Earth and the Sun is a centripetal force. When the distance decreases, the speed increases. (Imagine whirling a rock around your head in a horizontal circle. If you pull the string through your hand to shorten the distance between your hand and the rock, the rock speeds up.)

$$\frac{M_e v^2}{r} = \frac{G M_s M_e}{r^2} \quad \text{so} \quad v = \sqrt{\frac{G M_s}{r}}$$

Since the speed is greater in January, the distance must be less. This agrees with Kepler's second law (law of areal velocity).

33. The Earth's orbit is an ellipse, not a circle. Therefore, the force of gravity on the Earth from the Sun is not perfectly perpendicular to the Earth's velocity at all points. A component of the force will be parallel to the velocity vector and will cause the planet to speed up or slow down.

34. Standing at rest, you feel an upward force on your feet. In free fall, you don't feel that force. You would, however, be aware of the acceleration during free fall, possibly due to your inner ear.

35. If we treat $\vec{g}$ as the acceleration due to gravity, it is the result of a force from one mass acting on another mass and causing it to accelerate. This implies action at a distance, since the two masses do not have to be in contact. If we view $\vec{g}$ as a gravitational field, then we say that the presence of a mass changes the characteristics of the space around it by setting up a field, and the field then interacts with other masses that enter the space in which the field exists. Since the field is in contact with the mass, this conceptualization does not imply action at a distance.

36. You could take careful measurements of its position as a function of time in order to determine whether its tra-

jectory is an ellipse, a hyperbola, or a parabola. If the path is an ellipse, it will return, if its path is hyperbolic or parabolic, it will not return. Alternatively, by measuring its distance from the Sun, you can estimate the gravitational potential energy (per kg of its mass, and neglecting the planets) of the object, and by determining its position on several successive nights, the speed of the object can be determined [by applying conservation of mechanical energy principle, decrease/increase in PE = − increase/decrease in KE]. From this, its kinetic energy (per kg) can be determined. The sum of these two gives the comet's total energy (per kg) and if it is positive, it will likely swing once around the Sun and then leave the Solar System forever.

37. Newton's third law tells us that the forces are equal. They are also clearly equal when Newton's law of gravity is examined: $F_{12} = Gm_1 m_2/r^2$ has the same value whether $m_1 = $ Earth and $m_2 = $ sun or vice versa.

38. The force of the star on planet 1 is $F_1 = G\frac{M_s m_1}{r_1^2}$. The force of the star on planet 2 is $F_2 = G\frac{M_s m_2}{r_2^2}$. Since $m_2 = 2m_1$, and $r_2 = 2r_1$

$$\frac{F_2}{F_1} = \frac{G\frac{M_s(2m_1)}{(2r_1)^2}}{G\frac{M_s m_1}{r_1^2}} = \frac{2}{4} = \frac{1}{2}$$

39. (a) The force of the Earth on the first satellite is $F_1 = G\frac{M_e m_1}{r_1^2}$, while $F_2 = G\frac{M_e m_2}{r_2^2}$. Since $r_1 = r_2$,

$\frac{F_1}{F_2} = \frac{m_1}{m_2} = \frac{1000 \text{ kg}}{2000 \text{ kg}} = \frac{1}{2}$.

(b) With $F_{\text{net}} = ma$, $F_1 = m_1 a_1$ and $F_2 = m_2 a_2$, so

$$\frac{a_1}{a_2} = \left(\frac{F_1}{F_2}\right)\left(\frac{m_2}{m_1}\right) = \left(\frac{1}{2}\right)\left(\frac{2000 \text{ kg}}{1000 \text{ kg}}\right) = 1$$

The free-fall acceleration is the same for objects of different mass.

40. The astronauts can be "weightless" at any distance because an object is said to be weightless if it is in free fall (as in orbit). For the gravitational force to become zero, the spacecraft would have to be an infinite distance away.

41. The hammer will not fall to the Earth. The astronaut, shuttle, and hammer are all in free fall around the Earth (in an orbit), so the hammer has the same acceleration as the astronaut and does not move away from him.

42. The acceleration due to gravity is $g = G\frac{M}{R^2}$. With $g_1 = 20$ m/s^2, $M_2 = 2M_1$, and $R_2 = 2R_1$,

$$g_2 = G\frac{M_2}{R_2^2} = G\frac{(2M_1)}{(2R_1)^2} = G\frac{M_1}{R_1^2}\left(\frac{1}{2}\right) = \frac{1}{2}g_1 = 10 \text{ m/s}^2$$

43. The gravitational potential energy is negative because we choose to place the zero point of potential energy at infinity ($U_\infty = 0$). With this choice, the gravitational potential energy is negative because the conservative force of gravity is attractive. The two masses will gain kinetic energy as they approach each other.

44. You should fire the rocket in a direction to oppose the orbital motion of the satellite. As the satellite gets closer to Earth after the burn, the potential energy will decrease. However, the total mechanical energy will decrease due to the frictional drag forces transforming mechanical energy into thermal energy. The kinetic energy will increase until the satellite enters the atmosphere where the drag forces slow its motion.

45. The Earth's new orbital period would be about 11.9 years. For circular orbits, Kepler's third law relates orbital period to radius and can be written $T^2 = \left(\frac{4\pi^2}{GM}\right)r^3$. Here M is the mass of the body being orbited (the Sun in this problem.) The orbital period is independent of the mass of the orbiting body.

46. For a circular orbit, $v = \sqrt{GM/r}$. As r decreases, v increases, so the satellite speeds up as the satellite spirals inward.

47. A particle will be in equilibrium when the net force acting on it is equal to zero. Two particles under the action of their mutual gravitational force will be in equilibrium when they revolve around a common point under the influence of their mutual gravitational force of attraction. In this case, the gravitational pull is used up in providing the necessary centripetal force. Hence, the net force on the particles is zero and they are in equilibrium. This is also true for a three particle system.

48. Generally, weight of an object is the the force of gravity of earth on it, which we measure by spring or other balance. But when we say so about earth, whose force of gravity do we refer? and how do we measure it? It might be Sun, Moon, other planets.

So, instead the same thing is talked as force of gravitation between sun and earth, or earth and moon and so on.

49. The heavier bodies are attracted more strongly by the earth means the force applied by the earth is more. This force is the weight of the body = mg. Hence the acceleration of the body = force/mass = $mg/m = g$, which is independent of m. So all bodies fall with the same acceleration 'g'.

50. Any two massless particles will not exert gravitational force on each other. Such as two photons at rest (rest mass of photons= 0).

51. The Sun's gravity produces the same acceleration on both the Earth and the Moon. The Sun is pulling both of them along, but they are falling together.

You may imagine two skydiver jumping out of a plane at the same time (and we'd better ignore air resistance). They are subjected to gravitational forces from the Earth that vastly larger than the forces between them, but that doesn't pull them away from each other because they both experience the same acceleration.

52. The gravitational potential due to the system is given as

$$V = \frac{k}{r^2}$$

Gravitational field due to the system:

$$g = -\frac{dV}{dr}$$

$$\Rightarrow \quad g = -\frac{d}{dr}\left(\frac{k}{r^2}\right) = -\left(-\frac{2k}{r^3}\right)$$

$$\Rightarrow \quad g = \frac{2k}{r^3}$$

We can see that for this system, $g \propto \frac{1}{r^3}$

This type of system is not possible because F_g is always proportional to inverse of square of distance(experimental fact). If there were negative masses, then this type of system is possible. This system is a dipole of two masses, i.e., two masses, one positive and the other negative, separated by a small distance.

In this case, the gravitational field due to the dipole is proportional to $\frac{1}{r^3}$.

53. The gravitational potential energy of a two-particle system is given by $U = -\frac{Gm_1m_2}{r}$. This relation does not tell that the gravitational potential energy is zero at infinity. For our convenience, we choose the potential energies of the two particles to be zero when the separation between them is infinity.

No, if we suppose that the potential energy for $r = \infty$ is $30J$, then we need to modify the formula.

Now, potential energy of the two-particle system separated by a distance r is given by
$$[U(r)]_{\text{new}} = U(r) - U(\infty)$$
Given : $U(\infty) = 30J$
$$\therefore \qquad [U(r)]_{\text{new}} = -G\frac{m_1m_2}{r^2} - 30$$
This formula should be used to calculate the gravitational potential energy at separation r.

54. The weight of an object is more at the poles than that at the equator. In purchasing or selling goods, we measure the mass of the goods. The balance used to measure the mass is calibrated according to the place to give its correct reading. So, it is not beneficial to purchase goods at the equator and sell them at the poles. A beam balance measures the mass of an object, so it can be used here. For using a spring balance, we need to calibrate it according to the place to give the correct readings.

55. Acceleration due to gravity at surface,
$$g = \frac{GM}{R^2}$$
Taking log and differentiating,
$$\frac{\Delta g}{g} = -2\frac{\Delta R}{R}$$
Now as $\frac{\Delta R}{R} = -1\% \quad \Rightarrow \quad \frac{\Delta g}{g} = 2\%$
Hence, g is increased by 2%.

56. Yes. Because the centripetal force on the satellite is towards the center of the earth where the earth's mass is apparently considered to be concentrated.

57. yes, yes. See subsection 1.18.2.

58. Near the moon you would fire the rockets to accelerate the spacecraft with the thrust acting in the direction of your ship's velocity at the time. When the rockets have shut down, as you leave the moon, your kinetic energy will initially decrease (the moon's gravitational pull exceeds that of Earth), and your potential energy will increase. After a certain point, Earth's gravitational attraction would begin accelerating the ship and the kinetic energy would increase at the expense of the gravitational potential energy of the spacecraft-Earth-moon system. The spacecraft will enter Earth's atmosphere with its maximum kinetic energy. Eventually, landing in the ocean, the kinetic energy would be zero, the gravitational potential energy a minimum, and the total mechanical energy of the ship will have been surprisingly reduced due to air drag forces producing heat and light during re-entry.

59. A satellite can remain stationary above a point lying on the equatorial plane (geostationary). Since New Delhi does not lie on the equatorial plane it is not possible to have a geostationary satellite over New Delhi. Note that the time period of a geostationary satellite is 24h and its orbit lies in the equatorial plane. A satellite with a time period of 24 h that may or may not lie on an equatorial plane is called geosynchronous. All satellites in Indian Regional Navigation Satellite System (IRNSS) are geosynchronous and some of them are geostationary.

60. To feel the weightlessness in a circular orbit around the earth, the person should orbit with such a speed so that the centrifugal force must be equal to the apparent weight of the person.

For this condition $g = v^2/R$ at earth's surface. This gives,
$$v = \sqrt{gR} = \sqrt{(9.8 \times 6400000)} = 7920 \text{ m/s} = 7.92 \text{ km/s}$$
So only going in the circular orbit at the surface of the earth is not sufficient for the weightlessness but his speed also should be 7.92 km/s which is not possible for a person even in a vehicle or plane. Also, the resistance of air on the surface will make the vehicle burn at this speed.

61. As seen from the earth, the apparent time period for both the satellite will be different due to the rotation of the earth. The earth rotates from west to east. So, the satellite moving from east to west will have greater relative angular speed, hence, will have less time period.

62. It makes more sense to think of the Moon as orbiting the Sun, with the Earth providing a smaller force that makes the Moon "wobble" back and forth in its solar orbit.

63. The net force acting on the Moon is always directed toward the Sun, never away from the Sun. Therefore, the Moon's orbit must always curve toward the Sun. The path shown in the upper part of Figure 1.194, though it seems "intuitive," sometimes curves toward the Sun, sometimes away from the Sun. The correct path, shown in the lower part of Figure 1.194, curves sharply toward the Sun when both the Sun and the Earth pull inward on the Moon, and curves only slightly toward the Sun when the Moon is pulled in opposite directions by the Sun and the Earth.

1.26.7   Problems

1. **APPROACH** This is a direct application of the equation expressing Newton's law of gravitation:

SOLUTION By Newton's law of gravitation, we have
$$F = \frac{GMm}{r^2} = (6.67 \times 10^{-11} N \cdot m^2/kg^2)\frac{(1.50kg)(15.0 \times 10^{-3}\text{kg})}{(4.50 \times 10^{-2}\text{m})^2}$$
$$= 7.41 \times 10^{-10}\text{N}$$

2. For two 70 kg persons, modeled as spheres,
$$F_g = \frac{Gm_1m_2}{r^2}$$
$$= \frac{(6.67 \times 10^{-11} N \cdot m^2/kg^2)(70kg)(70kg)}{(2m)^2}$$
$$\sim 10^{-7} \text{ N}$$

3. (a) At the midpoint between the two objects, the forces exerted by the 200 kg and 500 kg objects are oppositely directed, and from $F_g = \frac{Gm_1m_2}{r^2}$, we have
$$\sum F = \frac{G(500 \text{ kg})(50.0 \text{ kg})}{(200 \text{ m})^2} - \frac{G(200 \text{ kg})(50.0 \text{ kg})}{(200 \text{ m})^2} -$$

or $\quad \sum F = \dfrac{G(50.0\text{ kg})(500\text{ kg}-200\text{ kg})}{(200\text{ m})^2}$
$$= 250 \times 10^{-7}\text{ N toward the 500 kg object.}$$

(b) At a point between the two objects at a distance d from the 500 kg object, the net force on the 50.0kg object will be zero when
$$\frac{G(50.0\text{ kg})(200\text{ kg})}{(4.00 m - d)^2} = \frac{G(50.0\text{ kg})(500\text{ kg})}{d^2}.$$
On solving for d, we get- $d = 2.45$ m from the 500 kg object toward the smaller object.

4. (a) $g = GM/r^2$, $\Delta g = -\left(2GM/r^3\right)\Delta r$.
In this case, $\Delta r = h$ and $M = 4\pi\rho r^3/3$, therefore,
$$\Delta W = m\Delta g = 8\pi G\rho mh/3.$$
(b) $\Delta W/W = \Delta g/g = 2h/r$.
So, an error of one part in a million will occur when h is one part in two million of r, or 3.2 meters.

5. $\quad F_{\text{net}} = mr\omega_s^2$, where ω_s is the rotational speed of the ship. But since the ship is moving relative to the earth with a speed v, we can write $\omega_s = \omega \pm v/r$, where the sign is positive if the ship is sailing east.
Therefore, $F_{\text{net}} = mr(\omega \pm v/r)^2$.
The scale measures a force W which is given by $mg - F_{\text{net}}$, or
$$W = mg - mr(\omega \pm v/r)^2.$$
Note that $W_0 = m\left(g - r\omega^2\right)$. Therefore,
$$W = W_0\frac{g - r(\omega \pm v/r)^2}{g - r\omega^2}$$
$$\approx W_0\left(1 \pm \frac{2\omega v}{1 - r\omega^2}\right)$$
$$\approx W_0(1 \pm 2v\omega/g)$$

6. (a) The Sun-Earth distance is 1.496×10^{11} m and the Earth-Moon distance is 3.84×10^8 m, so the distance from the Sun to the Moon during a solar eclipse is 1.496×10^{11} m $- 3.84 \times 10^8$ m $= 1.492 \times 10^{11}$ m
The mass of the Sun, Earth, and Moon are $M_s = 1.99 \times 10^{30}$ kg, $M_e = 5.98 \times 10^{24}$ kg and $M_m = 7.36 \times 10^{22}$ kg respectively, therefore, the gravitational force exerted by Sun on Moon
$$F_{SM} = \frac{GM_sM_m}{r_{sm}^2}$$
$$= \frac{(6.67\times10^{-11}\text{N}\cdot\text{m}^2/\text{kg}^2)(1.99\times10^{30}\text{kg})(7.36\times10^{22}\text{kg})}{(1.492\times10^{11}\text{m})^2}$$
$$= 4.39 \times 10^{20}\text{N}$$
(b) the gravitational force exerted by Earth on Moon
$$F_{EM} = \frac{GM_eM_m}{r_{em}^2}$$
$$F_{EM} = \frac{(6.67\times10^{-11}\text{N}\cdot\text{m}^2/\text{kg}^2)(5.98\times10^{24}\text{kg})(7.36\times10^{22}\text{kg})}{(3.84\times10^{8}\text{m})^2}$$
$$= 1.99 \times 10^{20}\text{N}$$
(c) the gravitational force exerted by Sun on Earth
$$F_{AE} = \frac{GM_sM_e}{r_{se}^2}$$
$$F_{SE} = \frac{(6.67\times10^{-11}\text{N}\cdot\text{m}^2/\text{kg}^2)(1.99\times10^{30}\text{kg})(5.98\times10^{24}\text{kg})}{(1.496\times10^{11}\text{m})^2}$$
$$= 3.55 \times 10^{22}\text{N}$$
(d) The force exerted by the Sun on the Moon is much stronger than the force of the Earth on the Moon. In a sense, the Moon orbits the Sun more than it orbits the Earth. The Moon's path is everywhere concave toward the Sun.

7. With one metric ton = 1000 kg,
$$F = m_1g = \frac{Gm_1m_2}{r^2}$$
$$g = \frac{Gm_2}{r^2}$$
$$= \frac{(6.67\times10^{-11}N\cdot m^2/kg^2)(4.00\times10^{7}kg)}{(100m)^2}$$
$$= 2.67 \times 10^{-7}m/s^2$$

8. The force exerted on the 4.00 kg mass by the 2.00 kg mass is directed upward and given by
$$\vec{F}_{21} = G\frac{m_2m_1}{r_{12}^2}\hat{j}$$
$$= (6.67 \times 10^{-11}\text{N}\cdot\text{m}^2/\text{kg}^2)\frac{(4.00\text{ kg})(2.00\text{ kg})}{(3.00\text{ m})^2}\hat{j}$$
$$= 5.93 \times 10^{-11}\hat{j}\text{ N}$$
The force exerted on the 4.00 kg mass by the 6.00 kg mass is directed to the left and it is given by:
$$\vec{F}_{23} = G\frac{m_2m_3}{r_{32}^2}(-\hat{i})$$
$$= (-6.67 \times 10^{-11}\text{N}\cdot\text{ m}^2/\text{kg}^2)\frac{(4.00\text{ kg})(6.00\text{ kg})}{(4.00\text{ m})^2}\hat{i}$$
$$= -10.0 \times 10^{-11}\hat{i}\text{N}$$
Therefore, the resultant force on the 4.00 kg mass is
$$\vec{F} = \vec{F}_{21} + \vec{F}_{23} = \left(5.93\hat{i} - 10.0\hat{i}\right) \times 10^{-11}\text{N}$$

9. The magnitude of the gravitational force is given by
$$F = \frac{Gm_1m_2}{r^2} = \frac{(6.672\times10^{-11}\text{ N}\cdot\text{m}^2/\text{kg}^2)(2.00\text{ kg})(2.00\text{ kg})}{(0.300\text{ m})^2}$$
$$= 2.97 \times 10^{-9}\text{ N}$$

10. Assume the masses of the sphere are the same. Using $F_g = \frac{Gm_1m_2}{r^2}$, we would find that the mass of a sphere is 1.22×10^5 kg. If the spheres have at most a radius of 0.500 m, the density of spheres would be at least $2.34 \times 10^5\text{kg/m}^3$, which is ten times the density of the most dense element, osmium.
The situation is impossible because no known element could compose the spheres.

11. We are given $m_1 + m_2 = 5.00$ kg, which means that $m_2 = 5.00$ kg $- m_1$.
Newton's law of universal gravitation then becomes
$$F = G\frac{m_1m_2}{r^2}$$
$$\Rightarrow \quad 1.00 \times 10^{-8}\text{N} = (6.67 \times 10^{-11}\text{N}\cdot\text{m}^2/\text{kg}^2)\frac{m_1(5.00\text{ kg}-m_1)}{(0200\text{ m})^2}$$
$$(5.00\ \text{ kg})m_1 - m_1^2 = \frac{(1.0.0 \times 10^{-8}\text{N})(0.0400\text{ m}^2)}{667 \times 10^{-11}\text{N}\cdot\text{ m}^2/\text{kg}^2} =$$
6.00 kg^2
Thus, $\quad m_1^2 - (5.00\text{ kg})m_1 + 6.00\text{ kg} = 0$
or $\quad (m_1 - 3.00\text{ kg})(m_1 - 2.00\text{ kg}) = 0$
giving $m_1 = 3.00$ kg, so $m_2 = (5.00 - 3.00)\text{kg} = 2.00$ kg.
The answer $m_1 = 2.00$ kg and $m_2 = 3.00$ kg is physically equivalent.

12. **APPROACH** We can determine the maximum range at which an object with a given mass can be detected by substituting the equation for the gravitational field in the expression for the resolution of the meter and solving for the distance. Differentiating $g(r)$ with respect to r, separating variables to obtain dg/g, and approximating Δr with dr will allow us to determine the vertical

change in the position of the gravity meter in Earth's gravitational field is detectable.

SOLUTION (a) Earth's gravitational field is given by:
$$g = \frac{GM}{R^2}$$
Express the gravitational field due to the mass m (assumed to be a point mass) of your friend and relate it to the resolution of the meter:
$$g(r) = \frac{Gm}{r^2} = 1.00 \times 10^{-11} g$$
$$= 1.00 \times 10^{-11} \frac{GM}{R^2}$$
Solving for r yields:
$$r = R\sqrt{\frac{100 \times 10^{11} m}{M}}$$
Substitute numerical values and evaluate r:
$$r = (6.37 \times 10^6 \text{m})\sqrt{\frac{100 \times 10^{11}(80 \text{ kg})}{598 \times 10^{24} \text{ kg}}} = 7.37 \text{ m}$$
(b) Differentiate g(r) and simplify to obtain:
$$\frac{dg}{dr} = \frac{-2Gm}{r^3} = -\frac{2}{r}\left(\frac{Gm}{r^2}\right) = -\frac{2}{r}g$$
Separate variables to obtain:
$$\frac{dg}{g} = -2\frac{dr}{r} = 10^{-11}$$
Approximating dr with Δr, evaluate Δr with $r = R$
$$|\Delta r| = |-\tfrac{1}{2}(1.00 \times 10^{-11})(6.37 \times 10^6 \text{m})|$$
$$= |31.9 \ \mu\text{m}|$$

13. Let θ represent the angle each cable makes with the vertical, L the cable length, x the distance each ball is displaced by the gravitational force, and $d = 1$ m the original distance between them. Then $r = d - 2x$ is the separation of the balls. For vertical equilibrium, we have
$$\sum F_y = 0: T\cos\theta - mg = 0$$
$$\Rightarrow \qquad T\cos\theta = mg \qquad\qquad \text{... (1)}$$
For horizontal equilibrium, we have

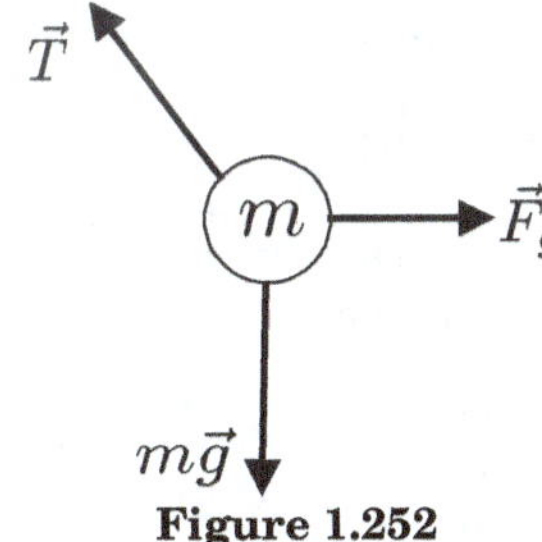

Figure 1.252

$$\sum F_x = 0: T\sin\theta - \frac{Gmm}{r^2} = 0$$
$$\Rightarrow \qquad T\sin\theta = \frac{Gmm}{r^2} \qquad\qquad \text{... (2)}$$
On dividing Eq.(2) by (1), we get
$$\tan\theta = \frac{Gmm}{r^2 mg}$$
or
$$\frac{x}{\sqrt{L^2 - x^2}} = \frac{Gm}{g(d - 2x)^2}$$
$$\Rightarrow \qquad x(d - 2x)^2 = \frac{Gm}{g}\sqrt{L^2 - x^2}$$
The factor Gm/g is numerically small. We expect that x is very small compared to both L and d, so we can treat the term $(d - 2x)$ as d, and $(L^2 - x^2)$ as L^2. We then have
$$x(1m)^2 = \frac{(6.67 \times 10^{-11} N \cdot m^2/kg^2)(100 kg)}{(9.80 m/s^2)}(45.00\ m)$$
or
$$x = 3.06 \times 10^{-8} m$$

14. The gravitational force on a small parcel of material

at the star's equator supplies the necessary centripetal acceleration:
$$\text{so,} \qquad \frac{GM_s\, m}{R_s^2} = \frac{mv^2}{R_s} = mR_s\omega^2$$
$$\omega = \sqrt{\frac{GM_s}{R_s^3}}$$
$$= \sqrt{\frac{(6.67 \times 10^{-11} \text{ N} \cdot \text{m}^2/\text{kg}^2)\left[2\left(1.99 \times 10^{30} \text{ kg}\right)\right]}{\left(10.0 \times 10^3 \text{ m}\right)^3}}$$
$$\omega = 1.63 \times 10^4 \text{rad/s}$$

15. The distance of the meteor from the center of Earth is $R + 3R = 4R$.
Therefore, the acceleration due to gravity at this distance.
$$g = \frac{GM}{r^2}$$
$$= \frac{(6.67 \times 10^{-11} \text{N} \cdot \text{m}^2/\text{kg}^2)(5.98 \times 10^{24}\text{kg})}{[4(637 \times 10^6 \text{m})]^2}$$
$$= 0.614 \text{ m/s}^2, \text{ toward Earth}$$

16. The gravitational field at the surface of the Earth or Moon is given by $g = \frac{GM}{R^2}$
The expression for density is, $\rho = \frac{M}{V} = \frac{M}{\frac{4}{3}\pi R^3}$
so, $\qquad M = \frac{4}{3}\pi\, \rho R^3$
and $\qquad g = \frac{G(\frac{4}{3}\pi\rho R^3)}{R^2} = \frac{4}{3}G\pi\rho R$
Noting that this equation applies to both the Moon and the Earth, and dividing the two equations,
$$\frac{g_M}{g_E} = \frac{\frac{4}{3}G\pi\rho_M R_M}{\frac{4}{3}G\pi\rho_E R_E} = \frac{\rho_M R_M}{\rho_E R_E}$$
Substituting for the fractions,
$$\frac{1}{6} = \frac{\rho_M}{\rho_E}\left(\frac{1}{4}\right) \quad \text{and} \quad \frac{\rho_M}{\rho_E} = \frac{4}{6} = \frac{2}{3}$$

17. From Fig.1.253, the gravitational field due to each individual mass is given by
(a) $g_1 = g_2 = \dfrac{MG}{r^2 + a^2}$
Along y axis-

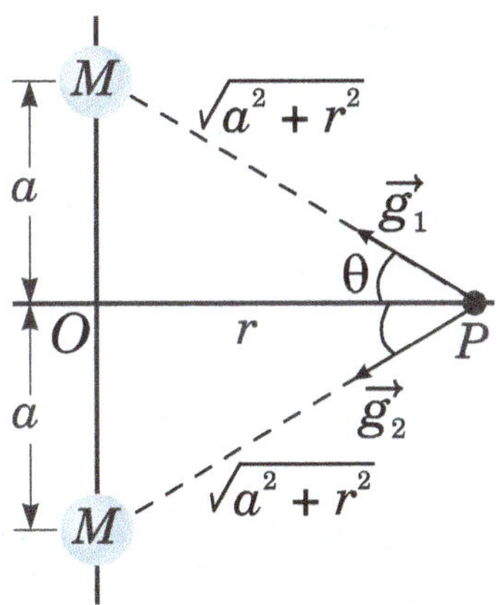

Figure 1.253

$$g_{1y} = -g_{2y}$$
Therefore, along y axis, the net gravitational field at point P:

$$g_y = g_{1y} + g_{2y} = 0$$

Along the negative direction of x axis-

$$g_{1x} = g_{2x} = g_2 \cos\theta$$

and from Fig.1.253, we have $\qquad \cos\theta = \dfrac{r}{(a^2+r^2)^{1/2}}$

Therefore, net field along the negative direction of x axis.

$$\vec{g} = 2g_{2x}(-\hat{i})$$

or $\qquad \vec{g} = \dfrac{2MGr}{(r^2+a^2)^{3/2}}$ toward the centre of mass

(b) At $r = 0$, the fields of the two objects are equal in magnitude and opposite in direction, to add to zero.

(c) As $r \to 0, 2MGr\left(r^2+a^2\right)^{-3/2}$ approaches $2MG(0)/a^3 = 0$

(e) As r becomes much larger than a, the expression approaches

$$2MGr\left(r^2+0^2\right)^{-3/2} = 2MGr/r^3 = 2MG/r^2 \text{ as required.}$$

18. If $\hat{e}_1$, $\hat{e}_2$ and $\hat{e}_3$ are unit vectors along OA, OB and OC directions respectively, then from Fig.1.254, net gravitational field at O will be given by:

$$\vec{g} = \dfrac{Gm}{l^2}\hat{e}_1 + \dfrac{Gm}{l^2}\hat{e}_2 + \dfrac{Gm}{(l\sqrt{2})^2}\hat{e}_3$$

Since, $\hat{e}_1 = \hat{i}$, $\hat{e}_2 = \hat{j}$, and $\hat{e}_3 = \cos 45.0°\hat{i} + \cos 45.0\hat{j}$, therefore, above equation gives:

$$\vec{g} = \dfrac{Gm}{l^2}\hat{i} + \dfrac{Gm}{l^2}\hat{j} + \dfrac{Gm}{2l^2}(\cos 45.0°\hat{i} + \cos 45.0\hat{j})$$

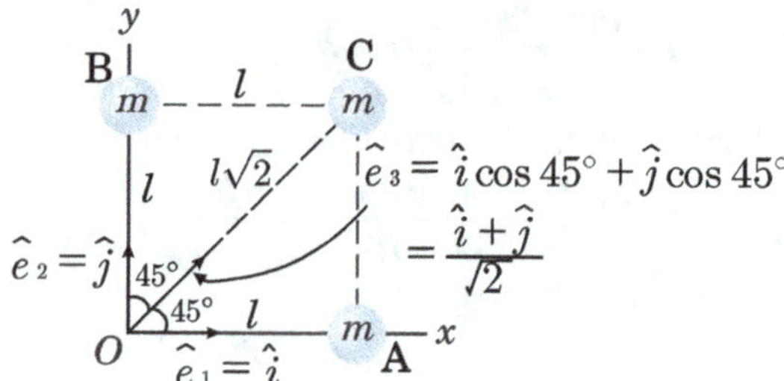

Figure 1.254

So, $\qquad \vec{g} = \dfrac{Gm}{l^2}\left(1+\dfrac{1}{2\sqrt{2}}\right)(\hat{i}+\hat{j})$

or $\qquad \vec{g} = \dfrac{Gm}{l^2}\left(\sqrt{2}+\dfrac{1}{2}\right)$, toward the opposite corner.

19. (a) By Newton's law of gravitation, we have

$$F = \dfrac{GMm}{r^2}$$

$$= \dfrac{(6.67 \times 10^{-11}\text{N}\cdot\text{m}^2/\text{kg}^2)[100(1.9.9 \times 10^{30}\text{kg})(10^3\text{kg})]}{(1.00 \times 10^4\text{m} + 500\text{m})^2}$$

$$= 1.31 \times 10^{17}\text{N}$$

(b) $\Delta F = \dfrac{GMm}{r^2_{front}} - \dfrac{GMm}{r^2_{back}}$

$$\Delta g = \dfrac{\Delta F}{m} = \dfrac{GM(r^2_{back} - r^2_{front})}{r^2_{front}r^2_{back}}$$

$$\Delta g = (6.67 \times 10^{-11}\text{N}\cdot\text{m}^2/\text{kg}^2)$$
$$\dfrac{[100(1.99 \times 10^{30}\text{kg})][(1.01 \times 1.0^4\text{m})^2 - (1.00 \times 10^4\text{m})^2]}{(1.00 \times 10^4\text{m})^2(101 \times 10^4\text{m})^2}$$

$$\Delta g = 2.62 \times 10^{12}\text{N/kg}$$

20. **APPROACH** Since, gravitational force of Sun on asteroid passes through the centre of Sun, therefore torque about the centre of Sun will be zero and hence the an-

gular momentum of asteroid about the centre of Sun will be conserved. So, we can use conservation of angular momentum to relate the asteroid's aphelion and perihelion distances.

(a) Using conservation of angular momentum, relate the angular momenta of the asteroid at aphelion and perihelion:

$$L_a - L_p = 0$$

or $\qquad mv_a r_a - mv_p r_p = 0 \quad \Rightarrow \quad \dfrac{r_a}{r_p} = \dfrac{v_p}{v_a}$

Substitute numerical values and evaluate the ratio of the asteroid's aphelion and perihelion distances:

$$\dfrac{r_a}{r_p} = \dfrac{20km/s}{14km/s} = 1.4$$

(b) It is farther from the Sun than Earth. Kepler's third law ($T^2 = Cr^3_{av}$) tells us that longer orbital periods together with larger orbital radii means slower orbital speeds, so the speed of objects orbiting the Sun decreases with distance from the Sun. The average orbital speed of Earth, given by $v = 2\pi r_{ES}/T_{ES}$, is approximately 30 km/s. Because the given maximum speed of the asteroid is only 20 km/s, the asteroid is farther from the Sun.

21. The gravitational force on satellite of mass m located at distance r from the centre of the Earth is

$$F_g = mg = GMm/r^2.$$

Thus, the acceleration of gravity at this location is

$$g = GM/r^2$$

If $g = 9.00$ m/s^2 at the location of the satellite, then, the radius of its orbit must be

$$r = \sqrt{\dfrac{GM}{g}} = \sqrt{\dfrac{(6.67 \times 10^{-11}\text{N}\cdot\text{m}^2/\text{kg}^2)(5.98 \times 10^{24}\text{kg})}{9.00\text{ m/s}^2}}$$
$$= 6.66 \times 10^6\text{m}$$

From Kepler's third law for Earth satellites,

$$T^2 = \dfrac{4\pi^2 r^3}{GM},$$

the period is found to be

$$T = 2\pi\sqrt{\dfrac{r^3}{GM}}$$
$$= 2\pi\sqrt{\dfrac{(666 \times 10^6\text{m})^3}{(667 \times 10^{-11}\text{Nm}^2/\text{kg}^2)(598 \times 10^{24}\text{kg})}}$$
$$= 5.41 \times 10^3\text{s}$$

or $\qquad T = (5.41 \times 10^3\text{s})\left(\dfrac{1h}{3600\text{ s}}\right)$
$$= 1.50\text{ h} = 90.0\text{ min}$$

22. (a) The desired path is an elliptical trajectory with the Sun at one of the foci, the departure planet at the perihelion, and the target planet at the aphelion [see Fig.1.255]. The perihelion distance r_D is the radius of the departure planet's orbit, while the aphelion distance r_T is the radius of the target planet's orbit. The semimajor axis of the desired trajectory is then

$$a = (r_D + r_T)/2.$$

If Earth is the departure planet,

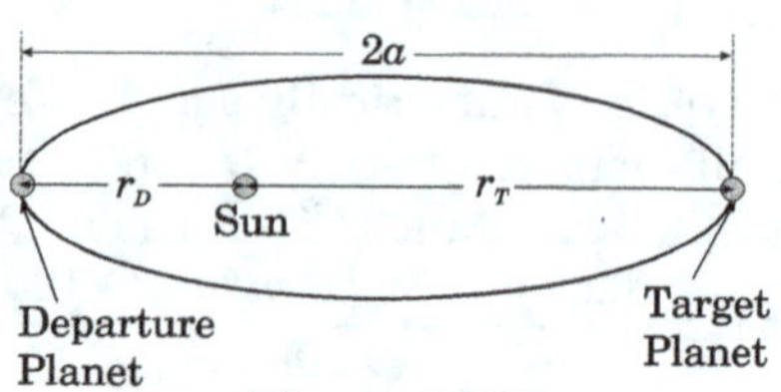

Figure 1.255

$$r_D = 1.496 \times 10^{11} \text{ m} = 1.00\text{AU}$$

With Mars as the target planet,
$$r_{\mathrm{T}} = 2.28 \times 10^{11} \text{ m} \left(\frac{1\text{AU}}{1.496 \times 10^{11} \text{ m}}\right) = 1.52\text{AU}$$
Thus, the semi-major axis of the minimum energy trajectory is
$$a = \frac{r_D + r_T}{2} = \frac{1.00\text{AU} + 1.52\text{AU}}{2} = 1.26\text{AU}$$
Kepler's third law (Eq.1.149), $T^2 = a^{39}$ then gives the time for a full trip around this path as
$$T = \sqrt{a^3} = \sqrt{(1.26 \text{ AU})^3} = 1.41 \text{ yr}$$
so the time for a one-way trip from Earth to Mars is
$$\Delta t = \frac{1}{2}T = \frac{1.41 \text{ yr}}{2} = 0.71 \text{ yr}$$
(b) This trip cannot be taken at just any time. The departure must be timed so that the spacecraft arrives at the aphelion when the target planet is located there.

23. (a) The particle does possess angular momentum, because it is not headed straight for the origin.
(b) Its angular momentum is constant. There are no identified outside influences acting on the object.
(c) Since speed is constant, the distance traveled between t_A and t_B is equal to the distance traveled between t_C and t_D. The area of a triangle is equal to one-half its (base) width across one side times its (height) dimension perpendicular to that side.
So, $\frac{1}{2}b_0(t_B - t_A) = \frac{1}{2}b_0(t_D - t_C)$
states that the particle's radius vector sweeps out equal areas in equal times.

24. Applying Newton's second law, $\sum F = ma$ yields $F_g = ma_c$ for each star:
$$\frac{GMM}{(2r)^2} = \frac{Mv^2}{r} \quad \text{or} \quad M = \frac{4v^2 r}{G}$$
We can write r in terms of the period, T, by considering the time and distance of one complete cycle. The distance travelled in one orbit is the circumference of the stars' common orbit, so
$$2\pi r = vT \quad \Rightarrow \quad r = \frac{vT}{2\pi}$$
Now, substituting this value of r in above expression, we get
$$M = \frac{4v^2 r}{G} = \frac{4v^2}{G}\left(\frac{vT}{2\pi}\right)$$

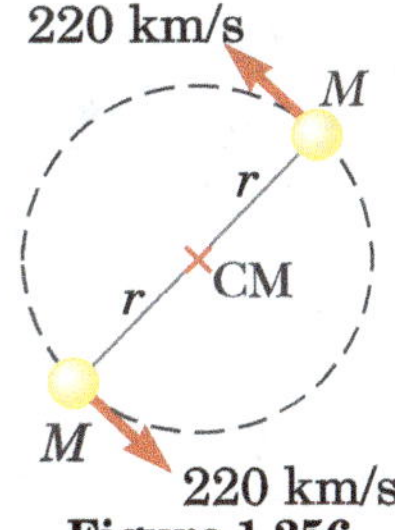

Figure 1.256

$$M = \frac{2v^3 \, T}{\pi G}$$

---

⁹If size of the orbit (a) is expressed in astronomical units (1 AU is equal to the average distance between the centre of earth and the centre of Sun) and period T is measured in years, then Kepler's third law: $T^2 = a^3$

$$= \frac{2\left(220 \times 10^3 \text{ m/s}\right)^3 (14.4 \text{ d})(86400 \text{ s/d})}{\pi \left(6.67 \times 10^{-11} \text{ N} \cdot \text{m}^2/\text{kg}^2\right)}$$
$$= 1.26 \times 10^{32} \text{ kg} = 63.3 \text{ solar masses}$$

25. To find the angular displacement of planet Y, we apply Newton's second law:
$$\sum F = ma: \quad \frac{Gm_{\text{planet}} M_{\text{star}}}{r^2} = \frac{m_{\text{planet}} v^2}{r}$$
Then, using $v = r\omega$,
$$\frac{GM_{\text{star}}}{r} = v^2 = r^2 \omega^2$$
$$GM_{\text{star}} = r^3 \omega^2 = r_x^3 \omega_x^2 = r_y^3 \omega_y^2$$
solving for the angular velocity of planet Y gives
$$\omega_y = \omega_x \left(\frac{r_x}{r_y}\right)^{3/2} = \left(\frac{90.0°}{5.00 \text{ yr}}\right)3^{3/2} = \frac{468°}{5.00 \text{ yr}}$$
Since, there are $360°$ in one revolution, therefore,

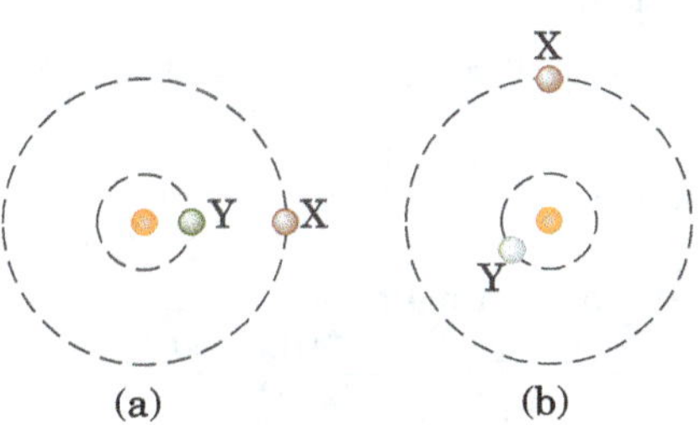

Figure 1.257

number of revolutions in $468° = \frac{468}{360}$ revolutions = 1.30 revolutions.

26. By Kepler's third law, $T^2 = ka^3$ (a = semimajor axis). For any object orbiting the Sun, with T in years and a in AU, $k = 1.00$. Therefore, for Comet Halley, and suppressing units [Fig.1.258],
$$(75.6)^2 = (1.00)\left(\frac{0.570 + x}{2}\right)^3$$
Therefore, the farthest distance the comet gets from the sun is given by
$$x = 2(75.6)^{2/3} - 0.570 = 35.2\text{AU}$$

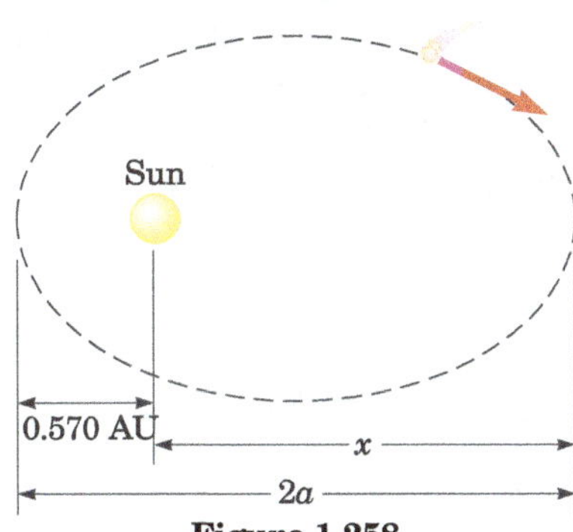

Figure 1.258

(out around the orbit of Pluto).

27. By conservation of angular momentum for the satellite,
$$r_p v_p = r_a v_a,$$
or
$$\frac{v_p}{v_a} = \frac{r_a}{r_p} = \frac{2289 \text{ km} + 6.37 \times 10^3 \text{ km}}{459 \text{ km} + 6.37 \times 10^3 \text{ km}} = \frac{8659 \text{ km}}{6829 \text{ km}} =$$
1.27
We do not need to know the period.

28. For an object in orbit about Earth, Kepler's third law gives the relation between the orbital period T and the average radius of the orbit ("semimajor axis") as We assume that the two given distances in the problem statements are the perigee and apogee, respectively. Thus, if the average radius is

$$r = \frac{r_{\min} + r_{\max}}{2} = \frac{6670 \text{ km} + 385000 \text{ km}}{2}$$
$$= 1.96 \times 10^5 \text{ km} = 1.96 \times 10^8 \text{ m}$$

The period (time for a round trip from Earth to the Moon) would be

$$T = 2\pi \sqrt{\frac{r^3}{GM}}$$

$$= 2\pi \sqrt{\frac{\left(1.96 \times 10^8 \text{ m}\right)^3}{\left(6.67 \times 10^{-11} \text{ N} \cdot \text{m}^2/\text{kg}^2\right)\left(5.98 \times 10^{24} \text{ kg}\right)}}$$

$$= 8.63 \times 10^5 \text{ s}$$

The time for a one-way trip from Earth to the Moon is then

$$\Delta t = \frac{1}{2} T = \left(\frac{8.63 \times 10^5 \text{ s}}{2}\right)\left(\frac{1 \text{ day}}{8.64 \times 10^4 \text{ s}}\right) = 4.99 \text{ days}$$

29. We find the satellite's altitude from

$$\frac{GM_J}{(R_J + d)^2} = \frac{4\pi^2 (R_J + d)}{T^2}$$

where d is the altitude of the satellite above Jupiter's cloud tops. Then,

$$GM_J T^2 = 4\pi^2 (R_J + d)^3$$

or $\left(6.67 \times 10^{-11} \text{ N} \cdot \text{m}^2/\text{kg}^2\right)\left(1.90 \times 10^{27} \text{ kg}\right)(9.84 \times 3600)^2$

$$= 4\pi^2 \left(6.99 \times 10^7 + d\right)^3$$

$$\Rightarrow \quad d = 8.92 \times 10^7 \text{ m} = 89200 \text{ km, above the planet}$$

30. (a) In $T^2 = 4\pi^2 a^3 / GM_{\text{central}}$ we take $a = 3.84 \times 10^8$ m.

$$M_{\text{central}} = \frac{4\pi^2 a^3}{GT^2}$$

$$= \frac{4\pi^2 \left(3.84 \times 10^8 \text{ m}\right)^3}{\left(6.67 \times 10^{-11} \text{ N} \cdot \text{m}^2/\text{kg}^2\right)(27.3 \times 86400 \text{ s})^2}$$

$$= 6.02 \times 10^{24} \text{ kg}$$

This is a little larger than 5.98×10^{24} kg.

Note: The Earth wobbles a bit as the Moon orbits it, so both objects move nearly in circles about their centre of mass, staying on opposite sides of it. The radius of the Moon's orbit is therefore a bit less than the Earth-Moon distance.

31. The speed of a planet in a circular orbit is given by

$$\Sigma F = ma: \quad \frac{GM_{\text{sun}} m}{r^2} = \frac{mv^2}{r} \Rightarrow v = \sqrt{\frac{GM_{\text{sun}}}{r}}$$

(a) For Mercury, the speed is

$$v_M = \sqrt{\frac{\left(6.67 \times 10^{-11} \text{ N} \cdot \text{m}^2/\text{kg}^2\right)\left(1.99 \times 10^{30} \text{ kg}\right)}{5.79 \times 10^{10} \text{ m}}}$$

$$= 4.79 \times 10^4 \text{ m/s}$$

and for Pluto,

$$v_P = \sqrt{\frac{\left(6.67 \times 10^{-11} \text{ N} \cdot \text{m}^2/\text{kg}^2\right)\left(1.99 \times 10^{30} \text{ kg}\right)}{5.91 \times 10^{12} \text{ m}}}$$

$$= 4.74 \times 10^3 \text{ m/s}$$

With greater speed, Mercury will eventually move far-

ther from the Sun than Pluto.

(b) With original distances r_P and r_M perpendicular to their lines of motion, they will be equally far from the Sun at time t, where

$$\sqrt{r_P^2 + v_P^2 t^2} = \sqrt{r_M^2 + v_M^2 t^2}$$

$$\Rightarrow \quad r_P^2 - r_M^2 = \left(v_M^2 - v_P^2\right) t^2$$

$$\Rightarrow \quad t = \sqrt{\frac{\left(5.91 \times 10^{12} \text{ m}\right)^2 - \left(5.79 \times 10^{10} \text{ m}\right)^2}{\left(4.79 \times 10^4 \text{ m/s}\right)^2 - \left(4.74 \times 10^3 \text{ m/s}\right)^2}}$$

$$= \sqrt{\frac{3.49 \times 10^{25} \text{ m}^2}{2.27 \times 10^9 \text{ m}^2/\text{s}^2}} = 1.24 \times 10^8 \text{ s} = 3.93 \text{ yr}$$

32. A satellite is in a circular orbit around the Earth (mass $= M$, radius $= R$) at an altitude of 2.80×10^6 m. Find (a) the period of the orbit, (b) the speed of the satellite, and (c) the acceleration of the satellite.

Sol (a) The radius of the satellite's orbit is

$$r = R + h = 6.37 \times 10^6 \text{ m} + 2.80 \times 10^6 \text{ m} = 9.17 \times 10^6 \text{ m}$$

Then, modifying Kepler's third law for orbital motion about the Earth rather than the Sun, we have

$$T^2 = \left(\frac{4\pi^2}{GM}\right) r^3 = \frac{4\pi^2 \left(9.17 \times 10^6 \text{ m}\right)^3}{\left(6.67 \times 10^{-11} \text{ N} \cdot \text{m}^2/\text{kg}^2\right)\left(5.98 \times 10^{24} \text{ kg}\right)}$$

$$= 7.63 \times 10^7 \text{ s}^2$$

or $T = \left(8.74 \times 10^3 \text{ s}\right)\left(\dfrac{1 \text{ h}}{3600 \text{ s}}\right) = 2.43$ h (b) The constant tangential speed of the satellite is

$$v = \frac{2\pi r}{T} = \frac{2\pi \left(9.17 \times 10^6 \text{ m}\right)}{8.74 \times 10^3 \text{ s}}$$

$$= 6.60 \times 10^3 \text{ m/s} = 6.60 \text{ km/s}$$

(c) The satellite's only acceleration is centripetal acceleration, so

$$a = a_c = \frac{v^2}{r} = \frac{\left(6.60 \times 10^3 \text{ m/s}\right)^2}{9.17 \times 10^6 \text{ m}} = 4.74 \text{ m/s}^2$$
$$\text{toward the Earth}$$

33. **APPROACH** While we could apply Newton's law of gravitation and second law of motion to solve this problem, we'll use Kepler's third law (derived from these laws) to find the period of Mimas and to relate the periods of the moons of Saturn to their mean distances from its centre. **SOLUTION** (a) By Kepler's third law, the period of Mimas having mean orbital radius r_M from the centre of Saturn:

$$T_M^2 = \frac{4\pi^2}{GM_s} r_M^3 \quad \Rightarrow \quad T_M = \sqrt{\frac{4\pi^2}{GM_s} r_M^3}$$

Substitute numerical values and evaluate T_M:

$$T_M = \sqrt{\frac{4\pi^2 (1.86 \times 10^8 m)^3}{(6.6726 \times 10^{-11} N \cdot m^2/kg^2)(5.69 \times 10^{26} kg)}}$$

$$= 8.18 \times 10^4 \text{ s} \approx 22.7 \text{ h}$$

(b) By Kepler's third law, the period of Titan, having mean orbital radius r_T from the centre of Saturn:

$$T_T^2 = \frac{4\pi^2}{GM_S} r_T^3 \Rightarrow r_T = \sqrt[3]{\frac{T_T^2 GM_S}{4\pi^2}}$$

Substitute numerical values and evaluate r_T:

$$r_T = \sqrt[3]{\frac{(1.38 \times 10^6 s)^2 (6.6726 \times 10^{-11} N \cdot m^2/kg^2)(5.69 \times 10^{26} kg)}{4\pi^2}}$$

$$= 1.22 \times 10^9 \text{m}$$

34. **APPROACH** While we could apply Newton's law of gravitation and second law of motion to solve this problem, we'll use Kepler's third law (derived from these laws) to relate the period of the moon to the mass of Earth and the mean Earth-moon distance.

SOLUTION By Kepler's third law, the period of the moon:

$$T_m^2 = \frac{4\pi^2}{GM_E} r_m^3 \Rightarrow M_E = \frac{4\pi^2}{GT_m^2} r_m^3$$

Substitute numerical values and evaluate M_E:

$$M_E = \frac{4\pi^2(3.84\times10^8 \text{ m})^3}{(6.6726\times10^{-11}\text{N·m}^2/\text{kg}^2)(273 \text{ days} \times \frac{24 \text{ h}}{\text{days}} \times \frac{3600 \text{ s}}{\text{h}})^2}$$

$$= 6.02 \times 10^{24} \text{ kg}$$

Remarks: This analysis neglects the mass of the moon; consequently the mass calculated here is slightly too large.

Note: In Problems 35 through 41, assume $U = 0$ at $r = \infty$.

35. (a) We compute the gravitational potential energy of the satellite Earth system from

$$U = -\frac{GMm}{r}$$

$$= -\frac{\left(6.67 \times 10^{-11} \text{ N·m}^2/\text{kg}^2\right)\left(5.98 \times 10^{24} \text{ kg}\right)(100 \text{ kg})}{(6.37 + 2.00) \times 10^6 \text{ m}}$$

$$= -4.77 \times 10^9 \text{ J}$$

(b), (c) The satellite and Earth exert forces of equal magnitude on each other, directed downward on the satellite and upward on Earth. The magnitude of this force is

$$F = \frac{GMm}{r^2}$$

$$= \frac{\left(6.67 \times 10^{-11} \text{ N·m}^2/\text{kg}^2\right)\left(5.98 \times 10^{24} \text{ kg}\right)(100 \text{ kg})}{\left(8.37 \times 10^6 \text{ m}\right)^2}$$

$$= 569 \text{ N}$$

36. **APPROACH** Let m represent the mass of the body that is projected vertically from the surface of Earth. To determine H', we apply the law of conservation of mechanical energy under the assumption that the gravitational field is constant. To find, H, we again apply the conservation of energy, with the zero of gravitational potential energy at infinity. Finally, we'll solve these two equations simultaneously to express H in terms of H'

SOLUTION When the gravitational field is constant: Let for constant gravitational field the reference potential energy level (i.e., zero potential energy level), is at the surface of Earth. Now, apply conservation of mechanical energy to relate the initial kinetic energy and the final potential energy of the object-Earth system, we get

$$U_f + K_f = U_i + Ki$$

Since, $K_f = U_i = 0$; therefore $U_f = K_i$

or $\quad mgH' = \frac{1}{2}mv^2$

$$\Rightarrow \quad H' = \frac{v^2}{2g} \quad \Rightarrow \quad v^2 = 2gH' \qquad \text{... (1)}$$

When Height Changes the Gravitational Field: Assuming the zero of gravitational potential energy be at infinity, we use conservation of mechanical energy to relate the initial kinetic energy and the final potential energy of the object-Earth system, i.e.

$$K_f + U_f = K_i + U_i$$

Since, $K_f = 0, \quad -K_i + U_f - U_i = 0$

Substitute for K_i, therefore, $U_f = K_i + U_i$

$$\Rightarrow \quad -\frac{GMm}{R+H} = \frac{1}{2}mv^2 - \frac{1}{2}mv^2 - \frac{GMm}{R}$$

On substituting the value of v from Eq.(1), we get

$$-\frac{GM}{R+H}\frac{1}{2}\left(2gH'\right) - \frac{GM}{R} = gH' - \frac{GM}{R}$$

or $\quad gH' = \frac{GM}{R} - \frac{GM}{R+H} = GM\left(\frac{H}{R(R+H)}\right)$

$$\Rightarrow \quad gH'R(R+H) = GMH$$

$$\Rightarrow \quad H(GM - gH'R) = gH'R^2$$

$$\Rightarrow \quad H = \frac{gH'R^2}{GM - gH'R}$$

Since, $GM = gR^2$, therefore, $H = \dfrac{gH'R^2}{gR^2 - gH'R}$

$$\Rightarrow \quad H' \frac{H'R^2}{R(R-H')} = \frac{H'R}{R - H'}$$

37. The work done by the Moon's gravitational field is equal to the negative of the change of potential energy of the meteor-Moon system:

$$W_{\text{grav}} = -\Delta U = -\left(\frac{-Gm_1m_2}{r} - 0\right)$$

$$W_{\text{grav}} = \frac{(6.67\times10^{-11} \text{ N·m}^2/\text{kg}^2)(7.36\times10^{22} \text{ kg})(1.00\times10^3 \text{ kg})}{1.74\times10^6 \text{ m}}$$

$$= 2.82 \times 10^9 \text{ J}$$

38. The energy required is equal to the change in gravitational potential energy of the object-Earth system:

$$U = -G\frac{Mm}{r} \quad \text{and} \quad g = \frac{G_E}{R_E^2} \quad \text{so that}$$

$$\Delta U = -GMm\left(\frac{1}{3R} - \frac{1}{R}\right) = \frac{2}{3}mgR$$

On substituting the numerical velue in above equation, we get

$$\Delta U = \frac{2}{3}(1000 \text{ kg})\left(9.80 \text{ m/s}^2\right)\left(6.37 \times 10^6 \text{ m}\right) = 4.17 \times 10^{10} \text{ J}$$

39. (a) The definition of density gives

$$\rho = \frac{M_S}{\frac{4}{3}\pi R_{\text{earth}}^2} = \frac{3\left(1.99 \times 10^{30} \text{ kg}\right)}{4\pi\left(6.37 \times 10^6 \text{ m}\right)^3} = 1.84 \times 10^9 \text{ kg/m}^3$$

(b) For an object of mass m on its surface, $mg = GM_S m/R_{\text{earth}}^2$. Thus,

$$g = \frac{GM_S}{R_{\text{earth}}^2}$$

$$= \frac{\left(6.67 \times 10^{-11} \text{ N·m}^2/\text{kg}^2\right)\left(1.99 \times 10^{30} \text{ kg}\right)}{\left(6.37 \times 10^6 \text{ m}\right)^2}$$

$$= 3.27 \times 10^6 \text{ m/s}^2$$

(c) Relative to $U_g = 0$ at infinity, the potential energy of the object-star system at the surface of the white dwarf is

$$U_g = -\frac{GM_S m}{R_{\text{earth}}}$$

$$= -\frac{\left(6.67 \times 10^{-11} \text{N·m}^2/\text{kg}^2\right)\left(1.99 \times 10^{30} \text{ kg}\right)(1.00 \text{ kg})}{6.37 \times 10^6 \text{ m}}$$

$$= -2.08 \times 10^{13} \text{ J}$$

40. (a) Energy conservation of the object-Earth system from release to radius r:

$$\left(K + U_g\right)_{\text{altitude } h} = \left(K + U_g\right)_{\text{radius } r}$$

$$\Rightarrow \quad 0 - \frac{GMm}{R+h} = \frac{1}{2}mv^2 - \frac{GMm}{r}$$

$$\Rightarrow \quad v = \left[2M\left(\frac{1}{r} - \frac{1}{R+h}\right)\right]^{1/2} = -\frac{dr}{dt}$$

(b) $\int_i^f dt = \int_i^f -\frac{dr}{v} = \int_f^i \frac{dr}{v}$.

The time of fall is,

$$\Delta t = \int_R^{R+h} \left[2GM\left(\frac{1}{r} - \frac{1}{R+h}\right)\right]^{-1/2} dr$$

or

$$\Delta t = \left(2 \times 6.67 \times 10^{-11} \times 5.98 \times 10^{24}\right)^{-1/2}$$

$$\times \int_{6.37 \times 10^6 \text{ m}}^{6.87 \times 10^6 \text{ m}} \left[\left(\frac{1}{r} - \frac{1}{6.87 \times 10^6 \text{ m}}\right)\right]^{-1/2} dr$$

Alternatively, to save time we can change variables to $u = \frac{r}{10^6}$. Then

$$\Delta t = \left(7.977 \times 10^{14}\right)^{-1/2} \int_{6.37}^{6.87} \left(\frac{1}{10^6 u} - \frac{1}{6.87 \times 10^6}\right)^{-1/2} 10^6 du$$

$$= 3.541 \times 10^{-8} \frac{10^6}{(10^6)^{-1/2}} \int_{6.37}^{6.87} \left(\frac{1}{u} - \frac{1}{6.87}\right)^{-1/2} du$$

On simplifying the integration of right hand side of above equation, we get

$$\int_{6.37}^{6.87} \left(\frac{1}{u} - \frac{1}{6.87}\right)^{-1/2} du = 9.596, \text{ therefore-}$$

$$\Delta t = 3.541 \times 10^{-8} \times 10^9 \times 9.596 = 339.8 = 340 \text{ s}$$

41. (a) Since the particles are located at the corners of an equilateral triangle, the distances between all particle pairs is equal to 0.300 m. The gravitational potential energy of the system is then

$$U_{\text{Tot}} = U_{12} + U_{13} + U_{23} = 3U_{12} = 3\left(-\frac{Gm_1 m_2}{r_{12}}\right)$$

$$U_{\text{Tot}} = -\frac{3\left(6.67 \times 10^{-11} \text{ N}\cdot\text{m}^2/\text{kg}^2\right)\left(5.00 \times 10^{-3} \text{ kg}\right)^2}{0.300 \text{ m}}$$

$$= -1.67 \times 10^{-14} \text{ J}$$

(b) Each particle feels a net force of attraction toward the midpoint between the other two. Each moves toward the center of the triangle with the same acceleration. They collide simultaneously at the centre of the triangle.

42. We use the isolated system model for energy: The principle of conservation of mechanical energy gives,

$$K_i + U_i = K_f + U_f$$

$$\frac{1}{2}mv_i^2 + GMm\left(\frac{1}{r_f} - \frac{1}{r_i}\right) = \frac{1}{2}mv_f^2$$

which becomes, $\quad \frac{1}{2}v_i^2 + GM\left(0 - \frac{1}{R}\right) = \frac{1}{2}v_f^2$

or

$$v_f^2 = v_1^2 - \frac{2GM}{R}$$

and

$$v_f = \left(v_1^2 - \frac{2GM}{R}\right)^{1/2}$$

$$v_f = \left[\left(2.00 \times 10^4 \text{ m/s}\right)^2 \right.$$

$$\left. - \left(\frac{2(6.67 \times 10^{-11} \text{ N}\cdot\text{m}^2/\text{kg}^2)(5.98 \times 10^{24} \text{ kg})}{6.37 \times 10^6 \text{ m}}\right)\right]^{1/2}$$

$$= 1.66 \times 10^4 \text{ m/s}$$

43. To determine the energy transformed to internal energy, we begin by calculating the change in kinetic energy of the satellite. To find the initial kinetic energy, we use

$$\frac{mv^2}{r_i} = \frac{GMm}{r^2}$$

or

$$\frac{v_i^2}{R+h} = \frac{GM_E}{(R+h)^2}$$

which gives

$$K_i = \frac{1}{2}mv_i^2 = \frac{1}{2}\left(\frac{GMm}{R+h}\right)$$

$$= \frac{1}{2}\left[\frac{\left(6.67 \times 10^{-11} \text{ N}\cdot\text{m}^2/\text{kg}^2\right)\left(5.98 \times 10^{24} \text{ kg}\right)\left(500 \text{ kg}\right)}{6.37 \times 10^6 \text{ m} + 0.500 \times 10^6 \text{ m}}\right]$$

$$= 1.45 \times 10^{10} \text{ J} \quad \text{Also,} \quad K_f = \frac{1}{2}mv_f^2 =$$

$\frac{1}{2}(500 \text{ kg})\left(2.00 \times 10^3 \text{ m/s}\right)^2 = 1.00 \times 10^9 \text{ J}$. The change in gravitational potential energy of the satellite-Earth system is $\Delta U = \dfrac{GMm}{r_i} - \dfrac{GMm}{r_f} = GMm\left(\dfrac{1}{r_i} - \dfrac{1}{r_f}\right)$

$$= \left(6.67 \times 10^{-11} \text{ N}\cdot\text{m}^2/\text{kg}^2\right)\left(5.98 \times 10^{24} \text{ kg}\right)\left(500 \text{ kg}\right)$$

$$\times \left(-1.14 \times 10^{-8} \text{ m}^{-1}\right) \times -2.27 \times 10^9 \text{ J}$$

The energy transformed into internal energy due to friction is then

$$\Delta E_{\text{int}} = K_i - K_f - \Delta U = (14.5 - 1.00 + 2.27) \times 10^9 \text{ J}$$

$$= 1.58 \times 10^{10} \text{ J}$$

44. To obtain the orbital velocity, we use

$$\sum F = \frac{GMm}{R^2} = \frac{mv^2}{R} \quad \Rightarrow \quad v = \sqrt{\frac{GM}{R}}$$

To escape the satellite, it's kinetic energy should be sufficient to make net mechanical energy zero, i.e.,

$$K + U = 0 \quad \Rightarrow \quad K = -U = -\left(-\frac{GMm}{R}\right)$$

$$\Rightarrow \quad \frac{1}{2}mv_{\text{esc}}^2 = \frac{GMm}{R} \quad \Rightarrow \quad v_{\text{esc}}^2 = \frac{2GM}{R}$$

$$\Rightarrow \quad v_{\text{esc}} = \sqrt{\frac{2GM}{R}} = \sqrt{2}v$$

45. (a) The total energy of the satellite-Earth system at a given orbital altitude is given by

$$E_{\text{tot}} = -\frac{GMm}{2r}$$

The energy needed to increase the satellite's orbit is then, suppressing units, $\quad \Delta E = \dfrac{GMm}{2}\left(\dfrac{1}{r_i} - \dfrac{1}{r_f}\right)$

$$= \frac{\left(6.67 \times 10^{-11}\right)\left(5.98 \times 10^{24}\right)}{2} \frac{10^3 \text{ kg}}{10^3 \text{ m}}\left(\frac{1}{6370 + 100}\right.$$

$$\left. - \frac{1}{6370 + 200}\right)$$

$$= 4.69 \times 10^8 \text{ J} = 469 \text{ MJ}$$

(b) Both in the original orbit and in the final orbit, the total energy is negative, with an absolute value equal to the positive kinetic energy. The potential energy is negative and twice as large as the total energy. As the satellite is lifted from the lower to the higher orbit, the gravitational energy increases, the kinetic energy decreases, and the total energy increases. The value of each becomes closer to zero. Numerically, the gravitational energy increases by 938MJ, the kinetic energy decreases by 469MJ, and the total energy increases by 469MJ.

46. (a) The major axis of the orbit is $2a = 50.5$ AU so $a = 25.25$ AU.

Further, $a + ae = 50$ AU, so $ae = 24.75$ AU. Then

$$e = \frac{24.75}{25.25} = 0.980$$

(b) In $T^2 = K_s a^3$ for objects in solar orbit, the Earth gives us

$$(1 \text{ yr})^2 = K_s(1\text{AU})^3 \quad \Rightarrow \quad K_s = \frac{(1 \text{ yr})^2}{(1 \text{ AU})^3}$$

Then, $T^2 = \frac{(1\text{yr})^2}{(1\text{AU})^3}(25.25\text{AU})^3 \Rightarrow T = 127$ yr

(c) $\quad U = -\frac{GMm}{r}$

$$= -\frac{(6.67\times10^{-11} \text{ N·m}^2/\text{kg}^2)(1.99\times10^{30} \text{ kg})(1.20\times10^{10} \text{ kg})}{50(1.496\times10^{11} \text{ m})}$$

$$= -2.13 \times 10^{17} \text{ J}$$

47. For a satellite in an orbit of radius r around the Earth, the total energy of the satellite-Earth system is $E = -\frac{GM_E}{2r}$. Thus, in changing from a circular orbit of radius $r = 2R_E$ to one of radius $r = 3R_E$, the required work is

$$W = \Delta E = E_f - E_i = -\frac{GMm}{2r_f} + \frac{GMm}{2r_i}$$

$$= GMm\left[\frac{1}{4R} - \frac{1}{6R}\right] = \frac{GMm}{12R}$$

48. (a) The work must provide the increase in gravitational energy:

$$W = \Delta U_g = U_{gf} - U_{gi}$$

$$= -\frac{GMm}{r_f} + \frac{GMm}{r_i}$$

$$= -\frac{GMm}{R+y} + \frac{GMm}{R}$$

$$= GMm\left(\frac{1}{R} - \frac{1}{R+y}\right)$$

$$= \left(\frac{6.67\times10^{-11} \text{ N·m}^2}{\text{kg}^2}\right)(5.98\times10^{24} \text{ kg})(100 \text{ kg})$$

$$\left(\frac{1}{6.37\times10^6 \text{ m}} - \frac{1}{7.37\times10^6 \text{ m}}\right)$$

$$= 850 \text{ MJ}$$

(b) In a circular orbit, gravity supplies the centripetal force:

$$\frac{GMm}{(R+y)^2} = \frac{mv^2}{R+y}$$

Therefore, $\quad \frac{1}{2}mv^2 = \frac{1}{2}\frac{GMm}{(R+y)}$

So, additional work = required kinetic energy, i.e.,

$$\Delta W = \frac{1}{2}\frac{(6.67\times10^{-11} \text{ N·m}^2/\text{kg}^2)(5.98\times10^{24} \text{ kg})(100 \text{ kg})}{7.37\times10^6 \text{ m}}$$

$$= 2.71 \times 10^9 \text{ J}$$

49. (a) The escape speed from the solar system, starting at Earth's orbit, is given by

$$v_{\text{solar escape}} = \sqrt{\frac{2M_{\text{Sun}}G}{R_{\text{Sun}}}}$$

$$= \sqrt{\frac{2\left(1.99\times10^{30} \text{ kg}\right)\left(6.67\times10^{-11} \text{ N·m}^2/\text{kg}^2\right)}{1.50\times10^9 \text{ m}}}$$

$$= 42.1 \text{ km/s}$$

(b) Let x represent the variable distance from the Sun. Then,

$$v = \sqrt{\frac{2M_{\text{Sun}}G}{x}} \quad \Rightarrow \quad x = \frac{v^2}{2M_{\text{Sun}}G}$$

If $v = \frac{125000 \text{ km}}{3600 \text{ s}} = 34.7 \text{ m/s}$, then

$$x = \frac{v^2}{2M_{\text{Sun}}G} = \frac{(34.7 \text{ m/s})^2}{2\left(1.99\times10^{30} \text{ kg}\right)\left(6.67\times10^{-11} \text{ N·m}^2/\text{kg}^2\right)}$$

$$= 2.20 \times 10^{11} \text{ m}$$

Note that at or beyond the orbit of Mars, 125000 km/h is sufficient for escape.

50. $F_C = F_G$ gives $\dfrac{mv^2}{r} = \dfrac{GMm}{r^2}$

which reduces to $v = \sqrt{\dfrac{GM}{r}}$

and period $= \dfrac{2\pi r}{v} = 2\pi r\sqrt{\dfrac{r}{GM}}$.

(a) $r = R + 200$ km $= 6370$ km $+ 200$ km $= 6570$ km. Thus, period,

$$T = 2\pi\left(6.57\times10^6 \text{ m}\right) \times \sqrt{\frac{6.57\times10^6 \text{ m}}{(6.67\times10^{-11} \text{ N·m}^2/\text{kg}^2)(5.98\times10^{24} \text{ kg})}}$$

$$= 5.30\times10^3 \text{ s} = 88.3 \text{ min} = 1.47 \text{ h}$$

(b) $v = \sqrt{\dfrac{GM}{r}}$

$$= \sqrt{\frac{(6.67\times10^{-11} \text{ N·m}^2/\text{kg}^2)\left(5.98\times10^{24} \text{ kg}\right)}{6.57\times10^6 \text{ m}}}$$

$$= 7.79 \text{ km/s}$$

(c) $K_f + U_f = K_i + U_i +$ Energy input, gives

Energy input $\quad = \dfrac{1}{2}mv_f^2 - \dfrac{1}{2}mv_i^2 + \left(\dfrac{-GMm}{r_f}\right)$

$$- \left(\frac{-GMm}{r_i}\right) \qquad \ldots (1)$$

here, $\quad r_i = R = 6.37\times10^6$ m

$$v_i = \frac{2\pi R}{86400 \text{ s}} = 4.63\times10^2 \text{ m/s}$$

$$r_f = 6570\times10^3 \text{ m}, \, v_f = 7.79\times \text{ m/s}$$

Substituting these values in Eq.(1), we get minimum energy input $= 6.43\times10^9$ J

51. The gravitational force supplies the needed centripetal acceleration.

Thus, $\quad \dfrac{GMm}{(R+h)^2} = \dfrac{mv^2}{R+h} \quad$ or $\quad v^2 = \dfrac{GM}{R+h}$

(a) $T = \dfrac{2\pi r}{v} = \dfrac{2\pi(R+h)}{\sqrt{\frac{GM}{(R+h)}}} = 2\pi\sqrt{\dfrac{(R+h)^3}{GM}}$

(b) $v = \sqrt{\dfrac{GM}{R+h}}$

(c) Minimum energy input is

$$\Delta E_{\text{min}} = \left(K_f + U_f\right) - \left(K_i - U_i\right)$$

This choice has the object starting with energy

$$K_i = \frac{1}{2}mv_i^2$$

with $v_i = \dfrac{2\pi R}{1.00 \text{ day}} = \dfrac{2\pi R}{86400 \text{ s}}$ and $U_i = -\dfrac{GMm}{R}$.

Thus, $\Delta E_{\text{min}} = \dfrac{1}{2}m\left(\dfrac{GM}{R+h}\right) - \dfrac{GMm}{R+h} - \dfrac{1}{2}m\left[\dfrac{4\pi^2R^2}{(86400 \text{ s})^2}\right] + \dfrac{GMm}{R}$

or $\quad \Delta E_{\text{min}} = GMm\left[\dfrac{R+2h}{2R(R+h)}\right] - \dfrac{2\pi^2R^2m}{(86400 \text{ s})^2}$

52. (a) Gravitational screening does not exist. The presence of the satellite has no effect on the force the planet exerts on the rocket. (b) The rocket has a gravitational potential energy with respect to Ganymede

$$U_1 = -\frac{Gm_1m_2}{r} = -\frac{(6.67\times10^{-11} \text{ N·m}^2)m_2(1.495\times10^{23} \text{ kg})}{(2.64\times10^6 \text{ m})\text{kg}^2}$$

$$U_1 = \left(-3.78\times10^6 \text{ m}^2/\text{s}^2\right)m_2$$

The rocket's gravitational potential energy with respect to Jupiter at the distance of Ganymede is

$$U_2 = -\frac{Gm_1m_2}{r} = -\frac{(6.67\times10^{-11}\ \text{N·m}^2)m_2(1.90\times10^{27}\ \text{kg})}{(1.071\times10^9\ \text{m})\text{kg}^2}$$

$$U_2 = \left(-1.18\times10^8\ \text{m}^2/\text{s}^2\right)m_2$$

To escape from both requires $\frac{1}{2}m_2v_{esc}^2 = +\left[(3.78\times10^6 + 1.18\times10^8)\ \text{m}^2/\text{s}^2\right]m_2$

$$v_{esc} = \sqrt{2\times\left(1.22\times10^8\ \text{m}^2/\text{s}^2\right)} = 15.6\ \text{km/s}.$$

53. (a) For the satellite $\sum F = ma$; $\frac{GMm}{r^2} = \frac{mv_i^2}{r}$ gives

$$v_i = \left(\frac{GM_E}{r}\right)^{1/2}$$

(b) Conservation of momentum in the forward direction for the exploding satellite gives: $(\sum mv)_i = (\sum mv)_f$

$$\Rightarrow \qquad 5mv_i = 4mv + m(0)$$

$$\Rightarrow \qquad v = \frac{5}{4}v_i = \frac{5}{4}\left(\frac{GM}{r}\right)^{1/2}$$

(c) With velocity perpendicular to radius, the orbiting fragment is at perigee. By conservation of angular momentum about the centre of earth, it's apogee distance and speed are related to r and v as:

$$4mrv = 4mr_fv_f \qquad v_f = \frac{rv}{r_f}$$

By conservation of mechanical energy , we have-

$$\frac{1}{2}mv^2 - \frac{GMm}{r} = \frac{1}{2}mv_f^2 - \frac{GMm}{r_f}$$

Substituting $v_f = \frac{vr}{r_f}$ we have $\frac{1}{2}v^2 - \frac{GM}{r} =$

$$\frac{1}{2}\frac{v^2r^2}{r_f^2} - \frac{GM}{r_f}$$

Further, substituting $v^2 = \frac{25}{16}\frac{GM}{r}$ gives

$$\frac{25}{32}\frac{GM}{r} - \frac{GM}{r} = \frac{25}{32}\frac{GMr}{r_f^2} - \frac{GM}{r_f}$$

$$\Rightarrow \qquad \frac{-7}{32r} = \frac{25r}{32r_f^2} - \frac{1}{r_f}$$

Clearing fractions we have

$$-7r_f^2 = 25r^2 - 32rr_f$$

or $\qquad 7\left(\frac{r_f}{r}\right)^2 - 32\left(\frac{r_f}{r}\right) + 25 = 0$

$$\Rightarrow \quad \frac{r_f}{r} = \frac{+32\pm\sqrt{32^2 - 4(7)(25)}}{14} = \frac{50}{14}\ \text{or}\ \frac{14}{14}$$

The latter root describes the starting point.

The outer end of the orbit has $\frac{r_f}{r} = \frac{25}{7} : r_f = \frac{25r}{7}$

54. From launch point to apogee at height h, conservation of energy gives

$$K_i + U_i + \Delta\text{E}_{mech} = K_f + U_f$$

$$\frac{1}{2}mv_i^2 - \frac{GMm}{R} + 0 = 0 - \frac{GMm}{R+h}$$

The mass of the projectile cancels out, giving

$$\frac{1}{2}v_i^2 - \frac{GM}{R} = -\frac{GM}{R+h}$$

$$\Rightarrow \qquad R+h = \frac{GM}{\frac{1}{2}v_i^2 - \frac{GM}{R}}$$

$$h = \frac{GM}{\frac{1}{2}v_i^2 - \frac{GM}{R}} - R$$

$$= \frac{(6.67\times10^{-11}\ \text{N·m}^2/\text{kg}^2)(5.98\times10^{24}\ \text{kg})}{\frac{1}{2}(10.0\times10^3\ \text{m/s})^2 - \frac{(6.67\times10^{-11}\ \text{N·m}^2/\text{kg}^2)(5.98\times10^{24}\ \text{kg})}{(6.37\times10^6\ \text{m})}}$$

$$= 2.2.52\times10^7\ \text{m}$$

55. **APPROACH** We can apply Newton's second law and the law of gravity to an object of mass m at the surface of the Sun and the neutron-Sun to find the ratio of the gravitational accelerations at their surfaces. Similarly, we can express the ratio of the corresponding expressions for the escape speeds from the two suns to determine their ratio.

SOLUTION (a) Express the gravitational force acting on an object of mass m at the surface of the Sun:

$$F_g = ma_g = \frac{GM_{\text{Sun}}m}{R_{\text{Sun}}^2}$$

Solving for a_g yields:

$$a_g = \frac{GM_{\text{Sun}}}{R_{\text{Sun}}^2} \qquad\qquad \ldots(1)$$

The gravitational force acting on an object of mass m at the surface of a neutron-Sun is:

$$F_g = ma'_g = \frac{GM_{\text{neutron-Sun}}m}{R_{\text{neutron-Sun}}^2}$$

Solving for a'_g yields:

$$a'_g = \frac{GM_{\text{neutron-Sun}}}{R_{\text{neutron-Sun}}^2} \qquad\qquad \ldots(2)$$

Dividing equation (2) by equation (1) to obtain:

$$\frac{a'_g}{a_g} = \frac{\frac{GM_{\text{neutron-Sun}}}{R_{\text{neutron-Sun}}^2}}{\frac{GM_{\text{Sun}}}{R_{\text{Sun}}^2}} = \frac{\frac{M_{\text{neutron-Sun}}}{R_{\text{neutron-Sun}}^2}}{\frac{M_{\text{Sun}}}{R_{\text{Sun}}^2}}$$

Because $M_{\text{neutron-Sun}} = M_{\text{Sun}}$.

$$\frac{a'_g}{a_g} = \frac{R_{\text{Sun}}^2}{R_{\text{neutron-Sun}}^2}$$

Substitute numerical values and evaluate the ratio a'_g/a_g

$$\frac{a'_g}{a_g} = \left(\frac{6.96\times10^8 m}{12.0\times10^3 m}\right)^2 = 3.36\times10^9$$

(b) The escape speed from the neutron-Sun is given by:

$$v'_e = \sqrt{\frac{GM_{\text{neutron-Sun}}}{R_{\text{neutron-Sun}}}}$$

The escape speed from the Sun is given by:

$$v_e = \sqrt{\frac{GM_{\text{Sun}}}{R_{\text{Sun}}}}$$

Dividing the first of these equations by the second and simplifying yields:

$$\frac{v'_e}{v_e} = \sqrt{\frac{R_{\text{Sun}}}{R_{\text{neutron-Sun}}}}$$

Substitute numerical values and evaluate v'_e/v_e :

$$\frac{v'_e}{v_e} = \sqrt{\frac{6.96\times10^8 m}{12.0\times10^3 m}} = 241$$

56. (a) When the rocket engine shuts off at an altitude of 250 km, we may consider the rocket to be beyond Earth's atmosphere. Then, its mechanical energy will remain constant from that instant until it comes to rest momentarily at the maximum altitude. That is,

$$K_f + U_f = K_i + U_i$$

or $\qquad 0 - \frac{GMm}{r_{\text{max}}} = \frac{1}{2}mv_i^2 - \frac{GMm}{r_i}$

or $\qquad \frac{1}{r_{\text{max}}} = -\frac{v_i^2}{2GM} + \frac{1}{r_i}$

With $r_1 = R + 250\ \text{km} = 6.37\times10^6\ \text{m} + 250\times10^3\ \text{m} = 6.62\times10^6\ \text{m}$ and $v_i = 6.00\ \text{km/s} = 6.00\times10^3\ \text{m/s}$, this gives

$$\frac{1}{r_{\text{max}}} = -\frac{\left(6.00\times10^3\ \text{m/s}\right)^2}{2\left(6.67\times10^{-11}\ \text{N·m}^2/\text{kg}^2\right)\left(5.98\times10^{24}\ \text{kg}\right)} + \frac{1}{6.62\times10^6\ \text{m}}$$

$$= 1.06\times10^{-7}\ \text{m}^{-1}$$

or $r_{\text{max}} = 9.44\times10^6\ \text{m}$. The maximum distance from

Earth's surface is then
$$h_{\max} = r_{\max} - R = 9.44 \times 10^6 \text{ m} - 6.37 \times 10^6 \text{ m}$$
$$= 3.07 \times 10^6 \text{ m}$$

(b) If the rocket were fired from a launch site on the equator, it would have a significant eastward component of velocity because of the Earth's rotation about its axis. Hence, compared to being fired from the South Pole, the rocket's initial speed would be greater, and the rocket would travel farther from Earth.

57. For a 6.00 km diameter cylinder, $r = 3000$ m and to simulate 1 g $= 9.80$ m/s^2
$$g = \frac{v^2}{r} = \omega^2 r$$
or $\qquad \omega = \sqrt{\frac{g}{r}} = 0.0572$ rad/s

The required rotation rate of the cylinder is $\dfrac{1 \text{ rev}}{110 \text{ s}}$.

58. To approximate the height of the sulfur, set $\frac{mv^2}{2} = mg_{10}\, h$, with $h = 70000$ m and $g_{10} = \frac{GM}{r^2} = 1.79$ m/s^2. This gives
$$v = \sqrt{2\, g_{10}\, h} = \sqrt{2\,(1.79 \text{ m/s}^2)\,(70000 \text{ m})}$$
$$\approx 500 \text{ m/s (over 1000 mi/h)}$$
We can obtain a more precise answer from conservation of energy:
$$\frac{1}{2}mv^2 - \frac{GMm}{r_1} = -\frac{GMm}{r_2}$$
$$\tfrac{1}{2}v^2 = \left(6.67 \times 10^{-11} \text{ N} \cdot \text{m}^2/\text{kg}^2\right)\left(8.90 \times 10^{22} \text{ kg}\right)$$
$$\times \left(\frac{1}{1.82 \times 10^6 \text{ m}} - \frac{1}{1.89 \times 10^6 \text{ m}}\right)$$
or $\qquad v = 492$ m/s

59. If one uses the result $v = \sqrt{\frac{GM}{r}}$ and the relation $v = (2\pi r/T)$, corresponding to $T = 1$ h $= 3600$ s one finds the radius of the orbit to be smaller than the radius of the Earth, so the spacecraft would need to be in orbit underground.

60. The acceleration of an object at the centre of the Earth due to the gravitational force of the Moon is given by
$$a = G\frac{M_M}{d^2}.$$
At the point A nearest the Moon,
$$a_+ = G\frac{M_m}{(d-R)^2}$$
At the point B farthest from the Moon,

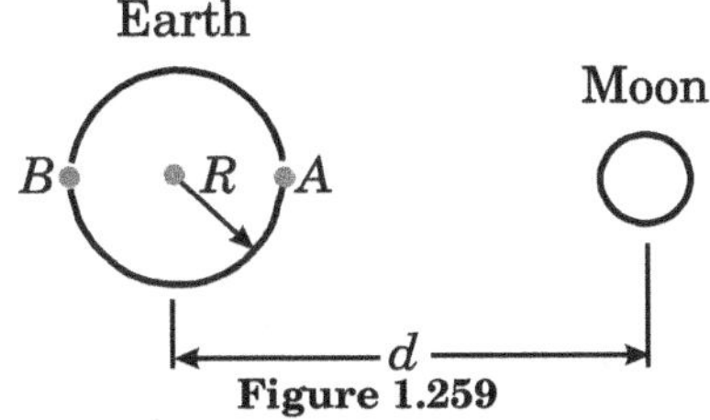

Figure 1.259

$$a_- = G\frac{M_m}{(d+R)^2}$$
From the above, we have
$$\frac{\Delta g_m}{g} = \frac{(a_+ - a)}{g} = \frac{GM_m}{g}\left[\frac{1}{(d-R)^2} - \frac{1}{(d+R)^2}\right]$$
Evaluating this expression, we find across the planet

$$\frac{\Delta g_m}{g} = \frac{\left(6.67 \times 10^{-11} \text{ N} \cdot \text{m}^2/\text{kg}^2\right)\left(7.36 \times 10^{22} \text{ kg}\right)}{9.80 \text{ m/s}^2}$$
$$\times \left[\frac{1}{\left(3.84 \times 10^8 \text{ m} - 6.37 \times 10^6 \text{ m}\right)^2} - \frac{1}{\left(3.84 \times 10^8 \text{ m} + 6.37 \times 10^6 \text{ m}\right)^2}\right]$$
$$= 2.25 \times 10^{-7}$$

61. (a) The only force acting on the astronaut is the normal force exerted on him by the "floor" of the cabin. The normal force supplies the centripetal force [Fig.1.260], i.e.,
$$F_c = \frac{mv^2}{r} \quad \text{and} \quad n = \frac{mg}{2}$$
This gives, $\qquad \dfrac{mv^2}{r} = \dfrac{mg}{2} \Rightarrow v = \sqrt{\dfrac{gr}{2}}$
$$\Rightarrow \qquad v = \sqrt{\frac{\left(9.80 \text{ m/s}^2\right)\left(10.0 \text{ m}\right)}{2}} \Rightarrow v = 7.00 \text{ m/s}$$
Since $v = r\omega$, we have
$$\omega = \frac{v}{r} = \frac{7.00 \text{ m/s}}{10 \text{ m}} = 0.7 \text{ rad/s}$$

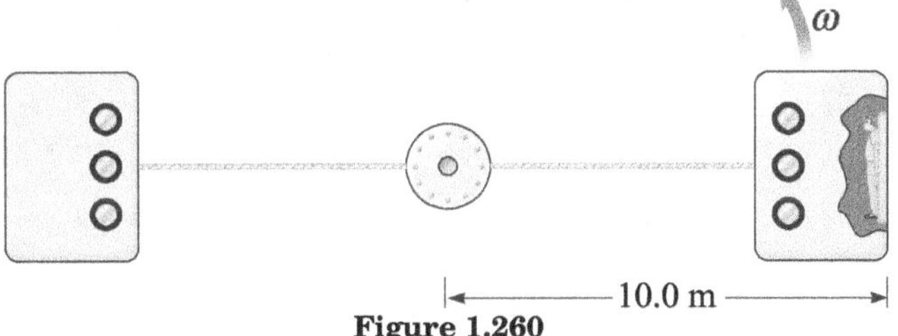

Figure 1.260

(b) Because his feet stay in place on the floor, his head will be moving at the same tangential speed as his feet. However, his feet and his head are travelling in circles of different radii.

(c) If he stands up without holding on to anything with his hands, the only force on his body is radial. Because the wall of the cabin near the traveller's head moves in a smaller circle, it moves at a slower tangential speed than that of the traveller's head so his head moves toward the wall-if he is not careful, there could be a collision. This is an example of the Coriolis force. Holding onto to a rigid support with his hands will provide a tangential force to the traveller to slow the upper part of his body down.

62. (a) Ignoring air resistance, the mechanical energy conservation for the object-Earth system from firing to apex is given by,
$$(K + U_g)_i = (K + U_g)_f$$
$$\frac{1}{2}mv_i^2 - \frac{GMm}{R} = 0 - \frac{GMm}{R+h}$$
where $\frac{1}{2}mv_{\text{esc}}^2 = \dfrac{GmM}{R}$. Then
$$\frac{1}{2}v_i^2 - \frac{1}{2}v_{\text{esc}}^2 = -\frac{1}{2}v_{\text{esc}}^2 \frac{R}{R+h}$$
$$\Rightarrow \qquad v_{\text{esc}}^2 - v_i^2 = \frac{v_{\text{esc}}^2 R}{R+h}$$
$$\Rightarrow \qquad \frac{1}{v_{\text{esc}}^2 - v_i^2} = \frac{R+h}{v_{\text{esc}}^2 R}$$

$$\Rightarrow \quad h = \frac{v_{esc}^2 R}{v_{esc}^2 - v_i^2} - R = \frac{v_{esc}^2 R - v_{esc}^2 R + v_i^2 R}{v_{esc}^2 - v_i^2}$$

$$\Rightarrow \quad h = \frac{R v_i^2}{v_{esc}^2 - v_i^2}$$

$$\Rightarrow \quad h = \frac{(6.37 \times 10^6 \text{ m})(8.76 \text{ km/s})^2}{(11.2 \text{ km/s})^2 - (8.76 \text{ km/s})^2} = 1.00 \times 10^7 \text{ m}$$

(b) The fall of the meteorite is the time-reversal of the upward flight of the projectile, so it is described by the same energy equation:

$$v_i^2 = v_{esc}^2 \left(1 - \frac{R}{R+h}\right) = v_{esc}^2 \left(\frac{h}{R+h}\right)$$

$$= (11.2 \times 10^3 \text{ m/s})^2 \left(\frac{2.51 \times 10^7 \text{ m}}{6.37 \times 10^6 \text{ m} + 2.51 \times 10^7 \text{ m}}\right)$$

$$= 1.00 \times 10^8 \text{ m}^2/\text{s}^2$$

$$v_i = 1.00 \times 10^4 \text{ m/s}$$

63. (a) Ignoring air resistance, the energy conservation for the object-Earth system from firing to apex is given by,

$$(K+U)_i = (K+U)_f$$

$$\Rightarrow \quad \frac{1}{2} m v_i^2 - \frac{GMm}{R} = 0 - \frac{GMm}{R+h}$$

where $\frac{1}{2} m v_e^2 = \frac{GMm}{R}$. Therefore,

$$\frac{1}{2} v_i^2 - \frac{1}{2} v_e^2 = -\frac{1}{2} v_e^2 \frac{R}{R+h}$$

$$\Rightarrow \quad v_e^2 - v_i^2 = \frac{v_e^2 R}{R+h}$$

$$\Rightarrow \quad \frac{1}{v_e^2 - v_i^2} = \frac{R+h}{v_e^2 R}$$

$$\Rightarrow \quad h = \frac{v_e^2 R}{v_e^2 - v_i^2} - R = \frac{v_e^2 R - v_e^2 R + v_i^2 R}{v_e^2 - v_i^2}$$

$$\Rightarrow \quad h = \frac{R v_i^2}{v_e^2 - v_i^2}$$

(b) The fall of the meteorite is the time-reversal of the upward flight of the projectile, so it is described by the same energy equation. From (a) above, replacing v_i with v_f, we have

$$v_f^2 = v_e^2 - v_e^2 \frac{R}{R+h}$$

$$\Rightarrow \quad v_f^2 = v_e^2 \left(1 - \frac{R}{R+h}\right) \quad \Rightarrow \quad v_f = v_e \sqrt{\frac{h}{R+h}}$$

(c) With $v_i << v_e$, $h \approx \frac{R v_i^2}{v_e^2} = \frac{R v_i^2 R}{2GM}$. But $g = \frac{GM}{R^2}$, so $h = \frac{v_i^2}{2g}$, in agreement with $0^2 = v_i^2 + 2(-g)(h - 0)$.

64. (a) If R represent the radius of the asteroid, then, its volume is $\frac{4}{3}\pi R^3$ and its mass is $\rho \frac{4}{3}\pi R^3$. For your orbital motion, $\sum F = ma$ gives

$$\frac{Gm_1 m_2}{R^2} = \frac{m_2 v^2}{R} \quad \Rightarrow \quad \frac{G\rho 4\pi R^3}{3R^2} = \frac{v^2}{R}$$

solving for R,

$$R = \left(\frac{3v^2}{G\rho 4\pi}\right)^{1/2}$$

$$= \left[\frac{3(8.50 \text{ m/s})^2}{(6.67 \times 10^{-11} \text{ N}\cdot\text{m}^2/\text{kg}^2)(1100 \text{ kg/m}^3) 4\pi}\right]^{1/2}$$

$$= 1.53 \times 10^4 \text{ m}$$

(b) $\rho \frac{4}{3}\pi R^3 = (1100 \text{ kg/m}^3)\frac{4}{3}\pi (1.53 \times 10^4 \text{ m})^3$

$$= 1.66 \times 10^{16} \text{ kg}$$

(c) $v = \frac{2\pi R}{T}$, $T = \frac{2\pi R}{v} = \frac{2\pi (1.53 \times 10^4 \text{ m})}{8.5 \text{ m/s}}$

$$= 1.13 \times 10^4 \text{ s} = 3.15 \text{ h}$$

(d) For an illustrative model, we take your mass as 90.0 kg and assume the asteroid is originally at rest. Angular momentum is conserved for you and asteroid system:

$$\sum L_i = \sum L_f$$

$$\Rightarrow \quad 0 = m_2 v R - L\omega$$

$$\Rightarrow \quad 0 = m_2 v R - \frac{2}{5} m_1 R^2 \frac{2\pi}{T_{asteroid}}$$

$$\Rightarrow \quad m_2 v = \frac{4\pi}{5} \frac{m_1 R}{T_{asteroid}}$$

$$\Rightarrow \quad T_{asteroid} = \frac{4\pi m_1 R}{5 m_2 v}$$

$$= \frac{4\pi (1.66 \times 10^{16} \text{ kg})(1.53 \times 10^4 \text{ m})}{5(90.0 \text{ kg})(8.50 \text{ m/s})}$$

$$= 8.37 \times 10^{17} \text{ s} = 26.5 \text{ billion years}$$

Thus your running does not produce significant rotation of the asteroid if it is originally stationary and does not significantly affect any rotation it does have.

This problem is realistic. Many asteroids, such as Ida and Eros, are roughly 30 km in diameter. They are typically irregular in shape and not spherical. Satellites such as Phobos (of Mars), Adrastea (of Jupiter), Calypso (of Saturn), and Ophelia (of Uranus) would allow a visitor the same experience of easy orbital motion.

65. (a) The two appropriate isolated system models are conservation of momentum and conservation of energy applied to the system consisting of the two spheres.

(b) Applying conservation of momentum to the system, we get

$$m_1 \vec{v}_{1i} + m_2 \vec{v}_{2i} = m_1 \vec{v}_{1f} + m_2 \vec{v}_{2f}$$

$$\Rightarrow \quad 0 + 0 = M \vec{v}_{1f} + 2M \vec{v}_{2f}$$

$$\Rightarrow \quad \vec{v}_{1f} = -2 \vec{v}_{2f}$$

(c) Applying conservation of energy to the system, we find

$$K_i + U_i + \Delta E = K_f + U_f$$

$$\Rightarrow \quad 0 - \frac{Gm_1 m_2}{r_i} + 0 = \frac{1}{2} m_1 v_{1f}^2 + \frac{1}{2} m_2 v_{2f}^2 - \frac{Gm_1 m_2}{r_f}$$

$$\Rightarrow \quad -\frac{GM(2M)}{12R} = \frac{1}{2} M v_{1f}^2 + \frac{1}{2}(2M) v_{2f}^2 - \frac{GM(2M)}{4R}$$

$$\Rightarrow \quad \frac{1}{2} M v_{1f}^2 = \frac{GM}{2R} - \frac{GM}{6R} - v_{2f}^2$$

$$\Rightarrow \quad v_{1f} = \sqrt{\frac{2GM}{3R} - 2v_{2f}^2}$$

(d) Combining the results for parts (b) and (c),

$$2v_{2f} = \sqrt{\frac{2GM}{3R} - 2v_{2f}^2}$$

$$\Rightarrow \quad 6v_{2f}^2 = \frac{2GM}{3R}$$

$$\Rightarrow \quad v_2 = \frac{1}{3}\sqrt{\frac{GM}{R}}$$

Therefore, $v_1 = \sqrt{\frac{2GM}{3R} - 2v_{2f}^2} = \frac{2}{3}\sqrt{\frac{GM}{R}}$

66. (a) At infinite separation $U = 0$ and at rest $K = 0$. Since the system is isolated, the energy and momentum of the two-planet system is conserved. We have

$$0 = \frac{1}{2} m_1 v_1^2 + \frac{1}{2} m_2 v_2^2 - \frac{Gm_1 m_2}{d} \qquad \ldots (1)$$

and $\qquad 0 = m_1 v_1 - m_2 v_2 \qquad \ldots (2)$

because the initial momentum of the system is zero.

On combining equations (1) and (2), we get

$$v_1 = m_2\sqrt{\frac{2G}{d(m_1+m_2)}} \quad \text{and} \quad v_2 =$$
$$m_1\sqrt{\frac{2G}{d(m_1+m_2)}}$$

The relative velocity is then

$$v_r = v_1 - (-v_2) = \sqrt{\frac{2G(m_1+m_2)}{d}}$$

(b) The instant before the collision, the distance between the planets is $d = r_1 + r_2$. Substitute given numerical values into the equation found for v_1 and v_2 in part (a) to find

$$v_1 = 1.03 \times 10^4 \text{ m/s and } v_2 = 2.58 \times 10^3 \text{ m/s}$$

Therefore,

$$K_1 = \frac{1}{2}m_1v_1^2 = 1.07 \times 10^{32} J \text{ and } K_2 = \frac{1}{2}m_2v_2^2 = 2.67 \times 10^{31} J$$

67. (a) The free-fall acceleration produced by the Earth is

$$g = \frac{GM}{r^2} = GMr^{-2} \text{ (directed downward)}$$

Its rate of change is

$$\frac{dg}{dr} = GM(-2)r^{-3} = -2GMr^{-3}$$

The minus sign indicates that g decreases with increasing height. At the Earth's surface,

$$\frac{dg}{dr} = -\frac{2GM}{R^3}$$

(b) For small differences,

$$\frac{|\Delta g|}{\Delta r} = \frac{|\Delta g|}{h} = \frac{2GM}{R^3}$$

Thus,

$$|\Delta g| = \frac{2GMh}{R^3}$$

(c) $|\Delta g| = \frac{2(6.67\times10^{-11}\text{N·m}^2/\text{kg}^2)(5.98\times10^{24}\text{ kg})(6.00\text{ m})}{(6.37\times10^6\text{ m})^3}$
$$= 1.85 \times 10^{-5} \text{ m/s}^2$$

68. (a) Each bit of mass dm in the ring is at the same distance from the object at A. The separate contributions $-\frac{Gmdm}{r}$ to the system energy add up to $-\frac{GmM_{\text{ring}}}{r}$. When the object is at A, this is

$$U_A = -\frac{(6.67\times10^{-11}\text{ N·m}^2/\text{kg}^2)(1000\text{ kg})(2.36\times10^{20}\text{ kg})}{\sqrt{(1.00\times10^8\text{ m})^2+(2.00\times10^8\text{ m})^2}}$$
$$= -7.04 \text{ N}$$

(b) When the object is at the center of the ring, the potential energy of the system is

$$U_B = -\frac{(6.67\times10^{-11}\text{ N·m}^2/\text{kg}^2)(1000\text{ kg})(2.36\times10^{20}\text{ kg})}{1.00\times10^8\text{ m}}$$
$$= -1.57 \times 10^5 \text{ J}$$

(c) Total energy of the object-ring system is conserved:

$$K_A + U_A = K_B + U_B$$
$$\text{or} \quad 0 - 7.04 \times 10^4 \text{ J} = \frac{1}{2}(1000\text{ kg})v_B^2 - 1.57 \times 10^5 \text{ J}$$
$$\text{or} \quad v_B = \left(\frac{2\times8.70\times10^4\text{ J}}{1000\text{ kg}}\right)^{1/2} = 13.2 \text{ m/s}$$

69. The original orbit radius is

$$r = a = 6.37 \times 10^6 \text{ m} + 500 \times 10^3 \text{ m} = 6.87 \times 10^6 \text{ m}$$

The original energy is

$$E_i = -\frac{GMm}{2a}$$
$$= -\frac{(6.67\times10^{-11}\text{ N·m}^2/\text{kg}^2)(5.98\times10^{24}\text{ kg})(10^4\text{ kg})}{2(6.87\times10^6\text{ m})}$$

$$= -2.90 \times 10^{11} \text{ J}$$

We assume that the perigee distance in the new orbit is 6.87×10^6 m. Then the major axis is $2a = 6.87 \times 10^6$ m $+ 2.00 \times 10^7$ m $= 2.69 \times 10^7$ m and the final energy is

$$E_f = -\frac{GMm}{2a}$$
$$= -\frac{(6.67\times10^{-11}\text{ N·m}^2/\text{kg}^2)(5.98\times10^{24}\text{ kg})(10^4\text{ kg})}{2.69\times10^7\text{ m}}$$
$$= -1.48 \times 10^{11} \text{ J}$$

The energy input required from the engine is

$$E_f - E_i = -1.48 \times 10^{11} \text{ J} - (-2.90 \times 10^{11} \text{ J}) = 1.42 \times 10^{11} \text{ J}$$

70. From the walk, $2\pi r = 25000$ m. Thus, the radius of the planet is

$$r = \frac{25000 \text{ m}}{2\pi} = 3.98 \times 10^3 \text{ m}$$

From the drop:

$$\Delta y = \frac{1}{2}gt^2 = \frac{1}{2} g(29.2 \text{ s})^2 = 1.40 \text{ m}$$

so, $\quad g = \frac{2(1.40 \text{ m})}{(29.2 \text{ s})^2} = 3.28 \times 10^{-3} \text{ m/s}^2 = \frac{MG}{r^2}$

which gives

$$M = 7.79 \times 10^{14} \text{ kg}$$

71. The quaternary star system is again shown in Fig.1.261b. The distance between the orbiting stars is $d = 2r\cos 30° = \sqrt{3}r$ since $\cos 30° = \frac{\sqrt{3}}{2}$. The net inward force on one orbiting star is

$$\frac{Gmm}{d^2}\cos 30° + \frac{GMm}{r^2}$$
$$+\frac{Gmm}{d^2}\cos 30° = \frac{mv^2}{r}$$
$$\Rightarrow \quad \frac{Gm2\cos 30°}{3r^2} + \frac{GM}{r^2} = \frac{4\pi^2 r^2}{rT^2}$$
$$\Rightarrow \quad G\left(\frac{m}{\sqrt{3}} + M\right) = \frac{4\pi^2 r^3}{T^2}$$

solving for the period gives

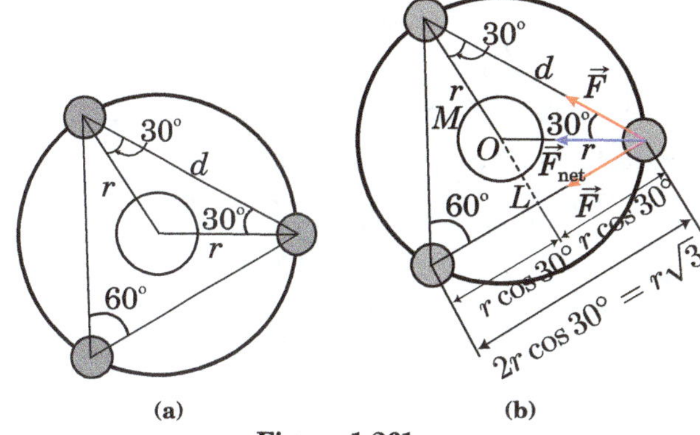

(a)   (b)

Figure 1.261

$$T^2 = \frac{4\pi^2 r^3}{G(M + m/\sqrt{3})}$$
$$T = 2\pi\left(\frac{r^3}{G(M + m/\sqrt{3})}\right)^{1/2}$$

72. (a) We find the period from

$$T = \frac{2\pi r}{V} = \frac{2\pi(30000 \times 9.46 \times 10^{15}\text{ m})}{2.50 \times 10^5 \text{ m/s}} = 7 \times 10^{15} \text{ s}$$

$$= 2 \times 10^8 \text{yr}$$

(b) We estimate the mass of the Milky Way from

$$M = \frac{4\pi^2 a^3}{GT^2} = \frac{4\pi^2 \left(30000 \times 9.46 \times 10^{15}\ \text{m}\right)^3}{\left(6.67 \times 10^{-11}\ \text{N} \cdot \text{m}^2/\text{kg}^2\right)\left(7.13 \times 10^{15}\ \text{s}\right)^2},$$

$$= 2.66 \times 10^{41}\ \text{kg}$$

or about 10^{41} kg

Note that this is the mass of the galaxy contained within the Sun's orbit of the galactic centre. Recent studies show that the true mass of the galaxy, including an extended halo of dark matter, is at least an order of magnitude larger than our estimate.

(c) A solar mass is about 1×10^{30} kg : $10^{41}/10^{30} = 10^{11}$
The number of stars is on the order of 10^{11}.

73. Energy conservation for the two-sphere system from release to contact:

$$-\frac{Gmm}{R} = -\frac{Gmm}{2r} + \frac{1}{2}mv^2 + \frac{1}{2}mv^2$$

or $\quad Gm\left(\dfrac{1}{2r} - \dfrac{1}{R}\right) = v^2 \quad \Rightarrow \quad v = \left(Gm\left[\dfrac{1}{2r} - \dfrac{1}{R}\right]\right)^{1/2}$

(a) The injected momentum is the final momentum of each sphere,

$$mv = m^{2/2}\left(Gm\left[\frac{1}{2r} - \frac{1}{R}\right]\right)^{1/2} = \left[Gm^3\left(\frac{1}{2r} - \frac{1}{R}\right)\right]^{1/2}$$

(b) If they now collide elastically each sphere reverses its velocity to receive impulse

$$mv - (-mv) = 2mv = 2\left[Gm^3\left(\frac{1}{2r} - \frac{1}{R}\right)\right]^{1/2}$$

74. (a) The net torque exerted on the Earth around Sun is zero. Therefore, the angular momentum of Earth is conserved. We use this to find the speed at aphelion:

$$mr_a v_a = mr_p v_p$$

and $\quad v_a = v_p\left(\dfrac{r_p}{r_a}\right) = \left(3.027 \times 10^4\ \text{m/s}\right)\left(\dfrac{1.471}{1.521}\right)$

$$= 2.93 \times 10^4\ \text{m/s}$$

(b) $K_p = \frac{1}{2}mv_p^2 = \frac{1}{2}\left(5.98 \times 10^{24}\ \text{kg}\right)\left(3.027 \times 10^4\ \text{m/s}\right)^2$

$$= 2.74 \times 10^{33}\ \text{J}$$

$$U_p = -\frac{GMm}{r_p}$$

$$= -\frac{\left(6.67 \times 10^{-11}\ \text{N} \cdot \text{m}^2/\text{kg}^2\right)\left(5.98 \times 10^{24}\ \text{kg}\right)\left(1.99 \times 10^{30}\ \text{kg}\right)}{1.471 \times 10^{11}\ \text{m}}$$

$$= -5.40 \times 10^{33}\ \text{J}$$

(c) Using the same form as in part (b),

$$K_a = 2.57 \times 10^{33}\ \text{J}$$

and $\quad U_a = -5.22 \times 10^{33}\ \text{J}$

(d) Compare to find that

$$K_p + U_p = -2.66 \times 10^{33}\ \text{J and } K_a + U_a = -2.65 \times 10^{33}\ \text{J}.$$

They agree, with a small rounding error.

75. For both circular orbits [Fig.1.262],

$$\Sigma F = ma: \quad \frac{GMm}{r^2} = \frac{mv^2}{r}$$

$$v = \sqrt{\frac{GM}{r}}$$

The original speed is

$$v_i = \sqrt{\frac{\left(6.67 \times 10^{-11}\ \text{N} \cdot \text{m}^2/\text{kg}^2\right)\left(5.98 \times 10^{24}\ \text{kg}\right)}{6.37 \times 10^6\ \text{m} + 2.00 \times 10^5\ \text{m}}}$$

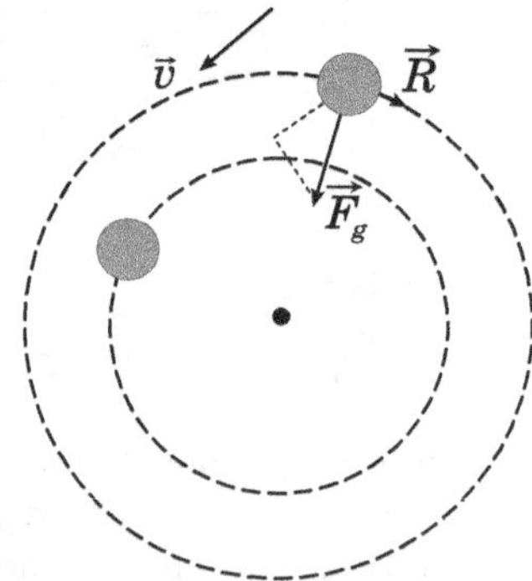

Figure 1.262

$$= 7.79 \times 10^3\ \text{m/s}$$

(b) The final speed is

$$v_i = \sqrt{\frac{\left(6.67 \times 10^{-11}\ \text{N} \cdot \text{m}^2/\text{kg}^2\right)\left(5.98 \times 10^{24}\ \text{kg}\right)}{6.47 \times 10^6\ \text{m}}}$$

$$= 7.85 \times 10^3\ \text{m/s}$$

The energy of the satellite-Earth system is

$$K + U = \frac{1}{2}mv^2 - \frac{GMm}{r} = \frac{1}{2}m\frac{GM}{r} - \frac{GM}{r} = -\frac{GMm}{2r}$$

(c) Originally,

$$E_i = -\frac{\left(6.67 \times 10^{-11}\ \text{N} \cdot \text{m}^2/\text{kg}^2\right)\left(5.98 \times 10^{24}\ \text{kg}\right)(100\ \text{kg})}{2\left(6.57 \times 10^6\ \text{m}\right)}$$

$$= -3.04 \times 10^9\ \text{J}$$

(d) Finally,

$$E_f = -\frac{\left(6.67 \times 10^{-11}\ \text{N} \cdot \text{m}^2/\text{kg}^2\right)\left(5.98 \times 10^{24}\ \text{kg}\right)(100\ \text{kg})}{2\left(6.47 \times 10^6\ \text{m}\right)}$$

$$= -3.08 \times 10^9\ \text{J}$$

(e) Thus the object speeds up as it spirals down to the planet. The loss of gravitational energy is so large that the total energy decreases by

$$E_i - E_f = -3.04 \times 10^9\ \text{J} - \left(-3.08 \times 10^9\ \text{J}\right) = 4.69 \times 10^7\ \text{J}$$

(f) The only forces on the object are the backward force of air resistance R, comparatively very small in magnitude, and the force of gravity. Because the spiral path of the satellite is not perpendicular to the gravitational force, one component of the gravitational force pulls forward on the satellite to do positive work and make its speed increase.

76. The centripetal acceleration of the blob comes from gravitational acceleration:

$$\frac{v^2}{r} = \frac{M_C G}{r^2} = \frac{4\pi^2 r^2}{T^2 r}$$

or $\quad G_C T^2 = 4\pi^2 r^3$

Solving for the radius gives

$$r = \left[\frac{\left(6.67 \times 10^{-11}\ \text{N} \cdot \text{m}^2/\text{kg}^2\right)(20)\left(1.99 \times 10^{30}\ \text{kg}\right)\left(5.00 \times 10^{-3}\ \text{s}\right)^2}{4\pi^2}\right]^{1/3}$$

or $\quad r_{\text{orbit}} = 119$ km

77. From Kepler's third law, minimum period means minimum orbit size. The "treetop satellite" in Problem 44 has minimum period. The radius of the satellite's circular orbit is essentially equal to the radius R of the planet.

$$\sum F = ma: \quad \frac{GMm}{R^2} = \frac{mv^2}{R} = \frac{m}{R}\left(\frac{2\pi R}{T}\right)^2$$

or $\qquad G\rho V = \dfrac{R^2\left(4\pi^2 R^2\right)}{R^2}$

or $\qquad G\rho\left(\dfrac{4}{3}\pi R^3\right) = \dfrac{4\pi^2 R^3}{T^2}$

$\Rightarrow \qquad T^2 G\rho = 3\pi \Rightarrow T = \sqrt{\dfrac{3\pi}{G\rho}}$

78. Let m represent the mass of the meteoroid and v_i its speed when far away [Fig.1.263]. No torque acts on the meteoroid about the centre of earth, so its angular momentum is conserved as it moves between the distant point and the point where it grazes the Earth, moving perpendicular to the radius:

$$L_i = L_f : m\vec{r}_i \times \vec{v}_i = m\vec{r}_f \times \vec{v}_f$$

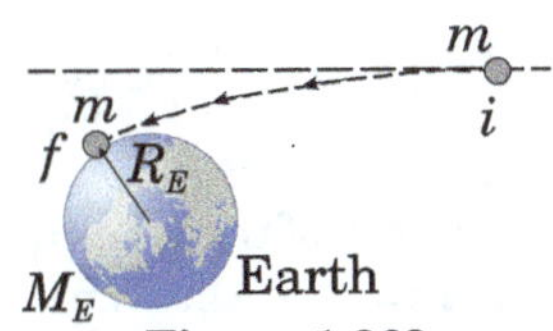
Figure 1.263

$\Rightarrow \quad m\left(3Rv_i\right) = mRv_f \quad \Rightarrow \quad v_f = 3v_i$

Now, the energy of the meteoroid-Earth system is also conserved:

$$K_i + U_i = K_f + U_f : \dfrac{1}{2}mv_i^2 + 0 = \dfrac{1}{2}mv_f^2 - \dfrac{GMm}{R}$$

$$\Rightarrow \qquad \dfrac{1}{2}v_i^2 = \dfrac{1}{2}\left(9v_i^2\right) - \dfrac{GM}{R}$$

or $\qquad \dfrac{GM}{R} = 4v_i^2 \quad \Rightarrow \quad v_i = \sqrt{\dfrac{GM}{4R}}$

79. If we choose the coordinate of the centre of mass at the origin [Fig.1.264], then by definition of centre of mass, we have

$$0 = \dfrac{\left(Mr_2 - mr_1\right)}{M+m} \quad \text{or} \quad mr_1 = Mr_2$$

or $\qquad \dfrac{r_1}{1/m} = \dfrac{r_2}{1/M} = \dfrac{\left(r_1 + r_2\right)}{\frac{1}{m} + \frac{1}{M}} = \left(r_1 + r_2\right)\dfrac{mM}{m+M}$

or $\qquad r_1 = \dfrac{M}{m+M}d \qquad\qquad \text{...(1)}$

and $\qquad r_2 = \dfrac{m}{m+M}d \qquad\qquad \text{...(2)}$

here, $d = \left(r_1 + r_2\right)$

The gravitational force of star of mass M on star of mass m is given by

$$F_g = \dfrac{GMm}{d^2}$$

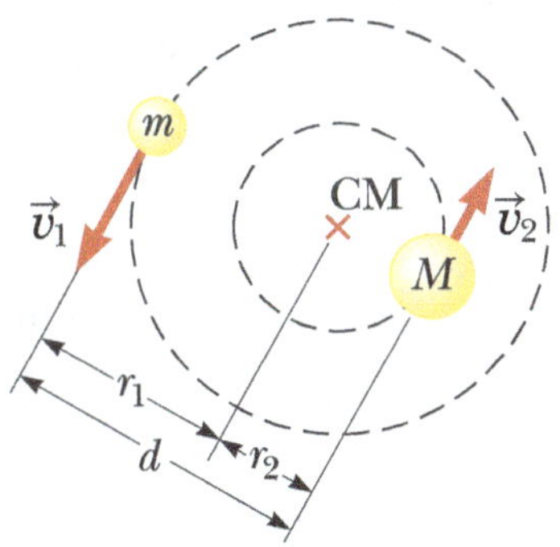
Figure 1.264

Since, there is no other force except gravitational force

on stars, therefore this force provides the required centripetal force to each star for circular motion about the centre of mass.

If ω is the angular frequency of each star about the centre of mass, then for circular motion of star of mass m, we can write

$$mr_1\omega^2 = F_g = \dfrac{GMm}{d^2}$$

or $\qquad r_1\omega^2 = \dfrac{GM}{d^2}$

or $\qquad \omega^2 = \dfrac{GM}{r_1 d^2} \qquad\qquad \text{...(3)}$

Substituting the value of r_1 from Eq.(1) in (3), we get

$$\omega^2 = \dfrac{GM}{\dfrac{M}{m+M}d^3} = \dfrac{G(m+M)}{d^3}$$

or $\qquad \omega = \sqrt{\dfrac{G(m+M)}{md^3}}$

If T is the time period of motion of each star, then

$$T = \dfrac{2\pi}{\omega} = \dfrac{2\pi}{\sqrt{\dfrac{G(m+M)}{md^3}}}$$

or $\qquad T^2 = \dfrac{4\pi^2 d^3}{G(M+m)}$

80. The gravitational forces the particles exert on each other are in the x direction. They do not affect the velocity of the centre of mass. Energy is conserved for the pair of particles in a reference frame coasting along with their center of mass, and momentum conservation means that the identical particles move toward each other with equal speeds in this frame:

$$U_i + K_{1i} + K_{2i} = U_f + K_{1f} + K_{2f}$$

$$\Rightarrow \quad -\dfrac{Gm_1 m_2}{r_i} + 0 = -\dfrac{Gm_1 m_2}{r_f} + \dfrac{1}{2}m_1 v^2 + \dfrac{1}{2}m_2 v^2$$

$$\Rightarrow \quad -\dfrac{\left(6.67\times10^{-11}\ \text{N·m}^2/\text{kg}^2\right)(1000\ \text{kg})^2}{20.0\ \text{m}}$$

$$= -\dfrac{\left(6.67\times10^{-11}\ \text{N·m}^2/\text{kg}^2\right)(1000\ \text{kg})^2}{2.00\ \text{m}} + 2\left(\dfrac{1}{2}\right)(1000\ \text{kg})v^2$$

$$\Rightarrow \quad \left(\dfrac{3.00\times10^{-5}\ \text{J}}{1000\ \text{kg}}\right)^{1/2} = v = 1.73\times10^{-4}\ \text{m/s}$$

Then their vector velocities are $\left(800 + 1.73\times10^{-4}\right)\hat{i}$ m/s and $\left(800 - 1.73\times10^{-4}\right)\hat{i}$ m/s for the trailing particle and the leading particle, respectively.

81. (a) The gravitational force exerted on m by the Earth (mass M) accelerates m according to $g_2 = \dfrac{GM}{r^2}$. The equal-magnitude force exerted on the Earth by m produces acceleration of the Earth given by $g_1 = \dfrac{Gm}{r^2}$. The acceleration of relative approach is then

$$g_2 + g_1 = \dfrac{Gm}{r^2} + \dfrac{GM}{r^2}$$

$$= \dfrac{\left(6.67\times10^{-11}\ \text{N·m}^2/\text{kg}^2\right)\left(5.98\times10^{24}\ \text{kg}+m\right)}{\left(1.20\times10^7\ \text{m}\right)^2}$$

$$= \left(2.77\ \text{m/s}^2\right)\left(1 + \dfrac{m}{5.98\times10^{24}\ \text{kg}}\right)$$

(b) and (c) Here $m = 5$ kg and $m = 2000$ kg are both negligible compared to the mass of the Earth, so the acceleration of relative approach is just 2.77 m/s^2. (d) Substituting m $= 2.00\times10^{24}$ kg into the expression for $(g_1 + g_2)$ above gives

$$g_1 + g_2 = 3.70\ \text{m/s}^2$$

(e) Any object with mass small compared to the Earth starts to fall with acceleration 2.77 m/s^2. As m increases to become comparable to the mass of the Earth, the acceleration increases, and can become arbitrarily large. It approaches a direct proportionality to m.

82. For the Earth, $\sum F = ma$: $\quad \frac{GM_s\,m}{r^2} = \frac{mv^2}{r} = \frac{m}{r}\left(\frac{2\pi r}{T}\right)^2$

Therefore, $GM_sT^2 = 4\pi^2 r^3$

Also, the angular momentum $L = mvr = m\frac{2\pi r}{T}r$ is a constant for the Earth. We eliminate $r = \sqrt{\frac{LT}{2\pi m}}$ between the equations:

$$GM_sT^2 \;=\; 4\pi^2\left(\frac{LT}{2\pi m}\right)^{3/2} \text{ gives } GM_sT^{1/2} \;=\;$$
$$4\pi^2\left(\frac{L}{2\pi m}\right)^{3/2}$$

Now the rates of change with time t are described by

$$GM_s\left(\frac{1}{2}\,T^{-1/2}\frac{dT}{dt}\right) + G\left(1\frac{dM_s}{dt}T^{1/2}\right) = 0$$

or $\quad \frac{dT}{dt} = -\frac{dM_s}{dt}\left(2\frac{T}{M_s}\right) \approx \frac{\Delta T}{\Delta t}$

which gives,

$$\Delta T \approx -\Delta t\frac{dM}{dt}\left(2\frac{T}{M_s}\right)$$
$$= -(5000\text{yr})\left(\frac{3.16\times10^7\text{ s}}{1\text{yr}}\right)\left(-3.64\times10^9\text{ kg/s}\right)$$
$$\Delta T = 5\left(2\frac{1\text{yr}}{1.99\times10^{30}\text{ kg}}\right)$$

83. Let m represent the mass of the spacecraft, r_E the radius of the Earth's orbit, and x the distance from Earth to the spacecraft.

The Sun exerts on the spacecraft a radial inward force of

$$F_s = \frac{GM_Sm}{(r_E - x)^2}$$

while the Earth exerts on it a radial outward force of

$$F_E = \frac{GM_Em}{x^2}$$

The net force on the spacecraft must produce the correct centripetal acceleration for it to have an orbital period of 1.000 year.

Thus, $F_S - F_E = \dfrac{GM_Sm}{(r_E - x)^2} - \dfrac{GM_Em}{x^2}$

$$= \frac{mv^2}{(r_E - x)} = \frac{m}{(r_E - x)}\left[\frac{2\pi(r_E - x)}{T}\right]^2$$

which reduces to

$$\frac{GM_S}{(r_E - x)^2} - \frac{GM_E}{x^2} = \frac{4\pi^2(r_E - x)}{T^2} \qquad \ldots(1)$$

Cleared of fractions, this equation would contain powers of x ranging from the fifth to the zeroth. We do not solve it algebraically. We may test the assertion that x is 1.48×10^9 m by substituting it into the equation, along with the following data: $M_S = 1.99 \times 10^{30}$ kg, $M_E = 5.974 \times 10^{24}$ kg, $r_E = 1.496 \times 10^{11}$ m, and $T = 1.000$ yr $= 3.156 \times 10^7$ s with $x = 1.48 \times 10^9$ m, the result is

6.053×10^{-3} m/s$^2 - 1.82 \times 10^{-3}$ m/s$^2 \approx 5.8708 \times 10^{-3}$ m/s^2

or $\quad 5.8709 \times 10^{-3}$ m/s$^2 \approx 5.8708 \times 10^{-3}$ m/s^2

To three-digit precision, the solution is 1.48×10^9 m.

As an equation of fifth degree, Eq.(1), has five roots. The Sun-Earth system has five Lagrange points, all revolving around the Sun synchronously with the Earth. The SOHO and ACE satellites are at one. Another is beyond the far side of the Sun. Another is beyond the night side of the Earth. Two more are on the Earth's orbit, ahead of the planet and behind it by 60°. The twin satellites of NASA's STEREO mission, giving three-dimensional views of the Sun from orbital positions ahead of and trailing Earth, passed through these Lagrange points in 2009. The Greek and Trojan asteroids are at the co-orbital Lagrange points of the Jupiter-Sun system.

84. (a) From the data about perigee, the energy of the satellite-Earth system is

$$E = \frac{1}{2}mv_p^2 - \frac{GM_Em}{r_p} = \frac{1}{2}(1.60\text{ kg})\left(8.23\times10^3\text{ m/s}\right)^2$$
$$-\frac{\left(6.67\times10^{-11}\text{ N}\cdot\text{m}^2/\text{kg}^2\right)\left(5.98\times10^{24}\text{ kg}\right)(1.60\text{ kg})}{7.02\times10^6\text{ m}}$$

or $E = -3.67 \times 10^7$ J

(b) $L = mvr\sin\theta = mv_pr_p\sin 90.0°$
$$= (1.60\text{ kg})\left(8.23\times10^3\text{ m/s}\right)\left(7.02\times10^6\text{ m}\right)$$
$$= 9.24\times10^{10}\text{ kg}\cdot\text{m}^2/\text{s}$$

(c) Since both the energy of the satellite-Earth system and the angular momentum of the Earth are conserved, at apogee we must have

$$\frac{1}{2}mv_a^2 - \frac{GMm}{r_a} = E$$

and $\quad mv_ar_a\sin 90.0° = L$

On substituting numeric values, we get

$$\tfrac{1}{2}(1.60\text{ kg})v_a^2 - \frac{\left(6.67\times10^{-11}\text{ N·m}^2/\text{kg}^2\right)\left(5.98\times10^{24}\text{ kg}\right)(1.60\text{ kg})}{r_a}$$
$$= -3.67\times10^7\text{ J}$$

and $\quad (1.60\text{ kg})v_ar_a = 9.24\times10^{10}\text{ kg}\cdot\text{m}^2/\text{s}$

Solving simultaneously, and suppressing units,

$$\frac{1}{2}(1.60)v_a^2 - \frac{\left(6.67\times10^{-11}\right)\left(5.98\times10^{24}\right)(1.60)(1.60)v_a}{9.24\times10^{10}}$$
$$= -3.67\times10^7$$

which reduces to

$$0.800v_a^2 - 11046v_a + 3.6723\times10^7 = 0$$

so, $\quad v_a = \dfrac{11046\pm\sqrt{(11046)^2 - 4(0.800)(3.6723\times10^7)}}{2(0.800)}$

This gives $v_a = 8230$ m/s or 5580 m/s.

The smaller answer refers to the velocity at the apogee while the larger refers to perigee.

Thus, $r_a = \dfrac{L}{mv_a} = \dfrac{9.24\times10^{10}\text{ kg}\cdot\text{m}^2/\text{s}}{(1.60\text{ kg})\left(5.58\times10^3\text{ m/s}\right)} = 1.04\times10^7$ m

(d) The major axis is $2a = r_p + r_{a'}$, so the semimajor axis is

$$a = \frac{1}{2}\left(7.02\times10^6\text{ m} + 1.04\times10^7\text{ m}\right) = 8.69\times10^6\text{ m}$$

and $\quad T = 8060$ s $= 134$ min

85. (a) Energy of the spacecraft-Mars system is conserved as the spacecraft moves between a very distant point and the point of closest approach:

$$0 + 0 = \frac{1}{2}mv_r^2 - \frac{GM_{\text{Mars}}\,m}{r}$$

or $\quad v_r = \sqrt{\dfrac{2G_{\text{Mars}}}{r}}$

After the engine burn, for a circular orbit we have $\sum F = ma$:

$$\frac{GM_{\text{Mars}}\,m}{r^2} = \frac{mv_0^2}{r}$$

or $\quad v_0 = \sqrt{\dfrac{GM_{\text{Mars}}}{r}}$

The percentage reduction from the original speed is

$$\frac{v_r - v_0}{v_r} = \frac{\sqrt{2}v_0 - v_0}{\sqrt{2}v_0} = \frac{\sqrt{2}-1}{\sqrt{2}} \times 100\% = 29.3\%$$

(b) The answer to part (a) applies with no changes , as the solution to part (a) shows.

1.26.8 Multiple Choice Assignments

Level 1

Q.No.	1	2	3	4	5	6	7	8	9	10
Ans.	A	A	D	A	C	B	C	C	A	C
Q.No.	11	12	13	14	15	16	17	18	19	20
Ans.	D	B	D	C	D	B	B	D	D	C
Q.No.	21	22	23	24	25	26	27	28	29	30
Ans.	A	A	A	A	C	D	B	C	A	C
Q.No.	31	32	33	34	35	36	37	38	39	40
Ans.	C	C	D	A	A	D	B	D	B	B
Q.No.	41	42	43	44	45	46	47	48	49	50
Ans.	A	C	D	C	C	D	B	B	D	B
Q.No.	51	52	53	54	55	56	57	58	59	60
Ans.	D	C	A	A	A	B	B	A	C	C
Q.No.	61	62	63	64	65	66	67	68	69	70
Ans.	B	A	D	B	D	B	C	B	D	D
Q.No.	71	72	73	74	75	76	77	78	79	80
Ans.	D	D	B	A	B	C	B	C	B	B
Q.No.	81	82								
Ans.	B	C								

Level 2

Q.No.	1	2	3	4	5	6	7	8	9	10
Ans.	C	A	D	B	A	B	C	A	C	A
Q.No.	11	12	13	14	15	16	17	18	19	20
Ans.	C	B	D	A	A	D	B	C	D	D
Q.No.	21	22	23	24	25	26	27	28	29	30
Ans.	B	D	C	C	A	A	C	B	A	B
Q.No.	31	32	33	34	35	36	37	38	39	
Ans.	C	A	C	D	D	A	B	B	B	

Level 3

Q.No.	1	2	3	4	5	6	7	8	9	10
Ans.	D	A	D	A	A	B	B	B	B	C
Q.No.	11	12	13	14	15	16	17	18	19	20
Ans.	C	C	A	B	B	A	B	A	C	C
Q.No.	21	22	23	24	25	26	27	28	29	30
Ans.	A	C	D	D	A	D	A	C	B	A

Column Matching

43	$A \to Q$	$B \to Q$	$C \to S$	$D \to T$
44	$A \to R$	$B \to R$		
45	$A \to Q$	$B \to P$	$C \to S$	$D \to Q$
46	$A \to R$	$B \to Q$	$C \to Q$	$D \to P$
47	$A \to R$	$B \to Q$	$C \to P$	
48	$A \to P$	$B \to P$	$C \to S$	
49	$A \to P$	$B \to Q$	$C \to Q$	$D \to Q$
50	$A \to S$	$B \to R$	$C \to P$	$D \to Q$

Level 4

SECTION - A

Q.No.	1	2	3	4	5	6	7	8	9	10
Ans.	C	C	A	C	B	B	B	D	A	B
Q.No.	11	12	13	14	15	16	17	18	19	20
Ans.	A	A	C	D	B	D	D	C	B	A
Q.No.	21	22	23	24	25	26	27	28	29	30
Ans.	A	C	D	D	B	A	C	B	D	D
Q.No.	31	32	33	34	35	36	37	38	39	40
Ans.	D	B	A	C	B	C	B	C	D	D
Q.No.	41	42	43	44	45	46	47	48	49	50
Ans.	C	A	C	C	D	B	D	C	C	A

SECTION - B

Q.No.	1	2	3	4	5	6	7	8	9	10
Ans.	D	A	B	A	C	A	A	D	C	B
Q.No.	11	12	13	14	15	16	17	18	19	
Ans.	D	C	A	A	B	B	C	B	2.3	

Note: In answer of Q. 19. of section B, the range lies in the interval $2.2 \to 2.4$.